GLOBAL STUDIES

MIDDLE EAST

ELEVENTH EDITION

Dr. William Spencer

OTHER BOOKS IN THE GLOBAL STUDIES SERIES
- Africa
- China
- Europe
- India and South Asia
- Japan and the Pacific Rim
- Latin America
- Russia, the Eurasian Republics, and
  Central/Eastern Europe

McGraw-Hill/Contemporary Learning Series
2460 Kerper Blvd., Dubuque, Iowa 52001
Visit us on the Internet—http://www.mhcls.com

Staff

Larry Loeppke	*Managing Editor*
Jill Peter	*Senior Developmental Editor*
Lori Church	*Permissions Assistant*
Maggie Lytle	*Cover*
Tara McDermott	*Designer*
Kari Voss	*Typesetting Supervisor/Co-designer*
Jean Smith	*Typesetter*
Sandy Wille	*Typesetter*
Karen Spring	*Typesetter*
Julie Keck	*Senior Marketing Manager*
Mary Klein	*Marketing Communications Specialist*
Alice Link	*Marketing Coordinator*
Tracie Kammerude	*Senior Marketing Assistant*

Sources for Statistical Reports

U.S. State Department, *Background Notes* (2000–2001).
The World Factbook (2001).
World Statistics in Brief (2001).
World Almanac (2001).
The Statesman's Yearbook (2001).
Demographic Yearbook (2001).
Statistical Yearbook (2001).
World Bank, World Development Report (2000–2001).

Copyright

Cataloging in Publication Data
Main entry under title: Global Studies: Middle East. 11th ed.
        1. Middle East—History. 2. Arab countries—History. 3. Israel—History.
I. Title: Middle East. II. Spencer, William, *comp.*
ISBN 97 800–07–340405–9   ISBN 0–07–340405–5   91–71258   ISSN 1098-3880

Eleventh Edition
Printed in the United States of America   1234567890QPDQPD09876   Printed on Recycled Paper

THE MIDDLE EAST

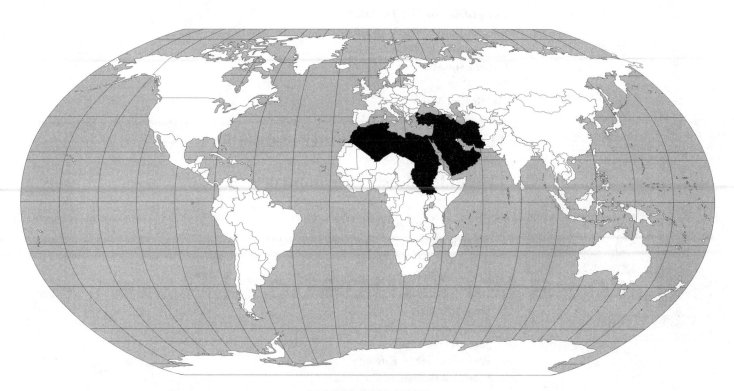

AUTHOR/EDITOR

Dr. William Spencer

The author/editor for *Global Studies: The Middle East* has specialized in the political and social history of the Middle East for over half a century. He retired from Florida State University in 1980 but has continued to be active in his field. He is currently an adjunct professor of history at Flagler College in St. Augustine, Florida, and is the author of many books in addition to the ten previous editions of this book. During his long professional career Dr. Spencer served as visiting professor at a number of colleges and universities in addition to field research and U.S. government and United Nations assignments. In his teaching, lecturing and service to school systems and community colleges he has made it his life's work helping students and educators develop a better understanding of this volatile region.

CONSULTANT

Elizabeth Bouvier Spencer

Elizabeth Bouvier Spencer is an artist and former teacher who traveled with her husband, Dr. William Spencer, on many research and professional trips to the Middle East. In addition to her contributions as grammarian and amanuensis she has been invaluable in the preparation of this and earlier editions in the areas of home and family life.

SERIES CONSULTANT

H. Thomas Collins

Washington, D.C.

Contents

The Middle East: Theater of Conflict   4

The Middle East: Heartland of Islam   19

Country Reports

Articles from the World Press

Using *Global Studies: The Middle East*

THE GLOBAL STUDIES SERIES

The Global Studies series was created to help readers acquire a basic knowledge and understanding of the regions and countries in the world. Each volume provides a foundation of information—geographic, cultural, economic, political, historical, artistic, and religious—that will allow readers to better assess the current and future problems within these countries and regions and to comprehend how events there might affect their own well-being. In short, these volumes present the background information necessary to respond to the realities of our global age.

Each of the volumes in the Global Studies series is crafted under the careful direction of an author/editor—an expert in the area under study. The author/editors teach and conduct research and have traveled extensively through the regions about which they are writing.

In *Global Studies: The Middle East*, the author/editor has written several regional essays and country reports for each of the countries included.

MAJOR FEATURES OF THE GLOBAL STUDIES SERIES

The Global Studies volumes are organized to provide concise information on the regions and countries within those areas under study. The major sections and features of the books are described here.

Regional Essays

For *Global Studies: The Middle East,* the author/editor has written several essays focusing on the religious, cultural, cosiopolitical, and exonomic differences and similarities of the countries and peoples in the various regions of the Middle East. Regional maps accompany the essays.

Country Reports

Concise reports are written for each of the countries within the region under study. These reports are the heart of each Global Studies volume. *Global Studies: The Middle East, Eleventh Edition,* contains 20 country reports.

The country reports are composed of five standard elements. Each report contains a detailed map visually positioning the country among its neighboring states; a summary of statistical information; a current essay providing important historical, geographical, political, cultural, and economic information; a historical timeline, offering a convenient visual survey of a few key historical events; and four "graphic indicators," with summary statements about the country in terms of development, freedom, health/welfare, and achievements.

A Note on the Statistical Reports

The statistical information provided for each country has been drawn from a wide range of sources. (The most frequently referenced are listed on page ii.) Every effort has been made to provide the most current and accurate information available. However, sometimes the information cited by these sources differs to some extent; and, all too often, the most current information available for some countries is somewhat dated. Aside from these occasional difficulties, the statistical summary of each country is generally quite complete and up to date. Care should be taken, however, in using these statistics (or, for that matter, any published statistics) in making hard comparisons among countries. We have also provided comparable statistics for the United States and Canada, which can be found on pages viii and ix.

World Press Articles

Within each Global Studies volume is reprinted a number of articles carefully selected by our editorial staff and the author/editor from a broad range of international periodicals and newspapers. The articles have been chosen for currency, interest, and their differing perspectives on the subject countries. There are 12 articles in *Global Studies: The Middle East, Eleventh Edition.*

The articles section is preceded by an annotated table of contents. This resource offers a brief summary of each article.

WWW Sites

An extensive annotated list of selected World Wide Web sites can be found on page vii in this edition of *Global Studies: The Middle East.* In addition, the URL addresses for country-specific Web sites are provided on the statistics page of most countries. All of the Web site addresses were correct and operational at press time. Instructors and students alike are urged to refer to those sites often to enhance their understanding of the region and to keep up with current events.

Glossary, Bibliography, Index

At the back of each Global Studies volume, readers will find a glossary of terms and abbreviations, which provides a quick reference to the specialized vocabulary of the area under study and to the standard abbreviations used throughout the volume.

Following the glossary is a bibliography that lists general works, national histories, and current-events publications and periodicals that provide regular coverage on China.

The index at the end of the volume is an accurate reference to the contents of the volume. Readers seeking specific information and citations should consult this standard index.

Currency and Usefulness

Global Studies: The Middle East, like the other Global Studies volumes, is intended to provide the most current and useful information available necessary to understand the events that are shaping the cultures of the region today.

This volume is revised on a regular basis. The statistics are updated, regional essays and country reports revised, and world press articles replaced. In order to accomplish this task, we turn to our author/editor, our advisory boards, and—hopefully—to you, the users of this volume. Your comments are more than welcome. If you have an idea that you think will make the next edition more useful, an article or bit of information that will make it more current, or a general comment on its organization, content, or features that you would like to share with us, please send it in for serious consideration.

Selected World Wide Web Sites for *Global Studies: The Middle East*

**All of these Web sites are hot-linked through the *Global Studies* home page: http://
www.mhcls.com/globalstudies (just click on a book).**

Some Web sites are continually changing their structure and content, so the information listed may not always be available.

GENERAL SITES

BBC News
http://news.bbc.co.uk/hi/english/world/middle_east/default.stm

Access current Middle East news from this BBC site.

CNN Interactive—World Regions: Middle East
http://www.cnn. com/WORLD/#mideast

This 24-hour news channel often focuses on the Middle East and is updated every few hours.

C-SPAN Online
http://www.c-span.org

See especially C-SPAN International on the Web for International Programming Highlights and archived C-SPAN programs.

Library of Congress
http://www.loc.gov

An invaluable resource for facts and analysis of 100 countries' political, economic, social, and national-security systems and installations.

ReliefWeb
http://www.reliefweb.int/w/rwb.nsf

UN's Department of Humanitarian Affairs clearinghouse for international humanitarian emergencies. It has daily updates, including Reuters and Voice of America.

United Nations
http://www.unsystem.org

The official Web site for the United Nations system of organizations. Everything is listed alphabetically, and data on UNICC and Food and Agriculture Organization are available.

UN Development Programme (UNDP)
http://www.undp.org

Publications and current information on world poverty, Mission Statement, UN Development Fund for Women, and much more. Be sure to see the Poverty Clock.

UN Environmental Programme (UNEP)
http://www.unep.org

Official site of UNEP with information on UN environmental programs, products, services, events, and a search engine.

U.S. Central Intelligence Agency Home Page
http://www.cia.gov/ index.htm

This site includes publications of the CIA, such as the World Factbook, Factbook on Intelligence, Handbook of International Economic Statistics, CIA Maps and Publications, and much more.

U.S. Department of State Home Page
http://www.state.gov/ www/ind.html

Organized alphabetically (i.e., Country Reports, Human Rights, International Organizations, and more).

World Health Organization (WHO)
http://www.who.ch

Maintained by WHO's headquarters in Geneva, Switzerland, the site uses Excite search engine to conduct keyword searches.

MIDDLE EAST SITES

Access to Arabia
http://www.accessme.com

Extensive information about traveling and working in the Arab world is presented on this Web site.

ArabNet
http://www.arab.net

This site is an extensive online resource for the Arab world in the Middle East and North Africa. There are links to every country in the region, covering current news, history, geography, culture, government, and business topics.

Arabia.On.Line
http://www.arabia.com

Discussions of Arab news, business, and culture are available at this site.

Camera Media Report
http://world.std.com/~camera/

This site is run by the Committee for Accuracy in Middle East Reporting in America, and it is devoted to fair and accurate coverage of Israel and the Middle East.

Center for Middle Eastern Studies
http://w3.arizona.edu/~cmesua/

This Web site is maintained by the University of Arizona Center for Middle Eastern Studies. The Center's mission is to further understanding and knowledge of the Middle East through education.

Center for Middle Eastern and Islamic Studies
http://www.dur. ac.uk/~dme0www/

The University of Durham in England maintains this site. It offers links to the University's extensive library of Middle East information; the Sudan Archive is the largest collection of documentation outside of Sudan itself.

The Middle East Institute
http://www.mideasti.org

The Middle East Institute is dedicated to educating Americans about the Middle East. The site offers links to publications, media resources, and other links of interest.

Middle East Internet Pages
http://www.middle-east-pages. com

A large amount of information on specific countries in the Middle East can be obtained on this site. Their engine allows you to browse through virtually every aspect of Middle East culture, politics, and current information.

Middle East Policy Council
http://www.mepc.org

The purpose of the Middle East Policy Council's Web site is to expand public discussion and understanding of issues affecting U.S. policy in the Middle East.

Middle East Times
http://metimes.com

The *Middle East Times* is a source for independent analysis of politics, business, religion, and culture in the Middle East.

Middle Eastern and Arab Resources
http://www.ionet.net/~usarch/WTB-Site.shtml

This omnibus site offers extensive information on each of the Middle Eastern countries. Scroll to their flags.

ISRAEL SITES

The Abraham Fund
http://www.coexistence.org

The goal of peaceful coexistence between Jews and Arabs is the theme of this site. Information to various projects and links to related sites are offered.

Zionist Archives
http://www.wzo.org.il/cza

This site is the official historical archives of the World Zionist Organization, the Jewish Agency, the Jewish National Fund, Karen Hayesod, and the World Jewish Congress.

We highly recommend that you review our Web site for expanded information and our other product lines. We are continually updating and adding links to our Web site in order to offer you the most usable and useful information that will support and expand the value of your book. You can reach us at: *http://www.mhcls.com*.

The United States (United States of America)

GEOGRAPHY

Area in Square Miles (Kilometers): 3,717,792 (9,629,091) (about 1/2 the size of Russia)

Capital (Population): Washington, DC (3,997,000)

Environmental Concerns: air and water pollution; limited freshwater resources, desertification; loss of habitat; waste disposal; acid rain

Geographical Features: vast central plain, mountains in the west, hills and low mountains in the east; rugged mountains and broad river valleys in Alaska; volcanic topography in Hawaii

Climate: mostly temperate, but ranging from tropical to arctic

PEOPLE

Population

Total: 280,563,000

Annual Growth Rate: 0.89%

Rural/Urban Population Ratio: 24/76

Major Languages: predominantly English; a sizable Spanish-speaking minority; many others

Ethnic Makeup: 77% white; 13% black; 4% Asian; 6% Amerindian and others

Religions: 56% Protestant; 28% Roman Catholic; 2% Jewish; 4% others; 10% none or unaffiliated

Health

Life Expectancy at Birth: 74 years (male); 80 years (female)

Infant Mortality: 6.69/1,000 live births

Physicians Available: 1/365 people

HIV/AIDS Rate in Adults: 0.61%

Education

Adult Literacy Rate: 97% (official)

Compulsory (Ages): 7–16; free

COMMUNICATION

Telephones: 194,000,000 main lines

Daily Newspaper Circulation: 238/1,000 people

Televisions: 776/1,000 people

Internet Users: 165,750,000 (2002)

TRANSPORTATION

Highways in Miles (Kilometers): 3,906,960 (6,261,154)

Railroads in Miles (Kilometers): 149,161 (240,000)

Usable Airfields: 14,695

Motor Vehicles in Use: 206,000,000

GOVERNMENT

Type: federal republic

Independence Date: July 4, 1776

Head of State/Government: President George W. Bush is both head of state and head of government

Political Parties: Democratic Party; Republican Party; others of relatively minor political significance

Suffrage: universal at 18

MILITARY

Military Expenditures (% of GDP): 3.2%

Current Disputes: various boundary and territorial disputes; "war on terrorism"

ECONOMY

Per Capita Income/GDP: $36,300/$10.082 trillion

GDP Growth Rate: 0%

Inflation Rate: 3%

Unemployment Rate: 5.8%

Population Below Poverty Line: 13%

Natural Resources: many minerals and metals; petroleum; natural gas; timber; arable land

Agriculture: food grains; feed crops; fruits and vegetables; oil-bearing crops; livestock; dairy products

Industry: diversified in both capital and consumer-goods industries

Exports: $723 billion (primary partners Canada, Mexico, Japan)

Imports: $1.148 trillion (primary partners Canada, Mexico, Japan)

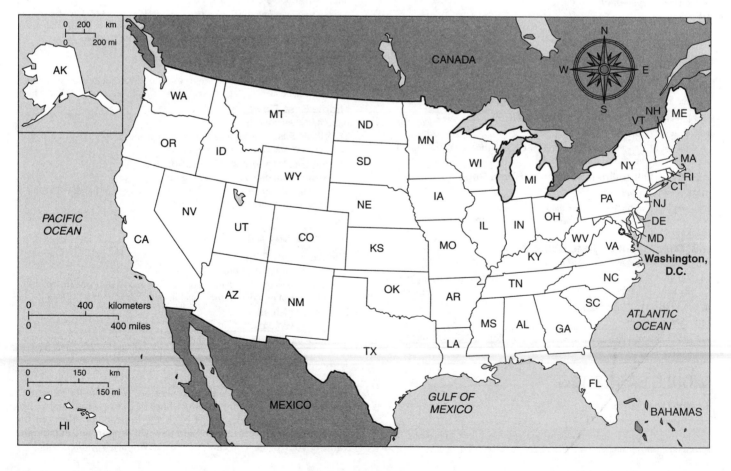

Canada

GEOGRAPHY

Area in Square Miles (Kilometers): 3,850,790 (9,976,140) (slightly larger than the United States)

Capital (Population): Ottawa (1,094,000)

Environmental Concerns: air and water pollution; acid rain; industrial damage to agriculture and forest productivity

Geographical Features: permafrost in the north; mountains in the west; central plains; lowlands in the southeast

Climate: varies from temperate to arctic

PEOPLE

Population

Total: 31,903,000

Annual Growth Rate: 0.96%

Rural/Urban Population Ratio: 23/77

Major Languages: both English and French are official

Ethnic Makeup: 28% British Isles origin; 23% French origin; 15% other European; 6% others; 2% indigenous; 26% mixed

Religions: 46% Roman Catholic; 36% Protestant; 18% others

Health

Life Expectancy at Birth: 76 years (male); 83 years (female)

Infant Mortality: 4.95/1,000 live births

Physicians Available: 1/534 people

HIV/AIDS Rate in Adults: 0.3%

Education

Adult Literacy Rate: 97%

Compulsory (Ages): primary school

COMMUNICATION

Telephones: 20,803,000 main lines

Daily Newspaper Circulation: 215/1,000 people

Televisions: 647/1,000 people

Internet Users: 16,840,000 (2002)

TRANSPORTATION

Highways in Miles (Kilometers): 559,240 (902,000)

Railroads in Miles (Kilometers): 22,320 (36,000)

Usable Airfields: 1,419

Motor Vehicles in Use: 16,800,000

GOVERNMENT

Type: confederation with parliamentary democracy

Independence Date: July 1, 1867

Head of State/Government: Queen Elizabeth II; Prime Minister Jean Chrétien

Political Parties: Progressive Conservative Party; Liberal Party; New Democratic Party; Bloc Québécois; Canadian Alliance

Suffrage: universal at 18

MILITARY

Military Expenditures (% of GDP): 1.1%

Current Disputes: maritime boundary disputes with the United States

ECONOMY

Currency ($U.S. equivalent): 1.46 Canadian dollars = $1

Per Capita Income/GDP: $27,700/$875 billion

GDP Growth Rate: 2%

Inflation Rate: 3%

Unemployment Rate: 7%

Labor Force by Occupation: 74% services; 15% manufacturing; 6% agriculture and others

Natural Resources: petroleum; natural gas; fish; minerals; cement; forestry products; wildlife; hydropower

Agriculture: grains; livestock; dairy products; potatoes; hogs; poultry and eggs; tobacco; fruits and vegetables

Industry: oil production and refining; natural-gas development; fish products; wood and paper products; chemicals; transportation equipment

Exports: $273.8 billion (primary partners United States, Japan, United Kingdom)

Imports: $238.3 billion (primary partners United States, European Union, Japan)

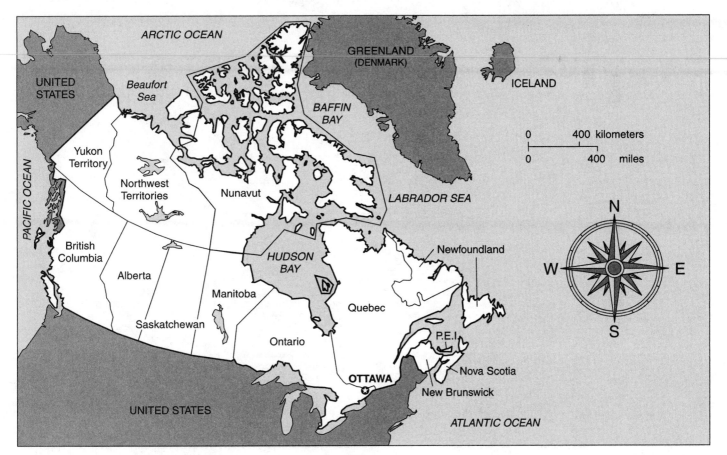

GLOBAL ✦ STUDIES

This map is provided to give you a graphic picture of where the countries of the world are located, the relationship they have with their region and neighbors, and their positions relative to major powers and power blocs. We have focused on certain areas to illustrate these crowded regions more clearly.

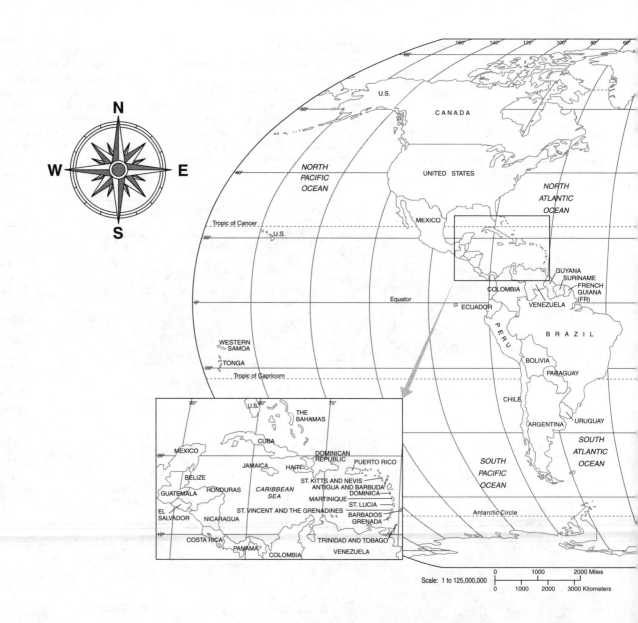

Scale: 1 to 125,000,000

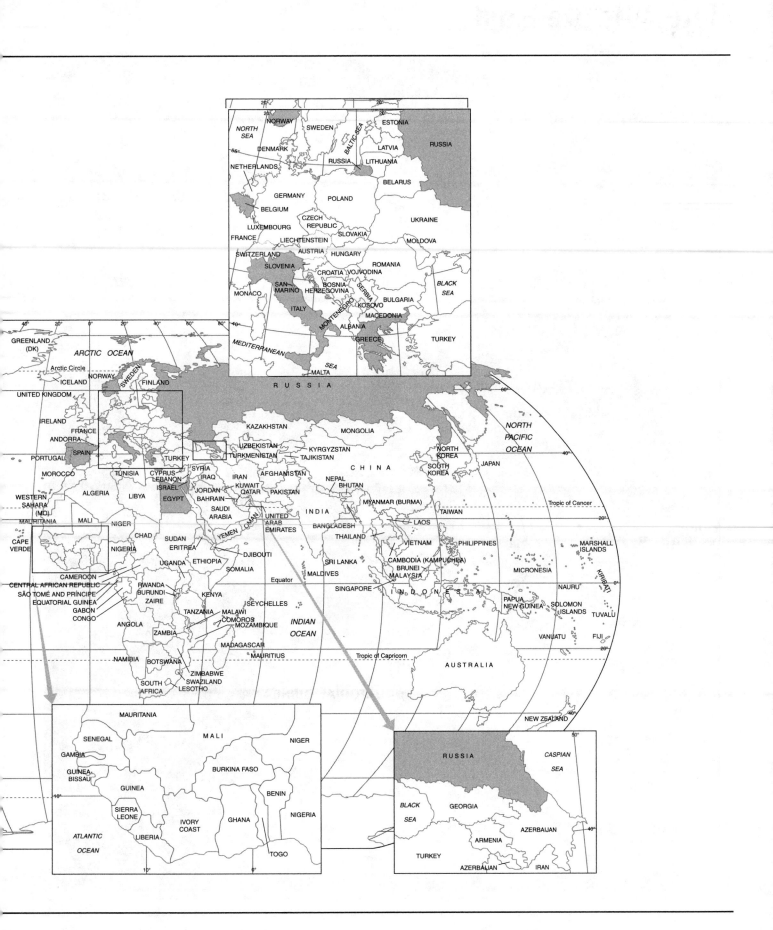

The Middle East

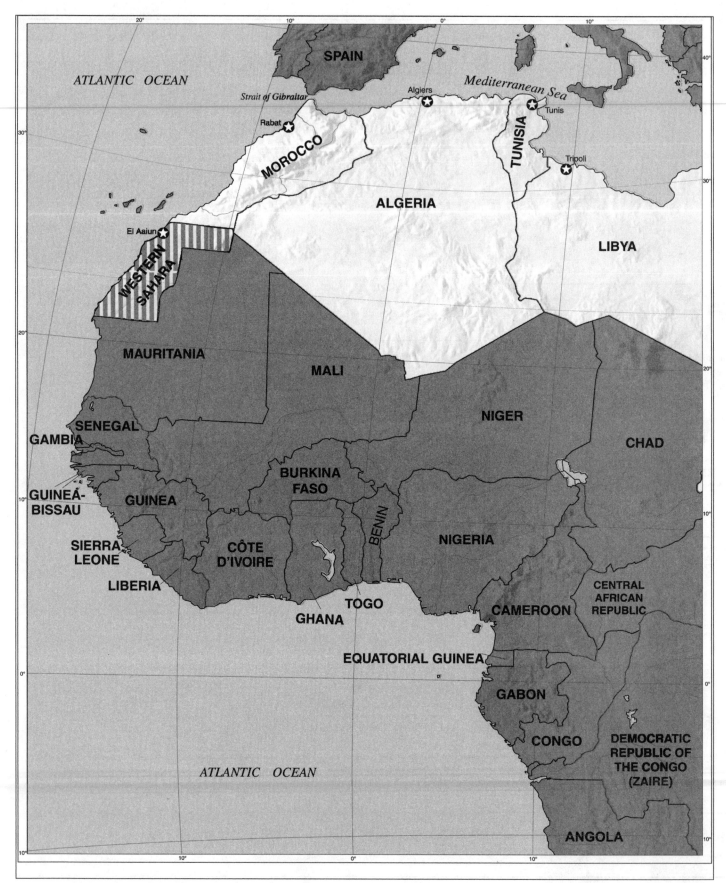

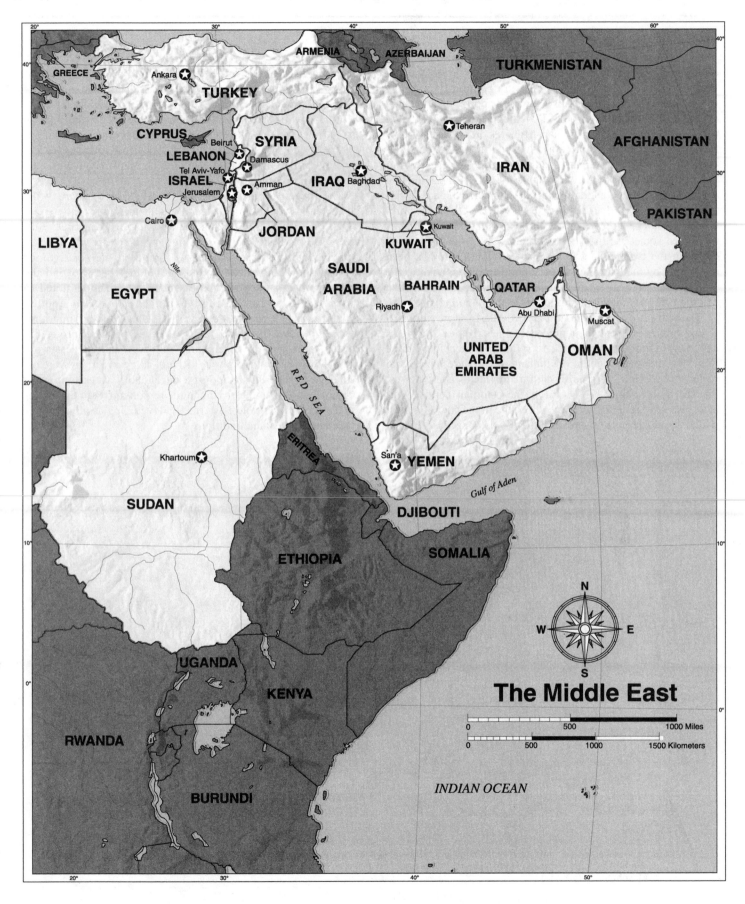

The Middle East

The Middle East: Theater of Conflict

The Middle East was formerly divided, in British and U.S. diplomatic parlance, into "Near" and "Middle" (plus the Far East) covering the rest of Asia, but in common present-day usage, it describes the geographical region approximately equal in size to the continental U.S. and extends from the Atlantic coast of Morocco to the Iranian-Afghan border. The Middle East is thus intercontinental rather than continental, with the diversity of topography, climate, and physical and social environments characteristic of the two continents, Africa and Asia, that define its territory. Geography and location have dictated a significant role in world affairs for the Middle East throughout recorded history; humankind's earliest cities, governments, organized societies, and state conflicts were probably located there. In the twentieth century, this traditional role has been confirmed by the exploitation of mineral resources vital to the global economy and by the rivalries of nations that regard the Middle East as strategically important to their national interests.[1]

The nations of the contemporary Middle East are very different, however, from their predecessors of 100 or 200 years ago. One important difference is political. When the United States became independent of England, there were three more or less "sovereign" Middle Eastern nation-states and empires: the Sherifian Sultanate of Morocco; the Ottoman Turkish Empire; and Iran, reunited by force under the new Qajar Dynasty, which would remain in power until it was succeeded by the Pahlavi Dynasty in the 1920s. These three states were still in place late in the nineteenth century, but European influence and control over their rulers had effectively robbed them of most of their independence. Since then—a process accelerated since World War II—the Middle Eastern map has been redrawn many times. The result of the redrawing process is the contemporary Middle East, 20 independent states with diverse political systems overlaying a pastiche of ethnic groups, languages, customs, and traditions.

The diversity of these states is compensated for, in part, by the cohesion provided by various unifying factors. One of these factors is geography. The predominance of deserts, with areas suitable for agriculture compressed into small spaces where water was available in dependable flow, produced the *oasis-village* type of social organization and agricultural life. Beyond the oases evolved a second type of social organization suited to desert life, a less settled lifestyle termed *nomadism*. Another type of village settlement evolved in plateau and mountain regions, wherever the topography afforded physical protection for the defense of the community. In Egypt, and to a lesser extent in the Tigris and Euphrates River Valleys, *villages* were established to take advantage of a dependable water supply for crop irrigation. Peoples living in the region mirrored these lifestyles, with the Middle Eastern city developing as an urban refinement of the same traditions.

Traditionally the world's earliest civilizations organized into societies with a social structure, governments, urban centers, agricultural surpluses, and such technological essentials as a written language, accounting procedures and a hierarchical system of divinities, developed first in Mesopotamia and then in China. However recent research has added the civilizations of Mesoamerica to the list.[1]

The broad set of values, traditions, historical experiences, kinship structures, and so on, usually defined as "culture," is a second cohesive factor for the Middle East's peoples. Islam, for example, is either the official state religion or the leading religion in all but one (Israel) of the states. The Arabic language, due to its identification with Islam, is a bond even for those peoples who use another spoken and/or written language (such as Turkish, Hebrew, or Farsi); and, in any case, the social order of Islam is another unifying force.

A third unifying factor, while it is intangible and difficult to define, is a common historical experience. Without exception, the states of the Middle East are the products of twentieth-century international politics and the conflict of interests of outside powers. Clashing national interests and external involvement in regional affairs have set the tone for the internal and regional conflicts of Middle Eastern states. Thus, the intercommunal violence in modern Lebanon has its roots in foreign (French and British) support for various communal groups in the 1860s, setting the groups against one another under the guise of protecting them from the Ottoman government. European intervention contributed significantly to the conflict process. Robert Aumann, the Israeli scholar who shared the Nobel Peace Prize in Economics with a U.S. scholar for their work on game theory, noted in his acceptance speech that the Arab-Israeli conflict had been going on for 80 years, roughly the period of European rule in the region. But that one is only among the more recent millennial conflicts, as invaders swept across the Middle East, conquering, being conquered by other invaders, eventually being absorbed into the larger society.

THE LAND ISLAND

Until recently, the Middle East was compartmentalized. Its peoples had little awareness of one another and even less of the outside world. Peoples who lived on or near the shores of the Mediterranean had more contact with each other and with Europeans, forming in general a broad trading community. But in the inland areas distances were great and communications poor. Often residents of one village knew nothing of those who lived in villages a few miles away. Travel throughout the region was slow and often dangerous. Caravans crossing the Sahara, for example, needed escorts from the nomadic Tuareg, the "Blue Men" so-called from their indigo robes, who would provide protection and guidance across the unmarked desert for a fee. It took several months, sometimes longer, for edicts from the Caliph in Baghdad to reach his governors in Spain or North Africa.

Consequently, the combination of vast distances, poor communications, and geographical isolation brought about the early development of subregions within the larger Middle East. As early as the tenth century A.D., three such subregions had been defined: North Africa, the Arab lands traditionally known to Europeans as the Near East, and the highland plateaus of Turkey and Iran. In the twentieth century, these three areas were further separated from one another by foreign political control with the establishment of League of Nations mandates over the Arab lands and recognition of French protectorates over Morocco and Tunisia. (Algeria, originally a French colony, became a department of metropolitan France in 1871.) With Egypt already a British protectorate (as of 1882), only the truncated Ot-

United Nations Photo (UN152627)

Life in the Sahara Desert is often perceived as nomadic, with the people living in tents and riding camels. To some extent this is still true, but there are many towns that offer a more settled way of life, such as this town in the Algerian Sahara.

toman empire and Qajar Iran remained as more or less sovereign countries on the eve of World War I.

In some respects the Middle East illustrates the theory developed by Admiral Alfred Thayer Mahan, distinguished naval officer and historian, many years ago. Mahan viewed the region from a naval perspective as a "land island" vulnerable to control by outside powers when their national interests seemed to be at stake. He based his theory on the nineteenth century struggle between Britain and Russia, the superpowers of the period, for control of greater Asia and by extension the Middle East. This struggle is often described historically as "The Great Game," with the two maneuvering as if on a checkerboard, or more romantically as "Tournament of Shadows" since it involved venturing into unknown areas and conducting warfare by proxies. Thus the British sought to protect India, their imperial prize, by a land barrier extending northward through Afghanistan to Central Asia. On their side the Russians worked to expand their land empire southward in search of a warm-water port. The overthrow of the Czarist empire and more recently the collapse of the Soviet Union essentially ended the Great Game, which was in any case unwinnable. But the extension of U.S. reach into its former Central Asian republics to ensure future oil needs and counter Russian and Iranian interests has in a sense revived it.[2]

The Ronald Reagan administration's commitment to a strongly anti-Communist policy revived the Great Game in new locations in the 1980s. Thus President Reagan insisted in 1983–1984 that American marines were in Lebanon to defend vital U.S. interests. In 1987, the United States accepted a request from Kuwait to "re-flag" Kuwaiti tankers in the Persian Gulf and provide them with naval protection, ostensibly to thwart Iran but also to forestall a Soviet move into the region. The U.S. "national interest" has been invoked on numerous occasions to justify military intervention or warnings thereof, most recently in Iraq in 2003, and with the Iraqi invasion and subjoined Israeli-Palestinian conflict it has become the primary focus of American foreign policy.

SUBREGIONAL CONFLICTS

The periodic outbreak of local or subregional conflicts characteristic of Middle Eastern societies, which stem from their tribal or ethnic origins, has brought the region to the forefront of world affairs in recent years. Several of these conflicts have generated international involvement, in particular the 1991 Gulf War and the long-running Arab-Israeli conflict. Others, such as the civil war in Lebanon and the conflict between North and South Sudan, remained essentially internal in nature.

Although thus far these Middle Eastern conflicts have been confined to their areas of origin or mediated by outside powers to reduce tension levels, some government policy makers continue to fear that they might spread and involve other nations in a wider war, possibly proving or at least demonstrating the effectiveness of the "domino theory" often invoked as a guide to modern international relations. The domino theory holds that tensions or unresolved disputes between two nations will widen as neighboring nations are drawn inevitably into the dispute, even without taking sides. The uninvolved nations will then become involved, as the particular dispute becomes buried in the rivalries of competing national interests. At some point, a specific incident ignites a general war, as nation after nation falls like a domino into the widening conflict. The classic example of the theory is World War I.

While the applicability of the domino theory to the Middle East has yet to be proven, regional conflict there thus far has not affected long-term global commerce or national survival, and international terrorist acts identified with Middle Eastern governments remain sporadic and uncoordinated. But there are very real limits to involvement or effective management, even by the superpowers. President Reagan recognized these limits implicitly by withdrawing American marines from Lebanon; and when Egypt's President Anwar al-Sadat ordered the withdrawal of all Soviet military advisers from his country some years ago, home they went.

A final point about these conflicts is that they are all direct results of European intervention in the Middle East. For much of its history, the Middle East was a region without defined borders, other than the intangible limits fixed for Muslims by their religion. Even the Ottoman Empire, the major power in the region for more than five centuries, did not mark off its territories into provinces with precise boundaries until well into the 1800s.

IRANIANS AND ARABS

Iranians (or Persians) and Arabs are nearly all Muslims. But they have very different ethnic origins, linguistic and geographical backgrounds, and histories.

The Iranians were a loosely organized nomadic people from Central Asia who migrated into the Iranian plateau some 3,000 years ago, settling in the province of Parsa (hence the name Persian) as sedentary farmers and herders. They gradually expanded their territory at the expense of the earlier inhabitants, and in about 600 B.C. they joined forces with the Medes, another tribal people, under the leadership of the Persian chief Cyrus. In a few years they conquered other Near Eastern peoples to form the world's first true empire, in the sense of many different peoples ruled by a single ruler. Cyrus's successors, the Achaemenian Dynasty, expanded Iranian territory east to the Indus River and westward to the edge of Asia on the Mediterranean Sea. Iranians have retained a lofty sense of their many important contributions to civilization beginning with this period. Except in Iraq and the Persian Gulf states, where there are significant Iranian communities, Iranians have largely remained in their country of origin.

The Arabs, whose Semitic language is different in roots and structure from Farsi (Persian) although they share alphabets, were the original population of the Arabian Peninsula. Historically they were divided into often-competing tribal clans and groups, the majority nomads (Bedouin) roaming the deserts but with a small urban element in trading cities such as Mecca. The element which eventually united them as a people was religion. Islam was brought to them by the Prophet Muhammad in the seventh century A.D. After his death, they expanded into other areas of the Middle East, first as conquerors, and then as settlers. Arabs form the majority in the broad area stretching from Morocco on the west, across North Africa, eastward to Egypt, Sudan, the "Arab states" and Arabia proper. There are also large Arab minorities in Iran and Israel.

But the European powers brought a different set of rules into the area. They laid down fixed borders sanctified by treaties, played ruler against ruler, divided and conquered. It was this European ascendancy, building on old animosities while creating new ones, that laid the groundwork for today's conflicts.

THE IRAN–IRAQ WAR: BATTLE OF ISLAMIC BROTHERS

The Iran–Iraq War broke out in September 1980, when Iraqi forces invaded Iran and occupied large portions of Khuzestan Province. This measure was in retaliation for Iranian artillery attacks across the border and efforts by Iranian agents to subvert the Iraqi Shia Muslim population, along with propaganda broadcasts urging Iraqis to overthrow the Iraqi regime of Saddam Hussein. But, as is the case with most Middle Eastern conflicts, the causes of the war are complex.

One factor is the ancient animosity between Iranians and Arabs, which dates back to the seventh century A.D., when invading Arab armies overran the once powerful Sassanid Empire of Iran, defeating the Iranian Army at the famous Battle of Qadisiya in 637. The Iranians were converted to Islam with relative ease, yet they looked down on the Arabs as uncivilized nomads who needed to be taught the arts of government and refined social behavior. The Arabs, in turn, despised the Iranians

Courtesy of informationwar.org (INFOWAR001)
A group of Iraqi prisoners being indoctrinated in Iran's Islamic revolution.

Saddam Hussain, like its predecessors, is paranoid about opposition in general, but about Shia opposition in particular.[4]

for what they considered their effeminateness—their love of gardens and flowers, their appreciation of wine and fine banquets. These attitudes have never entirely disappeared.[3] After the 1980 invasion, Iraq's government-controlled press praised it as Saddam Hussain's Qadisiya, reminding readers of the earlier Arab success.

Iran and Iraq have been at swords' points over a number of issues in recent years. One is occupation of three small islands at the mouth of the Persian Gulf. The British had included these islands in their protectorate over eastern Arabia and transferred them to the United Arab Emirates after that country became independent. But Shah Mohammed Reza Pahlavi, then the leader of Iran, contested the transfer, on the grounds that historically they had belonged to Iran. In 1971, an Iranian commando force seized the islands. Although the islands had never belonged to Iraq, the Iraqis denounced the occupation as a violation of *Arab* sovereignty and mounted a campaign among the predominantly Arab population of Iran's Khuzestan Province, adjacent to the border, to encourage them to revolt against the central government. The campaign failed, as would a later effort made by Saddam when his forces invaded Iran, and the islands remain under Iranian control.

Another issue was the shah's support for Kurdish guerrillas who had been fighting the Iraqi government for years to obtain autonomy for their mountain region. The shah also resented Iraq's grant of asylum to the Ayatollah Khomeini in 1963, because of the religious leader's continued anti-shah activities and propaganda broadcasts into Iran.

These disagreements intensified after the overthrow of the shah in 1979. Iraq accused the Khomeini regime of mistreatment of Khuzestan Arabs and of sending agents to incite its own Shia Muslim population to rebel. Iraqi governments have been dominated by the Sunni Muslim population since independence, although more than half of the population are Shia. The regime of

The personal hatred between Saddam and Khomeini certainly contributed to the war. The two had been bitter enemies since 1978, when Saddam ordered Khomeini expelled from Iraq and accused him of working with Iraqi Shia Muslim leaders to undermine the regime. But differences in their views on the nature of authority and of social development also set the two leaders in opposition. For Saddam Hussain, the development of Islamic society to the fullest is best achieved by a secular Socialist party (e.g., the Ba'th); Islam is tangential. Khomeini, in the republic that he fashioned for Iran based on his Islamic political philosophy, argued for authority to be vested in religious leaders like himself, since they are qualified by wisdom, moral uprightness, and insight to know what is best for the Islamic community.

One issue often overlooked as a cause of the war is a territorial dispute, dating back many centuries, that has been aggravated by European intervention in the Middle East. The dispute concerns the Shatt al-Arab, the 127-mile waterway from the junction of the Tigris and Euphrates Rivers south to the Persian Gulf. The waterway was a bone of contention between the Ottoman and Iranian Empires for centuries, due to its importance as a trade outlet to the Gulf. It came entirely under Ottoman control in the nineteenth century. But with the collapse of the Ottoman Empire in World War I, the new kingdom of Iraq, set up by Britain, came in conflict with a revitalized Iran over navigation and ownership rights. Iran demanded ownership of half the Shatt al-Arab under international law, which would mean to mid-channel at the deepest point. Iraq claimed the entire waterway across to the Iranian side. Conflict intensified as both countries built up their oil exports in the 1960s and 1970s. In 1969, the shah of Iran threatened to occupy Iran's side of the waterway with gunboats, and he began a program of military support to Kurdish (Sunni Muslim) rebels fighting the Iraqi government.

Iran was much wealthier and militarily stronger than Iraq at that time, and Iraq could do little about Iranian support for the Kurds. But the Iraqis did have the Shatt al-Arab as a bargaining chip, in that their rights to it were embodied in several treaties. In 1975, after lengthy negotiations, Houari Boumedienne, then the president of Algeria, interrupted an oil ministers' conference in Algiers to announce that "our fraternal countries Iran and Iraq have reached agreement on their differences."[5] Iraq agreed to recognize Iranian ownership of the Shatt from bank to mid-channel, and Iran agreed to stop supporting Kurdish rebels in Iraq.

The advantage to Iraq of bringing an end to the Kurdish rebellion was offset by the humiliation felt by Iraqi leaders because they had bartered away a part of the sacred Arab territory. Hussain considered the agreement a personal humiliation because he had been the chief negotiator. When he became president, he said that he had negotiated it under duress and that Iraq would one day be strong enough to revoke it.[6]

The overthrow of the Shah and the consequent weakening of his army after the establishment of the Islamic Republic seemed to Saddam Hussain an excellent opportunity to restore Iraqi pride at the expense of an inferior adversary. In September 1980, he announced that the 1975 treaty was null and void, and he demanded Iran's recognition of Iraqi sovereignty over the entire Shatt al-Arab. He also called for the return of the three islands seized by the shah's forces in 1971 and the transfer of predominantly Arab areas of Khuzestan to Iraqi control. Although the two countries were roughly equal in military strength at the time, purges in Iranian Army leadership, low morale, and lack of spare parts for weapons due to the U.S. economic boycott convinced Saddam that a limited attack on Iran would almost certainly succeed.

However, the quick and easy victory anticipated by the Iraqis did not materialize. Political expectations proved equally erroneous. Iraq had expected the Arabs of Khuzestan to support the invasion, but they remained loyal to Iran's Khomeini regime. The Iraqi forces failed to capitalize on their early successes and were stopped by determined Iranian resistance. The war quickly turned into a stalemate.

In 1981–1982, the momentum shifted strongly in Iran's favor. The war became a patriotic undertaking as thousands of volunteers, some barely teenagers, headed for the front. An Iranian operation, appropriately code-named Undeniable Victory, routed three Iraqi divisions. Iran's blockade of Iraqi oil exports put a severe strain on the Iraqi economy. After the defeat, Saddam withdrew all Iraqi forces from Iranian territory and asked for a cease-fire. But Iran refused; Khomeini set the ouster of "the traitor Saddam" as a precondition for peace.

Iraqi forces fared better on their own soil and threw back a number of large-scale Iranian assaults, with huge casualties. Subsequent Soviet deliveries of missiles and new aircraft gave Iraq total air superiority. In early 1985, the Iraqis launched a campaign of "total war, total peace," combining air raids on Iranian ports and cities with an all-out effort to bring international pressure on Iran to reach a settlement.

In March 1985, Iranian forces launched another major offensive toward Basra from their forward bases in the Majnoon Islands, deep in the marshes, which they had captured by surprise in 1984. Although they were driven back with heavy losses, a year later, the Iranian forces captured the Fao (Faw) Peninsula southeast of Basra in another surprise attack and moved to within artillery range of Iraq's second city.

With the ground war stalemated, conflict shifted in 1986 and 1987 to the sky and sea lanes. Iraq's vast air superiority enabled the country to carry the war deep into Iranian territory, with almost daily bombing raids on Iranian cities, industrial plants, and oil installations.

But the most dangerous aspect of the conflict stemmed from Iraqi efforts to interdict Iranian oil supplies in order to throttle its enemy's economy. The war had a high-risk potential for broader regional conflict from the start, and in 1984, Iraqi missile attacks on tanker traffic in the Persian Gulf came close to involving other states in the region in active participation.

The internationalization of the war, which had been predicted by many analysts, became a reality in its seventh year, like a plague of locusts. The secret dealings with the United States (revealed in the 1987 Iran–Contra hearings) had immeasurably strengthened Iran's air power and defenses; Iraq lost one fifth of its aircraft in a series of battles in the marshes. Iranian arms dealers were successful in purchasing weaponry from many sources. One of their major suppliers was China, from which they purchased a number of Silkworm missiles, which were installed at secret launching sites along the coast facing the Strait of Hormuz and the Fao Peninsula. At the same time, Iranian Revolutionary Guards established bases in various small harbors from whence they could mount missile and grenade attacks in fast patrol boats against ships passing in the Gulf. The government warned that tankers bound for Kuwait and other Gulf ports would be attacked if Iraq continued its air raids.

The direct cause of the internationalization of the war, however, was an Iraqi air raid on the U.S. naval frigate *Stark* on May 17, 1987. Thirty-seven American sailors were killed in the raid. (Although more than 200 ships had been attacked by Iraq or Iran since 1984, the *Stark* was the first warship attacked, and it suffered the heaviest casualties.) Saddam apologized, calling the attack a "tragic mistake." The United States drastically increased its naval forces in the Gulf and, in the following month, accepted a request from Kuwait for tanker protection under the American flag, along with naval escorts. In June 1987, the first convoy of "reflagged" Kuwaiti tankers traversed the Gulf without incident, escorted by U.S. warships and overflying jets from the aircraft carrier *Constellation*.

Predictably, Iran's threat to make the Gulf "safe for every one or no one," following the U.S. buildup in the region, affected nearby countries as well as international shipping. Saboteurs blew up oil installations and factories in the United Arab Emirates, Bahrain, and Saudi Arabia. Revolutionary Guardsmen carried out their earlier threats with hit-and-run strafing and grenade attacks on passing tankers. But the most serious danger came from floating mines strewn at random in shipping lanes. After a number of tankers had been damaged, the United States and several European countries previously uninvolved in the conflict, notably Italy, began sending minesweepers to the area.

With the Gulf in a state of high tension, the United Nations (UN) Security Council mounted a major effort to end the war. In July 1987, the Security Council unanimously approved Resolution 598. It called for an immediate cease-fire, the withdrawal of all forces to within recognized international boundaries, repatriation of all prisoners, and negotiations under UN auspices for a permanent peace settlement. Iraq accepted the resolution, but Iran temporized. Its president, Ali Khamenei, told the UN General Assembly: "The Security

Council's stance in relation to the war imposed on us has not changed up to this moment."[7]

A year later, though, Iran accepted *Resolution 598,* in an abrupt about-face. A number of factors combined to bring about this change, but the principal one was probably Iraqi success on the battlefield. Early in 1988, Republican Guard units specially trained in chemical warfare recaptured Fao in a massive assault, using nerve gases such as tabun and sarin along with mustard gas to thwart counterattacks. These chemical weapons worked with deadly effectiveness against the "human wave" tactics of teenage Iranian volunteers. Meanwhile, Saddam Hussain's crash program of development of long-range missiles enabled his forces to hit cities and military installations deep inside Iran and bring a further drop in Iranian morale.

Khomeini's death in June 1989 removed a major obstacle to peace negotiations. He had been persuaded only with great difficulty to approve the cease-fire, and his uncompromising hatred of Saddam Hussain was not shared by many of his associates.

A real peace settlement would enable both regimes to turn their full attention to the enormous problems of reconstruction. Unfortunately, their diametrically opposed positions on war gains worked against dialogue. Iran insisted on the withdrawal of Iraqi troops from its territory as a first step, while Iraq demanded that prisoner exchanges and clearing of the Shatt al-Arab should precede withdrawal.

The 1990 Iraqi invasion of Kuwait brought about an important change in the relationship. Urgently in need of allies, Saddam Hussain abruptly agreed to the original peace terms set by the United States and accepted by Iran. These terms required Iraqi troop withdrawal from occupied Iranian territory along with prisoner exchanges and clearance of mines and other obstacles from the Shatt al-Arab. Iran stayed neutral during the Gulf War and provided sanctuary for Iraqi pilots fleeing Allied air attacks, although it impounded their aircraft, which as of this writing have not been returned.

A formal peace treaty has yet to be signed between these long-time hostile neighbors, although they have reestablished diplomatic relations and exchanged some prisoners. Iran released 2,939 Iraqi POWs in May 2000; most of them had been held since the early days of the war. But a full exchange of prisoners and sharing of navigation rights in the Shatt al-Arab, the latter being one of the main causes for the war, have yet to be completed.

THE GULF WAR AND ITS AFTERMATH

On August 2, 1990, the Iraqi Army, which had been mobilized along the border, invaded and occupied Kuwait, quickly overcoming light resistance as the ruling Kuwaiti emir and his family escaped into exile. The invasion climaxed a long dispute between the two Arab neighbors over oil-production quotas, division of output from the oil fields of the jointly controlled Neutral Zone along the border, and repayment of Iraqi debts to Kuwait from the war with Iran. Saddam Hussain had criticized Kuwait for producing more than its quota as allotted by the Organization of Petroleum Exporting Countries (OPEC), thus driving down the price per barrel and costing Iraq $7 to $8 billion in lost revenues. The Iraqi leader also charged Kuwait with taking more than its share of the output

of the Neutral Zone. The Iraqi charges found considerable support from other Arab states, most of which consider the Kuwaitis to be greedy and arrogant. However, an Arab League summit meeting of oil ministers failed to resolve the dispute. Kuwait agreed only to a month-long adherence to its OPEC quota and continued to press for repayment of Iraqi debts owed for aid during the war with Iran.

What had been initially an inter-Arab conflict was globalized by the invasion. Although Iraq called its occupation a recovery of part of the Arab homeland, which had been "stolen" from the Arabs by the British and given its independence under false premises, the action was viewed as aggression by nearly all the countries in the world. The UN Security Council on August 6 approved *Resolution 660,* calling for an immediate withdrawal of Iraqi forces from Kuwait and restoration of the country's legitimate government. Pending withdrawal, a worldwide embargo would be imposed on Iraq, covering both exports and imports and including medical and food supplies as well as military equipment. A similar resolution approved by the League of Arab States denounced Iraq's aggression against the "brotherly Arab state of Kuwait" and demanded immediate Iraqi withdrawal and restoration of Kuwaiti independence.

The invasion divided the Arab states, as several, notably Yemen and Sudan, agreed with Iraq's contention that Kuwait was historically part of Iraq and that Kuwaiti arrogance was partly responsible for the conflict. Others took the opposite view. Egyptian president Hosni Mubarak accused Saddam Hussain of breaking a solemn pledge not to invade Kuwait. Saudi Arabia, fearing that it might be Iraq's next victim, requested U.S. help under the bilateral defense treaty to protect its territory. U.S. president George Bush and Soviet president Mikhail Gorbachev issued a joint pledge for action to expel Iraqi forces from Kuwait. A massive military buildup followed, largely made up of U.S. forces, but with contingents from a number of other countries, including several Arab states. Although led by U.S. military commanders, the collective force operated under the terms of UN *Resolution 660* and was responsible ultimately to the Security Council as a military coalition.

The UN embargo continued in effect for six months but failed to generate an Iraqi withdrawal from Kuwait, despite its severe impact on the civilian population. (The only concession made by Saddam Hussain during that period was the release of foreign technicians who had been working in Kuwait at the time of the invasion). As a result, the coalition forces launched the so-called Operation Desert Storm on January 16, 1991. With their total air superiority and superior military technology, they made short work of Iraq's army, as thousands of Iraqi soldiers fled into the desert or surrendered where they were. Subsequently, on the express orders of President George Bush, the campaign was halted on February 7, after Iraqi forces had been expelled from Kuwait. Yet Saddam remained in power, and uprisings of the Kurdish and Shia populations in Iraq were ruthlessly crushed by the reorganized Iraqi Army, which remained loyal to its leader. Although the Bush administration was unwilling to commit American forces to assist these populations and presumably risk significant casualties in unfamiliar territory, the United States and other members of the UN Security Council established "no-fly zones" north of the 36th parallel and south of the 33rd parallel of longitude, which would be off limits to Iraqi forces. The zones effectively limited Iraq's sovereignty to approximately two thirds of its own territory.

Saddam Hussain's running battle with the United Nations kept world attention focused on Iraq in 1992–1993. Despite his country's sound defeat in the Gulf War, the Iraqi dictator had gained the support not only of some other Arab states but also of many developing-world leaders, for what appeared to them to have been an infringement on Iraq's sovereignty by the United Nations in its zeal to destroy Iraq's weapons of mass destruction. But for the United Nations and the United States, the main issue involved Iraq's noncompliance with UN resolutions. Thus *Resolution 687* directed the country to destroy all its long-range ballistic missiles and dismantle its chemical- and nuclear-weapons facilities, while *Resolution 715* would establish a permanent UN monitoring system, with surveillance cameras, for all missile test sites and installations as well as nuclear facilities. Iraq's compliance with these resolutions would end the embargo imposed after the invasion of Kuwait and would enable Iraq to sell $1.6 billion in oil to finance imports of badly needed medicines, medical supplies, and foodstuffs. Iraqi representatives argued that their country had complied with *Resolution 687* by demolishing under international supervision the al-Atheer nuclear complex outside Baghdad and by opening all missile sites to UN inspectors. But they said that *Resolution 715* was illegal under international law, since it infringed on national sovereignty.

For the next four years Saddam held firm in his refusal to accept the resolution, along with a later one, *Resolution 986*, which would allow Iraq to sell 700,000 barrels per day (b/d) for six months, in return for compliance. His refusal placed a terrible burden on the Iraqi people, who suffered from extreme shortages of food, medical and other humanitarian supplies. In 1996, the Iraqi government accepted UN terms to allow it to sell $2 billion worth of oil every six months, in return for opening all missile-testing sites and biological- and chemical-weapons facilities to inspectors. The oil revenues would be used for purchases of critically needed food, medicines, and children's supplies. However, the UN Security Council was divided, with the United States insisting on Iraq's adherence to all its obligations specified in *Resolution 715,* while other Council members argued that the embargo was hurting the most vulnerable groups in the population without affecting the leadership.

The standoff hardened in late 1997, when Saddam ordered the American members of the inspection team to leave his country, saying that they were spies. The Clinton administration threatened to use force to compel their return and beefed up U.S. military strength in the Persian Gulf. A UN–sponsored mediation effort temporarily averted the threat. In extending the oil-for-food program in 1999, the Security Council replaced the Special Commission on Iraq with a new body, the UN Monitoring, Verification and Inspection Commission (UNMOVIC). Its members would be recruited from various member states, excluding the U.S. However its first two directors resigned, along with the head of the World Food Program in Iraq, to protest UN failure to alleviate the hardships imposed on Iraq's people by the sanctions.

Iraq's adamant refusal to accept renewed inspections along with the patrolling by U.S. and British aircraft of the "no-fly zones" led to increased attacks on the aircraft by Iraqi gunners. The United States and Britain, in response, began bombing military targets within the country, aimed particularly at Iraqi air-

defense and communications targets. The conflict escalated in December 1998, when some 100 such installations were either damaged or destroyed. The bombings were justified not only in response, but also to destroy installations used allegedly to produce weapons of mass destruction. In 2001, the attacks were scaled back. By that time Iraqi gunners were firing surface-to-air missiles (SAMS) at the aircraft. But in the course of 200,000 sorties none were hit. In August 2001 both overflights and Iraqi anti-aircraft fire increased significantly as the U.S. and Britain expanded their attacks on the country's air defense system. Overall there were 370 Iraqi violations of the no-fly zone in 2001, compared with 221 in 2000.

Saddam Hussain's determination to end the sanctions without preconditions, along with elimination of the no-fly zones as unwarranted interference in Iraq's internal affairs, generated a deadlock between the UN and its most recalcitrant member state. In December 2001 the Security Council renewed the sanctions for an additional six months, under a compromise resolution backed by the U.S. and Britain. It revised the oil-for-food program to establish a "goods review list." Items on the list that could be used for either civilian or military purposes would have to be approved by all Council members states before they could be imported into Iraq. Items not on the list could be imported without restrictions.

The September 11, 2001 attacks on mainland America changed the equation. President George Bush included Iraq in his "axis of evil" of states sponsoring terrorism as part of his global campaign against Osama bin Laden and his Al-Qaeda terrorist network. At the same time his administration began a massive buildup of military forces in the middle East in preparation for a presumed attack on Iraq as a part of that "axis".

The military buildup, among other factors, led Saddam to change his mind and admit the UNMOVIC inspection team. The inspectors then proceeded to Iraq, and this time received greater cooperation, even to opening of Saddam's palaces for inspection. Subsequently, in January 2003 Hans Blix (Norway), head of the team, reported that Iraq had destroyed 40 of its Al-Samoud 2 missiles, which have a range of 93-plus miles, above UN-set limits. The government stated that the remainder were in process of being destroyed. The team had found no evidence of weapons of mass destruction, he said. Saddam Hussain then announced a ban on production or importation of these weapons and the materials used to make them.

America Invades

Following this development the great majority of Security Council member states went on record as favoring continuation of the inspections until either evidence of the existence of Iraqi weapons emerged or there was absolute proof to the contrary. However the Bush administration's foreign policy architects had privately determined to invade Iraq, as the only feasible means of removing Saddam Hussain as a "threat to world peace". In October 2002 Congress, with little debate, approved a resolution to allow the president "to use the armed forces of the United States as necessary to defend the national interest against the continuing threat posed by Iraq". Efforts to develop a coalition similar to that set up by the first President Bush in the Security Council, however, foundered under the determined opposition of most member states, notably long-term allies France and Germany and Russia. Elsewhere, the Turkish parliament refused to allow the U.S. to use its bases for an attack on Iraq. Of all the Council's member states, only Britain supported an invasion and agreed to send troops.

By late February 2003 U.S. policy makers had concluded that the UN was incapable of taking action against Iraq. With the buildup of 140,000 U.S. forces in various Middle Eastern bases essentially complete, President Bush issued a warning to Saddam Hussain: go into exile within 48 hours or face the consequences. There was no response, and on March 19 Anglo-American forces invaded Iraq.

There is as yet little agreement between those who argue that Saddam Hussain needed to be removed as a threat to world peace or those who feel that the U.S. invasion was an arrant violation of international law governing the relations between sovereign states. It has also been said that by doing so, the U.S. tore apart the delicate fabric that has held the nation-state system together since the Peace of Westphalia in 1848. Other scholars have questioned the use of the "just war" theory first propounded by St. Augustine to support the U.S. invasion. But viewed in military terms, the invasion was a textbook example of technological perfection. With control of the skies, coalition aircraft carried out pinpoint bombing raids on strategic targets; they were followed by a land assault by tanks and mobile units and then by infantry moving swiftly through the desert. Basra, Iraq's southernmost city and only seaport, was taken first. Special teams captured key oil installations before they could be sabotaged. The Iraqi capital, Baghdad, offered little resistance, perhaps on Saddam's orders or those of his military commanders, the Iraqi army simply melted into the civilian population. Many Ba'th leaders were captured or killed in the weeks that followed. Their capture was made easier by the circulation by U.S. forces of a pack of 55 cards with their faces featured thereon, in an eerie duplication of Nuri al-Said's description of the Iraqi government as a "pack of cards" during the monarchy). Those killed included Saddam's two sons, Uday and Qusay, and "Chemical Ali," the architect of Anfal, the deadly campaign against the Kurds. Ultimately Saddam himself was captured, hiding in a small subterranean chamber in a house in his home town of Tikrit. His trial began in December 2005, and although he challenged its jurisdiction and declared that under Iraqi constitutional law and as head of state he could not be charged with genocide or any other purported crimes, the trial proceeded with intermittent delays, one of them caused by the assassination of members of his defense team.

Stumbling Toward Self-Government

On May 1, 2003, President Bush stood on the deck of the U.S.S. *Abraham Lincoln* and declared that the war in Iraq was over. It had taken Coalition forces eighteen days to capture Baghdad, the only slowdown from sandstorms in the desert. The absence of organized resistance and the terrifying amount of air and ground firepower loosed on Baghdad (which was more thoroughly destroyed than it had been by the Mongols) seemed to confirm the U.S. president's confident words. But in the ensuing years Bush's lofty assessment has morphed into a protracted long-term conflict, with no sure end yet in sight.

Coalition Forces In Iraq (as of 12/05)	
U.S.	148,000
Britain	9,000
South Korea	3,600
Italy	3,000
Poland	2,400
Ukraine	1,650
Netherlands	1,345
Romania	700
Denmark	525
Japan	500
Bulgaria	480
El Salvador	380
Georgia	300
Australia	250
Mongolia	173
Azerbaijan	151
Portugal	120
Latvia	110
Slovakia	105
Other countries	421 in all

In some respects the war has been internationalized, with the participation of troops from a large number of UN member states. Also a UN international election team monitoring the December elections pronounced them "transparent and credible", indicating that the final results would be endorsed when they were announced. In a sense the "miracle of Iraq" is to have reached this point in its reconstruction as a viable state.

Iraq's misfortune was to be seen as the first stage in the Bush administration's "crusade" against the terrorism let loose in 9/11 and coincidentally to establish democratic regimes where needed in the Middle East. But as a long-time scholar and observer of the region noted, "the establishment of democracy happens indigenously or it does not happen at all. It has never been imposed on governments or nations at the point of a bayonet."[8] Since the invasion was based on faulty intelligence (or the wilful disregard of warnings in sound intelligence reports), and the invaders arrived with no long-term plan for the political and economic reconstruction of a society battered by years of crippling sanctions and international isolation, U.S. forces face not only a hostile population but also a well-organized insurgency. Aside from billions in occupation costs and the prospect of an indefinite stay for U.S. troops, as of December 2005, two years after Bush declared the war ended, 2,163 U.S. military personnel had been killed along with 200 from other Coalition forces. According to the Iraq Body Count, which tracks deaths by news media reports, Iraqis themselves have suffered between 26,000 and 30,000 casualties, and the total may be far higher due to incomplete reportage.

It is a strange irony, that the U.S. invasion and occupation of Iraq were foreshadowed by the observations of a British general 88 years ago, then commanding British forces as they entered Baghdad in 1917. He wrote, "Our armies do not come into your cities and lands as conquerors but as liberators." An Iraqi voter echoed these very words, with the same theme, during the January, 2005 elections for a Transitional Assembly that would govern the country under U.S. control. He said: "The future of Iraq is a line that goes through the occupation. If you ask me why I was voting, it is because I want to find something to pull me out of this mud."[9]

THE ARAB–ISRAELI CONFLICT

Until very recently, the Arab–Israeli conflict involved two peoples: those grouped into the modern Arab states, for the most part products of European colonialism, and the modern State of Israel. Israel's military superiority over its Arab neighbors seemed to remove the likelihood of a sixth Arab–Israeli war, and the peace treaties with Jordan and Egypt were further deterrents to renewal of armed conflict. The Arab states that surround Israel, although politically new, are heirs to a proud and ancient tradition, reaching back to the period when Islamic–Arab civilization was far superior to that of the Western world. This tradition and the self-proclaimed commitment to Arab brotherhood, however, have yet to bring them together in a united front toward Israel. A major obstacle to Arab unity is the variety of political systems that exist in the individual Arab states. These range from patriarchal absolute rule in Saudi Arabia by a ruling family to the multiparty system of Lebanon. Other Arab states reflect a variety of political systems—constitutional and patriarchal monarchies, authoritarian single-party governments, and regimes dependent on a single individual, to name a few examples. The United Arab Emirates provides one model of successful unification, mainly because the individual emirates are patriarchally ruled; in addition, aside from large expatriate work forces their populations are ethnically and linguistically unified. The Yemen Arab Republic, the other successful example, resulted from unification of the Marxist People's Republic of Yemen (South Yemen) and the tribal Arab Republic of Yemen (North Yemen). Their unification nearly collapsed in 1994 due to civil war, but the triumph of northern over southern forces and subsequent coalition government, confirmed by national elections, seem to have assured its survival.

In the past several decades the conflict between the Arab states and Israel has shifted from an essentially external one involving sovereign entities to an internal Israeli-Palestinian one. In its essentials the conflict stems from opposing views of land ownership. The land in question is Palestine, ancient Judea and Samaria for Jews, claimed by modern Israel on historical, emotional, and symbolic grounds. The Jewish claim to possession is to fulfill God's original covenant with Abraham, patriarch of the ancient Jewish tribes. To the Jews, it is a sacred homeland. The modern Israelis are the returned Jews, immigrants from many lands, plus the small Jewish community that remained there during the centuries of dispersion (diaspora). The Palestinians, mostly descendants of peoples who settled there over the centuries, were formerly 80 percent Muslim and 20 percent Christian, but emigration and displacement under the Jewish state in its wars have reduced the latter to 2 percent of the population. For most of its history the territory was ruled by outside powers—Persians, Syrians, Rome, the Byzantine Empire, and from the 1400s to 1917 by sultans of the Ottoman Turkish Empire. Under Ottoman rule it was divided into two *vilayets* (provinces) plus the separate Sanjak of Jerusalem (a more or less self-governing district). When Britain was given the League of Nations mandate over the southern vilayet, named Acre from its principal town, the British named the territory "Palestine," possibly from biblical associations with the ancient Philistine inhabitants.

ZIONISM

Zionism may be defined as the collective expression of the will of a dispersed people, the Jews, to recover their ancestral homeland. This idealized longing was given concrete form by European Jews in the nineteenth century.

In 1882, a Jewish law student, Leon Pinsker, published *Auto-Emancipation,* a book that called on Jews, who were being pressed at the time between the twin dangers of anti-Semitism and assimilation into European society, to resist by establishing a Jewish homeland *somewhere.* Subsequently, a Viennese journalist, Theodor Herzl, published *Der Judenstaat (The Jewish State)* in 1896. Herzl argued that Jews could never hope to be fully accepted into the societies of nations where they lived; anti-Semitism was too deeply rooted. The only solution would be a homeland for immigrant Jews as a secular commonwealth of farmers, artisans, traders, and shopkeepers. In time, he said, it would become a model for all nations through its restoration of the ancient Jewish nation formed under a covenant with God.

Herzl's vision of the Zionist state would give equal rights and protection to people of other nationalities who came there. This secular view generated conflict with Orthodox Jews, who felt that only God could ordain a Jewish state and that, therefore, Zionism would have to observe the rules and practices of Judaism in establishing such a state.

In the twentieth century, the question of a Palestine homeland was given form and impetus by two nationalist movements: Zionism and Arab nationalism. *Zionism,* the first to develop political activism in implementation of a national ideal, organized large-scale immigration of Jews into Palestine. These immigrants, few of them skilled in agriculture or the vocations needed to build a new nation in a strange land, nevertheless succeeded in changing the face of Palestine. In a relatively short time, a region of undeveloped sand dunes near the coast evolved into the bustling city of Tel Aviv, and previously unproductive marshland was transformed into profitable farms and kibbutz settlements.

Arab nationalism, slower to develop, grew out of the contacts of Arab subject peoples in the Ottoman Empire with Europeans, particularly missionary-educators sent by their various churches to work with the Christian Arab communities. It developed political overtones during World War I, when British agents such as T. E. Lawrence encouraged the Arabs to revolt against the Turks, their "Islamic brothers." In return, the Arabs were given to understand that Britain would support the establishment of an independent Arab state in the Arab lands of the empire. An Anglo–Arab army entered Jerusalem in triumph in 1917 and Damascus in 1918, where an independent Arab kingdom was proclaimed, headed by the Emir Faisal, the leader of the revolt.

The Arab population of Palestine took relatively little part in these events. But European rivalries and conflicting commitments for disposition of the provinces of the defeated Ottoman Empire soon involved them directly in conflict over Palestine. The most important document affecting the conflict was the Balfour Declaration, a statement of British support for a Jewish homeland in Palestine in the form of a letter from Foreign Secretary Arthur Balfour to Lord Rothschild, a prominent Jewish banker and leader of the Zionist Organization.

Although the Zionists interpreted the statement as permission to proceed with their plans for a Jewish National Home in Palestine, neither they nor the Arabs were fully satisfied with the World War I peace settlement, in terms of the disposition of territories. The results soon justified their pessimism. The Arab kingdom of Syria was dismantled by the French, who then established a mandate over Syria under the League of Nations. The British set up a mandate over Palestine, attempting to balance support for Jewish aspirations with a commitment to develop self-government for the Arab population, in accordance with the terms of the mandate as approved by the League of Nations. Britain's rule was challenged by both Jews and Arabs, almost from the start of the mandate. Zionists and their supporters in England not only encouraged Jewish emigration to Palestine, but as early as the 1920s began smuggling guns there to protect the Jewish community from the Arabs as worked to build a Jewish state. On their side, the Palestinians engaged in often violent protests, first against the mandatory authorities and then against the Jews. These protests culminated in the 1929 riots, and later Palestinian leaders called a general strike, the so-called "Arab Revolt" of 1936–1939, against British rule.

On its part, Britain attempted to placate both communities at various times, placing limits on Jewish immigration and in 1939 halting it entirely. World War II, with its appalling slaughter by the Germans of the Jewish population of occupied Europe, made Britain's task even more difficult. Boatloads of desperate Jews escaping Nazism were turned back, a number of them sinking while in sight of the Palestinian shore. As violence between the Jewish and Arab communities became endemic, the British decided that the mandate was unworkable. They turned over the "Palestine problem" to the newly-formed United Nations, successor to the League. Britain set a departure date for its forces from Palestine on May 14, 1948. The UN had formed a committee, the Special Committee on Palestine (UNSCOP), in 1947 to deal with the issue, and it approved by majority vote a plan for the partition of Palestine into two states, one Arab and the other Jewish. On May 14 the last British soldier left Palestine, the Union Jack was lowered, and David Ben Gurion's Zionist Organization proclaimed the birth of the state of Israel.

Most state-to-state disputes are susceptible to arbitration and outside mediation, particularly when they involve borders or territory. But Palestine is a special case. Its location astride communication links between the eastern and western sections of the Arab world made it essential to the building of a unified Arab nation, the goal of Arab leaders since World War I. Its importance to Muslims as the site of one of their holiest shrines, the Dome of the Rock in Jerusalem, is underscored by Jewish control—a control made possible, in the Arab Muslim view, by the "imperialist enemies of Islam," and reinforced by the relatively lenient treatment given by an Israeli court to Jewish terrorists arrested for trying to blow up the shrines on the Dome and build a new temple on the site. Also, since they lack an outside patron, both the dispersed Palestinians and those remaining in Israel look to the Arab states as the natural champions of their cause.

Yet the Arab states have never been able to develop a coherent, unified policy toward Israel in support of the Palestinian cause. There are several reasons for this failure. One is the historic rivalry of Arab leaders—a competitiveness that has evolved

from ancient origins, strong individualism, and family pride. Other reasons include the overall immaturity of the modern Arab political system and the difficulty of distinguishing between rhetoric and fact. The Arabic language lends itself more to the former than the latter. Thus the repeated declarations of Arab leaders that with God's help "we will drive the Jews into the sea...." are not meant to be taken literally. But because rhetoric urges them to subscribe to the ideal of a single Arab nation, they are torn between the ideal of this nation and the reality of separate nations. Inasmuch as their populations are mostly Muslim, a further obstacle to their coming to terms with Israel is that of disagreements within their societies over the nature and purpose of "political Islam," interpreted by many Muslims as the restoration of the fundamentals of the religion in these states. The Arab states surrounding Israel see their regimes as under attack from fundamentalists, practitioners of political Islam. Israel's very existence and its development as a modern nation underline their own weakness and lack of political unity.

Another reason for Arab disunity stems from the relationship of the Arab states with the Palestinians. During the British mandate, the Arab Higher Committee—the nexus of what became the Palestine national movement—aroused the anger of Arab leaders in neighboring countries by refusing to accept their authority over the committee's policies in return for financial support. After the 1948 Arab–Israeli War, the dispersal of Palestinians into Arab lands caused further conflict. To this day the Palestinians are haunted by their forced departure from their homes and villages by that event, which they call *an-Naqba* ("the catastrophe"). Further disillusionment resulted from the poor performance of Arab armies in the wars with Israel. Constantine Zurayk of the American University of Beirut expressed their shame in his book *The Meaning of Disaster:*

> Seven Arab states declare war on Zionism, stop impotent before it and turn on their heels.... Declarations fall like bombs from the mouths of officials at meetings of the Arab League, but when action becomes necessary, the fire is still and quiet.

It should be noted that the aspirations of the Palestinians for a state of their own in Palestine were the focal point for Arab-Israeli conflict as this conflict developed after 1948. An important difference between the two communities was the degree of their organization. By the time of independence, the Israelis had developed all the necessary institutions of government plus a strong sense of national identity. The Palestinians had little sense of their identity as part of a Palestinian *nation*; they had always lived there, generation after generation, in compact self-supporting villages, authority (such as it was) held by a handful of leading families, one of whose members was the village *muhtar* (headman). A Palestinian national identity as such might have developed eventually, but its main cause was the establishment of a Jewish national identity and state in their midst.

The opposition of Israelis to any form of return of Palestinian refugees to their former lands, or to compensation thereof, has hardened in recent years to be essentially out of the question, especially as two Intifadas and a steady diet of suicide bombings set the two communities even farther apart. This was not always the case. Even before the establishment of Israel, many prominent Jews, Zionist and non-Zionist, sought an ac-

THE BALFOUR DECLARATION

The text of the Balfour Declaration is as follows: "I have much pleasure in conveying to you on behalf of His Majesty's Government the following declaration of sympathy with Jewish Zionist aspirations which has been submitted to and approved by the Cabinet:

"His Majesty's Government view with favor the establishment in Palestine of a National Home for the Jewish people and will use their best endeavors to facilitate the achievement of this project, it being clearly understood that nothing shall be done which may prejudice the civil and religious rights of existing non-Jewish communities in Palestine or the rights and political status enjoyed by Jews in any other country."

commodation of the two peoples. Chaim Weizmann, founder of the WZO and later Israel's first president, once expressed in a letter to a friend his belief that Palestine should be shared by the two "nations." Others, notably Dr. Judah Magnes and the brilliant theologian Martin Buber, argued tirelessly on behalf of Jewish-Arab harmony. On the eve of the UN partition resolution Buber warned: "What is really needed by the two peoples in Palestine is self-determination, autonomy."

More recently, Uri Avnery, a prominent Zionist and Knesset (Israeli Parliament) member, writing in the afterglow of Israel's triumph over the Arab states in the 1967 Six-Day War, said, "The government [should] offer the Palestine Arabs assistance in setting up a national republic of their own... [which] will become the natural bridge between Israel and the Arab world."

Israel and the Palestinians

The effort to distinguish between a rightful Jewish "homeland" and the occupied territories gained momentum in 1977 with the formation of Peace Now, a movement initiated by army officers who felt that the government of Menachem Begin should not miss the opportunity to negotiate peace with Egypt. Peace Now gradually became the engine of the Israeli peace movement, leading public opposition to invasion of Lebanon and establishment of Jewish settlements in the occupied territories. However, Peace Now's policy of "exchanging land for peace" was rejected by the Likud Bloc after it had defeated Labor in the 1996 elections and held power in the Knesset until 1999.

Domestic political pressures, the broad sympathy of Americans for Israel, and support for the country as a dependable ally in the Middle East have been passed along from one U.S. administration to another. This innate preference has not been helped by the position taken on the issue by the Palestine Liberation Organization (PLO), the international exponent organization of the Palestinian cause, which has sometimes resorted to terrorism. The PLO, upon its founding in 1964, issued a charter calling for the destruction of Israel and the establishment of a sovereign Palestinian Arab state. The PLO until recently was also ambivalent about its acceptance of UN *Resolutions 242* and *338,* which call for Israeli withdrawal from the West Bank and Gaza Strip as a prelude to peace negotiations.

For the first two decades of the Israeli occupation the Palestinians who remained in the territories were largely quiescent.

An entire generation grew up under Israeli control, living in squalid refugee camps or towns little changed since Ottoman times and deprived of even the elemental human rights supposedly guaranteed to an occupied population under international law. In December 1987, a series of minor clashes between Palestinian youths and Israeli security forces escalated into a full-scale revolt, or uprising, against the occupying power. This single event, called in Arabic the *intifada* (literally, "resurgence"), has changed the context of the Israeli–Palestinian conflict more decisively than any other in recent history.

The intifada caught not only the Israelis but also the PLO by surprise. Having lost their Beirut base due to the Israeli invasion of 1982, PLO leaders found themselves in an unusual situation, identified internationally with a conflict from which they were physically separated and could not control directly or even influence to any great degree. As more and more Palestinians in the territories were caught up in the rhythm of struggle, the routine of stone-throwings, waving of forbidden Palestinian flags, demonstrations, and cat-and-mouse games with Israeli troops, the PLO seemed increasingly irrelevant to the Palestinian cause.

In view of its isolation from the intifada, the only role the PLO could play was to keep the Palestinian cause prominently focussed in the global spotlight, via the media and before international organizations. Its leader, the late Yassir Arafat, ended a meeting of the PLO National Council, the organization's executive body, in Tunis, Tunisia, with the historic statement that, in addition to formal acceptance of *Resolutions 242* and *338* as the basis for peace negotiations, the PLO would recognize Israel's right to exist. Arafat amplified the statement at a special UN General Assembly session in Geneva, Switzerland, formally accepting Israeli sovereignty over its own territory and renouncing the use of terrorism by the PLO.

The PLO's return to the West Bank and Gaza Strip, and its reconstitution as the Palestine National Authority (PNA) in accordance with the Oslo Accords refocused the attention of the Arab states on the Palestine refugees, an Arab people exiled from their Arab homeland. The PNA now emerged as an international political actor in its own right. Nonetheless, its leadership faced the prospect of having to put years of armed struggle behind it and concentrate on state-building. This would be difficult under ordinary circumstances. But in the PNA's case it would have to be done in ways acceptable to Israel and the United States as Israel's principal sponsor.

The evidence of five wars and innumerable smaller conflicts suggests that the Arab–Israeli conflict will remain localized. Israel's invasion of Lebanon, like its predecessors, remained localized once the United States had intervened, and it proved only a temporary setback for the PLO, a displacement. The Arab states continue to be haunted by the Palestinians, an exiled, dispersed people who refuse to be assimilated into other populations or to give up their hard-won identity. Mohammed Shadid contends that "Palestine is the conscience of the Arab world and a pulsating vein of the Islamic world... perhaps the only issue where Arab nationalism and Islamic revivalism are joined."[14] In 1995, Libya's leader, Muammar al-Qadhafi, the Arab world's most fervent advocate of Arab unity since Gamal Abdel Nasser of Egypt, abruptly expelled 1,500 Palestinian workers that were long residents in his country. He did so, he said, to protest the Palestinian–Israeli peace agreements, which he called a sellout of Arab interests.

Libya is not the only Arab state affected by the Palestinians and their goal of return to their homeland. Israel's 1982 invasion of Lebanon, formulated and led by Ariel Sharon with the concurrence of Prime Minister Begin, was intended to drive the PLO leadership out of Beirut. It succeeded in that goal but led to the collapse of the Begin government after Christian militiamen, supposedly allied with Israelis, massacred Palestinians in the Sabra and Shatila refugee camps. An internal committee of inquiry placed the blame for the massacre on Sharon, because he had not prevented it. The other unintended result was the emergence in southern Lebanon of an anti-Israeli guerrilla force, Hizbullah, which eventually forced the Israelis to withdraw.

Aside from the immediate success of the Gulf War in its limited objectives, the one accomplishment of the George Bush administration vis-à-vis the Middle East political situation was the launching of direct peace talks among Arab, Israeli, and Palestinian representatives to establish a "total Middle East peace." These talks, begun in Madrid, Spain, in 1991, were continued at various locations for nearly a dozen rounds, but without much progress toward a solution.

Oslo and After

A major breakthrough took place in September 1993, as negotiations conducted in secret by Palestinian and Israeli negotiators in Norway, a neutral country, resulted in a historic agreement. The agreement, although it fell far short of Palestinian objectives of an independent state, provided for Israeli recognition of Palestinian territorial rights and acceptance of a Palestinian "mini-state" in the Gaza Strip and an area around the West Bank city of Jericho, which would be its capital. It would be governed by an elected Council and would have limited self-rule for a five-year transitional period, after which discussions would begin on its permanent status.

After many false starts and delays that were caused by extremist violence on both sides, Israel and the PLO reached an interim agreement for additional land transfers to Palestinian self-rule in 1995. However, the election of Benjamin Netanyahu as Israel's prime minister in that year put a hold on the process. Netanyahu did meet with Arafat at the Wye River Plantation in Maryland in October 1998, with President Bill Clinton and Jordan's King Hussein as mediators. Arafat and Netanyahu signed the "Wye Agreement," which provided for three more land transfers and a "safe passage route" for Palestinians between the Gaza Strip and the West Bank. In addition, 750 Palestinian prisoners held in Israel were released, bringing the number to 7,000 since 1993. Netanyahu later put a moratorium on the implementation of these provisions.

Ehud Barak's election as Israeli prime minister in May 1999 breathed new life into the peace process after a long hiatus under Netanyahu. In January 2000, Barak met with Syrian foreign minister Farouk al-Atassi in Sheperdstown, West Virginia, in a renewed effort to bring about a Syrian–Israeli peace treaty. President Clinton served as moderator and mediator. However, Israel and Syria remained far apart on issues of mutual concern, notably ownership of the Golan Heights, and the talks adjourned without agreement. The renewal of violent Israeli–Palestinian confrontation in 2000

made any further moves toward peace between Israel and its last hostile state neighbor impossible.

Barak brought partial peace with Lebanon by ordering the withdrawal of Israeli forces from their self-declared "security zone" just inside the Lebanese border in April 2000, ahead of schedule. Peacekeeping units of the United Nations Interim Force in Lebanon (UNIFIL) then moved up to the border. As a result, the occupied West Bank and Gaza Strip remained the only part of "Arab land" still under foreign control.

ISRAEL AND THE PALESTINIANS

With Israeli–Arab relations in a state of temporary peace, the focus of Middle East conflict centered on the Palestinian population of the occupied West Bank and Gaza Strip. Following Barak's election, several steps were taken to implement the 1993 Oslo Agreement. Another 12 percent of the West Bank was turned over to the Palestine National Authority in late 1999 and early 2000. In the summer of 2000, Barak met with PNA head Yassir Arafat at Camp David, Maryland, with President Clinton again serving as moderator. In what seemed at the time to be a generous proposal, Barak offered to turn over 94 percent of the West Bank, along with administrative control over the Dome of the Rock (exclusive of the Wailing Wall) and the Old City of Jerusalem, with its separate Jewish, Muslim, Christian, and Armenian quarters. Arafat, for his part, agreed that Jewish settlements outside of Jerusalem built illegally on Palestinian land would be exempt from Palestinian control. However, he insisted that the "right of return" of dispossessed Palestinians to their former homes and villages in what is now Israel be included in the agreement. This right of return has been an article of faith for them since the establishment of the Israeli state.

In retrospect, it seems doubtful that Barak's proposals would have been acceptable to the Israeli public and given the necessary approval by a bitterly divided Knesset. But in any case, they were doomed by Arafat's insistence on the right of refugee return. Subsequently, in 2001, Barak was defeated in elections for the office of prime minister, called prematurely after several no-confidence votes in the Knesset. His successor, former general Ariel Sharon, said he would not be bound by the proposals and would consider only interim agreements with the PNA, with no additional transfers of land to the Palestinians.

Intifada II

A recent profile of Sharon by a respected Israeli journalist described him as "as a man with no moral brakes; as Defense Minister and architect of the Lebanon invasion he said he had no idea that such a thing (as the massacres at the Sabra and Shatila refugee camps) could take place; he is known for deftly pretending to be doing what he is not doing."[10]

Intifada II (also called the Al-Aqsa Intifada, from its immediate cause) broke out in September 2000, after Sharon had made a widely-publicized visit to the Islamic shrines of the Dome of the Rock in Jerusalem. Although he was not a member of the government at that time, he said his visit was peaceful but was intended to emphasize Israel's right to sovereignty over its entire territory. But the visit enraged the Palestinians, who also hated Sharon for his indirect role in the massacres of Palestinians in the Sabra and Shatila refugee camps in Lebanon in 1982. It was as if a glove had been flung in their faces.

Intifada II also grew from the Palestinian conviction that Israel had no intention of honoring its commitment to a Palestinian sovereign state. From a psychoanalytic viewpoint it also drew on Palestinians' feelings of poor self-image—a people "despised and oppressed, without honor, rights and self-worth, and dependent on Israel for jobs, limited in their movements, and now cut off from each other by an Israeli concrete and barbed wire wall."[11]

Intifada II was marked by greater Palestinian use of lethal weaponry, including mortars and missiles, and the emergence of suicide bombers. The latter are called *shaheed* ("martyrs") because they willingly strap explosives around themselves and blow themselves up in public places, all for the good of the cause. This suicide bombing campaign, developed mainly by the militant Palestinian organizations Hamas and Islamic Jihad, has seriously destabilized Israeli society. On the Israeli side, demolition of homes suspected of harboring militants (now discontinued), periodic military incursions and targeting of Palestinian leaders for missile strikes, along with the frequent closure of borders, have had a devastating effect on the Palestinian economy. In May, 2003, following his public statement of support for a Palestinian state, Bush unrolled his "road map to peace" for the two adversaries. It would require commitments from both sides. The PNA would disarm Hamas militants and halt suicide bombings and other attacks on Israelis which have been its trademark. Israel would halt construction of new Jewish settlements in the West Bank and Gaza and dismantle the settlements already there. It would release Palestinian prisoners and withdraw its forces from the occupied territories.

However the Bush administration, already preoccupied with Iraq, has yet to put its full support behind the president's "road map to peace" and bring it to a successful conclusion. After Israel began targeting Hamas leaders and other Palestinian militants for assassination, killing Hamas spiritual leader Shaykh Ahmad Yasin and his successor in missile strikes, the Bush administration described the action as "unhelpful". However the U.S. subsequently vetoed a UN Security Council resolution which would have condemned them as violations of international law and the Geneva Convention governing the rights of peoples under occupation.

Israel's construction of a "security fence" that will eventually stretch for 370 miles along the border with the West Bank has created an additional obstacle on the road map to peace. It was described by Israeli officials as a necessary barrier to infiltration by suicide bombers and other attacks on its territory by Palestinian militants. But the net effect will be to isolate Palestinians from their farms, olive groves, even schools and hospitals, separating village from village as it sliced through Palestinian territory. Suits filed with the World Court to halt its completion were unsuccessful, as Israeli lawyers argued that it was essential to national security. However the Israeli cabinet, under prodding by the Supreme Court, approved in February 2005 an altered route that would incorporate 7 percent of the West Bank into its territory as opposed to the 16 percent envisaged in the original (2003) route. Even so, the fence has already created Palestinian enclaves separated from one another and surrounded by Jewish settlements linked by exclusive highways "off limits" to Palestinians.

One of the few bright spots in this dismal conflict was the evacuation of Israeli military forces and Jewish settlers from the Gaza Strip in late 2005. Despite outcries from the settler movement and a sort of "color war" within Israel between proponents of evacuation (blue t-shirts) and opponents wearing orange ones, the settler removal was accomplished without violence. Along with Gaza four small West Bank settlements were returned to Palestine control, the Gaza border with Egypt opened, and border control turned over to Egyptian troops. For the first time in 38 years, Gaza's people could inhale the fresh air of freedom.

The Israeli withdrawal from Gaza and removal of all Jewish settlements there has left its economy in shambles, unfortunately. The 400 hectares (800-plus acres) of greenhouses there, formerly owned and managed by Israeli settlers, were turned over miraculously intact to the Palestinians who had formerly worked for the owners. When they are fully operative they will bring substantial economic benefits to the Strip. But periodic border closures are a serious handicap to its economic revival. Thus the 130,000 Palestinians formerly working in Israel in the 1990s are down to a few thousand. By 2006 per capita income in Gaza had dropped by one-third, to $600, and despite foreign aid of $1 billion, highest per capita in the world, the poverty rate had doubled.

Two important developments in 2005–2006 have seriously altered the Palestinian-Israeli conflict equation with results as yet unpredictable. the first was the stroke suffered by Prime Minister Sharon in December 2005 and a more severe one in January 2006. As of this writing it appears that he will never return to political life, and certainly not before the March 2006 election. since Sharon had almost single-handedly brought about the evacuation of Gaza and removal of the Jewish settlements there, his sudden departure from the political scene suggested that the peace process would once again come to a dead end. However Acting Prime Minister Olmert has stated publicly that if his Kadima Party wins the election Israel will unilaterally evacuate 17 West Bank settlements and set its final borders over a 4-year period.

The second important development was the 2006 Palestinian elections for a new parliament. Hamas, the Islamist movement which has committed to the elimination of Israel, won a majority of 76 seats to 43 for Fatah, the secular movement which had dominated Palestinian politics for half a century behind the late Yassir Arafat.

Hamas's emergence, and the uncertainty about its policies once it has formed a government, placed Israel and the U.S. in a quandary. Initially Israel transferred $55 million in customs and tax duties to the Palestine National Authority as it has in the past. However, it said it would cease such transfers after the new parliament convenes on March 18. The cutoff would have a serious effect on the Palestinian economy. As presently established the Palestinian budget anticipates revenues of $70 million and expenditures of $330 million. Aid of $144 million pledged by the European Community, the World Bank and other sources still leaves an income gap of $112 million. With Israel already providing water and electricity to the Palestinian territories and deducting this cost from the now-withheld customs and tax payments, it may well be that simple economics will force a change in Hamas's charter if nothing else does.

SEPTEMBER 11, 2001

On September 11, 2001, a band of hijackers seized control of two U.S. commercial aircraft in flight and flew them deliberately into the twin towers of the World Trade Center in New York City. A third aircraft was hijacked and smashed into one side of the Pentagon near Washington, D.C.; while a fourth, perhaps intended for the U.S. Capitol Building, crashed in a field in Pennsylvania, well short of its objective. The hijackings not only caused significant loss of civilian lives and huge losses in property damages; they also brought global terrorism to American soil. For almost the first time in their history, Americans were rendered vulnerable by unknown, faceless enemies preaching a doctrine of hate.

The hijackings were a wake-up call for a nation that had always assumed that it was safe from the violence that had become a way of life elsewhere. The Bush administration launched a concerted effort, first to identify the hijackers and the network behind them and thereafter to bring those responsible to justice. Preliminary investigation indicated that the hijackers were Arabs from several Middle Eastern countries, mainly Egypt, Saudi Arabia, Sudan, and Yemen. They formed part of a terrorist network identified as *al-Qaeda* ("The Base" in Arabic).

Al-Qaeda's leader and founder is Osama bin Laden, a multi-millionaire member of a prominent Saudi Arabian family originally from the Hadhramaut, in southern Yemen. The family migrated to Saudi Arabia when he was young, and the senior bin Laden developed an enormously successful contracting business. Eventually the business became an international conglomerate, with interests and subsidiaries in many countries. But along the way, young Osama seemed to have developed an undying hatred of the United States and everything it stands for. He broke with his family, and, although an elitist by Saudi Arabian (for that matter universal) standards, chose first to join the mujahideen ("resistance" or "freedom fighters") in Afghanistan fighting the Soviet invasion, and then used that experience to build a secret terrorist network to carry out attacks on various perceived enemies. One end product was the September 11 terrorist attacks.

Since there has been as yet no repeat of the World Trade Center bombings, one might assume that either al-Qaeda's interests have shifted elsewhere or that the organization has been rendered ineffective by the U.S.-led campaign against international terrorism. But the reasons behind September 11, the "why" of hatred of the U.S. and responses thereto, require serious consideration. Why is this hatred so virulent that it motivated young men to take their own lives by attacking mainland America? Why is the U.S. perceived as the enemy of Islam by many Muslims, notably in the Middle East? Certainly part of the explanation is that the U.S. supports Israel in its occupation of Islamic lands (the West Bank and Gaza), thereby denying the Palestinians their right to a nation-state there.

Yet the Palestinian suicide bombers, young men and recently more and more young women, who attack targets in Israel are not doing so as Muslims. On a larger scale, anti-U.S. feeling has been enhanced by the presence of American military forces in the region, particularly on the "sacred Islamic soil" of Saudi Arabia—now mostly withdrawn—and more recently in Iraq. From the Islamic fundamentalist perspective also the U.S. is viewed as the main prop for repressive Muslim regimes, such as Egypt and Tunisia, that do not govern in accordance with Islamic law and practice.

Reaction in the Middle East to the September 11 attacks varied from country to country. Most national leaders condemned them as un-Islamic and contrary to the rules of jihad. Thus Saudi Arabia's highest-ranking legal scholar stated categorically that suicide bombers would not die as martyrs but, rather, as simple suicides, an action forbidden by the Koran. An emergency meeting of the 56-nation Organization of the Islamic Conference (OIC) criticized the Taliban regime in Afghanistan for sheltering the "heretic" bin Laden but warned against the "targeting of any Islamic or Arab state under the pretext of fighting terrorism." And the emir of Qatar told the conference that "we assert our utter rejection of these attacks and assert that those confronting them must not touch innocent civilians and must not extend beyond those who carry out those attacks."

Backing for the U.S.–led coalition against terrorism has varied widely. Britain and France promptly committed troops to the campaign in Afghanistan. Other European countries, notably Germany, Spain, and Italy, arrested numbers of members and alleged members of al-Qaeda. Indonesia, Malaysia, and Yemen, among other countries, pursued a roundup of militants and former "Afghan Arabs," returned veterans of the Soviet–Afghan War. In Yemen, a favorite hideout of anti–U.S. terrorism due to its geographical isolation and bin Laden's family background, American special forces were invited by the government in February 2002 to proceed there and train the Yemeni Army in counterterrorism.

One of the least desirable features in the "war on terror" of the Bush administration is that it has led the U.S. to adopt methods contrary to international law, human rights and the Geneva Conventions governing treatment of war prisoners, prisoners of conscience, etc. Apart from the well-documented practice of mistreatment of detainees in such prisons as Abu Ghraib and Guantananmo Bay, the worst feature of American policy under this administration has been the practice of "extraordinary rendition". Alleged terrorists and other suspects are arrested and then outsourced from the U.S. to countries such as Egypt, Syria or Jordan, which practice torture and other methods of interrogation. The Bush administration has justified the practice, which is managed primarily by the CIA, on grounds that the Geneva Conventions do not apply to terrorists since they do not represent a particular country, thus allowing the president to take whatever actions are necessary for national security.[12]

On the political side, in October 2001, Saudi Arabia and United Arab Emirates withdrew their recognition of the Taliban as the Afghan government. Iran closed its border with Afghanistan and provided arms to the Northern Alliance, while Iranian mediators in Germany helped unite the various Afghan warlords into an interim government for that war-torn country. But with bin Laden and leaders of his network possibly having escaped to Pakistan and with the technology available to al-Qaeda, the probability was that U.S. military success in Afghanistan was at best a first step in the long struggle to eliminate global terrorism.

NOTES

1. Charles C. Mann, *1491: New Revelations of the Americas Before Columbus* (New York: Knopf, 2005), pp. 176–177

2. Karl Meyer and Shareen Blair Brysac, *Tournament of Shadows: The Great Game and the Race for Central Asia* (Washington, D.C.: Counterpoint Press, 1999.) The term was coined by Count Nesselrode.

3. Terence O'Donnell, *Garden of the Brave in War* (New York: Ticknor and Fields, 1980), p. 19, states that in visits to remote Iranian villages, he was told by informants that the Arabs never washed, went around naked, and ate lizards.

4. Daniel Pipes, "A Border Adrift: Origins of the Conflict," in Shirin Tahir-Kheli and Shaheen Ayubi, eds., *The Iran-Iraq War: New Weapons, Old Conflicts* (New York: Praiger, 1983), pp. 10–13.

5. *Ibid.*, quoted on p. 20.

6. Stephen R. Grummon, *The Iran-Iraq War: Islam Embattled*, The Washington Papers/92, Vol. X (New York: Praeger, 1982), p. 10.

7. *The Christian Science Monitor* (September 23, 1987).

8. Gen. Stanley Maude, quoted in Anthony Shadid, *Night Draws Near: Iraq's People in the Shadow of America's War* (New York: Henry Holt, 2005, p. 388.

9. *Ibid.*, p. 396

10. David Shipler, *New York Times Week in Review*, June 1, 2003.

11. Avner Falk, *Fratricide in the Holy Land: A Psychoanalytic View of the Arab-Israeli Conflict* (Madison: University of Wisconsin Press, 2004)

12. See Jane Mayer, "Outsourcing Torture," *The New Yorker*, February 14, and 21, 2005.

The Middle East: Heartland of Islam

ISLAM IN FERMENT

"The world is no barrier to God,
He is visible in all that exists,
Remove yourself from yourself and Him,
Let Him speak to you as to Moses, from
    the burning bush"

Mehmed Esad Dede[1]

Until well after the Second World War, the world of Islam and its peoples, and particularly that part of the Islamic world centered in the Middle East and North Africa, were unknown to most Americans. Other than a handful of missionaries sent out by their churches to found schools and hospitals for their fellow-Christians and the occasional Biblically-inspired pilgrim to the Holy Land, few of us had visited the region and still fewer had any knowledge of its long-ago greatness. And unlike the First World War, when the area had been the scene of important conflicts, it was a backwater during World War II, except for Northern Africa. Young Americans including myself who passed through that region on their way to invade German-occupied Italy remembered it as a hot, dusty place, where men wearing what appeared to be bedsheets sat at small tables in fly-blown cafes, drinking small glasses of hot tea and speaking a strange guttural language accompanied by much waving of hands and shouting. This naive stereotype changed little in the intervening years, except for the addition of the State of Israel, whose Jewish peoples made the desert bloom and more than stood their ground militarily against the children of those men in "bed sheets."

Throughout the Middle East, in refugee camps, Islamic lands, towns in the Israeli-occupied West Bank, one hears five times a day the *azan*, the Islamic call to prayer. Five times a day the same message booms from the throats of the *muezzins*, professional "prayer-callers", often supplemented by loudspeakers or megaphones. The message is simple: God is Great! I testify that there is no God but God! I testify that Muhammad is the Messenger of God!" It is a rallying cry and order not only to the majority of Middle Easterners but to nearly half the world's population, believers in God, God's Holy Book, and God's Messenger or Prophet of Islam.

ISLAMIC ORIGINS

Islam, the third and youngest of the world's three great monotheistic religions after Judaism and Christianity, means "submission" or "surrender" in Arabic, its basic language. It developed among a particular people, the Arabs, living as many of them do today on the Arabian Peninsula. Fourteen centuries ago they were mostly illiterate, and more importantly they lacked any formal religion or system of gods and goddesses, except for belief in natural powers such as the wind and rain. They

United Nations Photo (UN143185)

The Middle East did not make a real impact on the American consciousness until 1979, when the followers of the Ayatollah Khomeini seized the U.S. Embassy in Teheran and held its occupants hostage for more than a year. The extent to which fundamentalist Shia Muslims would follow Khomeini, pictured on the placard displayed above, was little recognized before this event.

were also outsiders, not subject to any higher authority other than that of the chiefs of the various tribes they belonged to.

All this would change in the seventh century A.D., when a merchant named Muhammad had an incredible religious experience. He lived in the small but important trading city of Mecca, in southwest Arabia, and was well-respected as a mediator in disputes. In fact his nickname was al-Amin, "the Just." It was his practice from time to time to retire to a cave in the hills outside Mecca to meditate. And one evening while he was meditating there he received the first of many revelations from God, given him by the Angel Gabriel. These revelations form the basis for the new faith which he preached and propagated among his people. At first the revelations were given orally, but after his death they were written down in book form as the Koran (Qur'an, in Arabic), the Holy Book of Islam.

During Muhammad's lifetime, the various revelations he received were used to guide his followers along the "Way" of conduct (*Shari'a*, in Arabic) acceptable to God. The Arabs followed traditional religions in Muhammad's time, worshipping many gods. Muhammad taught belief in one God—Allah—and in the Word of God sent down to him as messenger. For this reason, Muhammad is considered the Prophet of Islam.

Muhammad's received revelations plus his own teachings issued to instruct his followers make up the formal religious system known as Islam. The word *Islam* is Arabic and has been translated variously as "submission," "surrender" (i.e., to God's

19

THE KORAN: THE HOLY BOOK OF ISLAM

Muslims believe that the Koran is the literal Word of God and that Muhammad was chosen to receive God's Word through the Angel Gabriel as a *rasul* (messenger). But the Koran does not cancel out the Bible and Torah, which preceded it. The Koran is viewed, rather, as providing a corrective set of revelations for these previous revelations from God, which Muslims believe have been distorted or not followed correctly. To carry out God's Word, as set down in the Koran, requires a constant effort to create the ideal Islamic society, one "that is imbued with Islamic ideals and reflects as perfectly as possible the presence of God in His creation."*

The Koran was revealed to Muhammad over the 22-year period of his ministry (A.D. 610–632). The revelations were of varying lengths and were originally meant to be committed to memory and recited on various occasions, in particular the daily prayers. Even today, correct Koranic practice requires memorization and recitation; during the fasting month of Ramadan, one section per day should be recited aloud.

In its original form, the Koran was either committed to memory by Muhammad's listeners or written down by one or more literate scribes, depending upon who was present at the revelation. The scribes used whatever materials were at hand: "paper, leather, parchment, stones, wooden tablets, the shoulder-blades of oxen or the breasts of men."** The first authoritative version was compiled in the time of the third caliph, Uthman, presumably on parchment. Since then, the Holy Book has been translated into many other languages as Islam has spread to include non-Arab peoples.

All translations stem from Uthman's text. It was organized into 114 *suras* (chapters), with the longest at the beginning and the shortest at the end. (The actual order of the revelations was probably the reverse, since the longer ones came mostly during Muhammad's period in Medina, when he was trying to establish guidelines for the community.)***

Many of the revelations provide specific guides to conduct or social relationships:

When ye have performed the act of worship, remember Allah sitting, standing and reclining.... Worship at fixed times hath been enjoined on the believers....

(*Sura IV*, 103)

Establish worship at the going down of the sun until the dark of night, and at dawn. Lo! The recital of the Koran at dawn is ever witnessed.

(*Sura XVII*, 78–79)

Make contracts with your slaves and spend of your own wealth that God has given you upon them....

(*Sura XXIV*, 33)

If you fear that you will be dishonest in regard to these orphan girls, then you may marry from among them one, two, three or four. But if you fear you will not be able to do justice among them, marry only one.

(*Sura IV*, 3)

Much of the content of the Koran is related to the ethical and moral. It is an Arab Koran, given to Arabs "in clear Arabic tongue" (*Sura XLI*, 44) and characterized by a quality of style and language that is essentially untranslatable. Muslim children, regardless of where they live, learn it in Arabic, and only then may they read it in their own language, and then always accompanied by the original Arabic version. Recitals of selections from the Koran are a feature of births, marriages, funerals, festivals, and other special events and are extraordinarily effective, whether or not the listener understands Arabic.****

* Peter Awn, "Faith and Practice," in Marjorie Kelly, ed., *Islam: The Religious and Political Life of a World Community* (New York: Praeger, 1984), pp. 2–7.
** *The Qur'an, The First American Version,* Translation and Commentary by T.B. Irving (Brattleboro, VT: Amana Books, 1985), Introduction, XXVII.
*** On this topic, see Fazlur Rahman, *Major Themes of the Qur'an* (Chicago: Bibliotheca Islamica, 1980), *passim.*
**** "The old preacher sat with his waxen hands in his lap and uttered the first Surah, full of the soft warm coloring of a familiar understanding.... His listeners followed the notation of the verses with care and rapture, gradually seeking their way together ... like a school of fish following a leader, out into the deep sea." Lawrence Durrell, *Mountolive* (London: Faber and Faber, 1958), p. 265.

---

will), and the fatalistic "acceptance." A better translation might be "receptiveness." Those who receive and accept the Word of God as transmitted to Muhammad and set down in the Koran are called Muslims.

The basic sources of Islam are the Koran, the *hadith* (sayings and instructions of Muhammad), the consensus of the community (*ijma'*), and inference or understanding by analogy (*qiyas*). The Koran, as the Word of God revealed to Muhammad, is the primary source, and as such it is considered divine, eternal and immutable. Its 114 Suras (chapters) were revealed to Muhammad over a 22-year period and were not set down in book form until long after his death. They were put together by scribes in the period of the Caliph Uthman, and this text is still considered the standard version.

Issues and problems not specifically addressed in the Koran are dealt with in the collected sayings and decisions of Muhammad. Since many of them derive from oral reports and were not collected until centuries later, inevitably what the Prophet said or decided may have errors. It was not until the late nineth century A.D. that an authoritative collection, the Six Books, was completed.

The third source, consensus, was used initially in the selection of caliphs as noted above. Increasingly, and especially for Shia Muslims, consensus has come to mean decisions or opinions rendered by their religious authorities.

The fourth source, inference by analogy, is used rarely and only when no clear examples are found in the Koran or the hadith and there is no agreement or consensus on an issue. It is similar to the use of precedent in Anglo-Saxon legal tradition.

Islam is essentially a simple faith. Five basic duties are required of the believer; they are often called the Five Pillars because they are the foundations of the House of Islam. They are:

1. The confession of faith: "I testify that there is no God but God, and Muhammad is the Messenger of God."

2. Prayer, required five times daily, facing in the direction of Mecca, the holy city.

3. Alms giving, a tax or gift of not less than 2½ percent of one's income, to the community for the help of the poor.

4. Fasting during the daylight hours in the month of Ramadan, the month of Muhammad's first revelations.
5. Pilgrimage, required at least once in one's lifetime, to the House of God in Mecca.

It is apparent from the above description that Islam has many points in common with Judaism and Christianity. All three are monotheistic religions, having a fundamental belief in one God. Muslims believe that Muhammad was the "seal of the Prophets," the last messenger and recipient of revelations. But they also believe that God revealed Himself to other inspired prophets, from Abraham, Moses, and other Old Testament (Hebrew Bible) figures down through history, including Jesus Christ. However, Muslims part company with Christians over the divinity of Jesus as the Son of God; the Resurrection of Jesus; and the tripartite division into Father, Son, and Holy Ghost or Spirit.

THE ISLAMIC CALENDAR

The Islamic calendar is a lunar calendar. It has 354 days in all, divided into 7 months of 30 days, 4 months of 29 days, and 1 month of 28 days. The first year of the calendar, A.H. 1 (Anno Hegira, the year of Muhammad's "emigration" to Medina to escape persecution in Mecca), corresponds to A.D. 622.

In the Islamic calendar, the months rotate with the Moon, coming at different times from year to year. It takes an Islamic month 33 years to make the complete circuit of the seasons. The fasting month of Ramadan moves with the seasons and is most difficult for Muslims when it takes place in high summer.

Although Muhammad is in no way regarded as divine by Muslims, his life is considered a model for their own lives. His *hadith* ("teachings" or "sayings") that were used to supplement Koranic revelations (or to deal with specific situations when no revelation was forthcoming) have served as guides to Muslim conduct since the early days of Islam. The Koran and hadith together form the *Sunna* (translated literally as "Beaten Path"), which provides an Islamic code of conduct for the believers.

The importance of Muhammad's role in Islam cannot be overemphasized. Among Muslims, his name is used frequently in conversation or written communication, always followed by "Peace Be Unto Him" (PBUH). A death sentence imposed on the writer Salman Rushdie by Iran's revolutionary leader Ayatollah Khomeini resulted from an unflattering portrait of Muhammad in Rushdie's novel *The Satanic Verses* (1988). And an Israeli woman's depiction of Muhammad as a pig writing in the Koran, on a poster displayed in Hebron, roused a storm of protest throughout the Muslim world. She was arrested by Israeli police and given a 21-year jail sentence for "harming religious sensibilities."[2]

The Five Pillars are not only required to be observed by Muslims, they are also specified in the Koran, even as to how they should be performed. The Confession of Faith (*Shahada*) is all that is required of a non-Muslim to become a convert. The ritual of prayer expands on the confession of faith, and its opening prayer comes directly from the first *Sura* ("chapter") of the Holy Book. Thereafter at each session the believers follow a set of ritual prayers (*rak'a*) and prostrations until the head has touched the floor. After each prayer the following creed is recited: "Peace be on thee, O Prophet! And the mercy of God and his blessings. Peace be on us and the righteous servants of God! I bear witness that none deserves to be worshipped but God, and I bear witness that Muhammad is His servant and apostle."

Almsgiving (*zakat*), the third Pillar of Islam, is enjoined upon the believer in the same Sura as that of prayer (Sura 31.2-5) Since God is the owner of all things, it is each Muslim's obligation to manage money and resources according to His will. This explains why Islam forbids the charging of interest and usury. However almsgiving by the wealthy for the benefit of the poor and needy may take the form of a permanent charitable donation (*waqf*) or endowment for the upkeep of a public institution such as a mosque, school, library or hospital. In general it should consist of not less than 2.5 percent of one's income.

Fasting (*sawm*), the fourth Pillar of Islam, is obligatory for Muslims in Ramadan, the ninth month of the Islamic calendar (see Box). God's first revelation to Muhammad was given on the 27th of that month, which is called *Dhu al-Qadir*) ("Night of Power"). Sura 97 of the Koran explains: "The Night of Power is better than a thousand months. Therein come down the angels and the Spirit, by God's permission ..."

It is a thirty-day fasting period, which begins, and ends, when a white thread cannot be distinguished from a black one. During the daylight hours Muslims must abstain from eating, drinking, smoking, sexual intercourse, even taking medicines. The end of the day's fast is called the *Iftar*, and at that time dinner may be served. The great value of the Ramadan fast is the discipline it imposes on the believers, helping them become more conscious of their dependence on God and more aware of the needs, the hunger and thirst, of the poor.

Pilgrimage (*Hijrah* or *Hajj*), the fifth Pillar, is incumbent on each Muslim at least once in his/her life, depending on health and financial circumstances. Those who cannot afford the journey will often borrow from others or go into debt in order to do so. The pilgrimage may be made at any time; this is called the lesser Hajj. The greater pilgrimage however takes place yearly on the first ten days of Dhu al-Hijrah, the pilgrimage month, and is sponsored and managed by the government of Saudi Arabia in its role as Guardian of the Holy Places (of Mecca and Medina). In past years, travel there was long and arduous, but today's pilgrims travel by jet and are escorted to Mecca from Jiddah airport in chartered buses, trucks and cars. Upon arrival they put on seamless white robes, and follow a series of prescribed rituals, circling seven times around the Ka'ba, kissing the Black Stone embedded in it, drinking water from the well of Zam Zam which supposedly nourished Hagar after Abraham cast her out and she wandered in the desert with her son Ishmael (Isaac's half-brother, in Biblical lore). Seven times they walk between the hills of Safa and Marwah, then on to Mina, where they symbolically throw stones to drive out Satan. A final seven-fold circumnavigation of the Ka'ba completes the pilgrimage ritual, and those who have taken part are known thereafter as *Hajjis*

ISLAMIC DIVISIONS: SUNNIS AND SHIAS

The great majority (90 percent) of Muslims are called Sunnis, because they follow the Sunna, observe the Five Pillars, and practice the rituals of the faith. They also interpret as correct the history of Islam as it developed after Muhammad's death, under a line of successors termed *caliphs* ("agents" or "deputies") who held spiritual and political authority over the Islamic community. However, a minority, while accepting the precepts and rituals of the faith, reject this historical process as contrary to what Muhammad intended for the community of believers. These Muslims are called Shias (commonly, but incorrectly, Shiites). The split between Sunnis and Shias dates back to Muhammad's death in A.D. 632.

Muhammad left no instructions as to a successor. Since he had said that there would be no more revelations after him, a majority of his followers favored the election of a caliph who would hold the community together and carry on his work. But a minority felt that Muhammad had intended to name his closest male blood relative, Ali, as his successor. Supporters of Ali declared that the succession to Muhammad was a divine right inherited by his direct descendants. Hence they are known as Shias ("Partisans") of Ali.

Initially the majority of the Islamic community, the *umma*, passed over Ali and by consensus elected Abu Bakr, Muhammad's father-in-law and the first convert after his wife Khadija., as Caliph. The next two caliphs were elected in the same manner, with Ali finally chosen as the fourth caliph after his predecessors had been killed by dissident Muslims. (In addition to the Sunni-Shia split, there were other deep divisions within the umma arising from traditional Arab tribal and family rivalries, as is the case in Iraq today). Ali himself was murdered in A.D. 661, and a rival family, the Umayyads, took over the office and moved Islam's capital from Mecca to Damascus, Syria.

Subsequently Ali's partisans put their support behind Husayn, his younger son, as the rightful caliph. However he and his family were ambushed by the army of the Umayyad caliph in A.D. 680 near the town of Karbala (Iraq), and his head sent to the caliph as proof of his death.

Husayn's murder widened the Sunni-Shia split, which has continued to the present day and is visible in the political conflicts raging in Iraq. Lacking a symbolic figure to turn to for leadership, Shia recognize a line of either seven or twelve Imams, descendants of Ali and Husayn, as their spiritual leaders. The Imams were not recognized as such by Sunni caliphs and were frequently persecuted, as were their followers. As a result the Shia developed the practice of *taqiya* ("dissimulation" or "concealment"). Outwardly they obeyed whatever religious or secular ruler was in power, but inwardly they continued to believe in the divine right of the Imams to rule the Islamic world and to anticipate the return of the last (12th) Imam from hiding as the *Mahdi* ("the Awaited One") to announce the Day of Judgment.

With the exception of Iran, the Shias remained a minority in the Islamic world and did not acquire political power. But in the early sixteenth century A.D. the leader of a religious brotherhood there, the Safavis, claimed descent from Ali and declared war on the Ottoman Turks, who ruled most of the Middle East at that time. In order to justify his war he made an agreement with Shia religious leaders—they would recognize him as head of state in return for his commitment to recognize Shia Islam as the official religion in Iran. Since then Iran has been a Shia nation both in population and in political power.

SHIA MUSLIMS AND MARTYRDOM

The murder of Husayn, far more than that of his father Ali, provided the Shia community with a martyr figure. This is due to the circumstances surrounding Husayn's death—the lingering image of Muhammad's grandson, with a small band of followers, surrounded in the waterless desert, to be cut down by the vastly superior forces of Yazid, has exerted a powerful influence on Shias. As a result, Shias often identify themselves with Husayn, a heroic martyr fighting against hopeless odds. For example, an important factor in the success of Iran in repelling the invasion of Iraqi forces in the bitter 1980–1988 Iran–Iraq War was the Basijis, teenage volunteers led into battle by chanters, in the firm belief that death at the hands of the Sunni Iraqi enemy was a holy action worthy of martyrdom.

'ASHURA

A special Shia Muslim festival not observed by Sunnis commemorates the 10 days of 'Ashura, the anniversary of the death of Husayn. Shia Muslims mark the occasion with a series of ritual dramas that may be compared to the Christian Passion Play, except that they may be performed at other times during the year. Particularly in Iran, the ritual, called Ta'zieyeh, is presented by strolling troupes of actors who travel from village to village to dramatize the story with songs, poetry, and sward dances. Ta'ziyeh also takes place in street parades in cities, featuring penitents who lash themselves with whips or slash their bodies with swards. Freya Stark, the great English travel writer, describes one such procession in her book *Baghdad Sketches*:

> All is represented, every incident on the fateful day of Karbala, and the procession stops at intervals to act one episode or another. One can hear it coming from far away by the thud of beaters beating their naked chests, a mighty sound like the beating of carpets, or see the blood pour down the backs of those who acquire merit with flails made of chains with which they lacerate their shoulders; and finally the slain body comes, headless, carried under a bloodstained sheet through wailing crowds.

The anniversary of Husayn's martyrdom is observed by Shia Muslims as 'Ashura, a 10-day festival featuring street parades with men slashing arms and backs with sharp knives in imitation of his death. Also featured is the so-called Passion Play series, with professional actors' troupes performing the event.

ISLAM AND EUROPE: CHANGING ROLES

The early centuries of Islam were marked by many brilliant achievements. An extensive network of trade routes linked the cities of the Islamic world. It was a high-fashion world in which the rich wore silks from Damascus ("damask"), slept on fine

United Nations Photo (UN152168)
The mosque at Khan El Khalili, Egypt.

sheets from Mosul ("muslin"), sat on couches of Morocco leather, and carried swords and daggers of Toledo steel. Islamic merchants developed many institutions and practices used in modern economic systems, such as banks, letters of credit, checks and receipts, and bookkeeping. Islamic agriculture, based on sophisticated irrigation systems developed in the arid Middle East, brought to Spain, in particular, a number of previously unknown food crops that include oranges, lemons, eggplant, radishes, and sugar. The very names of these foods are Arabic in origin, as are a great number of words in modern Spanish. Islamic civilization literally "flowered" in Cordoba, capital of Islamic Spain, and Baghdad, the caliph's eastern capital. The former has been well described as "the ornament of the world" with its magnificent gardens, hundreds of bookstores, palaces and sports arenas and of course the Great Mosque, at a time when Europe was medieval, dark and poor; and although it is hard to believe in the light of the present conflicts, Muslims, Christian and Jews lived there in harmony.[3]

Islamic medical technology reached a level of excellence in diagnosis and treatment unequaled in Europe until the nineteenth century. Muslim mathematics gave Europeans Arabic numerals and the concept of zero. Muslim navigators made possible Columbus's voyages, through their knowledge of seamanship and inventions such as the sextant and the compass. Their libraries were the most extensive in existence at that time.

The level of achievements of Islamic civilization from roughly 750 to 1200 was far superior to that of Europe. The first Europeans to come in direct contact with Islamic society were

Crusader knights, Christians who invaded the Middle East in order to recapture Jerusalem from its Muslim rulers. The Crusaders marveled at what they saw, even though they were the sworn enemies of Islam. This Christian occupation of the Holy Land, while short-lived (1099–1187), contributed significantly to the mutual hostility that has marked Christian–Muslim relations throughout their coexistence. (A retired German diplomat who became a Muslim in 1980 expressed the difference between Christianity and Islam in social terms: "The alternative to an increasingly amoral lifestyle in the West is Islam, but an Islam rigorously practiced and free from fanaticism, brutality, violence, violation of human rights and other practices erroneously associated in the Western mind with the religion.")[4]

The hostile relationship between Muslims and Christians resulting from the Crusades intensified with the rise to power in the Islamic world of the Ottoman Turks. In 1453 A.D. they captured Constantinople (modern Istanbul), capital of the East Roman or Byzantine empire. Soon Ottoman power extended across all of southeastern Europe. With the conquest of Egypt in 1517 the Ottomans captured the last caliph, who had taken refuge there after the seizure of Baghdad by the Mongols. He was "persuaded" to give up his title in return for security, and the Ottoman sultan now became the caliph.

The success of their armies and a belief that God had given Muslims, and particularly Ottomans, a superior way of life gave the Ottomans a feeling of superiority over their Christian European adversaries. Indeed this superiority lasted for several centuries. But gradually the roles were reversed. Militarily the Ottoman armies lost more and more battles, particularly to the expanding Russian power, and with defeat came the humiliation of being forced to sign peace treaties which cost them territory. Another shock came with the discovery that European cities had developed modern technology. They had electric lights, paved streets, telephone and telegraph lines, and other conveniences none of which existed in Ottoman/Middle Eastern cities.

Since that time the Ottoman empire has disappeared, its former Middle Eastern provinces being replaced by Islamic nations such as Syria, Lebanon, Jordan and Iraq that began their modern existence as dependencies of Christian European countries. Other Middle Eastern Islamic nations—Iran, Turkey and Egypt for example—have survived but in a different form politically from what they were before the Christian-Ottoman roles were reversed.

This change in status has caused much soul-searching in the Islamic community, particularly after the terrorist attacks of 9/11 in the United States. Some Islamic groups, such as Hamas in occupied Palestine and Osama Bin Laden's Al-Qaeda, argue that only through violence and the destruction of Israel and of non-representative Islamic regimes, especially those (i.e. Saudi Arabia) that have allowed non-Muslim forces to invade or use sacred Islamic territory, can the equation be balanced. A much larger but less articulate group favors accommodation with the West while holding fast to their faith, while a third, represented by the Wahhabi establishment in Saudi Arabia and possibly the clerical regime in Iran, argues for a return to the simple faith and rigid practice of the original Islam of Muhammad's day.

Unlike Christianity, Islam has never had a Reformation, which for Christians was a formalized rethinking that ended in the establishment of Protestant denominations. Today this pro-

S.M. Amin/Saudi Aramco World/PADIA (SA2282096)
A common concern among Muslims is how to achieve Islamic modernization, wherein social development considers the realities of the current era. This dichotomy is illustrated by these Egyptian women in traditional dress waiting for a bus, a modern convenience.

cess is getting underway, as Islamic "moderates" work to reform Islam from within. From Morocco to Iran, Turkey to Sudan, Lebanon to Yemen, they seek to build a progressive and effective framework for their faith, responding to the diversity of Muslim societies, the clash of authoritarian and democratic regimes and systems of government, the crisis generated by extremism, and the heavy hand of conservative religious leadership. Such a process took centuries to mature in Europe, and there is no reason not to expect a similar time span in the heartland of Islam.

ISLAMIC FUNDAMENTALISM

For a majority of Muslims the term "Islamic fundamentalism" would be better defined as "Islamic modernization," how to achieve political success and economic and social progress comparable to that of the West without sacrificing their commitment to the faith. In the absence of an Islamic Reformation, as mentioned above, advocates of modernization and conservative fundamentalists struggle for control in a number of Middle Eastern states, notably Iran, Turkey, Algeria and to a lesser extent Lebanon. Elsewhere regimes with varying degrees of authoritarianism remain in control but are increasingly challenged by a vocal opposition. On several occasions, in particular the Taliban in Afghanistan and the National Islamic Front in Sudan, such groups have been on the verge of seizing political control; indeed the Taliban did so before the U.S. invasion of that country. But thus far such groups have been thwarted, although in Turkey's case a revamped "Islamic" party won a majority in the legislature.

It should also be noted that interaction between the West and the Islamic Middle East in recent years has produced two or three generations of Western-educated, Western-oriented academic, professional and intellectual leaders. Even the structure of the Middle East's Islamic regimes is modeled on Western institutions. Bernard Lewis reminds us that the Islamic Republic of Iran—the prototype for putative Islamic regimes—has an elected Assembly and a written Constitution, "for which there is no precedent in the Islamic past."[5] Modern secularist Muslims and fundamentalists alike view Islam as a divinely ordained alternative to both communism and Western free-market capitalism. During the period of the last shah's rule, a prominent Iranian philosopher, Ali Shari'ati, coined the term "Westoxication" to describe the Muslim infatuation with the West, usually to the detriment of Islam. Islamic fundamentalists, rejecting this idea, would remove all elements of Western culture, even principles of law, ethics, politics and democratic practice, from Islamic societies, returning to the basic principles of the faith as the sole guiding force in society.

Muslims today, in the Middle East as elsewhere, are searching for ways to reassert their Islamic identity while reconciling a traditional Islamic way of life with the demands of the contemporary world. The commonality of the faith, its overarching cultural and social norms, is easily grasped and appeals to Muslims irrespective of their different backgrounds. One positive result of the fundamentalist movement is that it enables Muslims to take pride in their heritage. They no longer have to defend or apologize for their beliefs or practices. Islam, regardless of what often seems to be a bad press, has not only matured but also stands as a valid, legitimate institution, one that can be respected by adherents and non-adherents alike.

THE CONCEPT OF JIHAD

Jihad, often described as a "Sixth Pillar" of the House of Islam, may be defined as "sacred struggle" or "striving" (i.e of the individual to carry out God's will and encourage others to do so). It has also loosely but incorrectly been referred to as "holy war." In general it enjoins on the believer the obligation to lead a virtuous and sober life, and when necessary to help defend Islam against its enemies. It is this interpretation that has led to the division by Muslims of the world into the Dar al-Islam ("House of Islam or Peace") and the Dar al-Harb ("House of Dissidence"). Since the ongoing responsibility of all Muslims is to plant the flag of their faith everywhere in the world, jihad, in theory, requires them to propagate it everywhere and by whatever means are necessary, but peaceful ones.

According to the U.S. Commission on International Religious Freedom, Islam is specified as the state religion in the constitutions of 9 Middle Eastern Islamic countries: Algeria, Egypt, Jordan, Kuwait, Libya, Morocco, Qatar, Tunisia and the UAE. Bahrain, Oman, Saudi Arabia (which has no constitution) and Yemen do not, although describing themselves as "declared Islamic states". The Lebanese, Syrian and Sudanese constitutions do not make any such declaration.

Another definition of *jihad*, the striving of the individual for justice, is perhaps the most controversial. Islam teaches that if rulers—whether elected or appointed over some Islamic territory—become unjust, their subjects should bear the injustices

S.M. Amin/Saudi Aramco World/PADIA (SA3511024)

One of the Five Pillars of Islam, or five basic duties of Muslims, is to go on a pilgrimage to Mecca in Saudi Arabia once in their lifetime. There they circle seven times around the Black Box (the Ka'ba, pictured above), kiss the Black Stone, drink from the well of Zam Zam, and perform other sacred rites.

with fortitude; God will, in due course, reward their patience. Some Muslims interpret this injunction to mean that they should strive to help the leaders to see the error of their ways—by whatever action deemed necessary. Centuries ago, a secret society, the Hashishin ("Assassins"; so named because they reportedly were users of hashish), carried out many assassinations of prominent officials and rulers, claiming that God had inspired them to rid Islamic society of tyrants. Since Islam emphasizes the direct relationship of people to God—and therefore people's responsibility to do right in the eyes of God and to struggle to help other believers follow the same, correct path—it becomes most dangerous when individuals feel that they do not need to subject themselves to the collective will but, rather, to impose their own concepts of justice on others.

In our own day, jihad is often associated with the struggle of Shia Muslims for social, political, and economic rights within Islamic states. Inspired by the example of Iran, some seek to establish a true Islamic government in the House of Islam. But Iranian Muslims are not only militant, they are also strongly nationalistic. In this respect, they differ sharply in their approach to Islamic reform from the approaches of other militant Islamic groups, notably the Muslim Brotherhood. Its founder, Hassan al-Banna, stressed peaceful and gradual change in Islamic governments.: "The Brotherhood stands ready to assist [these] governments in the improvement of society through basic Islamification of beliefs, moral codes, and institutions."[6]

Osama bin Laden's al Qaeda terrorist network, in contrast, seeks the overthrow of these governments by violence, considering them unrepresentative and therefore un-Islamic.

No such strictures affect Iranians' view of jihad. An important element in their belief system derives from their special relationship with their religious leaders, particularly the late Ayatollah Khomeini. In his writings and sermons, Khomeini stressed the need for violent resistance to unjust authorities.

A strong example of "internal" Islamic jihad in recent years developed during the annual *Hajj* ("Great Pilgrimage") to Mecca in August 1987. Iranian pilgrims, taking literally Khomeini's injunction that the Hajj is the ideal forum for demonstration of the "proper use of Islam in politics," staged a political rally after midday prayer services. Demonstrators carrying posters of Khomeini shouted, "Death to America! Death to the Soviet Union! Death to Israel!" Saudi police attempting to control them were attacked, and the demonstration swiftly grew into a riot. When it was over, more than 400 people had been killed, including 85 policemen, and 650 people had been injured. "To take revenge for the sacred bloodshed is to free the holy shrines from the wicked Wahhabi," an Iranian government official told a crowd in Teheran.[7]

ISLAM IN THE TWENTY-FIRST CENTURY

The brief period of rule by the Taliban in Afghanistan, with its rigid interpretation of Islam based on Wahhabi principles, might suggest that extremist Islamic fundamentalist movements elsewhere in the Middle East can succeed by using similar tactics. Certainly the ideal of an Islamic community replicating that of the original one founded by Muhammad remains valid. However, the unique nature of the Taliban itself, and the special conditions that made its success possible, do not exist in North Africa or the rest of the Middle East. The Taliban's membership was recruited from Afghan refugee camps in Pakistan or remote villages within Afghanistan. Members, nearly all of them illiterate, were given only the rudiments of Koranic instruction, given by semiliterate mullahs, in so-called religious schools (*madrasas*) set up on a temporary basis in the camps or towns controlled by the organization. When they had completed this "education," they became *talibs* ("students"—hence the name) and were sent in a body to take part in a jihad in Afghanistan intended to restore law and order. Most Taliban members' exposure to the outside world—even the larger Islamic world—was nonexistent, and as a result they had no experience or model to turn to in their relation to women, to other cultures, or in state-building. Actions such as the destruction of ancient Buddhist statues, denial of job and education rights to women, and wholesale repression of normal social activities, eventually isolated them from the rest of the Islamic world.

As the millennium moves on, some seventeen centuries after the Prophet Muhammad (PBUH) announced his vision of a community of believers united in faith in One God, in prayer and in justice, Muslims everywhere are struggling to come to terms with a globalized world and to determine where and how their faith fits into this world. Long before there were separate nations with borders to cross or defend, nomadic peoples moved easily across the Middle East, giving visible substance to Ibn Khaldun's thesis that civilizations rise and fall according to a predictable pattern. Islam created one of those civilizations, and it may do so again. But increasingly in the region, borders are no longer barriers. They were formed artificially for nation-states, and they are crossed easily today, not so much by military forces as by technology, in the form of the Internet, satellite TV, cellular phones and the like.

In Europe particularly the 20-million Muslims community spread across the nations have struggled with problems of integrating themselves into European society while preserving their Islamic identity and values. Their difficulty was highlighted in 2005 by the infamous "cartoon crisis." On September 30 of that year a Danish newspaper, *Jyllands-Posten*, published 12 cartoon caricatures of the Prophet Muhammad. One showed him wearing a turban shaped as a bomb with a burning fuse. Another had him wearing a bushy grey beard and holding a sword, his eyes covered by a black rectangle, while a third pictured him in the desert, a middle-aged prophet walking in front of a donkey.

Aside from being in execrable taste and offensive to the values of a worldwide community, the cartoons seemed and were presented more as examples of European press freedom than as provocations. However, months after their publication they aroused a firestorm of protest, not only in Europe but more intensely in a number of Islamic countries. Danish and other European embassies in Syria and Lebanon, and elsewhere, were attacked and burned, and Muslim leaders in the Middle East called for a boycott of Danish goods.

What is notable about this crisis and Islamic reactions to it is that it was generated by radical Islamic groups not only to embarrass the European press but also those Middle Eastern governments that seem to be aligned with or supportive of the United States. The issue has led also to a more sober reappraisal of the faith by the great body of Muslims. This majority seeks to remain true to the ideals of Islam while accommodating itself to the multi-ethnic and economically and politically unequal world in which it lives. But the U.S. occupation of Islamic Iraq, its emphasis on the Islamic Middle East as the core of the "Axis of Evil", and U.S. support for Israel as an "alien presence" which denies the Palestinians their rightful place in a broad Islamic state, continue to fuel Muslim anger.

NOTES

1. R. Hrair Dekmejian, *Islam in Revolution* (Syracuse, NY: Syracuse University Press, 1985), p. 99.
2. Reported in *The Los Angeles Times* (July 3, 1997). Muslims follow Jewish dietary laws in regarding pigs as unclean animals, and Muslim scholars called her action "a declaration of war against Islam."
3. See Maria Rosa Menocal, *The Ornament of the World* (Boston: Little, Brown 2002). As is well known in history, Muslims and Jews were expelled from Spain after the Christian reconquest in 1492. In an ironic but purely Spanish touch, the first mosque built since then to serve returned Spanish Muslims opened its doors in January 2003.
4. Stephen King, in *The New York Times* (April 13, 1997).
5. Christine Eickelman, *Women and Community in Oman* (New York: New York University Press, 1984) pp. 80–111.
6. Cardamom, an aromatic spice ground from pods of a tall, palm-like plant, came originally from India. Today, the bulk of the crop is grown at high altitudes in Guatemala and exported to the Arab countries, which do not grow any. See Larry Luxner, "The Cardamom Connection," *Aramco World*, Vol. 48, No. 2 (March–April 1997).
7. A researcher at Neot Kedounim, a park in central Israel, has developed a cookbook with recipes for dishes eaten in King Solomon's time using plants that were grown there 2000 years ago and are being replicated in the park. In that long-ago time the foods we think of as "Mediterranean," tomatoes, eggplant, zucchini, rice, and oranges were unknown. The staple foods were wheat barley, lentils, beans and certain fruits, notably figs and dates.
8. Christine Bird, *Neither East Nor West* (New York: Pocket Books/Simon & Schuster, 2001), p. 24. Bird comments that "eating is a constant pastime in Iran" (p. 63).

Algeria (Peoples' Democratic Republic of Algeria)

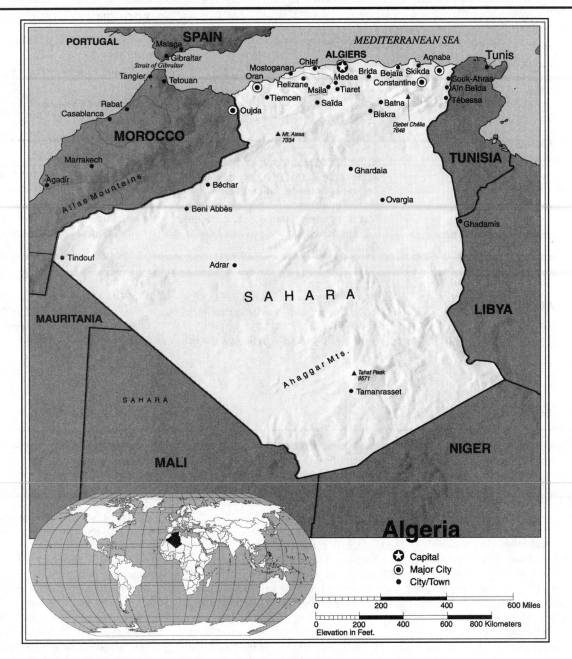

Algeria Statistics

GEOGRAPHY

Area in Square Miles (Kilometers):
919,352 (2,381,740) (about 3 1/2 times the size of Texas)
Capital (Population): Algiers (3,705,000)
Environmental Concerns: soil erosion; desertification; water pollution; inadequate potable water

Geographical Features: mostly high plateau and desert; some mountains; narrow, discontinuous coastal plain

Climate: arid to semiarid; mild winters and hot summers on coastal plain; less rain and cold winters on high plateau; considerable temperature variation in desert

PEOPLE

Population

Total: 32,531,853
Annual Growth Rate: 1.22%
Rural/Urban Population Ratio: 44/56
Major Languages: Arabic; Berber dialects; Ahaggar (Tuareg); French

Ethnic Makeup: 99% Arab-Berber; less than 1% European

Religions: 99% Sunni Muslim (Islam is the state religion); 1% Shia Muslim, Christian, and Jewish

Health

Life Expectancy at Birth: 73 years (male); 74.6 years (female)

Infant Mortality Rate (Ratio): 31/1,000 live births

Education

Adult Literacy Rate: 70%

Compulsory (Ages): 6–15

COMMUNICATION

Telephones: 2,199,600, plus 1,447,310 cell phones

Daily Newspaper Circulation: 52 per 1,000 people

Televisions: 71 per 1,000 people

Internet Users: 2 (2000)

TRANSPORTATION

Highways in Miles (Kilometers): 63,605 (102,424)

Railroads in Miles (Kilometers): 2,963 (4,772)

Usable Airfields: 137

Motor Vehicles in Use: 920,000

GOVERNMENT

Type: republic

Independence Date: July 5, 1962 (from France)

Head of State/Government: President Abdelaziz Bouteflika; Prime Minister Ahmed Ouyahia

Political Parties: National Liberation Front (FLN), majority party; National Democratic Rally (RND), National Reform Movement, chief minority parties; others include Movement for a Peaceful society; Islamic Salvation Front (FIS) outlawed since April 1992

Suffrage: universal at 18

MILITARY

Military Expenditures (% of GDP): 3.2%

Current Disputes: disputed southeastern border with Libya; Algeria supports Polisario Front which seeks to establish an independent Western Sahara, currently occupied by Morocco

ECONOMY

Currency ($U.S. Equivalent): 72 Algerian dinars = $1

Per Capita Income/GDP: $6,600/$177 billion

GDP Growth Rate: 6.1%

Inflation Rate: 3%

Unemployment Rate: 25.4%

Labor Force: 9,900,000

Natural Resources: petroleum; natural gas; iron ore; phosphates; uranium; lead; zinc

Agriculture: wheat; barley; oats; grapes; olives; citrus fruits; sheep; cattle

Industry: petroleum; natural gas; light industries; mining; electrical; petrochemicals; food processing

Exports: $32.16 billion (primary partners Italy, United States, France)

Imports: $15.25 billion (primary partners France, Italy, Germany)

Algeria Country Report

The modern state of Algeria occupies the central part of North Africa, a geographically distinctive and separate region of Africa that includes Morocco, Tunisia, and Libya. The name of the country comes from the Arabic word *al-Jaza'ir*, "the islands," because of the rocky islets along this part of the Mediterranean coast. The name of the capital, Algiers, has the same origin.

The official name of the state is the Democratic and Popular Republic of Algeria. It is the second-largest nation in Africa (after Sudan). The overall population density is low, but the population is concentrated in the northern third of the country. The vast stretches of the Algerian Sahara are largely unpopulated. The country had an extremely high birth rate prior to 1988, but government-sponsored family-planning programs have significantly reduced the rate.

GEOGRAPHY

Algeria's geography is a formidable obstacle to broad economic and social development. About 80 percent of the land is uncultivable desert, and only 12 percent is arable without irrigation. Most of the population live in a narrow coastal plain and in a fertile, hilly inland region called the Tell (Arabic for "hillock"). The four Saharan provinces have only 3 percent of the population but comprise more than half the land area.

The mineral resources that made possible Algeria's transformation in two decades from a land devastated by civil war to one of the developing world's success stories are all located in the Sahara. Economic growth, however, has been uneven, generally affecting the rural and lower-class urban populations unfavorably. The large-scale exodus of rural families into the cities, with consequent neglect of agriculture, has resulted in a vast increase in urban slums. Economic disparities were a major cause of riots in 1988, that led to political reforms and the dismantling of the socialist system responsible for Algerian development since independence.

Algeria is unique among newly independent Middle Eastern countries in that it gained its independence through a civil war. For more than 130 years (1830–1962), it was occupied by France and became a French department (similar to a U.S. state). With free movement from mainland France to Algeria and vice versa, the country was settled by large numbers of Europeans, who became the politically dominant group in the population although they were a minority. The modern Algerian nation is the product of the interaction of native Muslim Algerians with the European settlers, who also considered Algeria home.

Algeria's geography is a key to the country's past disunity. In addition to its vast Saharan territory, Algeria is broken up into discontinuous regions by a number of rugged mountain ranges. The Mediterranean coastline is narrow and is backed throughout its length by mountains, notably the imposing Kabyle range. The Algerian Atlas range, a continuation of the Moroccan Atlas, is a complex system of deep valleys, high plateaux, and peaks ranging up to 6,000 feet. In south central Algeria is the most impressive range in the country, the Aurès, a great mountain block.

The original inhabitants of the entire North African region were Berbers, a people of unknown origin grouped into various tribes. Berbers make up about 30 percent of the total population. The majority live in the eastern Kabylia region and the Aures (Chaouia), with a small, compact group in the five cities of the Mzab, in the Algerian Sahara. The Tuareg, a nomadic Berber people spread across southern Algeria, Mali, and Niger, are the only ones with a written script, called Tifinagh. In the past, they were literally "lords of the desert," patrolling the caravan routes on their swift camels and collecting tolls for safe passage as guides for caravaneers. They were a colorful sight in their tents with their indigo robes (which tinted their skin blue, hence the name for them, the "Blue Men"). But Saharan droughts, motorized transport, and the development of the oil industry have largely destroyed their traditional role and lifestyle. Today, Tuareg are more likely to be found pumping gas in cities or doing low-wage work in the oil fields than patrolling the desert.

The Arabs, who brought Islam to North Africa in the seventh century A.D., converted the Algerian Berbers after a fierce resistance. The Arabs brought their language as a unifying feature, and religion linked the Algerians with the larger Islamic world. Today, most follow Sunni Islam, but a significant minority, about 100,000, are Shia Muslims. They refer to themselves as *Ibadis*, from their observance of an ancient Shia rite, and live in five "holy cities" clustered in a remote Saharan valley where centuries ago they took refuge from Sunni rulers of northern Algeria. Their valley, the Mzab, has always maintained religious autonomy from Algerian central governments. The much larger Berber population of Kabylia has also resisted central authority, whether Islamic or French, throughout Algerian history. One of many pressures on the government today is that of an organized Kabyle movement, which seeks greater autonomy for the region and an emphasis on Berber language in schools, along with the revitalization of Kabyle culture.

HISTORY

The Corsair Regency

The foundations of the modern Algerian state were laid in the sixteenth century, with the establishment of the Regency of Algiers, an outlying province of the Ottoman Empire. Algiers in particular, due to its natural harbor, was developed for use by the Ottomans as a naval base for wars against European fleets in the Mediterranean. The Algerian coast was the farthest westward extent of Ottoman power. Consequently, Algiers and Oran, the two major ports, were exposed to constant threats of attack by Spanish and other European fleets. They could not easily be supported, or governed directly, by the Ottomans. The regency, from its beginnings, was a state geared for war.

The regency was established by two Greek-born Muslim sea captains, Aruj and Khayr al-Din (called Barbarossa by his European opponents because of his flaming red beard). The brothers obtained commissions from the Ottoman sultan for expeditions against the Spanish. They made their principal base at Algiers, then a small port, which Khayr al-Din expanded into a powerful fortress and naval base. His government consisted of a garrison of Ottoman soldiers sent by the sultan to keep order, along with a naval force called the corsairs.

Corsairing or piracy (the choice of term depended upon one's viewpoint) was a common practice in the Mediterranean, but the rise to power of the Algerine corsairs converted it into a more or less respectable profession.[1] The cities of Tetuan, Tunis, Salé (Morocco), and Tripoli (Libya) also had corsair fleets, but the Algerian corsairs were so effective against European shipping that for 300 years (1500–1800), European rulers called them the "scourge of the Mediterranean." One factor in their success was their ability to attract outstanding sea captains from various European countries. Renegades from Italy, Greece, Holland, France, and Britain joined the Algerian fleet, converted to Islam, and took Muslim names as a symbol of their new status. Some rose to high rank.

Government in Algiers passed through several stages and eventually became a system of deys. The deys were elected by the Divan, a council of the captains of the Ottoman garrison. Deys were elected for life, but most of them never fulfilled their tenure due to constant intrigue, military coups, and assassinations. Yet the system provided considerable stability, security for the population, and wealth and prestige for the regency. These factors probably account for its durability; the line of deys governed uninterruptedly from the late 1600s to 1830.

Outside of Algiers and its hinterland, authority was delegated to local chiefs and religious leaders, who were responsible for tax collection and remittances to the dey's treasury. The chiefs were kept in line with generous subsidies. It was a system well adapted to the fragmented society of Algeria and one that enabled a small military group to rule a large territory at relatively little cost.[2]

The French Conquest

In 1827, the dey of Algiers, enraged at the French government's refusal to pay an old debt incurred during Napoleon's wars, struck the French consul on the shoulder with a fly-whisk in the course of an interview. The king of France, Charles X, demanded an apology for the "insult" to his

representative. None was forthcoming, so the French blockaded the port of Algiers in retaliation. But the dey continued to keep silent. In 1830, a French army landed on the coast west of the city, marched overland, and entered it with almost no resistance. The dey surrendered and went into exile.[3]

The French, who had been looking for an excuse to expand their interests in North Africa, now were not sure what to do with Algiers. The overthrow of the despotic Charles X in favor of a constitutional monarchy in France confused the situation even further. But the Algerians considered the French worse than the Turks, who were at least fellow Muslims. In the 1830s, they rallied behind their first national leader, Emir Abd al-Qadir.

Abd al-Qadir was the son of a prominent religious leader and, more important, was a descendant of the Prophet Muhammad. Abd al-Qadir had unusual qualities of leadership, military skill, and physical courage. From 1830 to 1847, he carried on guerrilla warfare against a French army of more than 100,000 men with such success that at one point the French signed a formal treaty recognizing him as head of an Algerian nation in the interior. Abd al-Qadir described his strategy in a prophetic letter to the king of France:

> France will march forward, and we shall retire. But France will find it necessary to retire, and we shall return. We shall weary and harry you, and our climate will do the rest.[4]

In order to defeat Abd al-Qadir, the French commander used "total war" tactics, burning villages, destroying crops, killing livestock, and levying fines on peoples who continued to support the emir. These measures, called "pacification" by France, finally succeeded. In 1847, Abd al-Qadir surrendered to French authorities. He was imprisoned for several years, in violation of a solemn commitment, and was then released by Emperor Napoleon III. He spent the rest of his life in exile.

DEVELOPMENT

Algeria ranks fifth in the world in natural gas reserves and second in gas exports. The hydrocarbons sector generates 60 percent of revenues, 30 percent of GDP. In 2001 the country signed an association treaty with the European Union and has applied for membership in the World Trade Organization.

Although he did not succeed in his quest, Abd al-Qadir is venerated as the first

Algerian nationalist, able by his leadership and Islamic prestige to unite warring groups in a struggle for independence from foreign control. Abd al-Qadir's green and white flag was raised again by the Algerian nationalists during the second war of independence (1954–1962), and it is the flag of the republic today.

Algérie Française

After the defeat of Abd al-Qadir, the French gradually brought all of present-day Algerian territory under their control. The Kabyles, living in the rugged mountain region east of Algiers, were the last to submit. The Kabyles had submitted in 1857, but they rebelled in 1871 after a series of decrees by the French government had made all Algerian Muslims subjects but not citizens, giving them a status inferior to French and other European settlers.

The Kabyle rebellion had terrible results, not only for the Kabyles but for all Algerian Muslims. More than a million acres of Muslim lands were confiscated by French authorities and sold to European settlers. A special code of laws was enacted to treat Algerian Muslims differently from Europeans, with severe fines and sentences for such "infractions" as insulting a European or wearing shoes in public. (It was assumed that a Muslim caught wearing shoes had stolen them.)

In 1871, Algeria legally became a French department. But in terms of exploitation of natives by settlers, it may as well have remained a colony. One author notes that "the desire to make a settlement colony out of an already populated area led to a policy of driving the indigenous people out of the best arable lands."[5] Land confiscation was only part of the exploitation of Algeria by the *colons* (French settlers). They developed a modern Algerian agriculture integrated into the French economy, providing France with much of its wine, citrus, olives, and vegetables. Colons owned 30 percent of the arable land and 90 percent of the best farmland. Special taxes were imposed on the Algerian Muslims; the colons were exempted from paying most taxes.

The political structure of Algeria was even more favorable to the European minority. The colons were well represented in the French National Assembly, and their representatives made sure that any reforms or laws intended to improve the living conditions or rights of the Algerian Muslim population would be blocked.

In fairness to the colons, it must be pointed out that many of them had come to Algeria as poor immigrants and worked hard to improve their lot and to develop the country. By 1930, the centenary of the French conquest, many colon families had lived in Algiers for two generations or more. Colons had drained malarial swamps south of Algiers and developed the Mitidja, the country's most fertile region. A fine road and rail system linked all parts of the country, and French public schools served all cities and towns. Algiers even had its own university, a branch of the Sorbonne. It is not surprising that to the colons, Algeria was their country, "Algérie Française." Throughout Algeria they rebaptized Algerian cities with names like Orléansville and Philippeville, with paved French streets, cafes, bakeries, and little squares with flower gardens and benches where old men in berets dozed in the hot sun.

Jules Cambon, governor general of Algeria in the 1890s, once described the country as having "only a dust of people left her." What he meant was that the ruthless treatment of the Algerians by the French during the pacification had deprived them of their natural leaders. A group of leaders developed slowly in Algeria, but it was made up largely of *evolués*—persons who had received French educations, spoke French better than Arabic, and accepted French citizenship as the price of status.[6]

Other Algerians, several hundred thousand of them, served in the French Army in the two world wars. Many of them became aware of the political rights that they were supposed to have but did not. Still others, religious leaders and teachers, were influenced by the Arab nationalist movement for independence from foreign control in Egypt and other parts of the Middle East.

Until the 1940s, the majority of the evolués and other Algerian leaders did not want independence. They wanted full assimilation with France and Muslim equality with the colons. Ferhat Abbas, a French-trained pharmacist who was the spokesman for the evolués, said in 1936 that he did not believe that there was such a thing as an Algerian nation separate from France.

Abbas and his associates changed their minds after World War II. In 1943, they had presented to the French government a manifesto demanding full political and legal equality for Muslims with the colons. It was blocked by colon leaders, who feared that they would be drowned in a Muslim sea. On May 8, 1945, the date of the Allied victory over Nazi Germany, a parade of Muslims celebrating the event but also demanding equality led to violence in the city of Sétif. Several colons were killed; in retaliation, army troops and groups of colon vigilantes swept through Muslim neighborhoods, burning houses and slaughtering thousands of Muslims. From then on, Muslim leaders believed that independence through armed struggle was the only choice left to them.

The War for Independence

November 1 is an important holiday in France. It is called Toussaint (All Saints' Day). On that day, French people remember and honor all the many saints in the pantheon of French Catholicism. It is a day devoted to reflection and staying at home.

In the years after the Sétif massacre, there had been scattered outbreaks of violence in Algeria, some of them created by the so-called Secret Organization (OS), which had developed an extensive network of cells in preparation for armed insurrection. In 1952, French police accidentally uncovered the network and jailed most of its leaders. One of them, a former French Army sergeant named Ahmed Ben Bella, subsequently escaped and went to Cairo, Egypt.

As the day of Toussaint 1954 neared, Algeria seemed calm. But appearances were deceptive. Earlier in the year, nine former members of the OS had laid plans in secret for armed revolution. They divided Algeria into six *wilayas* (departments), each with a military commander. They also planned a series of coordinated attacks for the early morning hours of November 1, when the French population would be asleep and the police preparing for a holiday. Bombs exploded at French Army barracks, police stations, storage warehouses, telephone offices, and government buildings. The revolutionaries circulated leaflets in the name of the National Liberation Front (FLN), warning the French that they had acted to liberate Algeria from the colonialist yoke and calling on Algerian Muslims to join in the struggle to rebuild Algeria as a free Islamic state.

There were very few casualties as a result of the Toussaint attacks; for some time the French did not realize that they had a revolution on their hands. But as violence continued, regular army troops were sent to Algeria to help the hard-pressed police and the colons. Eventually there were 400,000 French troops in Algeria, as opposed to just 6,000 guerrillas. But the French consistently refused to consider the situation in Algeria a war. They called it a "police action." Others called it the "war without a name."[7] Despite their great numerical superiority, they were unable to defeat the FLN.

Elsewhere the French tried various tactics. They divided the country into small sectors, with permanent garrisons for each sector. They organized mobile units to track down the guerrillas in caves and hideouts. About 2 million villagers were moved

Agriculture is important in raising the living standards of Algeria. These farmers are harvesting forage peas, which will be used for animal feed.

into barbed-wire "regroupment camps," with a complete dislocation of their way of life, in order to deny the guerrillas the support of the population.

The war was settled not by military action but by political negotiations. The French people and government, worn down by the effects of World War II and their involvement in Indochina, grew sick of the slaughter, the plastic bombs exploding in public places (in France as well as Algeria), and the brutality of the army in dealing with guerrilla prisoners. A French newspaper editor expressed the general feeling: "Algeria is ruining the spring. This land of sun and earth has never been so near us. It invades our hearts and torments our minds."[8]

The colons and a number of senior French Army officers were the last to give up their dream of an Algeria that would be forever French. Together the colons and the army forced a change in the French government. General Charles de Gaulle, the French wartime resistance hero, returned to power after a dozen years in retirement. But de Gaulle, a realist, had no intention of keeping Algeria forever French. He began secret negotiations with FLN leaders for Algerian independence.

By 1961 the battlefield had extended into metropolitan France, with plastic bombs set off in cafés and other public places, killing hundreds of people. On its side, the French military routinely used torture and gang-style executions without trial to crush the rebellion. Some 3,000 of those

arrested simply disappeared.[9] Clashes between FLN fighters and those of its rival, the Algerian Nationalist Movement, caused further disruptions. In October, the shooting of Paris police officers led to the deaths of several hundred Algerians by the police during a peaceful protest march (an error not revealed by the French government until its archives for the period were opened in 1999).

Subsequently, colon and dissident military leaders united in a last effort to keep Algeria French. They staged an uprising against de Gaulle in Algiers, seizing government buildings and demanding his removal from office. But the bulk of the French Army remained loyal to him.

An attempted assassination of the French president in 1962 was unsuccessful. The colon–military alliance, calling itself the Secret Army Organization (OAS), then launched a savage campaign of violence against the Muslim population, gunning down people or shooting them at random on streets and in public markets. The OAS expected that the FLN would break the cease-fire in order to protect its own people. But it did not do so.

THE AGONY OF INDEPENDENCE

With the collapse of the OAS campaign against the FLN as well as its own government, the way was clear for Algeria to become an independent nation for the first time in its history. This became a reality on July 5, 1962, with the signing of a treaty

with France. However, few modern nations have become self-governing with so many handicaps. Several hundred thousand people—French, Algerian Muslims, men, women, and children—were casualties of the conflict. An even more painful loss was the departure of the entire European community. Panicked colons and their families boarded overcrowded ships to cross the Mediterranean, most of them to France, a land they knew only as visitors. Nearly all of the skilled workers, managers, landowners, and professionals in all fields were French, and they had done little to train Algerian counterparts.

The new Algerian government was also affected by factional rivalries among its leaders. The French writer Alexis de Tocqueville once wrote, "In rebellion, as in a novel, the most difficult part to invent is the end." The FLN revolutionaries had to invent a new system, one that would bring dignity and hope to people dehumanized by 130 years of French occupation and eight years of savage war.

The first leader to emerge from intraparty struggle to lead the nation was Ahmed Ben Bella, who had spent the war in exile in Egypt but had great prestige as the political brains behind the FLN. Ben Bella laid the groundwork for an Algerian political system centered on the FLN as a single legal political party, and in September 1963, he was elected president. Ben Bella introduced a system of *autogestion* (workers' self-management), by which tenant farmers took over the management

United Nations Photo (UN102110)
The rapid growth in the population of Algeria, coupled with urban migration, has created a serious housing shortage, as this crowded apartment building in Algiers testifies.

of farms abandoned by their colon owners and restored them to production as cooperatives. Autogestion became the basis for Algerian socialism—the foundation of development for decades.

Ben Bella did little else for Algeria, and he alienated most of his former associates with his ambitions for personal power. In June 1965, he was overthrown in a military coup headed by the defense minister, Colonel Houari Boumedienne. Ben Bella was sentenced to house arrest for 15 years; he was pardoned and exiled in 1980. While in exile, he founded the Movement for a Democratic Algeria, in opposition to the regime. In 1990, he returned to Algeria and announced plans to lead a broad-based opposition party in the framework of the multiparty system. He retired from political and public life in 1997, and the Movement was dissolved.

Boumedienne declared that the coup was a "corrective revolution, intended to reestablish authentic socialism and put an end to internal divisions and personal rule."[10] The government was reorganized under a Council of the Revolution, all military men, headed by Boumedienne, who subsequently became president of the republic. After a long period of preparation and gradual assumption of power by the re-

clusive and taciturn Boumedienne, a National Charter (Constitution) was approved by voters in 1976. The Charter defined Algeria as a socialist state with Islam as the state religion, basic citizens' rights guaranteed, and leadership by the FLN as the only legal political party. A National Popular Assembly (the first elected in 1977) was responsible for legislation.

In theory, the Algerian president had no more constitutional powers than the U.S. president. However, in practice, Boumedienne was the ruler of the state, being president, prime minister, and commander of the armed forces rolled into one. In November 1978, he became ill from a rare blood disease; he died in December. For a time, it appeared that factional rivalries would again split the FLN, especially as Boumedienne had named neither a vice-president nor a prime minister, nor had he suggested a successor.

The Algeria of 1978 was a very different nation from that of 1962. The scars of war had mostly healed. The FLN closed ranks and named Colonel Chadli Bendjedid to succeed Boumedienne as president for a five-year term. In 1984, Bendjedid was reelected. But the process of ordered socialist development was abruptly and forcibly interrupted in October 1988. A new genera-

tion of Algerians, who had come of age long after the war for independence, took to the streets, protesting high prices, lack of jobs, inept leadership, a bloated bureaucracy, and other grievances.

The riots accelerated the process of Algeria's "second revolution" toward political pluralism and dismantling of the single-party socialist system. President Bendjedid initially declared a state of emergency; and for the first time since independence, the army was called in to restore order. Some 500 people were killed in the rioting, most of them jobless youths. But the president moved swiftly to mobilize the nation in the wake of the violence. In a national referendum, voters approved changes in the governing system to allow political parties to form outside the FLN. Another constitutional change, also effective in 1989, made the cabinet and prime minister responsible to the National Assembly.

The president retained his popularity during the upheaval and was reelected for a third term, winning 81 percent of the votes. A number of new parties were formed in 1989 to contest future Assembly elections. They represented a variety of political and social positions. Thus, the People's Movement for Algerian Renewal advocated a "democratic Algeria, representative of moderate Islam," while the National Algerian Party, more fundamentalist in its views, had a platform of full enforcement of Islamic law and the creation of 2 million new jobs. The Socialist Forces Front (FFS), founded many years earlier by exiled FLN leader Hocine Ait Ahmed, resurfaced with a manifesto urging Algerians to support "the irreversible process of democracy."

For its part, the government sought to revitalize the FLN as a genuine mass party on the order of the Tunisian Destour, while insisting that it would not duplicate its neighbor country's *democratie de façade* but would instead embark on real political reforms. Recruitment of new members was extended to rural areas. Although press freedom was confirmed in the constitutional changes approved by the voters, control of the major newspapers and media was shifted from the government to the FLN, to provide greater exposure.

FOREIGN POLICY

During the first decade of independence, Algeria's foreign policy was strongly nationalistic and anti-Western. Having won their independence from one colonial power, the Algerians were vocally hostile toward the United States and its allies, calling them enemies of popular liberation. Algeria supported revolutionary movements all over the world, providing funds, arms, and training. The Palestine Liberation Organization, rebels against Portuguese colonial rule in Mozambique, Muslim guerrillas fighting the Christian Ethiopian government in Eritrea—all benefited from active Algerian support.

The government broke diplomatic relations with the United States in 1967, due to American support for Israel, and did not restore them for a decade. In the mid-1970s, Algeria moderated its anti-Western stance in favor of nonalignment and good relations with both East and West. Relations improved thereafter to such a point that Algerian mediators were instrumental in resolving the 1979–1980 American hostage crisis in Iran, since Iran regarded Algeria as a suitable mediator—Islamic yet nonaligned. However, Algeria's subsequent alignment with Iraq (in sympathy for Iraq as a fellow-Arab state) during the Iran–Iraq War caused a break in diplomatic relations with the Islamic Republic. They were not restored until 2000.

Until recently, Algeria's relations with Morocco were marked by suspicion, hostility, and periodic conflict. The two countries clashed briefly in 1963 over ownership of iron mines near Tindouf, on the border. Algeria also supported the Western Saharan nationalist movement fighting for independence for the former Spanish colony against Moroccan occupation. After Morocco annexed the territory, Algeria provided bases, sanctuary, funds, and weapons to the Polisario Front, the military wing of the movement. The Bendjedid government recognized the self-declared Sahrawi Arab Democratic Republic in 1980 and sponsored SADR membership in the Organization for African Unity (OAU).

Algeria's own economic difficulties plus the slow progress led by Bouteflika from military to civilian leadership and restoration of the parliamentary system helped reduce support for the Polisario. The Polisario office in Algiers was ordered closed. However the government continued its economic support for the 80,000 to 160,000 Sahrawis (even these figures are disputed) crowded into refugee camps in its Saharan territory.

THE ECONOMY

Algeria's oil and gas resources were developed by the French. Commercial production and exports began in 1958 and continued through the war for independence; they were not affected, since the Sahara was governed under a separate military administration. The oil fields were turned over to Algeria after independence but continued to be managed by French technicians until 1970, when the industry was nationalized.

Today, the hydrocarbons sector provides the bulk of government revenues and 90 percent of exports. New oil discoveries in 1996 and 2001 are expected to increase oil production, currently 852,000 barrels per day. Algeria provides 29 percent of the liquefied natural gas (LNG) imported by European countries, much of it through undersea pipelines to Italy and Spain.

During President Boumedienne's period in office, all sectors of the Algerian economy were governed under the 1976 National Charter. This document set forth provisions for national development under a uniquely Algerian form of state socialism. However, persistent economic difficulties caused by a combination of lower oil prices and global oversupply led the Bendjedid government to scrap the Charter. Since then, Algeria has borrowed heavily and regularly from international lenders to pay for continued industrial growth.

After a number of years of negative economic growth, the government initiated an austerity program in 1992. Imports of luxury products were prohibited and several new taxes introduced. The program was approved by the International Monetary Fund, Algeria's main source of external financing. In 1995, the IMF loaned $1.8 billion to cover government borrowing up to 60 percent under the approved austerity program to make the required "structural adjustment." In August of that year, the Paris Club—the international consortium that manages most of Algeria's foreign indebtedness—rescheduled $7 billion of the country's foreign debts due in 1996–1997, including interest payments, to ease the strain on the economy.

The agricultural sector employs 47 percent of the labor force and accounts for 12 percent of gross domestic product. But inasmuch as Algeria must import 70 percent of its food, better agricultural production is essential to overall economic development. Overall agricultural production growth has averaged 5 percent annually since 1990. The autogestion system introduced as a stop-gap measure after independence and enshrined later in FLN economic practice, when it seemed to work, was totally abandoned. In 1988, some 3,500 state farms were converted to collective farms, with individuals holding title to lands.

The key features of Bouteflika's economic reform program, one designed to attract foreign investment, include banking reforms, reduction of the huge government bureaucracy, favorable terms for foreign companies and privatization of state-owned enterprises. The telecommunications industry was privatized in 2000 and the government-owned cement and steel industries in 2002.

Privatization of state-owned enterprises is a key feature of the government's plan to attract foreign investment. The telecommunications sector was privatized in August 2000, and some 200 other public enterprises were in process of transfer to private ownership.

HEALTH/WELFARE

The 1984 Family Law improved women's rights in marriage, education and work opportunities. But professional women and, more recently, rural women and their children have become special targets of Islamic violence. Some 400 professional women were murdered in 1995 and more than 400 were killed in a one-day rampage in January 1998.

THE FUNDAMENTALIST CHALLENGE

Despite the growing appeal of Islamic fundamentalism in numerous Arab countries in recent years, Algeria until very recently seemed an unlikely site for the rise of a strong fundamentalist movement. The country's long association with France, its lack of historic Islamic identity as a nation, and several decades of single-party socialism militated against such a development. But the failure of successive Algerian governments to resolve severe economic problems, plus the lack of representative political institutions nurtured within the ruling FLN, brought about the rise of fundamentalism as a political. Fundamentalists took an active part in the 1988 riots; and with the establishment of a multiparty system, they organized a political party, the Islamic Salvation Front (FIS). It soon claimed 3 million adherents among the then 25 million Algerians.

FIS candidates won 55 percent of urban mayoral and council seats in the 1989 local and municipal elections. The FLN conversely managed to hold on to power largely in the rural areas. Fears that FIS success might draw army intervention and spark another round of revolutionary violence led the government to postpone for six months the scheduled June 1991 elections for an enlarged 430-member National People's Assembly. An interim government, under the technocrat prime minister Sid Ahmed Ghozali, was formed to oversee the transition process.

In accordance with President Bendjedid's commitment to multiparty democracy, the first stage of Assembly elections took place on December 26, 1991, with FIS candidates winning 188 out of 231 contested seats. But before the second stage could take place, the army stepped in. FIS leaders were arrested, and the elections were postponed indefinitely. President Bendjedid resigned on January 17, 1992, well ahead of the expiration (in 1993) of his third five-year term. He said that he did so as a sacrifice in the interest of restoring stability to the nation and preserving democracy. Mohammed Boudiaf, one of the nine historic chiefs of the Revolution, returned from years of exile in Morocco to become head of the Higher Council of State, set up by military leaders after the abortive elections and resignation of President Bendjedid. FIS headquarters was closed and the party declared illegal by a court in Algiers. Local councils and provincial assemblies formed by the FIS after the elections were dissolved and replaced by "executive delegations" appointed by the Higher Council.

Subsequently, Boudiaf named a 60-member Consultative Council to work with the various political factions to reach a consensus on reforms. However, the refusal of such leaders as former president Ben Bella and Socialist Forces Front (FFS) leader Hocine Ait Ahmed to participate limited its effectiveness. Boudiaf was also suspected of using it to build a personal power base. On June 29, 1992, he was assassinated, reportedly by a member of his own presidential guard.

With Boudiaf gone, Algeria's generals turned to their own ranks for new leadership. In 1994, General Liamine Zeroual, the real strongman of the regime, was named head of state by the Higher Council. Zeroual pledged that elections for president would be held in November 1995 as a first step toward the restoration of parliamentary government. The top FIS leaders, Abbas Madani and Ali Belhaj, who had been given 12-year jail sentences for "endangering state security" were released but had their sentences commuted to house arrest, on the assumption that in return for dialogue, they would call a halt to the spiraling violence.

However, the dialogue proved inconclusive, and Zeroual declared that the presidential elections would be held on schedule. Earlier, leaders of the FIS, FFS, FLN, and several smaller parties had met in Rome, Italy, under the sponsorship of Sant-Egidio, a Catholic service agency, and announced a "National Contract." It called for the restoration of FIS political rights in return for an end to violence, multiparty democracy, and exclusion of the military from government. The Algerian "personality" was defined in the Contract as Islamic, Arab, and Berber.

Military leaders rejected the National Contract out of hand, due to the FIS's participation. However, the November 1995 presidential election was held as scheduled, albeit under massive army protection—soldiers were stationed within 65 feet of every polling place. Zeroual won handily, as expected, garnering 61 percent of the votes. But the fact that the election was held at all, despite a boycott call by several party leaders and threats of violence from the Armed Islamic Group (GIA), was impressive.[11]

THE KILLING FIELDS

Algeria's modern history has been well described as one of excesses. Thus "the colonial period was unusually harsh, the war for independence particularly costly.... the insistence on one-party rule initially unwavering and the projects for industrialization overly ambitious," as specialists on the country have noted.[12] Extremes of violence are nothing new in Algerian life. But in addition to horrifying violence, the real tragedy of the conflict has been to pit "an inflexible regime and a fanatical opposition" against "innocent victims doomed by their secular lifestyle or their piety."[13]

Shortly before the 1995 election, Ahmed Ben Bella, Algeria's first president and the leader of the now-dissolved Movement for a Democratic Algeria, wrote a thoughtful analysis of the "dialogue at Rome" in which he had participated and that produced the National Contract. He noted: "A mad escalation of violence is the hallmark of everyday life. Nobody is safe: journalists, intellectuals, women, children and old people are all equally threatened. Yet the use of force, the recourse to violence, will not allow any of the protagonists to solve the problem to their advantage, and the solution must be a political one." The dialogue at Rome, he added, "was meant to lead to a consensus that would bring together everyone—including the regime in power—within the framework of the current Constitution, which stipulates political pluralism, democracy, respect for all human rights and freedoms."[14]

ACHIEVEMENTS

A new pipeline from the vast Hassi Berkine oil filed, the African continent's largest, went into production in 1998, with exports of 300,000 barrels per day. Increased foreign investment due to expansion of the hydrocarbons sector, privatization of state-owned enterprises, and favorable terms for foreign companies increased GDP growth to 5% in 2001.

The conflict between the armed wing of the FIS, the Armed Islamic Group (GIA), and the military regime reached a level of violence in the period after the 1995 "election" that left no room for compromise. The GIA targeted not only the army and police but also writers, journalists, government officials and other public figures, professional women, even doctors and dentists. Ironically, one of its victims was the head of the Algerian League for Human

Rights, which had protested the detention without trial of some 9,000 FIS members in roofless prisons deep in the Sahara, under unbelievably harsh conditions.

The GIA widened its circle of violence in the rest of the decade and on into the new century. In addition to Algerians it carried out attacks on foreigners, killing tourists as well as long-term foreign residents, notably Trappist monks. Rural villages were a favorite target, since they had neither police nor army protection. Entire village populations were massacred in a manner eerily reminiscent of the war for independence. The army and security forces did their share of killings, often arresting people and holding them indefinitely without charges. In 2003 the international organization Human Rights Watch reported that in addition to 120,000 deaths, some 7,000 persons had simply disappeared, never to be seen again by their families.

As the violence continued, the GIA also attacked foreigners, killing among others seven Trappist monks and the bishop of Oran. Rural villages were a favorite target since they lacked police or army protection. Men, women, and children in these villages were massacred under conditions of appalling brutality. A UN Human Rights subcommittee visited the country in 2000 and reported that the GIA and government troops were almost equally responsible for the casualties. By 2001, it was estimated that more than 120,000 people had been killed.

The violence tapered off in 2001, due in part to newly elected president Boutefli-kas's amnesty plan, withdrawal of popular support for the GIA, and loss of its main base in the Algiers Casbah. Unfortunately, few GIA fighters responded to the amnesty offer, as it would have required them to surrender their weapons. Violence broke out again in the coastal provinces. The amnesty offer expired in 2001, and with 5,000 guerrillas at large, the prospects for peace in Algeria seemed more remote than ever. In the first half of the year, there were more casualties (2,500) than in all of 1999.

WHAT PRICE DIALOGUE?

Algerians have been described as having one of two personality types: those "tolerant in matters of religion and way of life, multicultural in languages and traditions, open to the diversities of location at the great hub of the Mediterranean world" and those "secretive, violent, enemies of Islamic secularism, irreverence (toward Islam) and modernism." The establishment of the first group in control of the nation resulted largely from Zeroual's efforts. After his election in 1995, he ordered a referendum on key revisions to the Constitution. These included a ban on

religious, linguistic, regional, and gender-based political parties; a limit of two five-year terms for presidents; and commitments to Islam as the state religion and Arabic as its official language. Another revision established a bicameral legislature with an appointed upper house (the Council of the Nation) and a lower house, the National Assembly, popularly elected under a system of proportional representation. The revisions were approved in November by 85.8 percent of eligible voters.

Zeroual next set June 5, 1997, for elections to the 380-member Assembly, the first national election since the abortive 1992 one. Despite a meager 65.5 percent voter turnout due to fears of violence, Zeroual's newly formed party, the National Democratic Rally, won 115 seats. Along with the FLN's 64 seats, the results gave the regime a slim majority.

Two "moderate" Islamist parties (so called because they rejected violence) also participated in the elections. The Movement for a Peaceful Society ran second to the government party, with 69 seats; and An-Nahdah won 34 seats, giving at least a semblance of opposition in the Assembly. The first local and municipal elections in Algeria's modern history were held in 1997 as well, continuing the trend as government-backed candidates won the majority of offices.

In 1999, Zeroual resigned and scheduled open presidential elections for April. Seven candidates filed; they included former foreign minister Abdelaziz Bouteflika, who had lived in Switzerland for many years. Subsequently all the other candidates withdrew, citing irregularities in the election process. But the election went off as scheduled, and Bouteflika was declared the winner, with 74 percent of the vote. The names of the other candidates remained on the ballot (the two "moderate" Islamist candidates received 17 percent of the vote).

Despite the apparently endless violence, the government continues to move slowly toward restoration of representative government. The first local elections in the country's history were held in 1997. Elections for a 389-seat National Popular Assembly took place in May 2002, with candidates elected by popular vote. The FLN won the majority of seats, 199, followed by the RND with 48, the National Reform Movement with 43 and the Movement for a Peaceful Society with 38. Although the FIS continued to be excluded from political participation, its leaders, Madani and Belhaj, were released from house arrest in July 2003. They are still prohibited from political activity.

Timeline: PAST

1518–1520
Establishment of the Regency of Algiers

1827–1830
The French conquest, triggered by the "fly-whisk incident"

1847
The defeat of Abd al-Qadir by French forces

1871
Algeria becomes an overseas department of France

1936
The Blum-Viollette Plan, for Muslim rights, is annulled by colon opposition

1943
Ferhat Abbas issues the Manifesto of the Algerian People

1954–1962
Civil war, ending with Algerian independence

1965
Ben Bella is overthrown by Boumedienne

1976
The National Charter commits Algeria to revolutionary socialist development

1978
President Boumedienne dies

1980s
Land reform is resumed with the breakup of 200 large farms into smaller units; Arabization campaign

1990s
President Bendjedid steps down; the Islamic Salvation Front becomes a force and eventually is banned; the economy undergoes an austerity program; civil war

1999
Abdelaziz Bouteflika elected by voters as new president. Bicameral legislature elected with multi-party participation in 2001.

PRESENT

2000s
Efforts to restore the multiparty system

Continued civil conflict

2005
Bouteflika issues Charter for Peace and Reconciliation approved in national referendum

Improved security during Bouteflika's first term in office reduced the violence significantly, down from an average of 1,200 casualties a month in the early 1990s. Unfortunately few militants accepted the government's 1999 amnesty offer, but to the president's credit he has continued to work to restore the multiparty system and hold open Parliamentary elections. In January 2001, elections were held for the Senate, the upper house of the

Legislature. Its 144 seats are two-thirds elected and one-third appointed by the president. Suffrage was limited to the 15,000 members of communal and provincial popular assemblies elected in 1997. The National Rally for Democracy (RND), which holds a majority of seats in the lower house, won 78 Senate seats, to 31 for its main opposition, the Movement for a Peaceful Society (MSP).

Bouteflika's good intentions and early success became mired in 2000 and 2001, however, in a power struggle with military leaders. And with neither the campaign against the GIA nor moves toward reconciliation with the Islamists yielding much in the way of results, the Algerian people have become more disillusioned than ever.

Recent riots in Kabylia have compounded Bouteflika's difficulties. The Berber people of that region have chafed since independence against Arab political and economic domination, and have fought to preserve their culture and language. The death of a young Berber while in police custody sparked riots during April and May 2001, which soon spread over the whole country. Rioters clashed with police in Bejaia and Tizi Ouzou, the regional capitals, after heavy-handed police actions had killed some 80 persons. The crackdown generated public criticism abroad, particularly in France, where a government spokesman condemned "the violence of the repression" and urged a peaceful dialogue with the Berber population.

Another blow to the embattled regime came with the walkout from the Assembly of members of the Rally for Culture and Democracy (RCD), the main Berber party represented in the ruling coalition. Its leaders cited government failure to address Berber demands for official recognition of their language, affordable housing and job opportunities for youth as the main reasons for their withdrawal. Such issues as these, notably those of housing and jobs, affect the entire population. In August 2005 Bouteflika took another bold step toward restoration of civil peace by issuing a Charter for Peace and National Reconciliation.

It would be presented to the electorate in a national referendum. The Charter in essence asked the Algerian people to forget the past. It provided a little for everyone— amnesty for Islamic militants, for military and security forces for their extra-judicial killings, captures or torture of opposition elements, and compensation for families of victims of the decade-long violence. "Reconciliation, in my view, must protect us from experiencing once again the two evil phenomena of terrorist violence and extremism," Bouteflika said in his address to the nation.[15]

Reactions to the proposed Charter varied. Critics noted that it would strengthen presidential power, enabling him to set aside the constitution and continue as president for a third term. Others argued that it would effectively exonerate security forces and leave the sensitive issue of missing family members unresolved. But in a society fractured by political, linguistic, cultural and ethnic rivalries the Charter seemed the only way to restore unity and deal effectively with growing economic and social problems. A majority of voters thought so and voted Yes in the referendum. With the Charter now in effect, Bouteflika announced in March 2006 that the government would pardon some 2,100 militants. An additional 100 charged with more serious crimes would have their sentences reduced.

NOTES

1. See William Spencer, *Algiers in the Age of the Corsairs* (Norman, OK: University of Oklahoma Press, 1976), Centers of Civilization Series. "The corsair, if brought to justice in maritime courts, identified himself as *corsale* or *Korsan,* never as fugitive or criminal; his occupation was as clearly identifiable as that of tanner, goldsmith, potter or tailor," p. 47.
2. Raphael Danziger, *Abd al-Qadir and the Algerians* (New York: Holmes and Meier, 1977), notes that Turkish intrigue kept the tribes in a state of near-constant tribal warfare, thereby preventing them from forming dangerous coalitions, p. 24.
3. The usual explanation for the quick collapse of the regency after 300 years is that its forces were prepared for naval warfare but not for attack by land. *Ibid.,* pp. 36–38.
4. Quoted in Harold D. Nelson, *Algeria, A Country Study* (Washington, D.C.: American University, Foreign Area Studies, 1979), p. 31.
5. Marnia Lazreg, *The Emergence of Classes in Algeria* (Boulder, CO: Westview Press, 1976), p. 53.
6. For Algerian Muslims to become French citizens meant giving up their religion, for all practical purposes, since Islam recognizes only Islamic law and to be a French citizen means accepting French laws. Fewer than 3,000 Algerians became French citizens during the period of French rule. Nelson, *op. cit.,* pp. 34–35.
7. John E. Talbott, *The War Without a Name: France in Algeria, 1954–1962* (New York: Alfred A. Knopf, 1980).
8. Georges Suffert, in *Esprit,* 25 (1957), p. 819.
9. The opening of French historical archives in 1999 and recent interviews with leading generals in Algiers at the time, such as Jacques Massu, have reopened debate in France about the conduct of the war. In December 2000, members of the French Communist Party, which had backed the FLN, urged formation of a special commission to investigate charges of torture and provide compensation for victims' families. Suzanne Daley, in *The New York Times* (December 30, 2000). Retired general Paul Aussarress was fined $6,500 in 2001 for his 1999 book, *Algeria Special Forces 1955–1957,* in which he admitted the torture and execution of many Algerians and the disappearance of some 3,000 suspects while in custody; he was charged with "trying to justify war."
10. Nelson, *op. cit.,* p. 68.
11. Robert Mortimer, "Algeria: The Dialectic of Elections and Violence," *Current History* (May 1997), p. 232.
12. Frank Ruddy, who was assigned to Tindouf by the United Nations as a member of the observer group monitoring the referendum in the Western Sahara, comments on the town's national-history museum, "the one cultural attraction." However, "most of its space is devoted to especially grisly photos of terrible things the French did to Algerians during the Algerian war of independence." *The World & I* (August 1997), p. 138.
13. Robert Fisk, in *The Independent* (London) (March 16, 1995).
14. Ahmed Ben Bella, "A Time for Peace in Algeria," *The World Today* (November 1995), p. 209.
15. Michael Slackman, *New York Times,* September 26, 2005.

Bahrain (State of Bahrain)

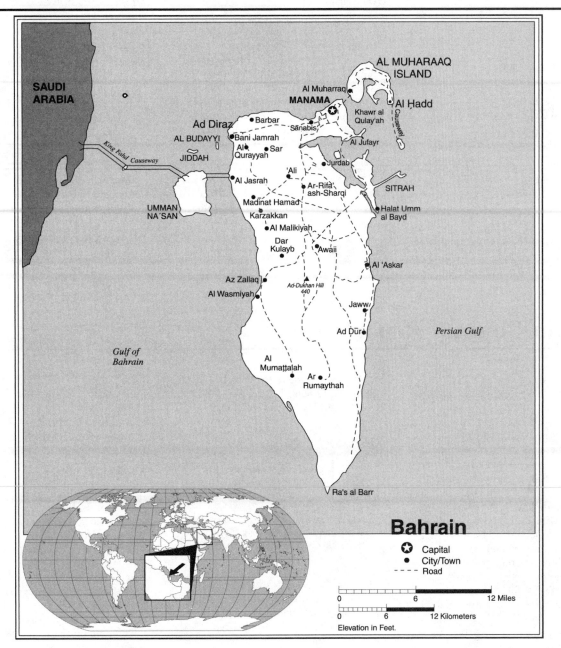

Bahrain Statistics

GEOGRAPHY

Area in Square Miles (Kilometers): 266 (688) (about 3 1/2 times the size of Washington, D.C.)

Capital (Population): Manama (166,200)

Environmental Concerns: desertification; coastal degradation resulting from oil spills and discharges from ships and industry; no natural freshwater

Geographical Features: mostly low desert plain, rising gently to low central escarpment

Climate: hot and humid summers; temperate winters

PEOPLE

Population

Total: 688,345 (235,108 non-nationals)

Annual Growth Rate: 1.51%

Rural/Urban Population Ratio: 9/91

Major Languages: Arabic; English

Ethnic Makeup: 63% Bahraini; 19% Asian; 10% other Arab; 8% Iranian

Religions: 70% Shia Muslim; 15% Sunni Muslim; 15% Bahai, Christian, and others

Health

Life Expectancy at Birth: 71 years (male); 76 years (female)
Infant Mortality Rate (Ratio): 17.27/1,000 live births

Education

Adult Literacy Rate: 89%
Compulsory (Ages): 6–17; free

COMMUNICATION

Telephones: 153,000 main lines
Daily Newspaper Circulation: 128 per 1,000 people
Televisions: 442 per 1,000 people
Internet Users: 1 (2000)

TRANSPORTATION

Highways in Miles (Kilometers): 1,927 (3,103)
Railroads in Miles (Kilometers): none
Usable Airfields: 3

Motor Vehicles in Use: 172,000

GOVERNMENT

Type: constitutional monarchy as of 2001
Independence Date: August 15, 1971 (from the United Kingdom)
Head of State/Government: King Hamad bin Isa al-Khalifa; Prime Minister Shaykh Salman bin Hamad al-Khalifa
Political Parties: none, but direct elections held in October 2002 for 40-member Chamber of Deputies (Parliament)
Suffrage: universal at 18

MILITARY

Military Expenditures (% of GDP): 6.3%
Current Disputes: none; dispute with Qatar resolved in 2001

ECONOMY

Currency ($U.S. Equivalent): 0.377 dinar = $1 (fixed rate)

Per Capita Income/GDP: $19,200/$13.1 billion
GDP Growth Rate: 5.6%
Inflation Rate: 2.1%
Unemployment Rate: 15%
Labor Force: 370,000 (4.470 non-nationals)
Natural Resources: oil; associated and non-associated natural gas; fish
Agriculture: fruits; vegetables; poultry; dairy products; shrimp; fish
Industry: petroleum processing and refining; aluminum smelting; offshore banking; ship repairing; tourism
Exports: $8.2 billion (primary partners India, United States, Saudi Arabia)
Imports: $5.87 billion (primary partners Saudi Arabia, Japan, Germany)

SUGGESTED WEBSITES

http://lcweb2.loc.gov/frd/cs/bhtoc.html
http://www.usembassy.com.bh

Bahrain Country Report

Bahrain is the smallest Arab state. It is also the only Arab island state, consisting of an archipelago of 33 islands, just five of them inhabited. The largest island, also named Bahrain (from the Arabic *bahrayn,* or "two seas"), has an area of 216 square miles.

Although it is separated from the Arabian mainland, Bahrain is not far away; it is just 15 miles from Qatar and the same distance from Saudi Arabia. A causeway linking Bahrain with mainland Saudi Arabia opened in 1986, technically ending its insular status. Improvements to the causeway in 2001 link the country even closer to its larger neighbor, as Saudis pour across the border to enjoy the movies, bars, and shopping boutiques of the freer Bahraini society.

Bahrain is unusual among the Persian Gulf states in that it started to diversify its economy early. Oil was discovered there in 1932. Its head start in the exportation of oil enabled the government to build up an industrial base over a long period and to develop a large, indigenous, skilled labor force. As a result, today about two thirds of the population are native-born Bahrainis.

HISTORY

Excavations by archaeologists indicate that roughly 5,000 years ago, Bahrain was the legendary *Dilmun,* "home of the gods" and the land of immortality in the Mesopotamian Epic of Gilgamesh. It had a fully urbanized society and was the center

of a far-flung trade network between Mesopotamia, Oman, the Arabian Gulf, and the Indus Valley cities farther east.

DEVELOPMENT

Bahrain's economy continues to develop and diversify. GDP growth held steady at 4.5–5 percent annually from 1998 to 2003, reaching 5.6 percent in 2005. In addition to a revived offshore banking industry it has the largest concentration of Islamic (non-interest charging) banks in the region.

During the centuries of Islamic rule in the Middle East, Bahrain (it was renamed by Arab geographers) became wealthy from the pearl-fishing industry. By the fourteenth century A.D., it had 300 villages. Bahraini merchants grew rich from profits on their large, lustrous, high-quality pearls. Bahraini sea captains and pearl merchants built lofty palaces and other stately buildings on the islands.

The Portuguese were the first Europeans to land on Bahrain, which they seized in the early sixteenth century as one of a string of fortresses along the coast to protect their monopoly over the spice trade. They ruled by the sword in Bahrain for nearly a century before they were ousted by Iranian invaders. The Iranians, in turn, were defeated by the al-Khalifas, a clan of the powerful Anaizas. In 1782, the clan leader, Shaykh Ahmad al-Khalifa, estab-

lished control over Bahrain and founded the dynasty that rules the state today. (The al-Khalifas belong to the same clan as the al-Sabahs, the rulers of Kuwait, and are distantly related to the Saudi Arabian royal family.)

A British Protectorate

In the 1800s, Bahrain came under British protection in the same way as other Gulf states. The ruler Shaykh Isa, whose reign was one of the world's longest (1869–1932), signed an agreement making Britain responsible for Bahrain's defense and foreign policy. He also agreed not to give any concessions for oil exploration without British approval. The agreement was important because the British were already developing oil fields in Iran. Control of oil in another area would give them an added source of fuel for the new weaponry of tanks and oil-powered warships of World War I. The early development of Bahrain's oil fields and the guidance of British political advisers helped prepare the country for independence.

INDEPENDENCE

Bahrain became fully independent in 1971. The British encouraged Bahrain to join with Qatar and seven small British-protected Gulf states, the Trucial States, in a federation. However, Bahrain and Qatar felt that they were more advanced economically, politically, and socially than were

United Nations Photo (UN152807)

Bahrain may be the first of the Gulf states to get out of the oil business, due to its dwindling reserves. Other income-generating industries are being explored, and diversification of the economy along with political stability make Bahrain a stable regional business center. The need for an effective educational system to supply an informed labor force is paramount, as these children and their teacher at a nursery school near Manama attest.

the Trucial States and therefore did not need to federate.

FREEDOM

Reinstatement of the 1973 Constitution and the Election of a new Chamber of Deputies in 2002 underlined the ruler's commitment to parliamentary democracy. The National Action Charter gives women the right to vote and run for public office; although none were elected to the Chamber, seven were subsequently appointed to the Cabinet of Ministers.

A mild threat to Bahrain's independence came from Iran. In 1970, Shah Mohammed Reza Pahlavi of Iran claimed Bahrain, on the basis of Iran's sixteenth-century occupation, plus the fact that a large number of Bahrainis were descended from Iranian emigrants. The United Nations discussed the issue and recommended

that Bahrain be given its independence, on the grounds that "the people of Bahrain wish to gain recognition of their identity in a fully independent and sovereign state."[1] The shah accepted the resolution, and Iran made no further claims on Bahrain during his lifetime.

The gradual development of democracy in Bahrain reached a peak after independence. Shaykh Khalifa (now called emir) approved a new Constitution and a law establishing an elected National Assembly of 30 members. The Assembly met for the first time in 1973, but it was dissolved by the emir only two years later.

What Had Happened?

Bahrain is an example of a problem common in the Middle East: the conflict between traditional authority and popular democracy. Fuad Khuri describes the problem as one of a "tribally controlled government that rules by historical right, opposed

to a community-based urban population seeking to participate in government through elections. The first believes and acts as if government is an earned right, the other seeks to modify government and subject it to a public vote."[2]

Governmental authority in Bahrain is defined as hereditary in the al-Khalifa family, according to the 1973 Constitution. The succession passes from the ruling emir to his eldest son. Since Bahrain has no tradition of representative government or political parties, the National Assembly was set up to broaden the political process without going through the lengthy period of conditioning necessary to establish a multiparty system. Members were expected to debate laws prepared by the Council of Ministers and to assist with budget preparation. But as things turned out, Assembly members spent their time arguing with one another or criticizing the ruler instead of dealing with issues. When the emir dissolved the Assembly, he said that it was preventing the government from doing what it was supposed to do.

Since that time, government in Bahrain has reverted to its traditional patriarchal-authority structure. However, Shia demands for reinstatement of the Assembly, a multiparty system with national elections, and greater representation for Shias in government have been met in part through changes in the governing structure. In 1993, the emir appointed a 30-member *Shura* (Council), composed of business and industry leaders along with members of the ruling family.

HEALTH/WELFARE

In January 2002 the American Mission Hospital, the first in Bahrain, marked its 100th anniversary. It was built in 1902 with a $6,000 gift from the Mason family, medical missionaries in Arabia. A 1993 labor law allows unions to organize and requires 60 percent local labor in new industries, both Bahraini and foreign-owned.

King Hamad, who changed his title as part of the country's move toward constitutional monarchy, has taken several long steps in that direction in recent years. In 2000 the Shura was enlarged to 40 members, including women and representatives of the large Shia community. The king also issued a "National Action Charter" which was endorsed in a referendum by 80 percent of voters. It provides for release of political prisoners, abolition of state security courts, and the annulment of a 1974 law which permitted imprisonment without trial or the right of appeal for persons charged with "political" crimes.

The region-wide anger of the Arab "street" over the U.S. invasion of Iraq and its continued support for Israel in its struggle with the Palestinians, reached Bahrain in 2005, as the Shia majority there staged the largest protest demonstration in memory. Earlier in the year, demands by this opposition movement's bloggers to use the Internet website to demand a new constitution, separation of powers, and greater political liberties resulted in arrests of several bloggers. The ministry of information then issued an edict requiring all websites to be registered with the government. One of them, Bahrain Online, had posted a UN report of governmental discrimination of the Shia majority.

FOREIGN RELATIONS

The U.S. involvement in Bahrain dates back over a century, to the establishment of the American Hospital in Manama in 1893.

American missionaries subsequently founded a number of schools, including the Bahrain-American School, which is still in existence. In recent years, Bahrain has become a key factor in U.S. Middle Eastern policy, due to its strategic location in the Persian Gulf. During the Iran–Iraq War, Bahrain's British-built naval base was a staging point for U.S. convoy escort vessels in the Gulf. Then, in the aftermath of the 1991 Gulf War, Bahrain became a "front-line state" in the American containment strategy toward Iraq. It is the permanent headquarters for the new U.S. Fifth Fleet, under a mutual defense agreement, with some 1,000 American military and naval personnel stationed there. Following the September 11, 2001 terrorist attacks in the United States, Bahrain strongly supported the international coalition against terrorism and granted use of its base for U.S. air strikes in Afghanistan.

Bahrain's only serious foreign-policy problem since its independence has involved neighboring Qatar. In 1992, the Qatari ruler unilaterally extended Qatar's territorial waters to include the islands of Hawar and Fishat al-Duble, which had been controlled by Bahrain since the 1930s, when both nations were British protectorates. Bahrain in turn demanded that Qatar recognize Bahrain's sovereignty over the al-Zeyara coastal strip, which adjoins Bahraini territorial waters. After considerable wrangling, the two countries agreed to take their dispute before the International Court of Justice (ICJ). In 2001, it confirmed Qatari sovereignty over Zabarah and Janan Islands. Bahrain was awarded sovereignty over Hawar and Qit'at Jaradeh Islands. The boundary between their maritime zones is to be drawn in accordance with the Court's decision.

In 1981, concern over threats by its more powerful neighbors led Bahrain to join with other Gulf states in forming the Gulf Cooperation Council, a regional mutual-defense organization. The other members are Kuwait, Oman, Qatar, Saudi Arabia, and the United Arab Emirates. In 2001, the members signed a formal defense pact, sponsored by the United States. The pact expanded the rapid-deployment force of 5,000 soldiers to 22,000, with members contributing troops in accordance with the size of their armed forces. The United States agreed to provide a $70 million early warning system to identify any chemical or biological weapons used by belligerent forces in an attack on GCC member states.

THREATS TO NATIONAL SECURITY

The 1979 Revolution in Iran caused much concern in Bahrain. The new Iranian gov-

ernment revived the old territorial claim, and a Teheran-based Islamic Front for the Liberation of Bahrain called on Shia Muslims in Bahrain to overthrow the Sunni regime of the emir. In 1981, the government arrested a group of Shia Bahrainis and others, charging them with a plot against the state, backed by Iran. The plotters had expected support from the Shia population, but this did not materialize. After seeing the results of the Iranian Revolution, few Bahraini Shia Muslims wanted the Iranian form of fundamentalist Islamic government. In 1982, 73 defendants were given prison sentences ranging from seven years to life. Bahrain's prime minister told a local newspaper that the plot didn't represent a real danger, "but we are not used to this sort of thing so we had to take strong action."[3]

Until the 1990s, the Shia community was politically inactive, and as a result, 100 Shia activists were pardoned by the emir and allowed to return from exile. However, the increase in fundamentalist activities against Middle Eastern regimes elsewhere has destabilized the island nation to some extent. Between 1994 and 1999, a low-level campaign of antigovernment violence claimed 30 lives and resulted in a number of sabotage incidents. In 1996, eight opposition leaders were arrested; they included Shaykh Abdul-Ameer al-Jamri, allegedly the head of the opposition movement. In July 1999, he was given a 10-year jail sentence and fined $14.5 million for having spied for a foreign power (unnamed), for inciting unrest, and for continuing to agitate illegally for political reform. But he was released the day after his sentencing, partly due to his ill health, but also due to criticism from the United States and other countries of Bahrain's failure to observe his civil and legal rights. The government also released some 320 other detainees.

ACHIEVEMENTS

In 2002 a female lawyer, Dr. Mariam bint Hassan Al-Khalifa, was appointed president of Bahrain University, the first to hold such a position in the Arab world. And in 2003 Bahrain's economy was named as the freest in the Arab Middle East by the Heritage Foundation.

AN OIL-LESS ECONOMY?

Bahrain was an early entrant in the oil business and may be the first Gulf state to face an oil-less future. Current production from its own oil fields is 42,000 barrels per day. The Bahrain Petroleum Company (Bapco) controls all aspects of production, refining, and export. However, Bapco must import

70,000 b/d from Saudi Arabia to keep its refinery operating efficiently.

In the past, Bahrain's economic development was characterized by conservative management. This policy changed radically with the accession of the current ruler. A January 2001 decree allows foreign companies to buy and own property, particularly for non-oil investment projects. (Oil currently accounts for 80 percent of exports and 60 percent of revenues.)

The slow decline in oil production in recent years has been balanced by expansion of the liquefied natural gas (LNG) sector. Current production is 170 million cubic feet per day. But with 9 billion cubic feet of proven reserves, production of LNG will long outlast oil production.

Aluminium Bahrain (ALBA), which accounts for 60 percent of Bahrain's non-oil exports, expanded its production in 2001 to become the world's largest aluminium smelter, with an annual output of 750,000 tons. Some 450,000 of this is exported. A seawater-desalination plant was completed in 1999. It uses waste heat from the smelter to provide potable water for local needs.

INTERNATIONAL FINANCE

During the Lebanese Civil War, Bahrain encouraged the establishment of "Offshore Banking Units" in order to replace Lebanon as a regional finance center. OBUs are set up to attract deposits from governments or large financial organizations such as the World Bank as well as to make loans for development projects. OBUs are "offshore" in the sense that a Bahraini cannot open a checking account or borrow money. However, OBUs bring funds into Bahrain without interfering with local growth or undercutting local banks.

The drop in world oil prices in the 1980s and the Iraqi occupation of Kuwait seriously disrupted the OBU system, and a number of offshore banks were closed. Increased oil production and higher world prices have revitalized the system, both for offshore and on-shore banks. In 2001, BNP Paribas, the sixth largest bank in the world, relocated its Middle East operations office to Bahrain, as did Turkey's Islamic Bank, an emerging giant in the Islamic banking system.

Bahrain's economic recovery has been enhanced by the Systems Development Council, a government body set up in 2000 to oversee development. The Bahrain Stock Exchange, opened in 1992, has revised its regulations to require foreign firms to employ 60 percent local labor, which strengthens Bahrainis' participation in the development of the emirate.

THE FUTURE

One key to Bahrain's future may be found in a Koranic verse (Sura XIII, II):

Lo! Allah changeth not the condition of a people until they first change what is in their hearts.

For a brief time after independence, the state experimented with representative government. But the hurly-burly of politics, with its factional rivalries, trade-offs, and compromises found in many Western democratic systems, did not suit the Bahraini temperament or experience. Democracy takes time to mature. Emile Nakhleh reminds us that "any serious attempt to democratize the regime will ultimately set tribal legitimacy and popular sovereignty on a collision course."[4]

The emir demonstrated his commitment to gradual democratization in 2000, issuing an edict, confirmed in a national referendum, that defines Bahrain as a constitutional monarchy ruled by a king. Municipal council elections were held in 2002, followed by those for a restored national parliament. Both male and female voters participated. But as noted earlier, these small steps toward representative government were felt to be inadequate by the large Shia population.

NOTES

1. UN Security Council *Resolution 287*, 1970. Quoted from Emile Nakhleh, *Bahrain* (Lexington, KY: Lexington Books, 1976), p. 9.
2. Fuad I. Khuri, *Tribe and State in Bahrain* (Chicago: University of Chicago Press, 1981), p. 219.
3. *Gulf Daily News* (May 15, 1982).
4. Nakhleh, *op. cit.,* p. 11.

Timeline: PAST

1602–1782
Periodic occupation of Bahrain by Iran after the Portuguese ouster

1783
The al-Khalifa family seizes power over other families and groups

1880
Bahrain becomes a British protectorate

1971
Independence

1973–1975
The new Constitution establishes a Constituent Assembly, but the ruler dissolves it shortly thereafter

1990s
Bahrain takes aggressive steps to revive and diversify its economy

PRESENT

2000s
Territorial disputes with Qatar resolved

Important changes in representation and participation in government

Egypt (Arab Republic of Egypt)

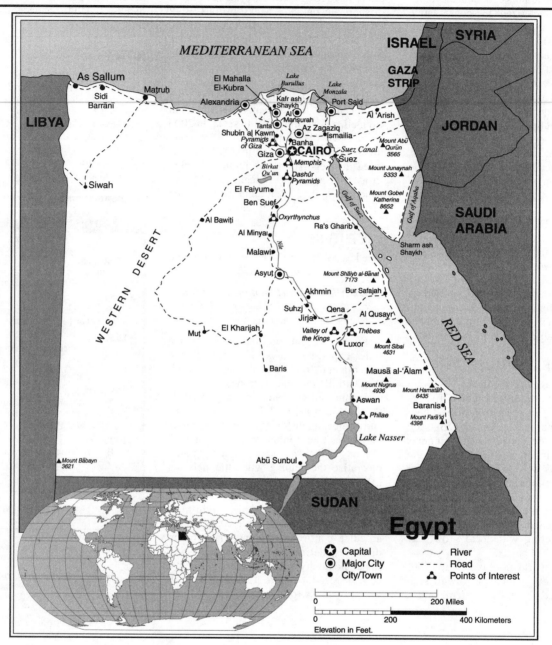

Egypt Statistics

GEOGRAPHY

Area in Square Miles (Kilometers):
386,258 (1,001,258) (about 3 times the size of New Mexico)

Capital (Population): Cairo (6,800,000)

Environmental Concerns: loss of agricultural land; increasing soil salinization; desertification; oil pollution threatening coral reefs and marine habitats; other water pollution; rapid population growth

Geographical Features: a vast desert plateau interrupted by the Nile Valley and Delta

Climate: desert; dry, hot summers; moderate winters

PEOPLE

Population

Total: 77,506,756

Annual Growth Rate: 1.78%

Rural/Urban Population Ratio: 55/45

Major Languages: Arabic; English

Ethnic Makeup: 99% Eastern Hamitic
(Egyptian, Bedouin, Arab, Nubian); 1%
others
Religions: 94% Muslim (mostly Sunni);
6% Coptic Christian and others

Health

Life Expectancy at Birth: 68.5 years
(male); 73 years (female)
Infant Mortality Rate (Ratio): 32.5/1,000
live births

Education

Adult Literacy Rate: 59.7%
Compulsory (Ages): for 5 years between 6
and 13

COMMUNICATION

Telephones: 9,600,000 main lines
Daily Newspaper Circulation: 43 per
1,000 people
Televisions: 110 per 1,000 people
Internet Users: 4,200,000

TRANSPORTATION

Highways in Miles (Kilometers): 39,744
(64,000)

Railroads in Miles (Kilometers): 2,973
(4,955)
Usable Airfields: 92
Motor Vehicles in Use: 1,703,000

GOVERNMENT

Type: republic
Independence Date: July 23, 1952, for the
republic; February 28, 1922, marking the
end of British rule
Head of State/Government: President
Mohammed Hosni Mubarak; Prime
Minister Ahmad Nazif
Political Parties: National Democratic
Party (NDP), majority party; others are
New Wafd; Tagammu (National
Progressive Unionist Group); Nasserist
Arab Democratic Party; Socialist Liberal
Party. NDP holds 88 percent majority in
Peoples Assembly
Suffrage: universal and compulsory at 18

MILITARY

Military Expenditures (% of GDP): 3.4%
Current Disputes: territorial dispute with
Sudan over the Hala'ib Triangle

ECONOMY

Currency ($U.S. Equivalent): 5.99
Egyptian pounds = $1
Per Capita Income/GDP: $4,200/$316
billion
GDP Growth Rate: 4.5%
Inflation Rate: 9.5%
Unemployment Rate: 10.9%
Labor Force: 20,600,000
Natural Resources: petroleum; natural gas;
iron ore; phosphates; manganese;
limestone; gypsum; talc; asbestos; lead; zinc
Agriculture: cotton; sugarcane; rice; corn;
wheat; beans; fruits; vegetables;
livestock; fish
Industry: textiles; food processing;
tourism; chemicals; petroleum;
construction; cement; metals
Exports: $11 billion (primary partners
European Union, Middle East, Afro-
Asian countries)
Imports: $19 billion (primary partners
European Union, United States, Germany)

SUGGESTED WEBSITE

http://lcweb2.loc.gov/frd/cs/
egtoc.htm

Egypt Country Report

The Arab Republic of Egypt is located at the extreme northeastern corner of Africa, with part of its territory—the Sinai Peninsula—serving as a land bridge to Southwest Asia. The country's total land area is approximately 386,000 square miles. However, 96 percent of this is uninhabitable desert. Except for a few scattered oases, the only settled and cultivable area is a narrow strip along the Nile River. The vast majority of Egypt's population is concentrated in this strip, resulting in high population density. Migration from rural areas to cities has intensified urban density; Cairo's population is currently 6.8 million, with millions more in the greater metropolitan area. It is a city that is literally "bursting at the seams."

Egypt today identifies itself as an Arab nation and is a founding member of the League of Arab States (which has its headquarters in Cairo). But its "Arab" identity is relatively new. It was first defined by the late president Gamal Abdel Nasser, who as a schoolboy became aware of his "Arabness" in response to British imperialism and particularly Britain's establishment of a national home for Jews in Arab Palestine. But Egypt's incredibly long history as a distinct society has given its people a sepa-

rate Egyptian identity and a sense of superiority over other peoples, notably desert people such as the Arabs of old.[1] Also, its development under British tutelage gave the country a headstart over other Arab countries or societies. Despite its people's overall low level of adult literacy, Egypt has more highly skilled professionals than do other Arab countries.

HISTORY

Although Egypt is a modern nation in terms of independence from foreign control, it has a distinct national identity and a rich culture that date back thousands of years. The modern Egyptians take great pride in their brilliant past; this sense of the past gives them patience and a certain fatalism that enable them to withstand misfortunes that would crush most peoples. The Egyptian peasants, the *fellahin,* are as stoic and enduring as the water buffaloes they use to do their plowing. Since the time of the pharaohs, Egypt has been invaded many times, and it was under foreign control for most of its history. When Nasser, the first president of the new Egyptian republic, came to power in 1954, he said that he was the first native Egyptian to rule the country in nearly 3,000 years.

DEVELOPMENT

Egypt's GDP growth rate, which held steady at 4–5 percent in the 90s, was affected by external events, notably the 9/11 terrorist attacks in the U.S. and the 2003 invasion of Iraq. The tourist industry, which generates 12 percent of revenues, saw a downturn but rebounded strongly in the early 2000s. However multiple bomb-blasts at the Sharm El-Shaikh Red Sea resort and in Cairo in 2004–2005, the first in 7 years, posed a new threat to the industry. The Sharm el-Shaikh attack in particular killed 88 persons and injured 119, mostly Egyptians but some foreigners.

It is often said that Egypt is the "gift of the Nile." The mighty river, flowing north to the Mediterranean with an enormous annual spate that deposited rich silt along its banks, attracted nomadic peoples to settle there as early as 6000 B.C. They developed a productive agriculture based on the river's seasonal floods. They lived in plastered mud huts in small, compact villages. Their villages were not too different from those one sees today in parts of the Nile Delta.

Each village had its "headman," the head of some family more prosperous or

These pyramids at Giza are among the most famous mementos of Egypt's past.

industrious (or both) than the others. The arrival of other nomadic desert peoples gradually brought about the evolution of an organized system of government. Since the Egyptian villagers did not have nearby mountains or wild forests to retreat into, they were easily governable.

The institution of kingship was well established in Egypt by 2000 B.C., and in the time of Ramses II (1300–1233 B.C.), Egyptian monarchs extended their power over a large part of the Middle East. All Egyptian rulers were called pharaohs, although there was no hereditary system of descent and many different dynasties ruled during the country's first 2,000 years of existence. The pharaohs had their capital at Thebes, but they built other important cities on the banks of the Nile. Recent research by Egyptologists indicate that the ancient Egyptians had an amazingly accurate knowledge of astronomy. The Pyramids of Giza, for example, were built so as to be aligned with true north. Lacking modern instruments, their builders apparently used two stars, Thaban and Draconis, in the Big Dipper, for their alignment, with a point equidistant from them to mark their approximation of true north. Only centuries later was the North Star identified as such. The pyramids and Sphinx were built with Egyptian labor and without more than rudimentary machinery. They underline the engineering expertise of this ancient people. The world's oldest irrigation canal, its base paved with limestone blocks, was unearthed near Giza in 1996; and in 1999, a 2,000-year-old cemetery, discovered by chance in Bahariya Oasis, was found to

contain rows of mummified men, women, and children along with wall murals showing funeral ceremonies, all in a remarkable state of preservation. The recent discovery of the bones of the second-largest dinosaur ever identified, in the same area, indicates that 90 million years ago, long before the pharaohs, Bahariya was a swampy tropical region similar to the Florida Everglades.

Another important discovery, in November 1999, was that of inscriptions on the walls of *Wadi Hoi* ("Valley of Terror") that may well be the world's oldest written language, predating the cuneiform letters developed by the Sumerians in Mesopotamia.

In the first century B.C., Egypt became part of the Roman Empire. The city of Alexandria, founded by Alexander the Great, became a center of Greek and Roman learning and culture. Later, it became a center of Christianity. The Egyptian Coptic Church was one of the earliest organized churches. The Copts, direct descendants of the early Egyptians, are the principal minority group in Egypt today. (The name Copt comes from *aigyptos,* Greek for "Egyptian.") The Copts welcomed the Arab invaders who brought Islam to Egypt, preferring them to their oppressive Byzantine Christian rulers. Muslim rulers over the centuries usually protected the Copts as "Peoples of the Book," leaving authority over them to their religious leaders, in return for allegiance and payment of a small tax. But in recent years, the rise of Islamic fundamentalism has made life more difficult for Egypt's Coptic minority. Coptic-Muslim friction declined significantly in the 1990s, but the 10 million Copts con-

tinue to complain of petty discrimination (such as university admission) and political disenfranchisement. To its credit, the government has eased building restrictions on new churches and allowed broadcasting of Christian services on state TV. Recently it restored property expropriated by the Nasser regime in 1950s to Coptic Church ownership.

THE INFLUENCE OF ISLAM

Islam was the major formative influence in the development of modern Egyptian society. Islamic armies from Arabia invaded Egypt in the seventh century A.D. Large numbers of nomadic Arabs followed, settling the Nile Valley until, over time, they became the majority in the population. Egypt was under the rule of the caliphs ("successors" of the Prophet Muhammad) until the tenth century, when a Shia group broke away and formed a separate government. The leaders of this group also called themselves caliphs. To show their independence, they founded a new capital in the desert south of Alexandria. The name they chose for their new capital was prophetic: *al-Qahira*—"City of War"—the modern city of Cairo.

In the sixteenth century, Egypt became a province of the Ottoman Empire. It was then under the rule of the Mamluks, originally slaves or prisoners of war who were converted to Islam. Many Mamluk leaders had been freed and then acquired their own slaves. They formed a military aristocracy, constantly fighting with one another for land and power. The Ottomans found it

United Nations Photo (UN152385)

Two young Egyptian boys gaze in wonder at the language of their ancestors, Karnak.

EGYPT ENTERS THE MODERN WORLD

At the end of the eighteenth century, rivalry between Britain and France for control of trade in the Mediterranean and the sea routes to India involved Egypt. The French general Napoleon Bonaparte led an expedition to Egypt in 1798. However, the British, in cooperation with Ottoman forces, drove the French from Egypt. A confused struggle for power followed. The victor was Muhammad Ali, an Albanian officer in the Ottoman garrison at Cairo. In 1805, the Ottoman sultan appointed him governor of Egypt.

Although he was not an Egyptian, Muhammad Ali had a vision of Egypt under his rule as a rich and powerful country. He began by forming a new army consisting of native Egyptians instead of mercenaries or slave-soldiers. This army was trained by European advisers and gave a good account of itself in campaigns, performing better than the regular Ottoman armies.[2] His successor, Ismail, went a step further by hiring some 50 demobilized veterans of the American Civil War, both Yankees and rebels, who brought discipline and military

experience to the training of Egyptian recruits. These mercenaries remained in Egypt after the end of the campaigns of Ismail and his successors in the Middle East, and when they died they were buried in the long-forgotten and neglected American cemetery in a corner of Cairo.

Muhammad Ali set up an organized, efficient tax-collection system. He suppressed the Mamluks and confiscated all the lands that they had seized from Egyptian peasants over the years, lifting a heavy tax burden from peasant backs. He took personal charge of all Egypt's exports. Cotton, a new crop, became the major Egyptian export and became known the world over for its high quality. Dams and irrigation canals were dug to improve cultivation and expand arable land. Although Muhammad Ali grew rich in the process of carrying out these policies, he was concerned for the welfare of the peasantry. He once said, "One must guide this people as one guides children; to leave them to their own devices would be to render them subject to all the disorders from which I have saved them."[3]

Muhammad Ali's successors were named *khedives* ("viceroys"), in that they ruled Egypt in theory on behalf of their superior, the sultan. In practice, they acted as independent rulers. Under the khedives, Egypt was again drawn into Euro-

pean power politics, with unfortunate results. Khedive Ismail, the most ambitious of Muhammad Ali's descendants, was determined to make Egypt the equal of Western European nations. His major project was the Suez Canal, built by a European company and opened in 1869. The Italian composer Verdi was invited to compose the opera *Aida* for its inauguration. He refused to do so at first, saying that Egypt was a country whose civilization he did not admire. Eventually he was persuaded (with the help of the then-princely bonus of $20,000!) and set to music the ancient Egyptian legend of imperialism and grand passion. Verdi's task was eased by the fact that Auguste Mariette, the preeminent Egyptologist of the period, wrote the libretto and designed sets and costumes with absolute fidelity to pharaonic times.

However, the expense of this and other grandiose projects bankrupted the country. Ismail was forced to sell Egypt's shares in the Suez Canal Company—to the British government!—and his successors were forced to accept British control over Egyptian finances. In 1882, a revolt of army officers threatened to overthrow the khedive. The British intervened and established a de facto protectorate, keeping the khedive in office in order to avoid conflict with the Ottomans.

United Nations Photo (UN67110)

In 1952, the Free Officers organization persuaded Egypt's King Farouk to abdicate. The monarchy was formally abolished in 1954, when Gamal Abdel Nasser (pictured above) became Egypt's president, prime minister, and head of the Revolutionary Command Council.

EGYPTIAN NATIONALISM

The British protectorate lasted from 1882 to 1956. An Egyptian nationalist movement gradually developed in the early 1900s, inspired by the teachings of religious leaders and Western-educated officials in the khedives' government. They advocated a revival of Islam and its strengthening to enable Egypt and other Islamic lands to resist European control.

During World War I, Egypt was a major base for British campaigns against the Ottoman Empire. The British formally declared their protectorate over Egypt in order to "defend" the country, since legally it was still an Ottoman province. The British worked with Arab nationalist leaders against the Turks and promised to help them form an independent Arab nation after the war. Egyptian nationalists were active in the Arab cause, and although at that time they did not particularly care about being a part of a new Arab nation, they wanted independence from Britain.

At the end of World War I, Egyptian nationalist leaders organized the *Wafd* (Arabic for "delegation"). In 1918, the Wafd presented demands to the British for the complete independence of Egypt. The British rejected the demands, saying that Egypt was not ready for self-government. The Wafd then turned to violence, organizing boycotts, strikes, and terrorist attacks on British soldiers and on Egyptians accused of cooperating with the British.

Under pressure, the British finally abolished the protectorate in 1922. But they retained control over Egyptian foreign policy, defense, and communications as well as the protection of minorities and foreign residents and of Sudan, which had been part of Egypt since the 1880s. Thus, Egypt's "independence" was a hollow shell.

Egypt did regain control over internal affairs. The government was set up as a constitutional monarchy under a new king, Fuad. Political parties were allowed, and in elections for a Parliament in 1923, the Wafd emerged as the dominant party. But neither Fuad nor the son who succeeded him, Farouk, trusted Wafd leaders. They feared that the Wafd was working to establish a republic. For their part, the Wafd leaders did not believe that the rulers were seriously interested in the good of the country. So Egypt waddled along for two decades with little progress.

THE EGYPTIAN REVOLUTION

During the years of the monarchy, the Egyptian Army gradually developed a corps of professional officers, most of them from lower- or middle-class Egyptian backgrounds. They were strongly patriotic and resented what they perceived to be British cultural snobbery as well as Britain's continual influence over Egyptian affairs.

The training school for these young officers was the Egyptian Military Academy, founded in 1936. Among them was Gamal Abdel Nasser, the eldest son of a village postal clerk. Nasser and his fellow officers were already active in anti-British demonstrations by the time they entered the academy. During World War II, the British, fearing a German takeover of Egypt, reinstated the protectorate. Egypt became the main British military base in the Middle East. This action galvanized the officers into forming a revolutionary movement. Nasser said at the time that it roused in him the seeds of revolt. "It made [us] realize that there is a dignity to be retrieved and defended."[4]

When Jewish leaders in Palestine organized Israel in May 1948, Egypt, along with other nearby Arab countries, sent troops to destroy the new state. Nasser and several of his fellow officers were sent to the front. The Egyptian Army was defeated; Nasser himself was trapped with his unit, was wounded, and was rescued only by an armistice. Even more shocking to the young officers was the evident corruption and weakness of their own government. The weapons that they received were inferior and often defective, battle orders were inaccurate, and their superiors proved to be incompetent in strategy and tactics.

Nasser and his fellow officers attributed their defeat not to their own weaknesses but to their government's failures. When they returned to Egypt, they were determined to overthrow the monarchy. They formed a secret organization, the Free Officers. It was not the only organization dedicated to the overthrow of the monarchy, but it was the best disciplined and had the general support of the army.

On July 23, 1952, the Free Officers launched their revolution. It came six months after "Black Saturday," the burning of Cairo by mobs protesting the continued presence of British troops in Egypt. The Free Officers persuaded King Farouk to abdicate, and they declared Egypt a republic. A nine-member Revolutionary Command Council (RCC) was established to govern the country.

EGYPT UNDER NASSER

In his self-analytical book *The Philosophy of the Revolution,* Nasser wrote, "I always imagine that in this region in which we live there is a role wandering aimlessly about in search of an actor to play it."[5] Nasser saw himself as playing that role. Previously, he had operated behind the scenes, but always as the leader to whom the other Free Officers looked up. By 1954, Nasser had emerged as Egypt's leader. When the monarchy was formally abolished in 1954, he became president, prime minister, and head of the RCC. Cynics said that Nasser came along when Egypt was ready for another king; the Egyptians could not function without one!

Nasser came to power determined to restore dignity and status to Egypt, to eliminate foreign control, and to make his country the leader of a united Arab world. It was an ambitious set of goals, and Nasser was only partly successful in attaining them. But in his struggles to achieve these goals, he brought considerable status to Egypt. The country became a leader of the "Third World" of Africa and Asia, developing nations newly freed from foreign control.

Nasser was successful in removing the last vestiges of British rule from Egypt. British troops were withdrawn from the Suez Canal Zone, and Nasser nationalized the canal in 1956, taking over the management from the private foreign company that had operated it since 1869. That action made the British furious, since the British government had a majority interest in the company. The British worked out a secret plan with the French and the Israelis, neither of whom liked Nasser, to invade Egypt and overthrow him. British and French paratroopers seized the canal in October

1956, but the United States and the Soviet Union, in an unusual display of cooperation, forced them to withdraw. It was the first of several occasions when Nasser turned military defeat into political victory. It was also one of the few times when Nasser and the United States were on the same side of an issue.

Between 1956 and 1967, Nasser developed a close alliance with the Soviet Union—at least, it seemed that way to the United States. Nasser's pet economic project was the building of a dam at Aswan, on the upper Nile, to regulate the annual flow of river water and thus enable Egypt to reclaim new land and develop its agriculture. He applied for aid from the United States through the World Bank to finance the project, but he was turned down, largely due to his publicly expressed hostility toward Israel. Again Nasser turned defeat into a victory of sorts. The Soviet Union agreed to finance the dam, which was completed in 1971, and subsequently to equip and train the Egyptian Army. Thousands of Soviet advisers poured into Egypt, and it seemed to U.S. and Israeli leaders that Egypt had become a dependency of the Soviet Union.

The lowest point in Nasser's career came in June 1967. Israel invaded Egypt and defeated his Soviet-trained army, along with those of Jordan and Syria, and occupied the Sinai Peninsula in a lightning six-day war. The Israelis were restrained from marching on Cairo only by a United Nations cease-fire. Nasser took personal responsibility for the defeat, calling it *al-Nakba* ("The Catastrophe"). He announced his resignation, but the Egyptian people refused to accept it. The public outcry was so great that he agreed to continue in office. One observer wrote, "The irony was that Nasser had led the country to defeat, but Egypt without Nasser was unthinkable."[6]

Nasser had little success in his efforts to unify the Arab world. One attempt, for example, was a union of Egypt and Syria, which lasted barely three years (1958–1961). Egyptian forces were sent to support a new republican government in Yemen after the overthrow of that country's autocratic ruler, but they became bogged down in a civil war there and had to be withdrawn. Other efforts to unify the Arab world also failed. Arab leaders respected Nasser but were unwilling to play second fiddle to him in an organized Arab state. In 1967, after the Arab defeat, Nasser lashed out bitterly at the other Arab leaders. He said, "You issue statements, but we have to fight. If you want to liberate [Palestine] then get in line in front of us."[7]

Inside Egypt, the results of Nasser's 18-year rule were also mixed. Although he

United Nations Photo (UN122838)

Nasser died in 1970 and was succeeded by Vice-President Anwar al-Sadat. Sadat, initially virtually unknown by the Egyptian people, took many bold steps in cementing his role as leader of Egypt.

talked about developing representative government, Nasser distrusted political parties and remembered the destructive rivalries under the monarchy that had kept Egypt divided and weak. The Wafd and all other political parties were declared illegal. Nasser set up his own political organization to replace them, called the Arab Socialist Union (ASU). It was a mass party, but it had no real power. Nasser and a few close associates ran the government and controlled the ASU. The associates took their orders directly from Nasser; they called him *El-Rais*—"The Boss."

As he grew older, Nasser, plagued by health problems, became more dictatorial, secretive, and suspicious. The Boss tolerated no opposition and ensured tight control over Egypt with a large police force and a secret service that monitored activities in every village and town.

Nasser died in 1970. Ironically, his death came on the heels of a major policy success: the arranging of a truce between the Palestine Liberation Organization and the government of Jordan. Despite his

health problems, Nasser had seemed indestructible, and his death came as a shock. Millions of Egyptians followed his funeral cortege through the streets of Cairo, weeping and wailing over the loss of their beloved Rais.

ANWAR AL-SADAT

Nasser was succeeded by his vice-president, Anwar al-Sadat, in accordance with constitutional procedure. Sadat had been one of the original Free Officers and had worked with Nasser since their early days at the Military Academy. In the Nasser years, Sadat had come to be regarded as a lightweight, always ready to do whatever The Boss wanted.

Many Egyptians did not even know what Sadat looked like. A popular story was told of an Egyptian peasant in from the country to visit his cousin, a taxi driver. As they drove around Cairo, they passed a large poster of Nasser and Sadat shaking hands. "I know our beloved leader, but who is the man with him?" asked the peasant. "I think he owns that café across the street," replied his cousin.

When Sadat became president, however, it did not take long for the Egyptian people to learn what he looked like. Sadat introduced a "revolution of rectification," which he said was needed to correct the errors of his predecessor.[8] These included too much dependence on the Soviet Union, too much government interference in the economy, and failure to develop an effective Arab policy against Israel. He was a master of timing, taking bold action at unexpected times to advance Egypt's international and regional prestige. Thus, in 1972 he abruptly ordered the 15,000 Soviet advisers in Egypt to leave the country, despite the fact that they were training his army and supplying all his military equipment. His purpose was to reduce Egypt's dependence on one foreign power, and as he had calculated, the United States now came to his aid.

A year later, in October 1973, Egyptian forces crossed the Suez Canal in a surprise attack and broke through Israeli defense lines in occupied Sinai. The attack was coordinated with Syrian forces invading Israel from the east, through the Golan Heights. The Israelis were driven back with heavy casualties on both fronts, and although they eventually regrouped and won back most of the lost ground, Sadat felt he had won a moral and psychological victory. After the war, Egyptians believed that they had held their own with the Israelis and had demonstrated Arab ability to handle the sophisticated weaponry of modern warfare. On the 25th anniversary of the

1973 October War, Egypt held its first military parade in 17 years, and 250 young couples were married in a mass public wedding ceremony at the Pyramids to remind the new generation—a third of the population are under age 15—of Egypt's great "victory."

Anwar al-Sadat's most spectacular action took place in 1977. It seemed to him that the Arab–Israeli conflict was at a stalemate. Neither side would budge from its position, and the Egyptian people were angry at having so little to show for the 1973 success. In November, he addressed a hushed meeting of the People's Assembly and said, "Israel will be astonished when it hears me saying … that I am ready to go to their own house, to the Knesset itself, to talk to them."[9] And he did so, becoming for a second time the "Hero of the Crossing,"[10] but this time to the very citadel of Egypt's enemy.

Sadat's successes in foreign policy, culminating in the 1979 peace treaty with Israel, gave him great prestige internationally. Receipt of the Nobel Peace Prize, jointly with Israeli prime minister Menachem Begin, confirmed his status as a peacemaker. His pipe-smoking affability and sartorial elegance endeared him to U.S. policymakers.

The view that more and more Egyptians held of their world-famous leader was less flattering. Religious leaders and conservative Muslims objected to Sadat's luxurious style of living. The poor resented having to pay more for basic necessities. The educated classes were angry about Sadat's claim that the political system had become more open and democratic when, in fact, it had not. The Arab Socialist Union was abolished and several new political parties were allowed to organize. But the ASU's top leaders merely formed their own party, the National Democratic Party, headed by Sadat. For all practical purposes, Egypt under Sadat was even more of a single-party state under an authoritarian leader than it had been in Nasser's time.

Sadat's economic policies also worked to his disadvantage. In 1974, he announced a new program for postwar recovery, *Infitah* ("Opening"). It would be an open-door policy, bringing an end to Nasser's state-run socialist system. Foreign investors would be encouraged to invest in Egypt, and foreign experts would bring their technological knowledge to help develop industries. Infitah, properly applied, would bring an economic miracle to Egypt.

Rather than spur economic growth, however, Infitah made fortunes for just a few, leaving the great majority of Egyptians no better off than before. Chief among those who profited were members of the Sadat family. Corruption among the small ruling class, many of its members newly rich contractors, aroused anger on the part of the Egyptian people. In 1977, the economy was in such bad shape that the government increased bread prices. Riots broke out, and Sadat was forced to cancel the increase.

On October 6, 1981, President Sadat and government leaders were reviewing an armed-forces parade in Cairo to mark the eighth anniversary of the Crossing. Suddenly, a volley of shots rang out from one of the trucks in the parade. Sadat fell, mortally wounded. The assassins, most of them young military men, were immediately arrested. They belonged to *Al Takfir Wal Hijra* ("Repentance and Flight from Sin"), a secret group that advocated the reestablishment of a pure Islamic society in Egypt—by violence, if necessary. Their leader declared that the killing of Sadat was an essential first step in this process.

Islamic fundamentalism developed rapidly in the Middle East after the 1979 Iranian Revolution. The success of that revolution was a spur to Egyptian fundamentalists. They accused Sadat of favoring Western capitalism through his Infitah policy, of making peace with the "enemy of Islam" (Israel), and of not being a good Muslim. At their trial, Sadat's assassins said that they had acted to rid Egypt of an unjust ruler, a proper action under the laws of Islam.

Sadat may have contributed to his early death (he was 63) by a series of actions taken earlier in the year. About 1,600 people were arrested in September 1981 in a massive crackdown on religious unrest. They included not only religious leaders but also journalists, lawyers, intellectuals, provincial governors, and leaders of the country's small but growing opposition parties. Many of them were not connected with any fundamentalist Islamic organization. It seemed to most Egyptians that Sadat had overreacted, and at that point, he lost the support of the nation. In contrast to Nasser's funeral, few tears were shed at Sadat's. His funeral was attended mostly by foreign dignitaries. One of them said that Sadat had been buried without the people and without the army.

MUBARAK IN POWER

Vice-President Hosni Mubarak, former Air Force commander and designer of Egypt's 1973 success against Israel, succeeded Sadat without incident. Mubarak dealt firmly with Islamic fundamentalism at the beginning of his regime. He was given emergency powers and approved death sentences for five of Sadat's assassins in 1982. But he moved cautiously in other areas of national life, in an effort to disassociate himself from some of Sadat's more unpopular policies. The economic policy of Infitah, which had led to widespread graft and corruption, was abandoned; stiff sentences were handed out to a number of entrepreneurs and capitalists, including Sadat's brother-in-law and several associates of the late president.

Mubarak also began rebuilding bridges with other Arab states that had been damaged after the peace treaty with Israel. Egypt was readmitted to membership in the Islamic Conference, the Islamic Development Bank, the Arab League, and other Arab regional organizations. In 1990, the Arab League headquarters was moved from Tunis back to Cairo, its original location. Egypt backed Iraq with arms and advisers in its war with Iran, but Mubarak broke with Saddam Hussein after the invasion of Kuwait, accusing the Iraqi leader of perfidy. Some 35,000 Egyptian troops served with the UN–U.S. coalition during the Gulf War; and as a result of these efforts, the country resumed its accustomed role as the focal point of Arab politics.

Despite the peace treaty, relations with Israel continued to be difficult. One bone of contention was removed in 1989 with the return of the Israeli-held enclave of Taba, in the Sinai Peninsula, to Egyptian control. It had been operated as an Israeli beach resort.

The return of Taba strengthened the government's claim that the 10-year-old peace treaty had been valuable overall in advancing Egypt's interests. The sequence of agreements between the Palestine Liberation Organization and Israel for a sovereign Palestinian entity, along with Israel's improved relations with its other Arab neighbors, contributed to a substantial thaw in the Egyptian "cold peace" with its former enemy. In March 1995, a delegation from Israel's Knesset arrived in Cairo, the first such parliamentary group to visit Egypt since the peace treaty.

But relations worsened after the election in 1996 of Benjamin Netanyahu as head of a new Israeli government. Egypt had strongly supported the Oslo accords for a Palestinian state, and it had set up a free zone for transit of Palestinian products in 1995. The Egyptian view that Netanyahu was not adhering to the accords led to a "war of words" between the two countries. Israeli tourists were discouraged from visiting Egypt or received hostile treatment when visiting Egyptian monuments, and almost no Egyptians opted for visits to Israel. The newspaper *Al-Ahram* even stopped carrying cartoons by a popular Israeli-American cartoonist because he had served in the Israeli Army. The two governments cooperated briefly in the return of a small Bedouin tribe, the Azazma, to its Egyptian home area in the Sinai. The tribe had fled into Israel following a dispute with another tribe that turned into open conflict.

The election of Ehud Barak as Israel's new prime minister was well received in Eygpt, notably due to his resumption of peace negotiations with the then-Palestinian leader Yassir Arafat. However the breakdown of those negotiations and Barak's defeat by Ariel Sharon in the 2000 Israeli elections re-established the "deep freeze" between the two countries. In 2004 President Mubarak declared Sharon incapable of making peace. However Israel's withdrawal from the Gaza Strip in 2005 helped improve Sharon's image in Egypt as peacemaker rather than butcher. One result was a prisoner swap. Mubarak released an Israeli Arab convicted of espionage in Egypt in return for 6 Egyptian students who had been studying in Israel.

FREEDOM

The Islamic fundamentalist challenge to Egypt's secular government has caused the erosion of many rights and freedoms enshrined in the country's Constitution. A state of emergency first issued in 1981 is still in effect; it was renewed in 2001 for a 3-year period. In June 2003 the Peoples' Assembly approved establishment of a National Council for Human Rights that would monitor violations or misuse of government authority.

Internal Politics

Although Mubarak's unostentatious lifestyle and firm leadership encouraged confidence among the Egyptian regime, the system that he inherited from his predecessors remained largely impervious to change. The first free multiparty national elections held since the 1952 Revolution took place in 1984—although they were not entirely free, because a law requiring political parties to win at least 8 percent of the popular vote limited party participation. Mubarak was re-

elected easily for a full six-year term (he was the only candidate), and his ruling National Democratic Party won 73 percent of seats in the Assembly. The New Wafd Party was the only party able to meet the 8 percent requirement.

New elections for the Assembly in 1987 indicated how far Egypt's embryonic democracy had progressed under Mubarak. This time, four opposition parties aside from his own party presented candidates. Although the National Democratic Party's plurality was still a hefty 69.6 percent, 17 percent of the electorate voted for opposition candidates. The New Wafd increased its percentage of the popular vote to 10.9 percent, and a number of Muslim Brotherhood members were elected as independents. The National Progressive Unionist Group, the most leftist of the parties, failed to win a seat.

Mubarak was elected to a fourth six-year term in September 1999, making him Egypt's longest-serving head of state in the country's independent history. His victory margin was 94 percent, two points less than in 1993, when as per usual he was the only candidate. Some 79 percent of Egypt's 24 million registered voters cast their ballots.

The Bush administration's drive to establish the Equivalent of American-style democracy in the Arab World, with freedom of press and assembly, political parties, human rights and particularly rights for women, may have promoted a small step in that direction in 2005. In February Mubarak announced the first direct, secret, multi-party presidential election in national history. It permitted other candidates to run against the incumbent. The National Assembly (Parliament), dominated by the President's National Democratic Party (NDP) approved a change in the constitution to allow such an election

AT WAR WITH FUNDAMENTALISM

Egypt's seemingly intractable social problems—high unemployment, an inadequate job market flooded annually by new additions to the labor force, chronic budgetary deficits, and a bloated and inefficient bureaucracy, to name a few—have played into the hands of Islamic fundamentalists, those who would build a new Egyptian state based on the laws of Islam. Although they form part of a larger fundamentalist movement in the Islamic world, one that would replace existing secular regimes with regimes that adhere completely to spiritual law and custom (*Shari'a*), Egypt's fundamentalists do not harbor expansionist goals. Their goal is to replace the Mubarak regime with a more purely "Islamic" one,

faithful to the laws and principles of the religion and dominated by religious leaders.

Egypt's fundamentalists are broadly grouped under the organizational name al-Gamaa al-Islamiya, with the more militant ones forming subgroups such as the Vanguard of Islam and Islamic Jihad, itself an outgrowth of al-Takfir wal-Hijra, which had been responsible for the assassination of Anwar Sadat. Ironically, Sadat had formed Al-Gamaa to counter leftist political groups. However, it differs from its parent organization, the Muslim Brotherhood, in advocating the overthrow of the government by violence in order to establish a regime ruled under Islamic law. During Mubarak's first term, he kept a tight lid on violence. But in the 1990s, the increasing strength of the Islamists and their popularity with the large number of educated but unemployed youth led to an increase in violence and destabilized the nation.

Violence was initially aimed at government security forces, but starting in 1992, the fundamentalists' strategy shifted to vulnerable targets such as foreign tourists and the Coptic Christian minority. A number of Copts were killed and many Copt business owners were forced to pay "protection money" to al-Gamaa in order to continue in operation. Subsequently the Copts' situation improved somewhat, as stringent security measures were put in place to contain Islamic fundamentalist violence. Gun battles in 1999 between Muslim and Copt villagers in southern Egypt resulted in 200 Christian deaths and the arrests of a number of Muslims as well as Copts. Some 96 Muslims were charged with violence before a state-security court, but only four were convicted, and to short jail terms. However, Muslim–Coptic relations remained unstable. Early in 2002, two Coptic weekly newspapers, *Al Nabaa* and *Akbar Nabaa*, were shut down after the Superior Press Council, a quasi-government body, filed a lawsuit charging them with "offending Egyptians and undermining national unity."

Islamic Jihad, the major fundamentalist organization and the one responsible for Sadat's assassination, subsequently shifted its locale and objectives in order to evade the repression of the Mubarak government. Many of its members joined the fighters in Afghanistan who were resisting the Soviet occupation of that country. After the Soviet withdrawal in 1989, some 300 of them remained, forming the core of the Taliban force that eventually won control of 90 percent of Afghanistan. In that capacity, they became associated with Osama bin Laden and his al-Qaeda international terror network. Two of their leaders, Dr. Ayman al-Zawahiri (a surgeon) and Muhammad Atif, are believed to have planned the September 11, 2001, terrorist bombings in the United States. However, Islamic Jihad's chief aim is the overthrow of the Mubarak government and its replacement by an Islamic one. Its hostility to the United States stems from American support for that government and for the U.S. alliance with Israel against the Palestinians.

In targeting tourism in their campaign to overthrow the regime, fundamentalists have attacked tourist buses. Four tourists were killed in the lobby of a plush Cairo hotel in 1993. In November 1997, 64 tourists were gunned down in a grisly massacre at the Temple of Hatshepsut near Luxor, in the Valley of the Kings, one of Egypt's prize tourist attractions. Aftershocks from the terrorist attacks on the United States have decimated the tourist industry, which is Egypt's largest source of income ($4.3 billion in 2000, with 5.4 million visitors in that year). Egyptair, the national airline, lost $56 million in October 2001 alone; and cancellations of package tours, foreign-airline bookings, and hotel reservations led to a 45 percent drop in tourist revenues.

One important reason for the rise in fundamentalist violence stems from the government's ineptness in meeting social crises. After the disastrous earthquake of October 1992, Islamic fundamentalist groups were first to provide aid to the victims, distributing $1,000 to each family made homeless, while the cumbersome, multilayered government bureaucracy took weeks to respond to the crisis. Similarly, al-Gamaa established a network of Islamic schools, hospitals, clinics, day-care centers, and even small industries in poor districts such as Cairo's Imbaba quarter.

The Mubarak government's response to rising violence has been one of extreme repression. The death penalty may be imposed for "antistate terrorism." The state of emergency that was established after Anwar Sadat's assassination in 1981 has been renewed regularly, most recently in 2001 for a three-year extension, over the vehement protests of opposition deputies in the Assembly. Some 770 members of the Vanguard of Islam were tried and convicted of subversion in 1993. The crackdown left Egypt almost free from violence for several years. But in 1996, al-Gamaa and two other hitherto unknown Islamic militant groups, Assiut Terrorist Organization and Kotbion (named for a Muslim Brotherhood leader executed in 1966 for an attempt to kill President Nasser), resumed terrorist activities. Eighteen Greek tourists were murdered in April, and the State Security Court sentenced five Assiut members to death for killing police and civilians in a murderous rampage. At their trial they chanted "God make a staircase of our skulls to Your glory," waving Korans in their cage, in an eerie replay of the trials of Sadat's assassins.

An unfortunate result of government repression of the militants is that Egypt, traditionally an open, tolerant, and largely nonviolent society, has taken on many of the features of a totalitarian state. Human rights are routinely suspended, the prime offenders being officers of the dreaded State Security Investigation (SSI). Indefinite detention without charges is a common practice, and torture is used extensively to extract "confessions" from suspects or their relatives. All of al-Gamaa's leaders are either in prison, in exile, or dead; and with 20,000 suspected Islamists also jailed, the government could claim with some justification that it had broken the back of the 1990s insurgency. Its confidence was enhanced in March 1999 when al-Gamaa said that it would no longer engage in violence. Two previous cease-fire offers had been spurned, but this newest offer resulted in the release of several hundred Islamists to "test its validity."

Due to the extremism of methods employed by both sides, the conflict between the regime and the fundamentalists has begun to polarize Egyptian society. As a prominent judge noted, "Islam has turned from a religion to an ideology. It has become a threat to Egypt, to civilization and to humanity."[11] The fundamentalists, in struggling to overthrow the regime and replace it with a more legitimately Islamic one (in their view) have at times attacked intellectuals, journalists, writers and others who do not openly advocate similar views or even oppose them. The novelist Farag Foda, who strongly criticized Egypt's "creeping Islamization" in his works, was killed outside of his Cairo home in the early 1990s, and Haguib Mahfouz, the Arab world's only Nobel laureate in literature, was critically wounded in 1994 by a Gamaa gunman.

In 1995, the regime imposed further restrictions on Egypt's normally freewheeling press and journalistic bodies. A law would impose fines of up to $3,000 and five-year jail sentences for articles "harmful to the state." The long arm of the law reached into the educational establishment as well. A university professor and noted Koranic scholar was charged with apostasy by clerics at Al-Azhar University, on the grounds that he had argued in his writings that the Koran should be interpreted in its historical/linguistic context alongside its identification as the Word of God. The charge came under the Islamic principle of *hisba*, "accountability." He was found guilty and ordered to divorce his wife, since a Muslim woman may not be married to an apostate. A 1996 law prohibited the use of hisba in the courts, but the damage had been done; the professor and his wife were forced into exile to preserve their marriage.

Another distinguished professor ran afoul of the government's Al-Azhar–imposed limits on free speech, as Saad Eddine Ibrahim, director of the American University at Cairo's Ibn Khaldoun Center for Democracy, was arrested in July 2000. He was charged with "defaming" Egypt abroad. The charge resulted from a documentary film produced at the Center which was critical of parliamentary electoral process for encouraging fraud. Due in part to widespread criticism in Europe and particularly the United States—Ibrahim holds dual Egyptian-American citizenship—he was released. In May 2001 he was rearrested, tried and convicted by a special court, and given a 7-year jail sentence. After strong international criticism and a warning by U.S. president Bush that his administration would oppose any increase in the $2 billion annual aid program to Egypt, the Court of Cassation, the country's highest court, threw out the conviction and ordered him released.[12] However, further international criticism followed, including a statement by President George Bush that his administration would oppose any increase in the $2 billion annual aid program to Egypt. The Court of Cassation, Egypt's highest court, subsequently threw out the conviction and ordered Ibrahim released.

ACHIEVEMENTS

Alexandria, founded in 332 B.C. by Alexander the Great, was one of the world's great cities in antiquity, with its Library, its Pharos (Lighthouse), its palaces, and other monuments. Most of them were destroyed by fire or sank into the sea long ago, as the city fell into neglect. Then, in 1995, underwater archeologists discovered the ruins of the Pharos; its location had not been known previously. Other discoveries followed—the palace of Cleopatra, the remains of Napoleon's fleet (sunk by the British in the Battle of the Nile), Roman and Greek trading vessels filled with amphorae, etc. The restoration of the Library was completed in 2000, with half of its 11 floors under the Mediterranean; visitors in the main reading room are surrounded by water cascading down its windows. After centuries of decay, Alexandria is again a magnet for tourists.

While Ibrahim's release suggests that the Egyptian government is sensitive to charges of misuse of human rights from abroad, its posture internally toward its citizens has changed little since emergency laws went into effect. Recently the head of the Group for Developing Democracy, a civil rights watchdog agency in Cairo, noted that "Egypt doesn't want real democracy. The state wants us puppets in its big show of paper democracy." And in July 2003 the director of the Cairo office of the UN Development Program told a reporter: "People seem to be accepting an immoral tradeoff between human rights and security."[13]

A small but significant step toward reawakening the country's moribund political system was taken in September 2005 with an open presidential election. The incumbent was opposed by 9 candidates, the most prominent being Ayman Nour, leader of the Al Ghad ("Tomorrow") opposition party. Mubarak was re-elected but won "only" 88.6 percent of the popular vote. However voting irregularities charged by opposition leaders led to a further crackdown. In December Ayman Nour was given a 5-year forced labor sentence for what U.S. officials later described as false charges. He had finished second to Mubarak in the popular vote.

The November 2005 parliamentary elections marked another small step forward politically for Egypt. The elections took place in 3 stages. For the first time Muslim Brotherhood members were allowed to run, although the organization is still unrecognized as a political party. At the end of the 3 stages they had won 88 seats. With the Brotherhood officially proscribed, the growing popularity of Kifaya, a secular pro-democracy movement, underscored the increasing disenchantment of Egyptians of all ages and social levels for the regime.

In a further effort to blunt the opposition, Mubarak in February 2006 abruptly postponed local council elections scheduled for April, canceling a promise he had made during the presidential campaign to promote greater democracy. While these councils have little actual power, a constitutional change in 2005 will give them control over nomination of candidates for president in 2011, when Mubarak is prohibited from running or serving.

A STRUGGLING ECONOMY

Egypt's economy rests upon a narrow and unstable base, due to rapid demographic growth and limited arable land and because political factors have adversely influenced national development. The country has a relatively high level of education and, as a result, is a net exporter of skilled labor to other Arab countries. But the overproduction of university graduates has produced a bloated and inefficient bureaucracy, as the government is required to provide a position for every graduate who cannot find other employment.

Agriculture is the most important sector of the economy, accounting for about one third of national income. The major crops are long-staple cotton and sugarcane. Egyptian agriculture since time immemorial has been based on irrigation from the Nile River. In recent years, greater control of irrigation water through the Aswan High Dam, expansion of land devoted to cotton production, and improved planting methods have begun to show positive results.

A new High Dam at Aswan, completed in 1971 upstream from the original one built in 1906, resulted from a political decision by the Nasser government to seek foreign financing for its program of expansion of cultivable land and generation of electricity for industrialization. When Western lending institutions refused to finance the dam, also for political reasons, Nasser turned to the Soviet Union for help. By 1974, just three years after its completion, revenues had exceeded construction costs. The dam made possible the electrification of all of Egypt's villages as well as a fishing industry at Lake Nasser, its reservoir. It proved valuable in providing the agricultural sector with irrigation water during the prolonged 1980–1988 drought, although at sharply reduced levels. However, the increased costs of land reclamation and loss of the sardine fishing grounds along the Mediterranean coast have made the dam a mixed blessing for Egypt.

Egypt was self-sufficient in foodstuffs as recently as the 1970s but now must import 60 percent of its food. Such factors as rapid population growth, rural-to-urban migration with consequent loss of agricultural labor, and Sadat's open-door policy for imports combined to produce this negative food balance. Subsidies for basic commodities, which cost the government nearly $2 billion a year, are an important cause of inflation, since they keep the budget continuously in deficit. Fearing a recurrence of the 1977 Bread Riots, the government kept prices in check. However, inflation, which had dropped to 8 percent in 1995 due to International Monetary Fund stabilization policies required for loans, rose to 37 percent in 1999 as the new free-market policy produced a tidal wave of imports. As a result, the foreign trade deficit increased drastically.

Egypt has important oil and natural-gas deposits, and new discoveries continue to strengthen this sector of the economy. Oil reserves increased to 3.3 billion barrels in 2001, due to new fields being brought on stream in the Western Desert. Proven natural-gas reserves are 51 trillion cubic feet, sufficient to meet domestic needs for 30 years at current rates of consumption.

A 2001 agreement with Jordan would guarantee Jordan's purchase of Egyptian natural-gas supplies, contingent on completion of the pipeline under the Red Sea from Al-Arish to Aqaba. But an earlier agreement with Israel, Egypt's closest and potentially most lucrative gas market, has been put on hold due to the renewed Palestinian–Israeli conflict. Under the agreement, Egypt would have provided $300 million a year in gas, meeting 15 percent of Israel's electric-power needs.

Egypt also derives revenues from Suez Canal tolls and user fees, from tourism, and from remittances from Egyptian workers abroad, mostly working in Saudi Arabia and other oil-producing Gulf states. The flow of remittances from the approximately 4 million expatriate workers was reduced and then all but cut off with the Iraqi invasion of Kuwait. Egyptians fled from both countries in panic, arriving home as penniless refugees. With unemployment already at 20 percent and housing in short supply, the government faced an enormous assimilation problem apart from its loss of revenue. The United States helped by agreeing to write off $4.5 billion in Egyptian military debts. However the imprisonment and unduly harsh sentence of dissident presidential candidate Ayman Nour led the U.S. in January 2006 to suspend a projected trade and tariff elimination agreement between the two countries.

One encouraging sign of brighter days ahead is the expansion of local manufacturing industries, in line with government efforts to reduce dependence upon imported goods. A 10-year tax exemption plus remission of customs duties on imported machinery have encouraged a number of new business ventures, notably in the clothing industry.

Unfortunately one of the few enterprises affecting Egypt's poor directly was literally "dumped" in January 2003 when the government stopped renewing licenses to the Zabbaleen, a 60,000-member Coptic community that traditionally collects a third of Cairo's 10,000 daily tons of garbage and trash. Future collections of all garbage and trash are to be made by foreign companies under contract. The new system would have certain advantages of the Zabbaleen system, mainly in terms of sanitation. But the economic impact on them will be severe. Over the years the Zabbaleen have used profits from trash collection to fund neighborhood improvements, schools and jobs for a great number of women.

In 1987, Mubarak gained some foreign help for Egypt's cash-strapped economy when agreement was reached with the International Monetary Fund for a standby credit of $325 million over 18 months to allow the country to meet its balance-of-pay-

Timeline: PAST

2500–671 B.C.
Period of the pharaohs

671–30 B.C.
The Persian conquest, followed by Macedonians and rule by Ptolemies

30 B.C.
Egypt becomes a Roman province

A.D. 641
Invading Arabs bring Islam

969
The founding of Cairo

1517–1800
Egypt becomes an Ottoman province

1798–1831
Napoleon's invasion, followed by the rise to power of Muhammad Ali

1869
The Suez Canal opens to traffic

1882
The United Kingdom establishes a protectorate

1952
The Free Officers overthrow the monarchy and establish Egypt as a republic

1956
Nationalization of the Suez Canal

1958–1961
Union with Syria into the United Arab Republic

1967
The Six-Day War with Israel ends in Israel's occupation of the Gaza Strip and the Sinai Peninsula

1970
Gamal Abdel Nasser dies; Anwar Sadat succeeds as head of Egypt

1979
A peace treaty is signed at Camp David between Egypt and Israel

1980s
Sadat is assassinated; he is succeeded by Hosni Mubarak; a crackdown on Islamic fundamentalists

1990s
The government employs totalitarian tactics in its battle with fundamentalists

PRESENT

2000s
Deep social and economic problems persist

ments deficit. The Club of Paris, a group of public and private banks from various industrialized countries, then rescheduled $12 billion in Egyptian external debts over a 10-year period.

Expanded foreign aid and changes in government agricultural policy required by the World Bank for new loans helped spur economic recovery in the 1990s, especially in agriculture. Production records were set in 1996 in wheat, corn, and rice, meeting 50 percent of domestic needs. The cotton harvest for that year was 350,000 tons, with 50,000 tons exported. However, a new agricultural law passed in 1992 but not implemented until 1997 ended land rents, allowing landlords to set their own leases and in effect reclaim their properties taken over by the government during the Nasser era. The purpose is to provide an incentive for tenant farmers to grow export crops such as cotton and rice. But as a result, Egypt's 900,000 tenant farmers have faced the loss of lands held on long-term leases for several generations.

However, these economic successes must be balanced against Egypt's chronic social problems and the lack of an effective representative political system. The head of the Muslim Brotherhood made the astute observation in a 1993 speech that "the threat is not in the extremist movement. It is in the absence of democratic institutions." Until such institutions are firmly in place, with access to education, full employment, broad political participation, civil rights, and the benefits of growth spread evenly across all levels of society, unrest and efforts to Islamize the government by force are likely to continue.

By 2000, the government's harsh repression had seriously weakened the fundamentalist movement, albeit at a heavy cost. Some 1,200 police officers and militants had been killed during the 1990s, and 16,000 persons remained jailed without charges on suspicion of membership in Islamic Jihad or other organizations. However, public disaffection continues to grow and to involve increasing numbers of nonfundamentalists. In May 2000, a demonstration by several hundred Al-Azhar students protesting the reprinting of a 1983 novel by Syrian writer Haider Haider was broken up by police. The demonstrators charged that the novel was insulting both to the Prophet Muhammad and to Islam; it had been reprinted as part of a Ministry of Culture project to promote modern Arabic literature. In 2001, the Cairo Book Fair, one of the Arab world's biggest cultural events, was boycotted by many Egyptian writers and editors, after its sponsor, the same Ministry of Culture, had banned a number of books. They were declared to be pornographic by Muslim Brotherhood members of the Legislature and criticized by Al-Azhar faculty as being offensive to Islam.

Egypt's own difficulties with fundamentalists caused some reluctance on its part when support for the U.S.–led international coalition against terrorism formed after the September 11, 2001, bombings of the World Trade Center in New York City and the Pentagon near Washington, D.C. The reluctance stemmed in part from public anger over continued American support for Israel against the Palestinians and the suffering of Iraq's fellow Arabs under the 11-year sanctions imposed on that country.

In March 2004 the government reached agreement with Israel to set up a number of Qualifying Industrial Zones (Q.I.Z.) in an effort to boost its flagging economy. Egyptian manufacturers, notable of textiles, will be able to export goods duty-free to the U.S. provided that 35 percent of goods exported were locally produced and a percentage reserved for Israeli products. Egypt's total exports to the U.S. of $3.3 billion included $336 million in textiles and clothing. The Q.I.Z.s will add significantly to this total.

Encouraged by a *fatwa* (Islamic religious ruling) on its lawfulness under Shari'a law given by a scholar at Al-Azhar University, Egypt's In Vitro Fertilization Center in Cairo became in October 2005 the first in the Arab world to conduct stem-cell research.

NOTES

1. Leila Ahmed, in *A Border Passage* (New York: Farrar, Strauss & Giroux, 1999), deals at length with Egyptian vs. Arab identity from the perspective of growing up in British-controlled Egypt.

2. An English observer said, "In arms and firing they are nearly as perfect as European troops." Afaf L. Marsot, *Egypt in the Reign of Muhammad Ali* (Cambridge, England: Cambridge University Press, 1984), p. 132.

3. *Ibid.*, p. 161.

4. Quoted in P. J. Vatikiotis, *Nasser and His Generation* (New York: St. Martin's Press, 1978), p. 35.

5. Gamal Abdel Nasser, *The Philosophy of the Revolution* (Cairo: Ministry of National Guidance, 1954), p. 52.

6. Derek Hopwood, *Egypt: Politics and Society 1945–1981* (London: George Allen and Unwin, 1982), p. 77.

7. Quoted in Vatikiotis, *op. cit.*, p. 245.

8. Hopwood, *op. cit.*, p. 106.

9. David Hirst and Irene Beeson, *Sadat* (London: Faber and Faber, 1981), p. 255.

10. "Banners slung across the broad thoroughfares of central Cairo acclaimed The Hero of the Crossing (of the October 1973 War)." *Ibid.*, pp. 17–18.

11. Said Ashmawy, quoted in "In God He Trusts," *Jerusalem Post Magazine* (July 7, 1995).

12. Neged Borat, quoted in *The New York Times* (May 21, 2001).

13. Reported in *The New York Times*, July 2003.

Iran (Islamic Republic of Iran)

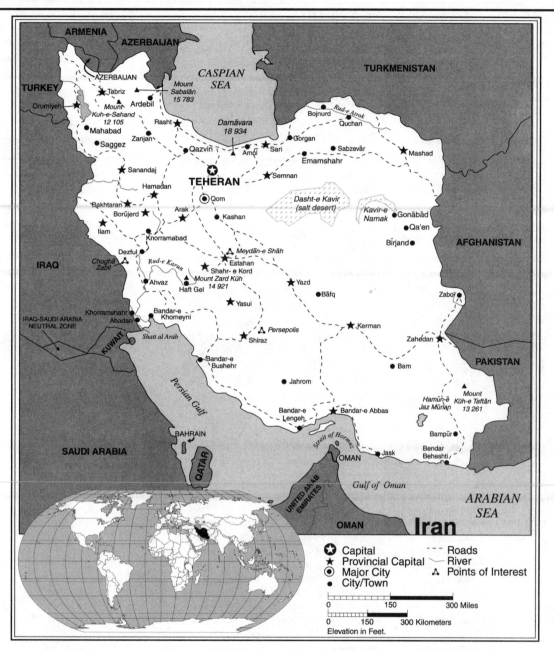

Iran Statistics

GEOGRAPHY

Area in Square Miles (Kilometers):
636,294 (1,648,000) (about the size of Alaska)

Capital (Population): Teheran (6,836,000)

Environmental Concerns: air and water pollution; deforestation; overgrazing; desertification; oil pollution; insufficient potable water

Geographical Features: a rugged, mountainous rim; a high central basin with deserts and mountains; discontinuous plains along both coasts

Climate: mostly arid or semiarid; subtropical along Caspian Sea coast

PEOPLE

Population

Total: 68,017,860

Annual Growth Rate: 0.86%

Rural/Urban Population Ratio: 40/60

Major Languages: Farsi (Persian); Azeri Turkish; Kurdish

53

Ethnic Makeup: 51% Persian; 24% Azeri; 8% Gilaki and Mazandarani; 7% Kurd; 10% others
Religions: 89% Shia Muslim; 10% Sunni Muslim; 1% Zoroastrian, Jewish, Christian, or Bahai

Health

Life Expectancy at Birth: 68 years (male); 71 years (female)
Infant Mortality Rate (Ratio): 41.58/1,000 live births

Education

Adult Literacy Rate: 79%
Compulsory (Ages): 6–10; free

COMMUNICATION

Telephones: 6,600,000 main lines
Daily Newspaper Circulation: 20 per 1,000 people
Televisions: 117 per 1,000
Internet Service Providers: 8 (2000)

TRANSPORTATION

Highways in Miles (Kilometers): 86,924 (140,200)
Railroads in Miles (Kilometers): 3,472 (5,600)

Usable Airfields: 305
Motor Vehicles in Use: 2,189,000

GOVERNMENT

Type: theocratic republic
Independence Date: April 1, 1979 (Islamic Republic of Iran proclaimed)
Head of State/Government: Supreme Guide Ayatollah Ali Hoseini-Khamenei; President Mahmud Ahmadi-Nejad
Political Parties: none registered as parties *per se* but the major ones are the Islamic Participation Front (IIPF), Executives of Construction, Solidarity, Islamic Labor, Mardom Salari, Militant Clerics Society, Mojahedin of the Islamic Revolution (MIRO) A new political group, Builders of Islamic Iran, won a majority of seats in the 2004 Majlis elections
Suffrage: universal at 15

MILITARY

Military Expenditures (% of GDP): 3.1%
Current Disputes: maritime boundary with Iraq in Shatt al-Arab not formally demarcated, and prisoner exchange not complete. Iran's occupation of Greater and Lesser Tunbs Islands disputed by UAE; Iran disagrees with Azerbaijan,

Russia, Kazakhstan and Turkmenistan over sharing of Caspian Sea waters.

ECONOMY

Currency ($U.S. Equivalent): 8,614 rials = $1 (changed from multi-exchange to fixed single rate in 2002)
Per Capita Income/GDP: $7,700/$516.7 billion
GDP Growth Rate: 6.3%
Inflation Rate: 15.5%
Unemployment Rate: 25%
Labor Force: 23,000,000
Natural Resources: petroleum; natural gas; coal; chromium; copper; iron ore; lead; manganese; zinc; sulfur
Agriculture: grains; sugar beets; fruits; nuts; cotton; dairy products; wool; caviar
Industry: petroleum; petrochemicals; textiles; cement and other construction materials; food processing; metal fabrication; armaments
Exports: $38.79 billion (primary partners Japan, Italy, United Arab Emirates)
Imports: $31.3 billion (primary partners Germany, France, Italy)

Iran Country Report

Iran is in many respects a subcontinent, ranging in elevation from Mount Demavend (18,386 feet) to the Caspian Sea, which is below sea level. Most of Iran consists of a high plateau ringed by mountains. Much of the plateau is covered with uninhabitable salt flats and deserts—the Dasht-i-Kavir and Dasht-i-Lut, the latter being one of the most desolate and inhospitable regions in the world. The climate is equally forbidding. The so-called Wind of 120 Days blows throughout the summer in eastern Iran, bringing dust and extremely high temperatures.

Most of the country receives little or no rainfall. Settlement and population density are directly related to the availability of water. The most densely populated region is along the Caspian Sea coast, which has an annual rainfall of 80 inches. The provinces of Azerbaijan in the northwest and Khuzestan along the Iraqi border, and the urban areas around Iran's capital, Teheran, are also heavily populated.

Water is so important to the Iranian economy that all water resources were nationalized in 1967. Lack of rainfall caused the development of a sophisticated system of

underground conduits, *qanats*, to carry water across the plateau from a water source, usually at the base of a mountain. Many qanats were built thousands of years ago and are still in operation. They make existence possible for much of Iran's rural population.

Until the twentieth century, the population was overwhelmingly rural; but due to rural–urban migration, the urban population has increased steadily. Nearly all of this migration has been to Teheran, whose metropolitan area now has a population of nearly 7 million, as compared to 200,000 in 1900. Yet the rural population has also increased. This fact has had important political consequences for Iran, when it was a monarchy as well as an Islamic republic. Attachment to the land, family solidarity, and high birth rates have preserved the strong rural element in Iranian society as a force for conservatism and loyalty to religious leaders, who then are able to influence whatever regime is in power. Indeed, the rural population strongly supported the first Islamic regime and contributed much of the volunteer manpower recruited to defend the country after the invasion by Iraqi forces in 1980.

ETHNIC AND RELIGIOUS DIVERSITY

Due to Iran's geographic diversity, the population is divided into a large number of separate and often conflicting ethnic groups. Ethnic Iranians constitute the majority. The Iranians (or *Persians,* from Parsa, the province where they first settled) are an Indo-European people whose original home was probably in Central Asia. They moved into Iran around 1100 B.C. and gradually dominated the entire region, establishing the world's first empire (in the sense of rule over various unrelated peoples in a large territory). Although its rulers failed to conquer Greece and thereby extend their empire into Europe, such achievements as their imperial system of government, the Persian language (Farsi), the monumental architecture of their capital at Persepolis, and their distinct cultural/historical heritage provide modern Iranians with pride in their ancient past and a national identity, unbroken to the present day.

The largest ethnic minority group is the Azeri (or Azerbaijani) Turks. The Azeris live in northwestern Iran. Their ethnic origin dates back to the ancient Persian Em-

Iranian society today has a considerable level of cultural conformity. Shia Islam is the dominant religion of Iran, and observance of this form of Islam permeates society, as this prayer meeting at Teheran University attests.

pire, when Azerbaijan was known as Atropene. The migration of Turkish peoples into this region in the eleventh and twelfth centuries A.D. encouraged the spread of the Turkish language and of Islam. These were reinforced by centuries of Ottoman rule, although Persian remained the written and literary language of the people.

Turkish dynasties originating in Azerbaijan controlled Iran for several centuries and were responsible for much of premodern Islamic Iran's political power and cultural achievements. In the late nineteenth and early twentieth centuries, Azeris were in the forefront of the constitutional movement to limit the absolute power of Iranian monarchs. They formed the core of the first Iranian Parliament. The Azeris have consistently fought for regional autonomy from the central Iranian government in the modern period and refer to their province as "Azadistan, Land of Freedom."

The Kurds are another large ethnic minority. Iran's Kurd population is concentrated in the Zagros Mountains along the Turkish and Iraqi borders. The Kurds are Sunni Muslims, as distinct from the Shia majority. The Iranian Kurds share a common language, culture, social organization, and ethnic identity with Kurds in Iraq, Turkey, and Syria. Kurds are strongly independent mountain people who lack a politically recognized homeland and who have been unable to unite to form one. The

Kurds of Iran formed their own Kurdish Republic, with Soviet backing, after World War II. But the withdrawal of Soviet troops, under international pressure, caused its collapse. Since then, Iranian Kurdish leaders have devoted their efforts toward greater regional autonomy. Kurdish opposition to the central Iranian government was muted during the rule of the Pahlavi dynasty (1925–1979), but it broke into the open after the establishment of the Islamic republic. The Kurds feared that they would be oppressed under the Shia Muslim government headed by Ayatollah Khomeini and boycotted the national referendum approving the republic. Central-government authority over the Kurds was restored in 1985. In 1992, Iraqi Kurdish leaders made an agreement with the Iranian government for deliveries of fuel and spare parts for their beleaguered enclave; in return, they pledged that the enclave would not be used by the People's Mujahideen or any other antigovernment group for military actions against Iran.

The Arabs are another important minority group (Iran and Turkey are the two Islamic countries in this region of the world with a non-Arab majority). The Arabs live in Khuzestan Province, along the Iraqi border. The Baluchi, also Sunni Muslims, are located in southeast Iran and are related to Baluchi groups in Afghanistan and Pakistan. They are semi-nomadic and have traditionally opposed any form of central-

government control. The Baluchi were the first minority to oppose openly the fundamentalist Shia policies of the Khomeini government. Non-Islamic minorities include Jews, Zoroastrians, and Armenians and other Christians. Altogether they make up 1 percent of the population. They are represented in the *Majlis* (Parliament) by two Armenian deputies and one each for Zoroastrians and Jews. Zoroastrians, about 30,000 in all, follow the ancient Persian religion preached by the prophet Zoroaster 2,500 years ago. Zoroaster defined life as a constant struggle between good (*Ahura Mazda*, "light") and evil (*Ahriman*, "darkness"). Zoroastrianism was the official religion of Iran during the pre-Islamic Sassanid empire. Its priests formed a privileged class, charged with responsibility for tending the sacred fire, which was (and still is) kept burning in the fire temples that are centers of faith and worship. Zoroastrians also traditionally buried their dead atop "towers of silence" rather than pollute the ground. Their religiously based customs and traditions have carried over into modern Iranian life. Thus Nowruz ("New Year"), the beginning of spring, the most important and popular of Iranian festivals, is Zoroastrian in origin, as is the solar calendar introduced by Reza Shah to replace the Arabic lunar one.

The Armenians, another small minority, are also protected under Article 13 of the republic's Constitution. Formerly, Armenians were important middlemen and traders in Iran. Armenian hairstylists were very popular before the Revolution, but the republic's strict Islamic code on male–female contacts forced them, as men, out of business. Armenian butcher shops are also required to post signs saying "Minority Religion" because they sell pork.

The Bahais, a splinter movement from Islam founded by an Iranian mystic called the *Bab* ("Door," i.e., to wisdom) and organized by a teacher named Baha'Ullah in the nineteenth century, are the largest non-Muslim minority group. Although Baha'Ullah taught the principles of universal love, peace, harmony, and brotherhood, his proclamations of equality of the sexes, ethnic unity, the oneness of all religions, and a universal rather than a Muslim God aroused the hostility of Shia religious leaders. Bahais in Iran were protected from Shia hostility during the Qajar and Pahlavi monarchy periods. But with the overthrow of Mohammed Reza Pahlavi, the religious leaders of the Islamic Republic undertook a campaign of persecution and mistreatment described by outside observers as "the genocide of a non-combatant people."[1]

Since 1979 over 200 Bahais have been executed and hundreds more jailed for various "offenses against Islam." Some 10,000 Bahais employed by the Shah's government were summarily terminated after the Revolution. Bahai youth may not attend Iranian universities, and the Bahai Institute of Higher Education, established as an alternative, was shut down in 1998. Bahai marriages are not recognized by the authorities and they are prohibited from owning property. The Bab's house in Shiraz, where the faith was founded in 1844, was razed along with other important Bahai sites in 2001. The 300,000 Bahais still resident in Iran have no legal rights; they are in essence a "non-people."

Despite Iran's official hostility toward Israel, its own Jewish population is recognized as a minority under the 1979 Constitution and has lived there for centuries. Some 100,000 Jews once lived in Iran, but emigration to Israel, the United States, and elsewhere has reduced their numbers. There are about 28,000 Jews in Iran today. They have been protected as "People of the Book" for generations; they observe their dietary laws without restriction, and their divorce, burial, and other laws are accepted in the Islamic courts. They also elect one deputy to the Majlis. However, the tensions in Iranian society between those favoring more openness to the outside world and hard-liners opposed to any accommodation, particularly with the United States and its Israeli ally, brought disaster to the Jewish community in Iran in 1999. Thir-

teen Jews were arrested and charged with spying for Israel. They included a rabbi, a university professor, office workers, and students. The trial began in May. Despite appeals from Jewish groups abroad and international criticism of the trial itself as being politically motivated, 10 of the defendants were given prison terms of up to 13 years; three were acquitted. The relative leniency of the sentences—the death sentence for espionage is normal in the Islamic Republic—underscored the trial's political nature. It met international demands for judicial fairness as well as the need to satisfy hard-liners. However, the Iranian Supreme Court denied an appeal by defense lawyers in 2001, saying that it had no legal basis for their consideration.

CULTURAL CONFORMITY

Despite the separatist tendencies in Iranian society caused by the existence of these various ethnic groups and religious divisions, there is considerable cultural conformity. Most Iranians, regardless of background, display distinctly Iranian values, customs, and traditions. Unifying features include the Farsi language, Islam as the overall religion, the appeal (since the sixteenth century) of Shia Islam as an Iranian nationalistic force, and a sense of nationhood derived from Iran's long history and cultural continuity.

DEVELOPMENT

Subsequent to the jump in global oil prices in 1999 Iran's economy has grown rapidly, from 5 percent in 2004 to the present 6.3 percent. Per capita income has kept pace. However it has yet to reach the levels of the Shah's regime. The oil and gas industries do not attract needed foreign investment, due largely to complicated contract requirements and the isolationist attitude of the conservative-dominated Majlis.

Iranians at all levels have a strongly developed sense of class structure. It is a three-tier structure, consisting of upper, middle, and lower classes. However, some scholars distinguish two lower classes: the urban wage earner, and the landed or landless peasant. The basic socioeconomic unit in this class structure is the patriarchal family, which functions in Iranian society as a tree trunk does in relation to its branches. The patriarch of each family is not only disciplinarian and decision maker but also guardian of the family honor and inheritance.

The patriarchal structure, in terms of the larger society, has defined certain behavioral norms. These include the seclusion of women, ceremonial politeness *(ta'aruf)*,

hierarchical authoritarianism with domination by superiors over subordinates, and the importance of face *(aberu)*—maintaining "an appropriate bearing and appearance commensurate with one's social status."[2] Under the republic, these norms have been increasingly Islamized as religious leaders have asserted the primacy of Shia Islam in all aspects of Iranian life.

HISTORY

Modern Iran occupies a much smaller territory than that of its predecessors. The Persian Empire (sixth–fourth centuries B.C.) included Egypt, the Arab Near East, Afghanistan, and much of Central Asia, prior to its overthrow by Alexander of Macedon. The Parthian and Sassanid monarchies were major rivals of Rome. Under the latter rulers (A.D. 226–651), Zoroastrianism became the state religion. The Sassanid administrative system, which divided its territory into provinces under a single central authority, was taken over intact by invading Arab armies bringing Islam to the land.

The establishment of Islam brought significant changes into Iranian life. Arab armies defeated the Sassanid forces at the Battle of Qadisiya (A.D. 637) and the later Battle of Nihavand (A.D. 641), which resulted in the death of the last Sassanid king and the fall of his empire. The Arabs gradually established control over all the former Sassanid territories, converting the inhabitants to Islam as they went. But the well-established Iranian cultural and social system provided refinements for Islam that were lacking in its early existence as a purely Arab religion. The Iranian converts to Islam converted the religion from a particularistic Arab faith to a universal faith. Islamic culture, in the broad sense—embracing literature, art, architecture, music, certain sciences and medicine—owes a great deal to the contributions of Iranian Muslims such as the poets Hafiz and Sa'di, the poet and astronomer Omar Khayyam, and many others.

Shia Muslims, currently the vast majority of the Iranian population and represented in nearly all ethnic groups, were in the minority in Iran during the formative centuries of Islam. Only one of the Twelve Shia Imams—the eighth, Reza—actually lived in Iran. (His tomb at Meshed is now the holiest shrine in Iran.) *Taqiya* ("dissimulation" or "concealment")—the Shia practice of hiding one's beliefs to escape Sunni persecution—added to the difficulties of the Shia in forming an organized community.

In the sixteenth century, the Safavids, who claimed to be descendants of the Prophet Muhammad, established control over Iran with the help of Turkish tribes. The first Safavid ruler, Shah Ismail, proclaimed Shiism as the official religion of his state and invited all Shias to move to Iran, where they would be protected. Shia domination of the country dates from this period. Shia Muslims converged on Iran from other parts of the Islamic world and became a majority in the population.

The Safavid rulers were bitter rivals of the Sunni Ottoman sultans and fought a number of wars with them. The conflict was religious as well as territorial. The Ottoman sultan assumed the title of caliph of Islam in the sixteenth century after the conquest of Egypt, where the descendants of the last Abbasid caliph of Baghdad had taken refuge. As caliph, the sultan claimed the right to speak for, and rule, all Muslims. The Safavids rejected this claim and called on Shia Muslims to struggle against him. In more recent years, the Khomeini government issued a similar call to Iranians to carry on war against the Sunni rulers of Iraq, indicating that Shia willingness to struggle and, if necessary, incur martyrdom was still very much alive in Iran.

King of Kings

The Qajars, a new dynasty of Turkish tribal origin, came to power after a bloody struggle at the end of the eighteenth century. They made Teheran their capital. Most of Iran's current borders were defined in the nineteenth century by treaties with foreign powers—Britain (on behalf of India), Russia, and the Ottoman Empire. Due to Iran's military weaknesses, the agreements favored the outside powers and the country lost much of its original territory.

Despite Iran's weakness in relation to foreign powers, the Qajar rulers sought to revive the ancient glories of the monarchy at home. They assumed the old Persian title *Shahinshah*, "King of Kings." At his coronation, each ruler sat on the Peacock Throne, the gilded, jewel-encrusted treasure brought to Iran by Nadir Shah, conqueror of northern India and founder of the short-lived Iranian Afshar dynasty. They assumed other grandiose titles, such as "Shadow of God on Earth" and "Asylum of the Universe." A shah once told an English visitor, "Your King, then, appears to be no more than a first magistrate. I, on the other hand, can elevate or degrade all the high nobles and officers you see around me!"[3]

Qajar pomp and grandeur were more illusion than reality, as was shown in a recent exhibit of Qajar art. Thus portraits of Fath Ali Shah, the second Qajar ruler, show him receiving the bows of European envoys, in poses meant to duplicate those of the great Sassanid rulers of the pre-Islamic Iranian past. Unfortunately, Iran's

grandeur had passed. Its strategic location between British-ruled India and the Russian Empire that extended across Central Asia guaranteed that it would become a pawn in the contest for control, usually referred to as the "Great Game." Under these difficult circumstances, Qajar rulers survived mainly by manipulating tribal leaders and other groups against one another with the tacit support of the mullahs (religious leaders).

Nasr al-Din Shah, Iran's ruler for most of the nineteenth century, was responsible for a large number of concessions to European bankers, promoters, and private companies. His purpose was to demonstrate to European powers that Iran was becoming a modern state and to find new revenues without having to levy new taxes, which would have aroused more dangerous opposition. The various concessions helped to modernize Iran, but they bankrupted the treasury in the process. The shah realized that the establishment of a trained professional army would not only defend Iran's territory but would also demonstrate to the European powers that the country was indeed "modern." To that end he reached agreement with the Russian Czar to send officers from the Cossacks, a traditionally warlike tribe in Russia, to train the core of such an army. This unit subsequently became known as the Cossack Brigade, Iran's only elite military force.

In the mid-nineteenth century, the shah was encouraged by European envoys to turn his attention to education as a means of creating a modern society. In 1851, he opened the Polytechnic College, with European instructors, to teach military science and technical subjects. The graduates of this college, along with other Iranians who had been sent to Europe for their education, and a few members of aristocratic families, became the nucleus of a small but influential intellectual elite. Along with their training in military subjects, they acquired European ideas of nationalism and progress. They were "government men" in the sense that they worked for and belonged to the shah's government.

But they also came to believe that the Iranian people needed to unite into a nation, with representative government and a European-style educational system, in order to become a part of the modern world. The views of these intellectuals put them at odds with the shah, who cared nothing for representative government or civil rights, only for tax collection. The intellectuals also found themselves at odds with the mullahs, who controlled the educational system and feared any interference with their superstitious, illiterate subjects.

The intellectuals and mullahs both felt that the shah was giving away Iran's assets and resources to foreigners. For a long time the intellectuals were the only group to complain; the illiterate Iranian masses could not be expected to protest against actions they knew nothing about. But in 1890, the shah gave a 50-year concession to a Briton named Talbot for a monopoly over the export and distribution of tobacco. Faced with higher prices for the tobacco they grew themselves, Iranians staged a general strike and boycott, and the shah was forced to cancel the concession. A similar pattern of protests that evolved into mass opposition to arbitrary rule marked the Constitutional Revolution of 1905, the 1979 revolution that overthrew Shah Mohammed Reza Pahlavi, and most recently the present contest between "reformers" advocating a more liberal Islamic government and the Khomeini regime.

By the end of the nineteenth century, the people were roused to action, the mullahs had turned against the ruler, and the intellectuals were demanding a constitution that would limit his powers. One of the intellectuals wrote, "It is self-evident that in the future no nation—Islamic or non-Islamic—will continue to exist without constitutional law.... The various ethnic groups that live in Iran will not become one people until the law upholds their right to freedom of expression and the opportunity for [modern] education."[4] One century and two revolutions later, Iran is still struggling to put this formula into operation.

According to Roy Mottahedeh, "the bazaar and the mosque are the two lungs of public life in Iran."[5] The bazaar, like the Greek agora and the Roman forum, is the place where things are bought, deals are consummated, and political issues are aired for public consideration or protest. The mosque is the bastion of religious opinion; its preachers can, and do, mobilize the faithful to action through thundering denunciations of rulers and government officials. Mosque and bazaar came together in 1905 to bring about the first Iranian Revolution, a forerunner, at least in pattern, of the 1979 revolt. Two sugar merchants were bastinadoed (a punishment, still used in Iran, in which the soles of the feet are beaten with a cane) because they refused to lower their prices; they complained that high import prices set by the government gave them no choice. The bazaar then closed down in protest. With commercial activity at a standstill, the shah agreed to establish a "house of justice" and to promulgate a constitution. But six months later, he still had done nothing. Then a mullah was arrested and killed for criticizing the ruler in a Friday sermon. Further protests were met with mass arrests and then gunfire; "a river of blood now divided the court from the country."[6]

In 1906, nearly all of the religious leaders left Teheran for the sanctuary of Qum, Iran's principal theological-studies center. The bazaar closed down again, a general strike paralyzed the country, and thousands of Iranians took refuge in the British Embassy in Teheran. With the city paralyzed, the shah gave in. He granted a Constitution that provided for an elected Majlis, the first limitation on royal power in Iran in its history. Four more shahs would occupy the throne, two of them as absolute rulers, but the 1906 Constitution and the elected Legislature survived as brakes on absolutism until the 1979 Revolution. In this sense, the Islamic Republic is the legitimate heir to the constitutional movement.

The Pahlavi Dynasty

Iran was in chaos at the end of World War I. British and Russian troops partitioned the country, and after the collapse of Russian power due to the Bolshevik Revolution, the British dictated a treaty with the shah that would have made Iran a British protectorate. Azeris and Kurds talked openly of independence; and a Communist group, the Jangalis, organized a "Soviet Republic" of Gilan along the Caspian coast.

The only organized force in Iran at this time was the Cossack Brigade. Its commander was Reza Khan, a villager from an obscure family who had risen through the ranks on sheer ability. In 1921, he seized power in a bloodless coup, but he did not overthrow the shah. The shah appointed him prime minister and then left the country for a comfortable exile in Europe, never to return.

Turkey, Iran's neighbor, had just become a republic, and many Iranians felt that Iran should follow the same line. But the religious leaders wanted to keep the monarchy, fearing that a republican system would weaken their authority over the illiterate masses. The religious leaders convinced Prime Minister Reza that Iran was not ready for a republic. In 1925, Reza was crowned as shah, with an amendment to the Constitution that defined the monarchy as belonging to Reza Shah and his male descendants in succession. Since he had no family background to draw upon, Reza chose a new name for his dynasty: Pahlavi. It was a symbolic name, derived from an ancient province and language of the Persian Empire.

Reza Shah was one of the most powerful and effective monarchs in Iran's long history. He brought all ethnic groups under the control of the central government and established a well-equipped standing army

to enforce his decrees. He did not tamper with the Constitution; instead, he approved all candidates for the Majlis and outlawed political parties, so that the political system was entirely responsible to him alone.

Reza Shah's New Order

Reza Shah wanted to build a "new order" for Iranian society, and he wanted to build it in a hurry. He was a great admirer of Mustafa Kemal Ataturk, founder of the Turkish Republic. Like Ataturk, Reza Shah believed that the religious leaders were an obstacle to modernization, due to their control over the masses. He set out to break their power through a series of reforms. Lands held in religious trust were leased to the state, depriving the religious leaders of income. A new secular code of laws took away their control, since the secular code would replace Islamic law. Other decrees prohibited the wearing of veils by women and the fez, the traditional brimless Muslim hat, by men. When religious leaders objected, Reza Shah had them jailed; on one occasion, he went into a mosque, dragged the local mullah out in the street, and horse whipped him for criticizing the ruler during a Friday sermon.

In 1935, a huge crowd went to the shrine of Imam Reza, the eighth Shia Imam, in Meshad, to hear a parade of mullahs criticize the shah's ruthless reform policies. Reza Shah ringed the shrine with troops. When the crowd refused to disperse, they opened fire, killing a hundred people. It was the first and last demonstration organized by the mullahs during Reza Shah's reign. Only one religious leader, a young scholar named Ruhollah al-Musavi al-Khomeini, consistently dared to criticize the shah, and he was dismissed as being an impractical teacher.

Iran declared its neutrality during the early years of World War II. But Reza Shah was sympathetic to Germany; he had many memories of British interference in Iran. He allowed German technicians and advisers to remain in the country, and he refused to allow war supplies to be shipped across Iran to the Soviet Union. In 1941, British and Soviet armies simultaneously occupied Iran. Reza Shah abdicated in favor of his son, Crown Prince Mohammed, and was taken into exile on a British warship. He never saw his country again.

Mohammed Reza Pahlavi

When the new shah came to the throne, few suspected that he would rule longer than his father and hold even more power. Mohammed Reza Pahlavi was young (22) and inexperienced, and he found himself ruling a land occupied by British and Soviet troops and threatened by Soviet-sponsored separat-

ist movements in Azerbaijan and Kurdistan. Although these movements were put down, with U.S. help, a major challenge to the shah developed in 1951–1953.

A dispute over oil royalties between the government and the Anglo-Iranian Oil Company (AIOC) aroused intense national feeling in Iran. Mohammed Mossadegh, a long-time Majlis member and ardent nationalist, was asked by the shah in 1951 to serve as prime minister and to implement the oil-nationalization laws passed by the Majlis. The AIOC responded by closing down the industry, and all foreign technicians left the country. The Iranian economy was not affected at first, and Mossadegh's success in standing up to the company, which most Iranians considered an agent of foreign imperialism, won him enormous popularity.

Mossadegh served as prime minister from 1951 to 1953, a difficult time for Iran due to loss of oil revenues and internal political wrangling. Although his policies embroiled him in controversy, Mossadegh's theatrical style—public weeping when moved, fainting fits, a preference for conducting public business from his bed while dressed in pajamas, and a propensity during speeches to emphasize a point by ripping the arm from a chair—enhanced his appeal to the Iranian people. His radio "fireside chats" soon won him a mass following; he became more popular than the shy, diffident young shah, and for all practical purposes he ruled Iran.

By 1953, Iran's economy was in bad shape. With no oil revenues coming in, there was mass unemployment and high inflation. However, political factors rather than economic ones led to Mossadegh's overthrow. By then he had suspended the Majlis, and ruled by decree. As opposition to his decrees increased, Mossadegh and his Tudeh allies responded with arrests of key opponents. Newspapers were closed if they dared to criticize the government, in an eerie mirroring of the Iran of today. Many of the prime minister's supporters had broken with him due to what they considered his unwise leadership and dictatorial policies. As a result, his main support came from the Tudeh ("Masses"), the Soviet-aligned Iranian Communist Party. The Cold War confrontation between the United States and the Soviet Union was then at its height, and the Dwight Eisenhower administration feared that Mossadegh's policies would lead to a Communist takeover of Iran.

The shah's advisers convinced him to dismiss Mossadegh and appoint General Fazlollah Zahedi, minister of interior, in his place. When the decree was announced, Mossadegh's supporters, along with Tudeh members, took to the streets to protest the dismissal. The shah then fled the country, for the first time in his life. But after a confused series of demonstrations and counter-demonstrations, a popular uprising, which started in the poor sections of Iran where the ruler had his greatest support, overthrew Mossadegh. He was arrested as he tried to climb the back fence of his house. On August 19, the shah returned to his country in triumph. He then gradually gathered all authority in his hands and developed the vast internal security network that eliminated parliamentary opposition. Mossadegh was kept in prison for three years, all the while loudly protesting his innocence. He was then tried and sentenced to house arrest at his estate at Ahmedabad, west of Teheran. He died there in March 1967.[7]

By the 1960s, the shah felt that he was ready to lead Iran to greatness. In 1962, he announced the Shah–People Revolution, also known as the White Revolution. It had these basic points: land reform, public ownership of industries, nationalization of forests, voting rights for women, workers' profit sharing, and a literacy corps to implement compulsory education in rural areas. The plan drew immediate opposition from landowners and religious leaders. But only one spoke out forcefully against the shah: Ayatollah Ruhollah Khomeini, by then the most distinguished of Iran's religious scholars. "I have repeatedly pointed out that the government has evil intentions and is opposed to the ordinances of Islam," he said in a public sermon. His message was short and definite: The shah is selling out the country; the shah must go.

Khomeini continued to criticize the shah, and in June 1963, he was arrested. Demonstrations broke out in various cities. The shah sent the army into the streets, and again a river of blood divided ruler from country. Khomeini was released, re-arrested, and finally exiled to Iraq. For the next 15 years, he continued attacking the shah in sermons, pamphlets, and broadsides smuggled into Iran through the "bazaar network" of merchants and village religious leaders. Some had more effect than others. In 1971, when the shah planned an elaborate coronation at the ancient Persian capital of Persepolis to celebrate 2,500 years of monarchy, Khomeini declared, "Islam is fundamentally opposed to the whole notion of monarchy. The title of King of Kings ... is the most hated of all titles in the sight of God.... Are the people of Iran to have a festival for those whose behavior has been a scandal throughout history and who are a cause of crime and oppression ... in the present age?"[8]

Yet until 1978, the possibility of revolution in Iran seemed to be remote. The shah controlled all the instruments of power. His secret service, SAVAK, had informers everywhere. The mere usage of a word such as "oppressive" to describe the weather was enough to get a person arrested. Whole families disappeared into the shah's jails and were never heard from again.

The public face of the regime, however, seemed to indicate that Iran was on its way to wealth, prosperity, and international importance. The shah announced a 400 percent increase in the price of Iranian oil in 1973 and declared that the country would soon become a "Great Civilization." Money poured into Iran, billions of dollars more each year. The army was modernized with the most sophisticated U.S. equipment available. A new class of people, the "petro-bourgeoisie," became rich at the expense of other classes. Instead of the concessions given to foreign business firms by penniless Qajar shahs, the twentieth-century shah became the dispenser of opportunities to business people and bankers to develop Iran's great civilization with Iranian money—an army of specialists imported from abroad.

FREEDOM

Iran's Constitution calls its political system a "religious democracy," leaving the term ambiguous. The Council of Constitutional guardians, an unelected body appointed by the Supreme legal Guide, controls the national election process through its approval (or rejection) or candidates for public office. Two other unelected bodies, the Expediency Council and The Assembly of Experts, serve as checks and balances on each other and to a limited degree on the CCG. Otherwise personal, press and other freedoms seesaw between hard and fast rules and more relaxed ones, as has been the case since the Revolution.

In 1976, the shah seemed at the pinnacle of his power. His major adversary, Khomeini, had been expelled from Iraq and was now far away in Paris. U.S. president Jimmy Carter visited Iran in 1977 and declared, "Under your leadership (the country) is an island of stability in one of the more troubled areas of the world." Yet just a month later, 30,000 demonstrators marched on the city of Qum, protesting an unsigned newspaper article (reputed to have been written by the shah) that had attacked Khomeini as being anti-Iranian. The police fired on the demonstration, and a massacre followed.

Gradually, a cycle of violence developed. It reflected the distinctive rhythm of Shia Islam, wherein a death in a family is followed by 40 days of mourning and every death represents a martyr for the faith. Massacre followed massacre in city after city. In spite of the shah's efforts to modernize his country, it seemed to more and more Iranians that he was trying to undermine the basic values of their society by striking at the religious leaders. Increasingly, marchers in the streets were heard to shout, "Death to the shah!"

Even though the shah held absolute power, he seemed less and less able or willing to use his power to crush the opposition. It was as if he were paralyzed. He wrote in his last book, "A sovereign may not save his throne by shedding his countrymen's blood.... A sovereign is not a dictator. He cannot break the alliance that exists between him and his people."[9] The shah vacillated as the opposition intensified. His regime was simply not capable of self-reform nor of accepting the logical consequences of liberalization, of free elections, a return to constitutional monarchy, and the emergence of legitimate dissent.[10]

THE ISLAMIC REPUBLIC

The shah and his family left Iran for good in January 1979. Ayatollah Ruhollah Khomeini returned from exile practically on his heels, welcomed by millions who had fought and bled for his return. The shah's Great Civilization lay in ruins. Like a transplant, it had been an attempt to impose a foreign model of life on the Iranian community, a surgical attachment that had been rejected.

In April 1979, Khomeini announced the establishment of the Islamic Republic of Iran. He called it the first true republic in Islam since the original community of believers was formed by Muhammad. Khomeini said that religious leaders would assume active leadership, serve in the Majlis, even fight Iran's battles as "warrior mullahs." A "Council of Guardians" was set up to interpret laws and ensure that they were in conformity with the sacred law of Islam.

Khomeini, as the first Supreme Guide, embodied the values and objectives of the republic. Because he saw himself in that role, he consistently sought to remain above factional politics yet to be accessible to all groups and render impartial decisions. But the demands of the war with Iraq, the country's international isolation, conflicts between radical Islamic fundamentalists and advocates of secularization, and other divisions forced the aging Ayatollah into a day-to-day policy-making role. It was a role that he was not well prepared for, given his limited experience beyond the confines of Islamic scholarship. (Quite possibly the war

United Nations Photo (UN109371)

Iran has a long Islamic tradition. This worshipper is praying at the Shah Mosque, which makes up one side of the magnificent Royal Square of Shah Abbas in Isfahan.

with Iraq, for example, could have been settled earlier if it were not for Khomeini's vision of a pure Shia Iran fighting a just war against the atheistic, secular regime of Saddam Hussein.)

A major responsibility of the Council of Guardians was to designate—with the Ayatollah's approval—a successor to Khomeini as Supreme Legal Guide. In 1985, the Council chose Ayatollah Hossein Ali Montazeri, a former student and close associate of the bearded patriarch. Montazeri, although politically inexperienced and lacking Khomeini's charisma, had directed the exportation of Iranian Islamic fundamentalist doctrine to other Islamic states after the Revolution, with some success. This responsibility had identified him abroad as the architect of Iranian-sponsored terrorist acts such as the taking of hostages in Lebanon. But during his brief tenure as Khomeini's designated successor, he helped make changes in prison administration, revamped court procedures to humanize the legal system and reduce prisoner mistreatment, and urged a greater role for opposition groups in political life. However, Montazeri resigned in March 1989 after publishing an open letter, which aroused Khomeini's ire, criticizing the mistakes made by Iranian leaders during the Revolution's first decade.

The Islamic Republic staggered from crisis to crisis in its initial years. Abol Hassan Bani-Sadr, a French-educated intellectual who had been Khomeini's right-hand man in Paris, was elected president in 1980 by 75 percent of the popular vote. But it was one of the few post revolutionary actions that united a majority of Iranians. Although the United States, as the shah's supporter and rescuer in his hour of exile, was proclaimed the "Great Satan" and thus helped to maintain Iran's revolutionary fervor, the prolonged crisis over the holding of American Embassy hostages by guards who would take orders from no one but Khomeini embarrassed Iran and damaged its credibility more than any gains made from tweaking the nose of a superpower. Although the hostages were held for over a year and so damaged the credibility of President Carter as to cost him the 1980 presidential election, one group of American Embassy diplomats managed to escape and take refuge in the Canadian Embassy in Teheran. Subsequently they escaped from the country disguised as members of a Canadian film company that had supposedly produced a film on Iran, thus being the only Americans to leave Iran in the year of the hostage crisis.

Historically, revolutions often seem to end by devouring those who carry them out. A great variety of Iranian social groups had united to overthrow the shah. They had different views of the future; an "Islamic republic" meant different things to different groups. The Revolution first devoured all those associated with the shah, in a reign of terror intended to compensate for 15 years of repression. Islamic tribunals executed thousands of people—political leaders, intellectuals, and military commanders.

The major opposition to Khomeini and his fellow religious leaders came from the radical group Mujahideen-i-Khalq. The group favored an Islamic socialist republic and was opposed to too much influence on government by religious leaders. However, the Majlis was dominated by the religious leaders, many of whom had no experience in government and knew little of politics beyond the village level. As the conflict between these groups sharpened, bombings and assassinations occurred almost daily.

The instability and apparently endless violence during 1980–1981 suggested to the outside world that the Khomeini government was on the point of collapse. Iraqi president Saddam Hussein thought so, and in September 1980, he ordered his army to invade Iran—a decision that proved to be a costly mistake. President Bani-Sadr was dismissed by Khomeini after an open split developed between him and religious leaders over the conduct of the war; he escaped to France subsequently and has remained out of politics. In his absence the Mujahideen continued their assaults against the regime, and in 1981 carried out a series of bomb attacks on government leaders. One in particular killed more than 70 of them, and the present Supreme Legal Guide, Ali Khamenei, lost an arm in another attack.

The Khomeini regime showed considerable resilience in dealing with its adversaries. In 1983, Mujahideen leaders were hunted down and killed or imprisoned. Their organization had been the vanguard in the Iranian Revolution, but their Marxist, atheist views caused them to be viewed by the clerical regime as an enemy. The Mujahdeen leader, Massoud Rajavi, escaped to France, where he was given asylum. While there he organized the National Council of Resistance as the Mujahideen's political and PR arm. However the French government expelled him in 1986, in a move to improve relations with Iran. Rajavi then found refuge in Iraq, where he was welcomed by Khomeini's old adversary Saddam Hussein. Saddam found him a useful ally, and provided the Mujahideen with money, weapons and bases along the border from whence to launch raids and intelligence-gathering activities into Iran.

Toward the end of the war the Mujahideen took advantage of Iraqi successes to seize several towns inside Iran, freeing political prisoners and executing minor officials, such as prison wardens, without trial. But the organization had little internal support in the country. Its Marxist views were not shared by the majority of people, and Mujahideen claims of 90,000 executions and more than 150,000 political prisoners held by the regime were believed to be wildly exaggerated.

The U.S. invasion and occupation of Iraq set up an awkward situation both for the Mujahideen and for U.S. forces. Although it continued to be labeled as a terrorist organization by the U.S. State Department, its opposition to the Iranian regime indicated to U.S. policy makers that it might serve as a useful ally in the event of a future confrontation. Rajavi went into hiding after the occupation, leaving the organization's leadership in the hands of his wife Maryam.

In 2003 then-President Khatami issued an amnesty offer to the Mujahidin in Iraq; they could return if they repented of their past political acts. Subsequently the Iraq Governing Council appointed by the U.S. ordered the organization to leave its territory. Some 250 militants out of 3,500 accepted the offer, returning to families who thought they would never see them again. The return was arranged by the Red Cross under guarantees of their safe treatment.

Otherwise the republic's government continues to treat the Mujahidin as an enemy. In 1990, Rajavi's brother was killed by unknown gunmen in Geneva, Switzerland. In August 1991, Shahpour Bakhtiar, the last prime minister under the monarchy, was murdered under similar circumstances in Paris, where he had been living in exile. Although the elderly Bakhtiar had been opposed to the shah as well as to Khomeini, fellow exile and former Iranian president Bani-Sadr charged that the regime had a "hit list" of opponents, including himself and the former prime minister, slated for execution.

The other main focus of opposition was the Tudeh (Masses) Party. Although considered a Communist party, its origins lay in the Iranian constitutional movement of 1905–1907, and it had always been more nationalistic than Soviet-oriented. The shah banned the Tudeh after an assassination attempt on him in 1949, but it revived during the Mossadegh period of 1951–1953. After the shah returned from exile in 1953, the Tudeh was again banned and went underground. Many of its leaders fled to the Soviet Union. After the 1979 Revolution, the Tudeh again came out into the open and collaborated with the Khomeini regime. It was tolerated by the religious leaders for its nationalism, which made its Marxism acceptable. Being militarily weak at that time, the regime also wished to remain on good terms with its Soviet neighbor. However, the rapprochement was brief. In 1984, top Tudeh leaders were arrested in a series of surprise raids and were given long prison terms.

INTERNAL POLITICS

What may be described as the "surreal world" of Iranian politics is largely the result of institutions grafted onto the structure of the pre-revolutionary government by clerical leaders. The clerical regime preserved that structure, which consisted of the Majlis, cabinet of ministers, civil service and the armed forces. SAVAK was replaced by a similar and equally repressive security and intelligence called SAVAM. Then a parallel structure of government was formed under the authority and leadership of the Supreme Legal Guide, responsible only to him as the final repository of Islamic law.

This structure consists of a 12-member Council of Constitutional Guardians (CCG), an 88-member Assembly of Experts elected for 8-year terms, and a 31-member appointed Expediency Council. These three bodies function within the system as a sort of checks and balances on each other. Thus the Assembly of Experts, based in Qum and popularly elected, chooses the Supreme Legal Guide when there is a vacancy. All Assembly members must be clerics and after being elected must be "approved" by the CCG before taking their seats. The die is further loaded by the fact that half of the members of the CCG must also be clerical leaders.

The Expediency Council, currently headed by former President Ali Akbar Hashimi Rafsanjani, was set up to resolve disputes over legislation between the Majlis and the CCG and to verify whether laws passed by the Majlis are compatible or not with Islamic law. The Supreme Legal Guide also controls the judiciary, the Radio-TV Ministry, the chiefs of the armed forces, the Revolutionary Guard and SAVAM and the

police. In the light of the present conflict over reform, pitting advocates of more openness in the system against those determined at all cost to preserve the Islamic republic as it stands, this parallel structure represents a formidable obstacle.

The Revolution that overturned one of the most ruthless authoritarian regimes in history has been in effect long enough to provide some clues to its future direction. One clue is the continuity of internal politics. Despite wreaking savage vengeance on persons associated with the shah's regime, Khomeini and his fellow mullahs preserved most of the Pahlavi institutions of government. The Majlis, civil service, secret police, and armed forces were continued as before, with minor modification to conform to strict Islamic practice. The main addition was a parallel structure of revolutionary courts, paramilitary Revolutionary Guards (Pasdaran), workers' and peasants' councils, plus the Council of Guardians as the watchdog over legislation.

An important change between the monarchy and the republic concerns the matter of appropriate dress. Decrees issued by Khomeini required women to wear the enveloping chador and *hijab* (headscarf) in public. Painted nails or too much hair showing would often lead to arrests or fines, sometimes jail. The decrees were enforced by Revolutionary Guards and *komitehs* ("morals squads") patrolling city streets and urban neighborhoods. Also, the robe and turban worn by Ayatollah Khomeini and his fellow clerics were decreed as correct fashion, preferred over the "Mr. Engineer" business suit and tie of the shah's time. The necktie in particular was considered a symbol of Western decadence and derided as a "donkey's tail" by the country's new leaders. In subsequent years, the dress code and other restrictions on behavior have seesawed between extremes. In Khatami's first term, most restrictions were lifted; the necktie even staged a comeback among professionals—doctors, lawyers, and others. Then, in 2001, following his reelection, the so-called hard-line conservatives who control the judiciary issued new regulations, banning neckties, wearing of heavy makeup in public places such as restaurants, and other forms of "un-Islamic" social behavior. Shopping malls were ordered not to play music or display women's underwear in their shop windows; shopkeepers and restaurateurs were warned of "heavy consequences" for violations.

The 1985 presidential election in Iran continued the secular trend. The ruling Islamic Republican Party (since dissolved) nominated President Ali Khamenei for a second term, against token opposition.

However, Mehdi Bazargan, the republic's first prime minister, who subsequently went into opposition and founded the Freedom Movement, announced that he would be a candidate. The Council of Guardians vetoed his candidacy on the grounds that his opposition to the war with Iraq, although well publicized, would be damaging to national solidarity if he ran for president. Khamenei won reelection handily, but nearly 2 million votes were cast for one of the two opposition candidates, a religious leader. The relatively high number of votes for this candidate reflected increasing dissatisfaction with the Khomeini regime's no-quarter policy toward Iraq, rather than opposition to the regime itself.

Relations with other countries have varied between cooperation and hostility, perhaps the only constant being Iran's determination to play an important regional role. Iranian negotiators helped privately to arrange the release of Western hostages in Lebanon in 1990–91. In 2005 the government reopened the case of an Iranian-Canadian journalist killed while in custody in 2003. Her death had seriously affected relations with Canada.

The death sentence issued in 1989 by Khomeini on the Anglo-Pakistani author Salman Rushdie for his novel *The Satanic Verses* further estranged Iran from Western countries. Khomeini called the book sacrilegious because it allegedly defamed the Prophet Muhammad. The *fatwa* (religious edict) imposing sentence was lifted in 1998 and Britain then re-established diplomatic relations.

The regime has had difficulties in ensuring popular support and making use of its majority in the fractious Majlis to carry out necessary economic reforms. In the 1992 Majlis elections, two political groups presented candidates: the Society of Radical Clergy (*Ruhaniyat,* loosely but incorrectly translated as "moderates") and the Combatant Religious Leaders (*Ruhaniyoun,* "hard-liners"). All candidates had to be approved by the Council of Constitutional Guardians (CCG), a 12-member body of senior religious scholars, to ensure that their views were compatible with Islam.

Recent elections for the Majlis reflect the surreal political world in which Iranians live and vote. In the 1992 elections two groups presented candidates vetted by the CCG, the Society of Radical Clergy and the Combatant Religious Leaders. An outsider would have been hard-pressed to distinguish between them. In the 1996 elections, the distinction between Ruhaniyat and Ruhaniyoun became even more blurred, reflecting the arcane nature of Iranian politics. The former, renamed the Conservative Combatant Clergy Society, was now opposed by the Servants of Iran's Construction, a coalition of centrist supporters of then–president Rafsanjani. The Freedom Movement, now headed by former prime minister Ibrahim Yazdi, was banned by the CCG. In October 2001, 30 of its members, excluding Yazdi, were arrested and held for trial on charges of "diverting" the Islamic Revolution from its true course.

In the 2000 Majlis elections, held in three stages, new political groupings emerged, corresponding to the reformist and hard-line elements in the population. The Islamic Iran Participation Front, a coalition of disparate groups united mainly in their goal of political reform, won 200 of the seats in the Legislature, to 55 for the Executives of Construction, composed of clerical leaders and government officials opposed to reform.

IRAN AFTER KHOMEINI

In June 1989, Ayatollah Ruhollah Khomeini died of a heart attack in a Teheran hospital. He was 86 years old and had struggled all his life against the authoritarianism of two shahs.

The Imam left behind a society entirely reshaped by his uncompromising Islamic ideals and principles. Every aspect of social life in republican Iran is governed by these principles, from prohibition of the production and use of alcohol and drugs to a strict dress code for women outside the home, compulsory school prayers, emphasis on theological studies in education, and required fasting during Ramadan. One positive result of this Islamization program has been a renewed awareness among Iranians of their cultural identity and pride in their heritage.

Khomeini also bequeathed many problems to his country. The most immediate problem concerned the succession to him as Supreme Legal Guide.

A separate body of senior religious leaders and jurists, the Assembly of Experts, resolved the succession question by electing President Khamenei as Supreme Guide. However, the choice emphasized Khomeini's unique status as both political and spiritual leader. As a *Hojatulislam* (a lower-ranking cleric), Khamenei lacked the credentials to replace the Ayatollah. But he was an appropriate choice, having served as a part of the governing team. Also, he was the most available religious leader. He had completed two terms as president and was ineligible for reelection.

THE PRESIDENCY

The chief executive's powers in the Iranian system were greatly strengthened by a constitutional amendment abolishing the office of prime minister, approved by voters in a July 1989 referendum. But concern for a smooth transfer of power after Khomeini's death prompted the government to advance the date for electing a new president to succeed Ali Khamenei from October to July. There were only two candidates: Majlis speaker Ali Akbar Hashemi Rafsanjani, Khomeini's right-hand man almost from the start of the Revolution; and Agriculture Minister Abbas Sheibani, a political unknown. As anticipated, Rafsanjani won handily, with 95 percent of the 14.1 million votes cast.

Rafsanjani's first cabinet consisted mostly of technocrats, suggesting some relaxation of the policy of exporting the Revolution and supporting revolutionary Islamic groups outside the country. At home, there was also a slight relaxation of the strict enforcement of Islamic codes of behavior enforced by the morals squads and security police. But the easing was temporary. In 1992, the new Supreme Guide issued an edict ordering these codes enforced. A similar edict was issued for arts and culture, and the minister of culture and Islamic guidance was forced to resign after clerical leaders charged him with being "too permissive" in allowing concerts and films of a "non-Islamic" nature to be presented to the public. (His name: Mohammed Khatami.)

Rafsanjani was reelected in 1993, but by a much smaller margin of the electorate, 63 percent. The decline in his popularity was due mainly to his failure to improve the economy, but also for continuation of the strict Islamic dress and behavior codes. When his term ended, he kept his seat in the Majlis. In the February 2000 Majlis elections, he finished 29th in the contest for Teheran's 30 seats, barely avoiding a runoff. However, Ayatollah Khamenei appointed him chairman of the powerful Expediency Council, which serves as the final advisory body to the Supreme Legal Guide in political and legislative matters. In this capacity, he continued to play an important role in Iran's evolving political process.

FOREIGN POLICY

The end of the war with Iraq left a number of issues unresolved. A prisoner exchange was agreed to in principle, but it has yet to be completed. Under the terms of an agreement negotiated by the International Red Cross, Iran released 1,999 Iraqi prisoners in April 2000, and an additional 930 in May, for a total of 2,929. Many of them had been held since the end of the war in 1988, and others since 1983. As of December 2002 an estimated 1,000 Iranian POWS remained in Iraq, although they have probably been released since the fall of Saddam Hussein. On its part Iran has refused to return 100 Iraqi warplanes flown there and impounded since the 1991 Gulf War.

A formal peace treaty ending the war and among other issues defining navigation rights in the Shatt al-Arab has yet to be signed. However relations between the two neighbor countries have improved significantly since the overthrow of Saddam Hussein. His regime had been the principal backer of the Mujahidin-I-Khalq.

The unwillingness of European countries to go along with secondary sanctions on Iran, due to its economic and trade importance to them, brought an improvement in Iranian–European relations in the mid-1990s. In July 1996, Total of France was given a concession to develop newly discovered oil fields near Sirri Island in the Gulf. The concession had been granted previously to the U.S. Conoco Oil Company, but Conoco withdrew under pressure from the Clinton administration to comply with sanctions regulations.

The conviction of four Iranians in a German court in 1997 for the 1992 assassinations of Iranian Kurd opposition leaders in Berlin temporarily halted the increase in European–Iranian contacts. More than 100,000 people marched in Teheran to protest the decision, and the European Union suspended the "dialogue" with Iran on ways to sanctions. But the Iranian market remains too attractive to Europe in terms of investment. Between 1998 and 2001, the Khatami regime negotiated $13 billion in foreign investment, most of it from European firms.

The long and often hostile relationship with Russia, dating back to the years of gradual acquisition of Iranian territory by advancing Russian armies and settlers in the 1800s, entered a new phase in the new millennium with a joint agreement for training of Iranian officers at Russian bases. Russia also agreed to complete the reparation of Iran's German-built nuclear reactor at Bushehr, badly damaged in the war with Iraq. Also, over vehement U.S. objections, Russia sold Kilo-class submarines and other weapons to Iran, including parts for a new ballistic missile.

Iran's relations with the United States remained glacially frozen, at least on the official level. For more than two decades, the only official connections had been through the Swiss Embassy in Teheran and the Pakistan Embassy in Washington, D.C. However, an indirect channel through the U.S.–Iran Claims Tribunal set up in the Hague has been quite successful. It was set up to adjudicate claims by U.S. firms for work contracted and equipment delivered to the shah's regime but not paid for due to the Revolution. The tribunal is also responsible for examining Iranian claims against U.S. companies. To date it has awarded $6 billion to U.S. claimants and $4 billion to Iranian claimants. In 2003, in a separate court action, a U.S. district judge charged Iran, as sponsor of the Lebanese Hezbollah organization, with responsibility for the 1983 truck bombing that killed 241 Marines in Beirut. Their families would be compensated from Iranian assets frozen in the U.S. once Congress passes the necessary legislation.

Although Iranians continue to view the American people favorably, the clerical regime and its hard-line supporters insist that the U.S. government is the cause of their failure to establish a sound economy, and an obstacle to improved relations with the outside world. In November 2001, the 22nd anniversary of the occupation of the U.S. Embassy in Teheran and the holding of 52 Americans as hostages for 444 days, the embassy building was reopened as a museum, with an exhibit of alleged "crimes against the Iranian people" committed by America. One exhibit showed photos of the U.S. helicopters that had crashed in the desert on a failed rescue mission.

ACHIEVEMENTS

Despite periodic restrictions on dress and public behavior imposed by the regime and enforced by *Basiji* (Revolutionary Guards), women in Iran participate more actively in national life than their sisters in other Islamic countries. They drive their own cars, work outside the home, and are represented in the Majlis. A 2003 law gives them the right to initiate divorce. Iranian women teams compete in international sports events, and in 2005 a female mountaineering group scaled Mount Everest.

Prior to the election of a new Iranian president in 1997, the U.S.–Iranian relationship became truly ice-bound. The Clinton administration imposed a trade embargo in 1995, calling the country a major sponsor of international terrorism. Congress then passed a bill penalizing companies that invest $40 million or more in Iran's oil and gas industry. The ban would apply equally to U.S. and foreign companies.

As might have been expected, the bill aroused a storm of protest in Europe. It was viewed as unwarranted interference in the internal affairs of European countries, and the European Union denounced it as a violation of the principles of international free trade. It was patently unenforceable outside U.S. borders, and in May 1998, the administration approved a waiver to allow the consortium formed to exploit the former Conoco concession. With U.S companies excluded from the Iranian market, Russia has become the country's main source of weaponry as well as

technical aid for its nuclear power program. Russian experts began building the first nuclear power plant, at Bushire, in 1995. When it is fully operational, scheduled for 2005, it will meet 20 percent of national electricity needs. A law issued by the Guardian Council in 2005 requires the government to develop nuclear technology, including uranium enrichment, despite warnings by the U.S. and other Western countries that the action would not be in compliance with IAEA strictures.

President Khatami's reelection prompted a reassessment of U.S. policy toward Iran. Former secretary of state Madeleine Albright acknowledged some "mistakes of the past" shortly after the February 2000 Majlis elections. They included the overthrow of Mossadegh, U.S. support for Iraq in the war with Iran, and alignment with the shah's regime despite its brutal suppression of dissent.

There remains a considerable reservoir of good will among ordinary Iranians toward the U.S. After the southeastern city of Bam, with its ancient citadel, was leveled by a massive earthquake with some 45,000 casualties, the U.S. sent a medical disaster team along with volunteers from many humanitarian agencies (including a team from Alabama who brought Maxwell House coffee to a land of tea-drinkers and cooked meals daily for the survivors!) The door between the two countries opened a crack wider when Iran agreed to participate in an international donors' conference devoted to Iraq's reconstruction, agreeing to provide potable water and electricity.

Iran's foreign policy in recent years has been essentially regional. In 2001, it signed an agreement with Saudi Arabia for joint efforts to combat terrorism and drug trafficking. Iran's relations with the newly independent Islamic republics of Central Asia expanded rapidly in the late 1990s, both in diplomatic relations and exchanges of trade. However, a dispute with Azerbaijan over oil exploration rights in the Caspian Sea—which holds an estimated 200 billion barrels of offshore oil and 600 billion cubic meters of natural gas—came to a head in July 2001, when an Iranian warship halted exploration in the offshore Alov field by Azerbaijani research vessels.

The dispute over this field reflects the larger issue of Caspian Sea boundaries. After the collapse of the Soviet Union in 1991, the five countries bordering the Caspian—Iran, Azerbaijan, Kazakhstan, Russia, and Turkmenistan—signed an agreement that gave Iran 12 percent of shoreline but left open the issue of oil and gas development. With exploration under way, Iran is demanding control of 20 percent, including the Alov field, which lies 60 miles north of its current territorial waters.

In a related dispute, Iran has objected strenuously to the U.S.–sponsored plan for a pipeline to carry Azerbaijani oil from Baku to the Turkish port of Ceyhan (Adana) on the Mediterranean, thus bypassing Iranian territory. For its part, the Azerbaijan government accused Iran of funding new mosques and paying their prayer leaders to support pro-Iranian subversive activities.

After the September 11, 2001, terrorist attacks in the United States, the government condemned the action and the killing of civilians. In a mosque sermon, Expediency Council president Rafsanjani stated that "despite all our differences we are willing to join the U.S.–international coalition against terrorism under the umbrella of the United Nations, if America does not impose its own view." Iran closed its border with Afghanistan to Afghan refugees, although it allowed private relief organizations to continue making food deliveries to that war-ravaged country.

An important element in Iran's willingness to cooperate with the United States against the Afghanistan-based al-Qaeda network and its leader, Osama bin Laden, stemmed from drug smuggling. The Taliban regime that controlled 90 percent of that country until its overthrow in late 2001 had encouraged drug cultivation as a means of income and turned a blind eye to smuggling, much of it through Iran. Since 1996, some 3,000 Iranian border guards have been killed by smugglers, and easy access to drugs has resulted in a large number of Iranian addicts.

But any further thawing of U.S.-Iranian relations was quickly refrozen when Bush included Iran in his "axis of evil" covering countries supposedly sponsoring terrorism. The charge was made more explicit in August 2003 with revelations that the country was "well advanced" toward a nuclear weapons capability. Iranian officials insist that extraction of newly-discovered uranium resources near the town of Natanz, along with the heavy water plant at Arak, both serve its intention to develop peaceful uses for nuclear energy.

Disclosure of Iran's secret nuclear enrichment program with its implications of acquisition of nuclear weapons, a program which has been going on for 20 years, led to a major confrontation with European countries and by implication the U.S. in 2004–2006. As a signatory to the Nuclear Non-Proliferation Treaty (NPT) and a member of the International Atomic Energy Agency (IAEA) of the United Nations, the country is required to disclose its program and to invite IAEA inspectors to examine its facilities. In late 2004 Iran suspended its enrichment program to comply with these requirements, and inspectors proceeded to examine the Natanz enrichment plant. However the country insisted it had the right to develop nuclear energy for peaceful uses on its own soil. This insistence was a major obstacle to a resolution of the negotiations. Several compromises were floated including one that would have Russia produce the necessary centrifuges for delivery to Iran, but as of this writing there has been no settlement. Several members of the UN security Council including the U.S. have threatened to bring the issue before the Council and impose sanctions on Iran, but thus far China and Russia, both of them heavy importers of Iranian oil, have indicated they would veto the proposal. In January 2006 Iran resumed production at the Natanz plant and announced that it would open its own stock exchange (bourse) to sell its oil in Euros rather than dollars, the traditional medium of currency for oil sales worldwide.

The nuclear issue ironically united all Iranians in a new-found pride in their national identity. "We will definitely not stop our nuclear activities; it is our red line"— so read the lettering on a banner carried in a recent demonstration in Tehran.[11] This new unity also has made it more problematic for the U.S. to seek regime change as a prelude to normalization of relations. In February 2006 Secretary of State Condoleezza Rice requested funding from Congress of $75 million to support "pro-democracy forces" within the country. But as observers pointed out, such forces only exist outside Iran and consist mainly of discredited groups such as the Iraq-based Peoples' Majahideen.

AN ELECTION SURPRISE

In May 1997, Iranian voters went to the polls to elect a new president. Four candidates had been cleared and approved by the CCG: a prominent judge; a former intelligence-agency director; Majlis speaker Ali Akbar Nalegh-Nouri; and Mohammed Khatami, the former minister of culture and Islamic guidance, a more or less last-minute candidate, since he had been out of office for five years and was not well known to the public. Speaker Nalegh-Nouri was the choice of the religious leaders and the Majlis and was expected to win easily.

In Iranian politics, though, the devil is often in the details; the unexpected may be the rule. With 25 million out of Iran's 33 million potential voters casting their ballots, Khatami emerged as the winner in a startling upset, with 69 percent of the votes as compared to 25 percent for Nalegh-Nouri. Support for the new president came mainly from women, but he was also backed by the large number of Iranians un-

der age 25, who grew up under the republic but are deeply dissatisfied with economic hardships and Islamic restrictions on their personal freedom.

Khatami took office in August. Despite some opposition in the Majlis, all 22 of the ministers in his cabinet were approved, although his first nominee for minister of interior was later impeached by a 137–128 vote. (Khatami then assigned him to a sub-cabinet post.)

During the short-lived openness of Iran's 1997 version of "Prague Spring," civic and press freedom flourished, with many new newspapers and magazines appearing on the streets. *Society,* a new magazine, even dared to publish articles mildly critical of the regime and photographs of unveiled women on its front pages. Another harbinger of change was the impressive victory of Khatami supporters in the local and municipal elections, the first since the 1979 Revolution. They won the great majority of the 200,000 seats on town and city councils, including 15 seats on the influential Teheran City Council. The 2003 municipal elections brought similar results, with reform candidates garnering the majority of seats.

As noted earlier, the president's powers are limited. His limitations were underscored in 1998 by the murders of prominent Iranian writers and dissidents. A public outcry resulted in the uncovering of a "rogue squad" in the intelligence service responsible for these and other murders of prominent Iranians over the years—intellectual, political, and other leaders. Members of the squad were tried and convicted in 1998 by a military court, but the Iranian Supreme Court overturned the convictions on the grounds of investigative irregularities. In August 2001, a new trial of 15 "rogue agents" began in Teheran. By that time, the highest-ranking official involved, Deputy Minister Saeed Emani, had died in custody. Supporters of President Khatami had charged a cover-up in order to bring about a new trial, claiming that death squads had attempted to silence opponents of the clerical regime for more than a decade.

The contest between "hard-liners" holding fast to the Islamic structure as laid out by Khomeini and implemented by his successor has intensified in recent years. In February 1998, the popular mayor of Teheran, Gholarn Hossein Karbaschi, was arrested and charged with corruption in office. He had won popular favor with his beautification and cleanup program for the capital and had been active in Khatami's election campaign. During his trial, the embattled mayor described it as "politically motivated on the part of enemies of reform and openness." Nonetheless, he received a

two-year sentence; however, in January 2000, Khamenei pardoned him in another of the unexpected twists that give Iranian politics its unique flavor.

THE ECONOMY

Iran's bright economic prospects during the 1970s were largely dampened by the 1979 Revolution. Petroleum output was sharply reduced, and the war with Iraq crippled industry as well as oil exports. Ayatollah Khomeini warned Iranians to prepare for a decade of grim austerity before economic recovery would be sufficient to meet domestic needs. After the cease-fire with Iraq, Khomeini enlarged upon his warning, saying that the world would be watching to see if the Revolution would be destroyed by postwar economic difficulties.

Iran's remarkable turnaround since the end of the war with Iraq, despite the U.S.–imposed trade restrictions, suggests that the late Ayatollah Khomeini was a better theologian than economist. The country's foreign debts were paid off by 1990. Since then, however, loans for new development projects and purchase of equipment, including a nuclear reactor for peaceful uses set up by Russian technicians in 1998, along with reduced oil revenues, have generated foreign debts of $30 billion.

Iran was self-sufficient in food until 1970. The White Revolution redistributed a considerable amount of land, most of it from estates that Reza Shah had confiscated from their previous owners. But the new owners, most of them former tenant farmers, lacked the capital, equipment, and technical knowledge needed for productive agriculture. The revolutionary period caused another upheaval in agriculture, as farmers abandoned their lands to take part in the struggle, and fighting between government forces and ethnic groups disrupted production. Production dropped 3.5 percent in 1979–1980, the first full year of the Islamic Republic of Iran, and continued to drop at the same rate through 1982. The war with Iraq caused another upheaval. Rural youths, traditionally Khomeini's strongest supporters, flocked to join the army, most of them as *basijis,* "volunteers," advancing ahead of regular troops in human wave tactics that wore down the Iraqis but caused enormous Iranian casualties, and as a consequence decimating the able-bodied rural population.[12]

These difficulties, plus large-scale rural–urban migration, have hampered development of the agricultural sector, which accounts for 20 percent of gross domestic product. Formerly self-sufficient in food, Iran is now the world's biggest wheat importer. In the non-oil sector, Iran is the world's major producer

United Nations Photo (UN16049_11.1986)
Today Iran is still in a state of internal unrest. The country's future will be in the hands of the young and how well they are educated.

of pistachio nuts. (Former president Rafsanjani is himself a pistachio farmer in his home province of Rafsanjan, the center of production.) Caviar exports from Iran's Caspian Sea waters, formerly significant, have been reduced due to too much fishing and pollution from on-shore development. The country is also the number-one world producer of saffron; unfortunately, much of the crop is stolen by poachers and smuggled into Europe, where it sells for up to $272 per pound!

Petroleum is Iran's major resource and the key to economic development. Oil was discovered there in 1908, making the Iranian oil industry the oldest in the Middle East. Until 1951, the Anglo-Iranian Oil Company produced, refined, and distributed all Iranian oil. After the 1951–1953 nationalization period, when the industry was closed down, a consortium of foreign oil companies—British, French, and American—replaced the AIOC. In 1973, the industry was again nationalized and was operated by the state-run National Iranian Oil Company.

After the Revolution, political difficulties affected oil production, as the United States and its allies boycotted Iran due to the hostage crisis. Other customers balked at the high prices ($37.50 per barrel in 1980, as compared to $17.00 per barrel a year earlier). The war with Iraq was a further blow to the industry. Japan, Iran's biggest customer, stopped purchases entirely in 1981–1982. War damage to the important Kharg Island terminal reduced Iran's export capacity by a third, and the Abadan refinery was severely crippled. Periodic Iraqi raids on other Iranian oil terminals in more distant places such as Lavan and Qeshm, reachable by longer-range aircraft, seriously decreased Iran's export output.

The instability of world oil prices, along with domestic energy subsidies (gasoline prices in Iran are about 10 percent lower than world prices) has led the country to boost natural gas production from huge reserves. However, difficulties of access to foreign markets and the hard bargaining involved for foreign companies have kept production low. Iran presently produces a bare 300 cubic meters per day, and would need to produce 700 cm^3 to increase exports to an acceptable level.

In addition to its oil and gas reserves, Iran has important bauxite deposits, and in 1994, it reported the discovery of 400 million tons of phosphate rock to add to its mineral resources. It is now the world's sixth-largest exporter of sulfur. However, oil and gas remain the mainstays of the economy. Oil reserves are 88 billion barrels; with new gas discoveries each year, the country sits astride 70 percent of the world's reserves.

Iran's great natural resources, large population, and strong sense of its international importance have fueled its drive to become a major industrial power. The country is self-sufficient in cement, steel, petrochemicals, and hydrocarbons (as well as sugar—Iranians are heavy users). Production of electricity meets domestic needs. The nuclear power plant at Bushire, begun by German engineers, is now in the last stages of completion after numerous construction delays, with the work being done by some 3,700 Russian technicians, and should be operative in late 2006. It could meet 20 percent of energy needs.

Since the breakup of the Soviet Union into independent states, Iran has been active in developing trade and economic links with the mainly Islamic republics of Central Asia. A rail link from the port of Bandar Abbas to Turkmenistan was completed in 1996, giving that landlocked country access to the outside world by a watery "Silk Road" to India and the Far East.

Aside from the hydrocarbons industry much of Iran's economy is controlled by the state, either directly or through the network of *bazaaris* (urban merchants and business owners) and semi-public foundations called *bonyads*. The support of the bazaaris was essential to the Revolution, while the bonyads were set up after 1979 ostensibly to administer properties confiscated by the republic from the monarchy. They are accountable only to the Supreme Legal Guide and control the bulk of the non-oil economy.

To their credit, the reformers backing the Khatami government have set out to reform the economy. In June 2001, the Majlis approved a 26-article law to stimulate economic growth. Among other elements, it would allow imports of modern machinery to replace Iran's aging industrial equipment. But the Guardian Council, which has final authority over legislation, rejected the law, arguing that it would discriminate against domestic investment and pave the way for foreign control of the economy. It was approved and resubmitted by the Majlis, but in December 2001 the Council ruled that it was not consistent with Islamic law. Constitutional restrictions on foreign ownership of oil or other concessions and profit-sharing agreements also militate against government-planned reforms. A new foreign investment law was passed by the Majlis in 2002 and approved by the CCG, but heavy taxes, extensive bargaining over agreements and the high costs of doing business continue to deter foreign companies from investing in Iran's potentially huge market.

A REVOLUTION FROM WITHIN?

Two decades after the Revolution that brought the first Islamic republic into existence, a debate is still under way to determine how "Islamic" Iranian society should be. The debate is between those Iranians who advocate strict adherence to Islamic law and those who would open up the society to diverse social behavior and norms. As noted earlier in this report, Iran's new

president was the choice of women and young people, two groups who feel they have benefited little from the Revolution. As their candidate, he may expect to come under increasing pressure from this new and largely disadvantaged constituency. As one Iranologist noted, "Having united religion and politics, the regime now has to face antagonisms directed at the clerics for failing to deliver on lofty promises. Pressure from Islamic radicals to push for further purification of social and political practices has alienated important elements in society."[12]

On June 8, 2001, Mohammed Khatami was reelected handily for a second term as Iran's president, garnering 77 percent of the 28.2 million votes cast. His closest opponent, the former minister of labor, won 16.5 percent of the vote. The remainder was divided among eight other candidates. Voter turnout was 68 percent, lower than in 1997, but Khatami's victory margin was higher.

Normally support of such magnitude would give an elected leader a clear mandate for carrying out his or her programs.in most countries. But politically support of such magnitude would give the winner a clear mandate for carrying out his or her programs. But politically, the Islamic Republic of Iran is a unique institution. Its population is sharply divided between "hard-liners" and "reformers," terms that do little to explain Iran's political complexities. The former, generally speaking, are those who would preserve at all costs the theocratic rule and Islamic values bequeathed to the republic by its founders. The latter, centered on Khatami as the "great white hope" in achieving their goals, seek to reshape Iran as a more open society, committed to justice and the rule of law and with personal freedom and rights guaranteed. The slogan "Iran for all Iranians" adopted by the reformers during the February 2000 Majlis elections perhaps best describes—and symbolizes—these goals. The imbalance between the "two Irans" became apparent during Khatami's first term, which was marred by the murders of leading writers and intellectuals, and by several violent confrontations of university students with security police, the former being at the forefront of resistance to hard-liners. A police raid on a Teheran University dormitory in 1999 resulted in a number of deaths and arrests of student leaders. Further demonstrations generated scenes reminiscent of the last days of the monarchy before the Revolution.

The election of a reformer-dominated Majlis in February 2000, and that of Khatami for a second term as president in June 2001, seemed to promise success for his reform program. Known as the Second

Khordad Movement because of its date (that of his election), it would move Iranian society away from the strict Islamic interpretation of social and political behavior of the clerical regime, toward a more open society receptive to ideas and influences from the outside world, yet without compromising its Islamic nature.

However, the reform movement very quickly ran afoul of hard-line elements supporting and enforcing the rules of the regime. Freedom House, a New York-based human rights watchdog organization, reported in a survey in 2000 that "the state continues to maintain control through terror: arbitrary detention, torture, disappearance, summary trial and execution are commonplace."[13] In 2000 in particular, new newspapers deemed critical of these policies were closed almost before they began publication, and their reporters and editors arrested. A group of Iranian journalists attending a conference in Berlin on human rights were given jail sentences on their return; it was estimated that Iran had more journalists in prison than any other country in the world.

The hard-liners also used public floggings as a visible method of slowing the clock of political and social change. In August 2001, 13 young men were given 80 lashes each for "offenses against public order," notably use of alcohol or being seen in public with women to whom they were not related. The head of the judiciary defended the floggings as necessary to combat un-Islamic behavior and rising rates of crime and drug use.

Its control over the levers of power enabled the regime to repress the reform movement effectively during Khatami's first term. The second term brought a very different story. In June 2003, a reform bill was passed by the Majlis which would strip the CCG of its power to veto candidates for public office. As expected the CCG rejected the bill under its role of approving (or disapproving) bills before they become law.

The 2004 Majlis elections brought to a head the evolving conflict between harliners committed to all-out support of the clerical regime and moderates advocating a more secular society with fewer limits on personal behavior and dress and freer mixing of genders. Prior to the election the CCG had disqualified some 3,000 of the 8,200 candidates on grounds their "speech and behavior" were disloyal to Islam and to the constitution. As a result, one-third of the 290 deputies resigned in protest, while others joined the public in mass sit-ins before the Majlis building. The Interior Ministry warned that the election would have to be postponed unless the banned candidates were made eligible, noting that the action violated election law. Subsequently the CCG reinstated 200 candidates but refused to do so for the remainder. President Khatami then asked Ayatollah Khamenei to intervene, in his capacity as supreme Legal Guide. However, he took no action, and the election was held as scheduled in February. To no one's surprise the hardliners won a huge majority of seats, including 30 of 38 in Tehran, ensuring the continuation of what one observer called the "dictatorship of the mullahs."

Timeline: PAST

551–331 B.C.
The Persian Empire under Cyrus the Great and his successors includes most of ancient Near East and Egypt

A.D. 226–641
The Sassanid Empire establishes Zoroastrianism as the state religion

637–641
Islamic conquest at the Battles of Qadisiya and Nihavard

1520–1730
The Safavid shahs develop national unity based on Shia Islam as the state religion

1905–1907
The constitutional movement limits the power of the shah by the Constitution and Legislature

1925
The accession of Reza Shah, establishing the Pahlavi Dynasty

1941
The abdication of Reza Shah under Anglo–Soviet pressure; he is succeeded by Crown Prince Mohammed Reza Pahlavi

1951–1953
The oil industry is nationalized under the leadership of Prime Minister Mossadegh

1962
The shah introduces the White Revolution

1979–1980
Revolution overthrows the shah; Iran becomes an Islamic republic headed by the Ayatollah Khomeini

1980s
The Iran–Iraq war; Khomeini dies

1990s
Iran's economy begins to recover; foreign relations improve; debate over how Islamic the Islamic Republic should be

PRESENT

2000s
President Mohammed Khatami wins reelection handily
U.S. president George W. Bush calls Iran part of "evil axis" of terror, reigniting tensions with Iran

The June 2005 election for president produced further surprises. Our of over 1,000 candidates for the office only eight were approved by the Guardian Council. The favorite and front-runner, former president Rafsanjani, campaigned on the basis of liberalizing the economy and normalizing relations with the U.S. His chief rival, the former minister of science, was favored by the youth population for his platform emphasizing intellectual freedom and cultural diversity. However the unexpected winner was former Tehran mayor Ahmad Ahmadinejad (Ahmadi-Nejad, in Farsi). A man of humble origins (his father was a blacksmith) and a "success story" in U.S. terms, he was elected largely due to campaign promises to deal vigorously with public corruption, economic problems and the disparity between rich and poor.

The new president however wasted no time before engaging in inflammatory rhetoric, probably reflecting his lack of foreign experience. In a November speech he called for Israel to be "wiped off the map," and in another he declared that the Holocaust never existed. Although his initial ministerial appointments were rejected by the Majlis on grounds of professional inexperience, in other ways the new president reflects the conservative thrust and national pride which has resulted from the nuclear standoff with Europe. The majority of his predecessor's diplomatic and gubernatorial appointees were dismissed after his election as being "too liberal". In a recent speech Ahmadinjad stated: "Iran has the right to peaceful nuclear technology and scientific progress. With high esteem and self-confidence our nation now is moving in the direction of development and perfection ..."[14]

It would appear that the national pride and unity revived under the new president has sharply reduced the division between "moderates" and "conservatives" (or soft and hard-liners) in Iranian society during the Khatami regime. But the Iranians are an extraordinary people, gifted, technically proficient, used to the practice of dissimulation (Taqiya) which has always characterized Shia Muslims as an often-persecuted minority. It would be premature to forecast the demise of this unique experiment in theocratic government. But as one scholar observes, "civil society here ... has not been crushed, as it has by authoritarian leaders in the Arab world. On many levels and for important issues leaders must draw consensus from the different levels of power ..."[15] This struggle, between conservative and reformism, idealism and pragmatism, the old guard and the new generation, would seem to encourage the active participation of non-political, non-governmental groups in the construction of an effective civil society in Iran. Without such participation, supplemented by wise leadership, the country's unique experiment in clerical authority is unlikely to succeed.

NOTES

1. A 1991 memo from Ayatollah Khamenei ordered that Bahais should be expelled from prevented from attaining "positions of influence," and denied employment and access to education. The memo is in sharp contrast to UN General Assembly *Resolution 52/142,* which calls on Iran to "emancipate" its Bahai population.

2. Golamreza Fazel, "Persians," in Richard V. Weekes, ed., *Muslim Peoples: A World Ethnographic Survey,* 2nd ed. (Westport, CT: Greenwood Press, 1984), p. 610. "Face-saving is in fact one of the components of *Ta'aruf,* along with assertive masculinity *(gheyrat)."*

3. John Malcolm, *History of Persia* (London: John Murray, 1829), Vol. II, p. 303.

4. Behzad Yaghmaian, *Social Change in Iran* (Albany, NY: State University of New York Press, 2001), p. 127.

5. Roy Mottahedeh, *The Mantle of the Prophet* (New York: Simon & Schuster, 1985), p. 52.

6. *Ibid.,* p. 34.

7. Until recently, the CIA had always been credited with engineering Mossadegh's overthrow and had made no effort to deny this charge. Publication recently of the agency's secret history of "Operation Ajax" clearly emphasized its limited role and lack of effectiveness. The unpublished memoirs of Ardeshir Zahedi, the general's son ("Five Decisive Days, August 14–18, 1953") indicate that U.S. involvement was incidental to a genuine popular uprising.

8. Imam Khomeini, *Islam and Revolution,* transl. by Hamid Algar, tr. (Berkeley, CA: Mizan Press, 1981), p. 175.

9. Mohammed Reza Pahlavi, Shah of Iran, *Answer to History* (New York: Stein and Day, 1980), pp. 152–153.

10. Sepehr Zabih, *Iran's Revolutionary Upheaval: An Interpretive Essay* (San Francisco, CA: Alchemy Books, 1979), pp. 46–49.

11. Scott Peterson, in *The Christian Science Monitor,* March 10, 2005.

12. "Wearing red headbands and inspired by professional chanters before battle, their heads were filled with thoughts of death and martyrdom and going to Paradise." V. S. Naipaul, "After the Revolution," *The New Yorker* (May 26, 1997), p. 46.

13. "A Survey of Iran," *The Economist,* January 18, 2003.

14. Evan Osnos, *Chicago Tribune,* February 23, 2006.

15. Michael Slackman, "Letter from Iran", *The New York Times,* February 15, 2006.

Iraq (Republic of Iraq)

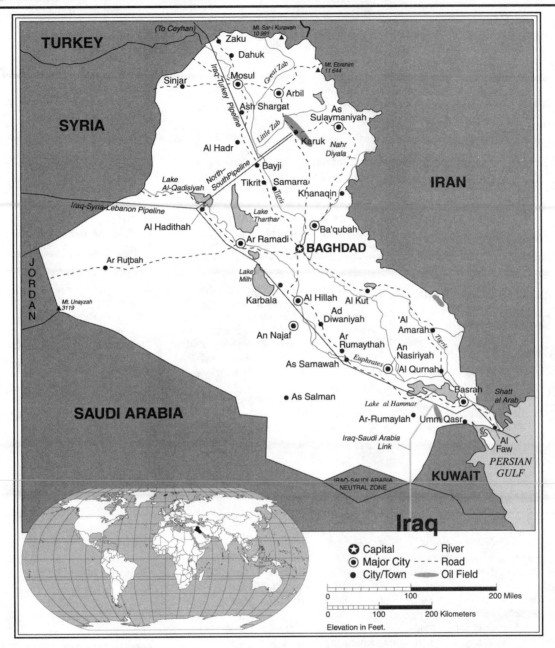

Iraq Statistics

GEOGRAPHY

Area in Square Miles (Kilometers):
168,710 (437,072) (about twice the size of Idaho)

Capital (Population): Baghdad (3,842,000)

Environmental Concerns: draining of marshes near An Nasiriyah has affected wetlands and destroyed ecosystems with heavy impact on wildlife; inadequacy of potable water; air and water pollution in cities; soil degradation due to excess salinity and erosion; desertification

Geographical Features: broad plains shading to desert in central and south; mountains in north and northwest; hilly north of Baghdad. The vast southern marshlands, covering some 3,000 square miles and isolated for centuries, were largely drained by Saddam Hussein's government for an irrigation and land reclamation project. Since the U.S. occupation they have been about 40 percent restored.

Climate: dry and extremely hot, with very short winters, except in northern mountains which have cold winters with much snow and temperate summers

PEOPLE

Population

Total: 26,074,906
Annual Growth Rate: 2.78%
Rural/Urban Population Ratio: 25/75
Major Languages: Arabic; Kurdish
Ethnic Makeup: 75% Arab; 20% Kurdish;
5% Turkoman, and others
Religions: 60% Shia Muslim; 35% Sunni
Muslim; 5% others (Christian, Yazidi,
Sabaean)

Health

Life Expectancy at Birth: 67 years (male);
69 years (female)
Infant Mortality Rate: 50/1,000 live births

Education

Adult Literacy Rate: 40.4% (males 55.9%,
females 24%) (*Note:* Iraq had a much
higher rate prior to Saddam's wars. War
disruption and UN sanctions have
lowered the literacy rate significantly. In
the 1970s Iraq had the highest level in the
Arab world.
Compulsory (Ages): 6–12; free

COMMUNICATION

Telephones: 675,000 main lines in use
Daily Newspaper Circulation: n/a (same
reason)
Televisions: n/a (same reason)
Internet Users: 1 (2000)

TRANSPORTATION

Highways in Miles (Kilometers): 29,435
(47,400)
Railroads in Miles (Kilometers): 1,262
(2,032)
Usable Airfields: 109
Motor Vehicles in Use: 1,040,000

GOVERNMENT

Type: Presently in transition from single-
party to multi-party republic. A
Transitional National Assembly was
elected in January 2005 and drafted a
constitution which was approved in a
nationwide plebiscite. In December 2005
voters elected a permanent National
Assembly. However members have not
yet agreed on the country's first full-term
four-year government.
Independence Date: October 3, 1932,
from Britain; July 14, 1958, as republic
after the overthrow of the monarchy
Head of State/Government: President
Jalal Talabani (Kurd), whose term will
expire with the formation of the new
government. Prime Minister Ibrahim al-
Jaafari, appointed in the January 2005
election, was reconfirmed in March by a
1-vote margin by the Shia majority in the
Assembly.
Political Parties: United Iraqi Alliance
(Shia), The Iraqi List (Sunni-Shia),
Kurdish Alliance List, largest Parties;

others are Assembly of Independent
Democrats (Sunni-Shia), National
Democratic Party (Sunni), National
Rafidain List (Christian)
Suffrage: universal at 18

MILITARY

Military Expenditures (% of GDP): n/a
($1.3 billion)
Current Disputes: continued insurgency
since the U.S. invasion and occupation

ECONOMY

Currency ($U.S. Equivalent): 1,890 new
dinars = $1
Per Capita Income/GDP: $2,100/$54
billion
GDP Growth Rate: 5.2%
Inflation Rate: 25.4%
Unemployment: 25–30%
Labor Force: 6,700,000
Natural Resources: petroleum; natural
gas; phosphates; sulfur; lead; gypsum;
iron ore
Agriculture: wheat; barley; rice;
vegetables; dates; cotton; sheep; cattle
Industry: petroleum; chemicals; textiles;
construction materials; food processing
Exports: $10 billion, exclusively crude oil
(primary partners United States, France,
Russia)
Imports: $9.9 billion (primary partners
France, Australia, Italy)

Iraq Country Report

The Republic of Iraq is a young state in a very old land. In ancient times, its central portion was called *Mesopotamia,* a word meaning "land between the rivers." Those rivers are the Tigris and the Euphrates, which originate in the highlands of Turkey and flow southward for more than a thousand miles to join in an estuary called the Shatt al-Arab, which carries their joint flow into the Persian (or, to Iraqis, the Arab) Gulf.

The fertility of the land between the rivers encouraged human settlement and agriculture from an early date. The oldest farming community yet discovered anywhere was unearthed near Nineveh, capital of the Assyrian Empire, in 1989; it dates back to 9000 B.C. Other settlements grew in time into small but important cities with local governments, their economies based on trade and crafts production in addition to agriculture. And the process of using a written alphabet with characters rather than symbols probably originated here. In 1999, President Saddam Hussein announced an

international festival for the year 2000 to mark the 5,000th anniversary of this Meso-potamian invention (although recent ar-chaeological discoveries in Egypt suggest that writing may have developed there even earlier).

Present-day Iraq (*Iraq* is an Arabic word meaning "cliff" or, less glamorously, "mud bank") occupies a much larger territory than the original Mesopotamia. Iraqi territory also includes a Neutral Zone on the border with Saudi Arabia. Iraq's other borders are with Turkey, Syria, Jordan, Kuwait, and Iran. These borders were established by the British on behalf of the newly formed Iraqi government, which they controlled after World War I. Disagreement with Kuwait over oil production and allocation from their shared Rumaila field was a factor in Iraq's 1990 invasion of Kuwait.

In 1994, Iraq accepted the border demar-cated under United Nations *Resolutions 687, 773,* and *883,* formally relinquishing its claims to Kuwait and the islands of Bubiyan

and Warbah. The new border, realigned northward by an international commission, removed 1,870 feet from Iraqi jurisdiction.

Iraq's other border in question, disputed with Iran, is in the Shatt al-Arab (Arab Delta), a broad, navigable estuary extending from the confluence of the Tigris and Euph-rates Rivers down to the Persian Gulf. Dur-ing the years of the British mandate and early independence, Iraq claimed ownership from the west to east banks. Iran's claim ex-tended from its own (east) bank to mid-channel. Iraq recognized the Iranian claim in a 1975 agreement; in return, Iran with-drew support from Kurds in northern Iraq who were seeking autonomy within the Iraqi state. The unilateral abrogation of this agreement by Saddam Hussein was one fac-tor in the 1980–1988 Iran–Iraq War.

History

The "land between the rivers" has had many occupiers in its long history as a settled area, and it has seen many kingdoms and empires rise and fall. Over the centuries many peoples—Sumerians, Babylonians, Assyrians, Persians, Arabs and others added layer upon layer to the mix of Mesopotamian civilization. The world's first cities probably began there, as did agriculture, the growing of food crops, made possible by an ingenious irrigation system developed by Sumerian "engineers" to bring Tigris-Euphrates water from those rivers to their fields. Since theirs was essentially an agricultural society, the Sumerians developed a system of recording land ownership and grain sales on clay tablets, in what is generally considered the world's first written alphabet. Their transactions were recorded in cube-shaped letters called cuneiform, rather than the picture-words (hieroglyphs) used in ancient Egypt which were much more cumbersome.

The Sumerians seem to have disappeared as a people around 2,200 B.C.E., absorbed into other more powerful peoples, though there is some evidence that their descendants may still be found in the marshlands of southern Iraq, where they survived as "people of the reeds" until the 1990s, when their habitat was largely destroyed by the government, partly for land reclamation but also to pursue Shia rebels who had taken refuge there.

Successor Mesopotamian peoples also contributed much to our modern world. The Babylonian king Hammurabi developed the first code of laws; there are 282 of them in all, inscribed on steles (pillars) placed at strategic points in his kingdom to warn people what they should or should not do, and the consequences thereof. A later Babylonian king, Nebuchadnezzar, built the world's first capital city, at Babylon. Its Hanging Gardens, a series of overhanging terraces planted with flowers and trees and watered by hidden waterwheels, was considered one of the Seven Wonders of the ancient world. The king had constructed it to please his queen, who came from far-off mountains and pined for her homeland.

Other contributions have come down to us from this ancient land. The first political system, based on sovereign city-states, began there. Each Mesopotamian city was built around a central sacred space, the earthly home of a god. His shrine was the *ziggurat*, a stepped pyramid reaching many stories toward the sky and from whence he could descend to earth when needed.

One would think that the modern Iraqis would take great pride in their storied heritage. But unlike the Iranians and the Chinese, whose history goes back as far and who have an innate sense of their grandeur as peoples, until recently this glorious past was largely forgotten, unimportant except to archaeologists, most of them foreigners. Gertrude Bell, Leonard Woolley and many others spent years literally digging up Ur, Babylon, Nineveh and other ancient Mesopotamian cities, collecting priceless artifacts which were kept on display in the National Museum in Baghdad. One of the first horror stories of the overthrow of Saddam Hussein by coalition forces in 2003 was the looting of the museum. Thus far some 3,411 priceless artifacts have been recovered or returned, but at least 10,000 pieces including the Akkadian Bassetki copper statue, all 330 pounds of it, are still missing. (Ironically Saddam in his megalomaniac view of himself as a world leader liked to remind the Iraqis of their past greatness, rebuilding the city of Babylon with his own hands and calling himself a modern-day Hammurabi or Nebuchadnessar).

Although occupying peoples and their rulers have made Mesoporamia a place of many-layered civilizations, the most important influence in Iraqi social and cultural life today comes from the conquest of the region by Islamic Arabs. In A.D. 637, an Arab army defeated the Persians, who were then rulers of Iraq, near the village of Qadisiya, not far from modern Baghdad, a victory of great symbolic importance for Iraqis today. Arab peoples settled the region and intermarried with the local population, producing the contemporary Iraqi–Arab population.

Aside from Saddam's myth-making the most important influence on Iraq's people has been that of Islam, brought to their ancestors by Arabs from Arabia in the seventh century A.D. In 637 an invading Arab army defeated the Persians, who ruled Mesopotamia at that time, in the battle of Qadisiyya, near modern Baghdad. The invaders established a string of military bases to protect their new conquest, and these bases gradually grew into cities as Arab tribal groups settled them. Thus an Arab mix was added to the population to produce the modern Iraqis.

During the early years of Islam, Iraq played an important role in Islamic political history. Shia Muslims who believe in the right of Ali, Muhammad's son-in-law, and his descendants to serve as caliphs formed a majority there, as they do today. The tombs of Ali, Muhammad's son-in-law and the fourth and last leader of a united caliphate, and his son Husayn, martyred in a power struggle with his Damascus-based rival Yazid, are both in Iraq (at Najaf and Karbala, respectively).

In the period of the Abbasid caliphs (A.D. 750–1258), Iraq was the center of a vast Islamic empire stretching from Morocco on the west to the plains of India. Caliph al-Mansur laid out a new capital for the world of Islam, some 60 miles from the ruins of Babylon. He named his new capital Baghdad, possibly derived from a Persian word for "garden," and, according to legend, laid bricks for its foundations with his own hand.

Baghdad was a round city, built in concentric circles, each one walled, with the caliph's green-domed palace and mosque at the center. It was the world's first planned city, in the sense of having been laid out in a definite urban configuration and design. Under the caliphs, Baghdad became a center of science, medicine, philosophy, law, and the arts, at a time when London and Paris were mud-and-wattle villages. The city became wealthy from the goods brought by ships from Africa, Asia, and the Far East, since it was easily reachable by shallow-draught boats from the Gulf and the Indian Ocean moving up the Tigris to its harbor.

In the thirteenth century A.D. the Mongols, a group of tribes in Central Asia near the Chinese border, were united under the leadership of Genghis Khan and swept across the Middle East, conquering the Islamic lands as far as Egypt as well as present-day Russia. A Mongol army led by Genghis Khan's son Hulagu captured Baghdad in 1258. The city was not destroyed as completely as other Islamic cities but it suffered extensive damage.

That year also marked the end of the Abbasid Caliphate. The ruling caliph was seized and executed by the non-Muslim, shaman-worshipping Mongols. A more serious result of the Mongol invasion was their destruction of the ancient irrigation system that had made agriculture not only possible but also extremely productive. However, after their invasion and occupation the Mongols were converted to Islam, settled down and became peaceful farmers and herdsmen, as their descendants are today.

After the fall of Baghdad, Iraq came under the rule of various local princes and dynasties. In the sixteenth century, it was included in the expanding territory of the Safavid Empire of Iran. The Safavid shah championed the cause of Shia Islam; as a result, the Ottoman sultan, who was Sunni, sent forces to recover the area from his hated Shia foe. Possession of Iraq went back and forth between the two powers, but the Ottomans eventually established control until the twentieth century.

Iraq was administered as three separate provinces under appointed Ottoman governors. The governors paid for their appointments and were interested only in recovering their losses. The result was heavy taxation and indifference to social and economic needs. The one exception was the province of Baghdad. It was governed by a man whom today we would call an enlightened administrator. This governor, Midhat Pasha, set up a provincial newspaper, hospitals, schools, munitions factories, and a fleet of barges to carry produce down river to ports on the Gulf. His administration also ensured public security and an equitable taxation system. Midhat Pasha later became the grand vizier (prime minister) of the Ottoman Empire and was the architect of the 1876 Constitution, which limited the powers of the sultan.

The British Mandate

World War I found England and France at war with Germany and the Ottoman Empire. British forces occupied Iraq, which they rechristened Mesopotamia, early in the war. British leaders had worked with Arab leaders in the Ottoman Empire to launch a revolt against the sultan; in return, they promised to help the Arabs form an independent Arab state once the Ottomans had been defeated. A number of prominent Iraqi officers who were serving in the Ottoman Army then joined the British and helped them in the Iraqi campaign.

The British promise, however, was not kept. The British had made other commitments, notably to their French allies, to divide the Arab provinces of the Ottoman Empire into British and French "zones of influence." An independent Arab state in those provinces was not in the cards.

The most that the British (and the French) would do was to organize protectorates, called mandates, over the Arab provinces, promising to help the population become self-governing within a specified period of time. The arrangement was approved by the new League of Nations in 1920. Iraq then became a British mandate, with Faisal ibn Hussein as its king, but with British advisers appointed to manage its affairs. (Faisal had been ruler of the short-lived Arab kingdom of Syria set up after the war, but he was expelled when the French occupied Damascus and established their mandate.)

The British kept their promise with the mandate. They worked out a Constitution for Iraq in 1925 that established a constitutional monarchy with an elected Legislature and a system of checks and balances. Political parties were allowed, although most of them were groupings around prominent personalities and had no platform other than independence from Britain. In 1932, the mandate formally ended, and Iraq became an independent kingdom under Faisal. Iraq became an independent kingdom under Faisal. It also joined the League of Nations, forerunner of the UN, as the first Middle Eastern nation to do so. Although its sovereignty was limited in certain areas, especially in British control of military bases and management of foreign policy, by that time Iraq had reached the goal of Arab nationalists of a sovereign state.

The Iraqi Monarchy: 1932–1958

The new kingdom cast adrift on perilous international waters was far from being a unified nation. It was more of a patchwork of warring and competing groups. The Muslim population was divided into Sunni and Shia, as it is today, with the Sunnis forming a minority but controlling the government and business and dominating urban life. The Shias, although a majority, were mostly rural peasants and farmers, many of them migrants to the cities, where they formed a large underclass.

The country also had large Christian and Jewish communities, the latter tracing its origins back several thousand years to the exile of Jews from Palestine to Babylonia after the conquest of Jerusalem by Nebuchadnezzar. The Assyrians formed the largest Christian group. Formerly residents in Ottoman Turkey, they supported the British as "our smallest ally" in World War I and, as a result, were allowed to resettle in Iraq after the establishment of the Turkish Republic. The British protected them, recruiting Assyrians as guards for military and air bases and the British-controlled police force. As one observer noted, "with their slouch hats and red and white hackles, they became a symbol for British rule in Iraq.

However, when they pushed their luck to the point of demanding autonomy in 1933, the British-trained Iraqi Army moved against them, destroying villages and massacring their Assyrian inhabitants.[1] Many fled into exile, forming large communities (particularly in Detroit and San Diego). A small minority remained in Iraq, but most of them have now left, due to the country's economic difficulties. A group of 172 Assyrians took asylum in Mexico in September 2000. Some 6,000 Assyrians resident in southeastern Turkey since World War I have been allowed to practice their faith without hindrance, and in 2005 the Turkish government approved their public celebration of their annual festival, Akito

The new state also included other non-Muslim communities, such as the Yazidis (called "devil-worshippers," from their religious practice) and the Sabaeans, descended from the ancient Babylonians. These social and religious divisions in the population plus great economic disparities made the new state almost impossible to govern or develop politically.[2]

King Faisal I was the single stabilizing influence in Iraqi politics, so his untimely death in 1933 was critical. His son and successor, Ghazi, was more interested in racing cars than anything else and was killed at the wheel of one in 1939. Ghazi's infant son succeeded him as King Faisal II, while Ghazi's first cousin became regent until the new ruler came of age.

After King Faisal's death and during the minority of his young son, the regent, Abd al-Ilah, governed the country along with a constantly shifting coalition of landowners, merchants, tribal leaders, and urban politicians. As a result, there was little political stability or progress toward national unity. Between 1933 and 1936, for example, 22 different cabinets held office. The late Nuri al-Said, who served as prime minister a number of times, once compared the Iraqi government to a pack of cards. You must shuffle them often, he said, because the same faces keep turning up.[3]

THE REVOLUTION OF 1958

To their credit, the king's ministers kept the country's three broad social divisions—the Kurdish north, the Sunni Arab center, and the Shia Arab south—in relative balance and harmony. Oil revenues were channeled into large-scale development projects. The formation of a modern school system with a Western-model curriculum, along with adult literacy programs, establishment of a national army, and opportunities for its officers to attend British military academies such as Sandhurst, gave Iraq a head start toward self-government, well ahead of other Arab States. Education was promoted strongly, and one of the sad results of Saddam Hussein's wars and UN sanctions has been a sharp decline in the literacy rate, formerly the highest in the Arab world. The press was free, and, though it had a small and ingrown political elite, there was much participation in legislative elections. Despite its legitimate Arab credentials as one of the successor states fashioned by the British after World War I, however, a new generation of pan-Arab nationalist Iraqis viewed the royal regime as a continuation of foreign rule, first Turkish and then British.

Resentment crystallized in the Iraqi Army. On July 14, 1958, a group of young officers overthrew the monarchy in a swift, predawn coup. The king, regent, and royal family were killed. Iraq's new leaders proclaimed a republic that would be reformed, free, and democratic, united with the rest of the Arab world and opposed to all foreign ideologies, "Communist, American, British or Fascist."

Iraq has been a republic since the 1958 Revolution, and July 14 remains a national

Homer Sykes/Katz/Woodfin (SYKES001)

The image of Saddam Hussain has become part of the Iraqi landscape; his portrait appears in public buildings, at the entrances to cities, in homes, even on billboards along highways.

holiday. But the republic has passed through many different stages, with periodic coups, changes in leadership, and political shifts, most of them violent. Continuing sectarian and ethnic hatreds, maneuvering of political factions, ideological differences, and lack of opportunities for legitimate opposition to express itself without violence have created a constant sense of insecurity among Iraqi leaders. A similar paranoia affects Iraq's relations with its neighbors. The competition for influence in the Arab world and the Persian/Arab Gulf and other factors combine to keep the leadership constantly on edge.

This pattern of political instability showed itself in the coups and attempted coups of the 1960s. The republic's first two leaders were overthrown after a few years. Several more violent shifts in the Iraqi government took place before the Ba'th Party seized control in 1968. Since that time, the party has dealt ruthlessly with internal opposition. A 1978 decree outlawed all political activity outside the Ba'th for members of the armed forces. Many Shia clergy were executed in 1978–1979 for leading antigovernment demonstrations after the Iranian Revolution; and following Saddam Hussain's rise to the presidency, he purged a number of members of the Revolutionary Command Council (RCC), on charges that they were part of a plot to overthrow the regime.

THE BA'TH PARTY IN POWER

The Ba'th Party in Iraq began as a branch of the Syrian Ba'th founded in the 1940s by two Syrian intellectuals: Michel Aflaq, a Christian teacher, and Salah al-Din Bitar, a Sunni Muslim. Like its Syrian parent, the Iraqi Ba'th was dedicated to the goals of Arab unity, freedom, and socialism. However, infighting among Syrian Ba'th leaders in the 1960s led to the expulsion of Aflaq and Bitar. Aflaq went to Iraq, where he was accepted as the party's true leader. Eventually, he moved to Paris, where he died in 1989. His body was brought back to

Iraq for burial, giving the Iraqi Ba'th a strong claim to legitimacy in its struggle with the Syrian Ba'th for hegemony in the movement for Arab unity.

Prior to the 2003 U.S. invasion and overthrow of Saddam Hussein the country was governed by a Provisional Constitution issued by the Revolutionary Command Council of the Ba'th, the party's decision-making body. It defined Iraq, without elaboration, as a sovereign peoples' democratic republic. Other clauses provided for a National Assembly responsible only for ratification of Ba'th-issued laws and RCC decisions.

The Iraqi Communist Party, founded in the 1930s, survived the ins and outs of interparty conflict during the monarchy and gained a pre-eminent position after the 1958 coup, being the only one with a strong central organization even though it represented an alien ideology. It was banned in 1978 and went underground, continuing to oppose Saddam Hussein on a sub rosa basis. After his overthrow, its secretary-general, Hamid Majid Moussa, was appointed as a member of the Iraq Governing Council set up by U.S. authorities to prepare for a transfer of power.

An abortive coup in 1973, which pitted a civilian faction of the Ba'th against the military leadership headed by President Ahmad Hasan al-Bakr, stirred party leaders to attempt to broaden their base of popular support. They reached agreement with the Iraqi Communist Party to set up a Progressive National Patriotic Front. Other organizations and groups joined the Front later. Although the Iraqi Communist Party had cooperated with the Ba'th on several occasions, the agreement marked its first legal recognition as a party. However, distrust between the two organizations deepened as Ba'th leaders struggled to mobilize the masses. The Communists withdrew from the Front in 1979 and refused to participate in parliamentary elections. Their party was declared illegal in 1980.

SADDAM HUSSAIN

Politics in Iraq have always been marked by secrecy and behind-the-scenes maneuvering among its factions and groups. This situation became more pronounced after the overthrow of the monarchy in 1958. From then until the Ba'th's accession to power, authority was fragmented. Iraq also has always lacked any tradition of accountability on the part of its leaders or public pressure to bring them to account. As a result national politics became completely screened from public or international view, One writer's question, "Will the Iraqi ruling class please stand up?" is apt.[4] But in the late 1970s and early 1980s, one of its leaders, Saddam Hussain, emerged from the pack to become an absolute ruler.

Saddam Hussain's early history did not suggest such an achievement. He was born in 1937 in the small town of Tikrit, on the Tigris halfway between Baghdad and Mosul. Tikrit's chief claim to fame, until the twentieth century, was that it was the birthplace of Saladin, hero of the Islamic world in the Middle Ages against the Crusaders. (The Iraqi leader has at times identified himself with Saladin as another great Tikriti, although Saladin was a Kurd and Saddam Hussain's distrust of Kurds is well known.) His family belonged to the Begat clan of the Al Bu Nasser tribe, settled around Tikrit. They lived in a nearby village and farmed some 12 acres of land. According to accounts, Saddam was a bully and street fighter from early childhood. He spent little time in school and was hired out from time to time as a shepherd. As a teenager he left home for Baghdad, lived with an uncle, and joined the Ba'th Party. He played a very minor role in the Ba'th's attempted assassination in 1959 of Abd al-Karim Qassem, leader of the 1958 Revolution, and escaped to Egypt in disguise. He returned to Iraq after Qassem's overthrow and execution and gradually worked his way up through the ranks of the Ba'th. Eventually he became vice-chairman, and then chairman, of the Revolutionary Command Council, the party's ruling body. As chairman, he automatically became president of Iraq under the 1970 Constitution. As there are no constitutional provisions limiting the terms of office for the position, the National Assembly named him president-for-life in 1990.[5]

As one might expect from a leader whose political experience was limited to intra-party intrigue and anti-government plots, Saddam Hussein came to office with none of the attributes needed for leader and statesman. He had never served in the army, traveled only as far as Egypt, and had little knowledge of foreign affairs or non-Arab peoples. His first effort was to instill in the Iraqi people a "climate of fear" which would enable him to govern unopposed. Thus almost immediately after he became president he declared that a plot by other Ba'th leaders to overthrow him had been discovered. He called an emergency meeting of the RCC, read out the names of the conspirators, had them called forward to "confess," and then taken out and shot. Finally, he sent a videotape of the proceedings to other Arab heads of state "so they would understand the need for the Ba'th to destroy enemies within its ranks."[6]

The Iran–Iraq War of the 1980s was a severe test for the Ba'th and its leader. A series of Iraqi defeats with heavy casualties in the mid-1980s suggested that the Iranian demand for Saddam Hussain's ouster as a precondition for peace might ignite a popular uprising against him. But Iranian advances into Iraqi territory, and in particular the capture of the Fao Peninsula and the Majnoon oil fields, united the Iraqis behind Saddam. For one of the few times in its history, the nation coalesced around a leader and a cause.

DEVELOPMENT

The Security Council formally ended sanctions on Iraq in May 2003. Aside from their devastating impact on the population, the 13-year sanctions had severely limited oil production and led to the deterioration of oil refinery installations, industries and service systems. Oil production was 3.5 million barrels per day (b/d) prior to 1991 but dropped to 2.6 million b/d during the sanctions period and stopped entirely during the U.S. invasion. Since the occupation began, sabotage has further delayed economic recovery. Insurgent attacks on oil pipelines and refineries in 2004–2005 have frequently reduced the country's normal output of 2 million barrels by as much as 100,000 barrels per day (bpd). The pipeline to Turkey, Iraq's major export outlet, was completely shut down on several occasions.

Thereafter Saddam combined ferocious ruthlessness toward real (or imagined) opponents of his regime with a sort of "father of his country" image, taking part in family events, visiting farms, etc., always in the appropriate dress besides his trademark beret. Posters of him sprouted with ever-increasing frequency on public buildings, in banks, schools and larger-than-life highway billboards. His numerous presidential palaces, most of them occupying choice real estate, reminded the Iraqis of his power on a daily basis as they walked, drove or did business. One of the very few jokes that circulated in the fear-ridden Saddam years was: "What is Iraq's population? Twenty-one million Iraqis and twenty-one million portraits of Saddam Hussain."

RECENT DEVELOPMENTS

The end of the war with Iran and Saddam Hussain's popularity as the heroic defender of the Iraqi Arab nation against the Shia Iranian enemy prompted a certain lifting of Ba'thist repression and authoritarian rule. Emergency wartime regulations in force since 1980 were relaxed in 1989, and an amnesty was announced for all political exiles except "agents of Iran."

In July 1990, the RCC and the Arab Ba'th Regional Command, the party's governing body, approved a draft constitution to replace the 1970 provisional one. The new Constitution "legalized" the formation of political parties other than the Ba'th, as long as they conformed to Ba'thist principles. It also established freedom of the press and other civil rights, although again in conformity with Ba'thism.

The 1990 Iraqi occupation of Kuwait and the ensuing Gulf War halted even these small steps toward representative government. The draft Constitution remained in suspension; it was not issued unilaterally by the regime nor submitted to voters in a referendum. In the 1990s the regime implemented a law in the provisional constitution to allow elections for local and municipal councils. Candidates were required to be members of the Ba'th, or if they weren't, to run as independents. The councils were given the authority under a 1995 law to administer programs in health, education, and economic development in their respective localities.

As a survivor of many party conflicts, one would expect Saddam Hussein to make bold moves and take great risks when his survival seemed to be at stake. Such moves also characterized his foreign policy. To further his goal of establishing Iraq as a major regional power and threatening its enemies, notably Israel, he undertook a large-scale project of building an arsenal of

weapons of mass destruction (WMD), mainly nuclear, chemical and biological. According to his chief "bomb maker," in 1971 he directed a team of scientists and engineers to build a nuclear bomb equal to that exploded by the United States over Hiroshima in 1945. The first one made by Iraq would be dropped unannounced on Israel. Such a bomb was tested in 1987, but the project was halted in 1990 by the invasion of Kuwait. By that time, it had employed 12,000 engineers and scientists and cost more than $10 billion.[7]

THE KURDS

The Kurds, the largest non-Arab minority in Iraq today, form a relatively compact society in the northern mountains. Formerly the Ottoman province (*vilayet*) of Mosul, the territory was occupied by British troops after World War I and included in the mandate over Iraq by the League of Nations, despite angry protests by the Turks demanding its inclusion in the new Republic of Turkey. The Kurds living there agitated for self-rule periodically during the monarchy; for a few months after World War II, they formed their own republic in Kurdish areas straddling the Iraq–Iran and Iraq–Turkey borders.

In the 1960s, the Kurds rebelled against the Iraqi government, which had refused to meet their three demands (self-government in Kurdistan, use of Kurdish in schools, and a greater share in oil revenues). The government sent an army to the mountains but was unable to defeat the Kurds, masters of guerrilla warfare. Conflict continued intermittently into the 1970s. Although the 1970 Constitution named Arabs and Kurds as the two nationalities in the Iraqi nation and established autonomy for Kurdistan, the Iraqi government had no real intention of honoring its pledges to the Kurds.

A major Iraqi offensive in 1974 had considerable success against the Kurdish *Pesh Merga* ("Resistance"), even capturing several mountain strongholds. At that point, the shah of Iran, who had little use for Saddam Hussein, began to supply arms to the Pesh Merga. The shah also kept the Iraq–Iran border open as sanctuary for the guerrillas.

In 1975, a number of factors caused the shah to change his mind. He signed an agreement with Saddam Hussain, redefining the Iran–Iraq border to give Iran control over half of the Shatt al-Arab. In return, the shah agreed to halt support for the Kurds. The northern border was closed and, without Iranian support, Kurdish resistance collapsed. A similar fate befell the Assyrian community. Entire villages were destroyed, and the surviving population was resettled farther south as Saddam Hussein pressed his drive to "purify" the Iraqi nation and preserve Sunni Arab minority authority.

In 1986, Iran resumed support for the Pesh Merga, to use its warriors as an auxiliary force against the Iraqis. Kurdish forces carried out a number of raids into northern Iraq.

But with the end of the war with Iran in 1988, the Iraqi Army turned on the Kurds in a savage and deliberate campaign of genocide. Operation *Anfal* ("spoils," in Arabic) involved the launching of chemical attacks on such villages as Halabja and the forced deportation of Kurdish villagers from their mountains to detention centers in the flatlands. Anfal was directed by Saddam's paternal first cousin Majid, a Tikriti and member of the Ba'th inner circle. Under his personal direction 4,000 Kurdish villages were destroyed, and 5,000 Kurds, mostly old men, women and children were killed in a cyanide gas attack on the border town of Halabya. He became notorious as "Chemical Ali" due to his Anfal leadership, and was thought to have been killed in an air strike on his villa after the U.S. invasion but was subsequently captured. His capture, however, was of little comfort to the families of the 182,000 Kurds slaughtered during the anti-Kurdish campaign.[8]

A second exodus of Kurdish refugees took place in 1991, after uprisings of Kurdish rebels in northern Iraq were brutally suppressed by the Iraqi Army, which had remained loyal to Saddam Hussein. The United States and its allies sent troops and aircraft to the Iraqi–Turkish border and barred Iraq from using its own air space north of the 36th Parallel, the main area of Kurdish settlement. Several hundred thousand refugees subsequently returned to their homes and villages.

FREEDOM

Although violence has continued and even accelerated since the occupation, with not only U.S. soldiers but ordinary Iraqis in the newly-formed police force, Shias and even Sunnis accused of collaborating with Coalition forces as targets, progress toward building a new Iraqi nation under representative government continues, albeit at a slow pace. In September 2003 the U.S. administrator formed a 25-member Governing Council with representatives from women, Kurds, Sunni and Shia communities, Turkoman tribes and Assyrian Christians. In March 2004 the Council reached agreement on an interim constitution as the basis for electing a National Assembly in 2005. The constitution includes a 13-article Bill of Rights.

Under this umbrella of air protection and the exclusion of Iraqi forces from the Kurdish region, the Iraqi Kurds moved toward self-rule in their region. The two main factions—the Kurdish Democratic Party (KDP), led by Massoud al-Barzani, and the Patriotic Union of Kurdistan (PUK) of Jalal al-Talabani—agreed to the formation of a joint Parliament elected by the Kurdish population. This new Parliament, which was divided equally between KDP and PUK members, approved a law defining a federal relationship with Iraq, providing for internal autonomy for the Kurdish region. Kurdish was confirmed as the official language, and a Kurdish university was established.

But the tragedy of the Kurds has always been their inability to unite unless there is an external threat. With the Iraqi regime effectively removed from Kurdistan, the traditional cleavages and inner conflicts of Kurdish society came to the surface. A new Kurdish Parliament was scheduled to be elected in September 1995. However, clashes between the two factions broke into open conflict before the elections could take place. By 1996, the PUK controlled two thirds of the region, including the major cities of Irbil and Sulaymaniyah. Barzani's KDP, although it controlled only one third, held an economic advantage over its rival because of its control over the main source of Kurdish revenues. The lion's share of these revenues came from trade (and smuggling) across the Turkish border.

The bell rang for another tragic hour in Iraqi Kurdistan in September 1996. Barzani's KDP struck a deal with Saddam Hussein to help him unseat the rival PUK; and KDP forces, backed by 30,000 to 40,000 Iraqi troops with tanks and artillery, swept down on Irbil and Sulaymaniyah to drive the PUK from its strongholds. The KDP success was brief; the PUK withdrew into the mountains to regroup and then launched a counteroffensive, which recovered all its lost territories, except Irbil, by mid-October. Saddam Hussein withdrew his forces after a blunt warning from the United States, but not before rounding up opposition dissidents who had remained there after the Gulf War and were supported by the U.S. Central Intelligence Agency to form the anti-Saddam Iraqi National Congress (INC).

Operation Provide Comfort, which was set up after the Gulf War to give the Kurds in northern Iraq a "safe haven" under U.S. and British air protection, enabled the Kurdish region to develop institutions of self-government and preserve its culture while remaining part of the Iraqi state. In 2001 reconciliation between the KDP and PUK ended nearly a decade of party conflict. The establishment of an elected parliament, a school system and bureaucracy with Kurd-

ish as the primary language of instruction and government documents, a university and police force, and an economy based in part on oil revenues from northern Iraqi fields all underscore the emergence of a viable autonomous Iraqi Kurdistan.

The U.S.–U.K. air "umbrella" helped lessen the impact on the Iraqi Kurds of UN sanctions, compared with the rest of the population. A substantial share of income form the "oil-for-food" program was sent there directly, while significant revenues accrued to the Kurds from cross-border smuggling of oil across the Turkish border.

In spite of the Iraqi Kurds' success in developing a viable self-governing region in Iraq, collectively the Kurds remain caught in a bind. It is a double bind, part internal, part international. Even such legendary Kurdish leaders as Mullah Mustafa Barzani (whose son Massoud serves as president pro tem of the Kurdish parliament) were never able to promote the larder interests of Iraqi Kurds beyond tribal and family alliances.[9] Their international bind results from the division of the Kurdish homeland among modern nation-states. As a result, the Kurds remain as almost the only people without a homeland and sovereignty of their own.

THE OPPOSITION

Saddam Hussein's ruthless repression of opposition, made possible by his control of security services, the brutality of his sons and his legion of spies and informers, made sure that no organized group of opponents would emerge to challenge his rule. Those who could to so left the country for exile. But the Iraqi defeat in the Gulf War, followed by the Kurdish and Shia popular uprisings—although they were unsuccessful—suggested that Saddam might be overthrown from abroad. The success of Iraq's Kurds in establishing de facto autonomy under UN–U.S. protection also encouraged the opposition, as did the continuing UN–imposed limitations on Iraqi sovereignty. In June 1992, representatives of some 30 opposition groups met in Vienna, Austria, to form the Iraqi National Congress.

Composed of various opposition groups, the INC described its purpose as the overthrow of Saddam Hussein and the Ba'th Party and their replacement by a secular Islamic regime. The state would be governed under a constitution providing specific guarantees of human rights, protection of minorities, and a multiparty political system. The United States then began funneling funds through its Central Intelligence Agency to the INC, while other CIA operatives worked with its rival anti-Saddam organization, the Iraq National Accord (INA), based in Jordan. In 1994, a team of CIA officers went to

Iraqi Kurdistan to establish a base for the INC as the starting point for a coup against the Iraqi dictator.

However, the conflict between Kurdish rival groups described above encouraged the Iraqi Army to invade their territory. The CIA/INC base was overrun. Those of its members not arrested or executed by the Iraqis were evacuated to the United States after the 1996 presidential election. A similar fate befell the INA, which was potentially more dangerous to Saddam because it was centered in units that formed the core of his support, the Republican Guards and the Security Service (the dreaded *Mukhabarat*). In June 1996, the officers involved were arrested, tortured, and executed as they prepared to stage their coup.[10]

OTHER COMMUNITIES

The Shia community, which forms approximately two-thirds of the total population of Iraq, has been ruled by the Sunni minority since independence. Shias have been consistently underrepresented in successive Ba'thist governments and are the most economically deprived component of the population. However, they remained loyal to the regime (or at least quiescent) during the war with Iran. In a belated attempt to undo decades of deprivation and assure their continued loyalty, the government invested large sums in the rehabilitation of Shia areas in southern Iraq after the war ended. Roads were built, and sacred Shia shrines were repaired.

Long-held Shia grievances against Ba'thist rule erupted in a violent uprising after Iraq's defeat in the Gulf War. The uprising was crushed, however, as Iraqi troops remained loyal to Saddam Hussain. Some 600 troops were killed in an Alamo-type siege of the sacred shrines, which were badly damaged. A few rebels escaped into the almost impenetrable marshlands of southern Iraq. But their expectations of U.S. support proved illusory. The Clinton administration feared that Iran would intervene on behalf of the Iraqi Shias. U.S. helicopters did nothing except to overfly Iraqi gunships as they strafed columns of fleeing refugees. In a half-hearted policy, the Clinton administration eventually declared a no-fly zone south of the 32nd Parallel, off limits to Iraqi aircraft. But by then, the uprising had been crushed.

The rebels' retreat into the marshlands served as an excuse for the Iraqi regime to bring another distinctive community—the Marsh Arabs—under centralized government control. This community, believed by some to be descended from the original inhabitants of southern Iraq and by others to be descended from slaves, has practiced for centuries a unique way of life based on fishing and hunting in the marshes, living in papyrus-and-mud houses and traveling in reed boats through the maze of unmarked channels of their watery region. Prior to 1990, they numbered about 750,000. The late intrepid English explorer Sir Wilfred Thesiger (died 2003) lived with them in the 1950s and chronicled their lifestyle in his eloquent book *The Marsh Arabs*. Tragically this lifestyle was the victim of perhaps Saddam Hussain's most ruthless and repressive actions toward his people in his decades in power. In 1991 he sent troops and artillery ferried by helicopters into the marshes in pursuit of Shia rebels who had fled there after their failed uprising.

HEALTH/WELFARE

Manpower losses in two wars plus violence since the 2003 U.S. occupation have resulted in a surplus of women; they make up approximately 60 percent of the population today. However the UN sanctions and lack of hospital care and obstetrics training have adversely affected women, particularly pregnant mothers. In 2003, the most recent year of accurate measurements of mortality, some 370 out of 100,000 women died in childbirth. This is triple 1990 rates and 31 times the U.S. rate of 12/100,000. The continuing insurgency forces many women to give birth at home, usually attended by an untrained midwife or obstetrical assistant. A government program to train Iraqi doctors, nurses, and midwives begun in 2005 has had some success but is severely limited by the violence.

After destroying a number of villages and removing much of the population for resettlement elsewhere, Saddam Hussain followed the military campaign with a vast reclamation project. It involved construction of a "third river," a canal between the Tigris and the Euphrates Rivers which would draw water from both rivers through a network of small dams, levees, and diked ponds to irrigate new land being brought under cultivation. A body of 6,000 workers drawn from surplus Iraqi labor, working around the clock, completed the project in 3 months in 1993. And a decade later 93–95 percent of the 3,000-square mile marshes, larger than the state of Maryland, lay bone-dry and almost uninhabited.[11]

THE ECONOMY

Iraq's economy since independence has been based on oil production and exports. The country also has large natural-gas reserves as well as phosphate rock, sulfur, lead, gypsum, and iron ore. Ancient Mesopotamia was probably the first area in the world to develop agriculture, using the fertile soil nourished by the Tigris and Euphrates Rivers. Until recently, Iraq was the world's largest exporter of dates. However, by 1999 the UN embargo and the longest drought in a century had brought food production to a near standstill. An estimated 70 percent of wheat and barley crops, mainstays of agriculture, were lost, and government officials described the situation as a "food catastrophe" comparable to the collapse of the health-care system. The Ba'th economic policies emphasized state control of the economy while the party was in power, under the Ba'thist rubric of guided socialism. In 1987, the regime began a major economic restructuring program. More than 600 state organizations were abolished, and young technocrats replaced many senior ministers. In 1988, the government began selling off state-run industries, reserving only heavy industry and hydrocarbons for state operation. Light industries such as breweries and dairy plants would henceforth be run by the private sector.

The oil industry was developed by the British during the mandate but was nationalized in the early 1970s. Nationalization and price increases after 1973 helped to accelerate economic growth. The bulk of Iraqi oil shipments are exported via pipelines across Turkey and Syria. During the war with Iran, the Turkish pipeline proved essential to Iraq's economic survival, since the one across Syrian territory was closed and Iraq's own refineries and ports were put out of commission by Iranian attacks. Turkey closed this pipeline during the 1990–1991 Gulf crisis, a decision that proved a severe strain for the Iraqi economy (not to mention a huge sacrifice for coalition-member Turkey).

Iraq has proven oil reserves of some 100 billion barrels, the fifth largest in the world, and new discoveries continue to augment the total. Oil output was cut to 2 million barrels per day in 1986–1987, in accordance with quotas set by the Organization of Petroleum Exporting Countries, but was increased to 4.5 million b/d in 1989 as the country sought to recover economically from war damage.

The economic impact of the eight-year Iran–Iraq War was heavy, causing delays in interest payments on foreign loans, defaults to some foreign contractors, and postponement of major development projects except for dams, deemed vital to agricultural production. The war also was a heavy drain on Iraqi finances; arms purchases between 1981 and 1985 cost the government $23.9 billion. By 1986, the external debt was $12 billion. By 1988, the debt burden had gone up to nearly $60 billion, although half this total had been given by the Arab Gulf states as war aid and was unlikely ever to be repaid.

United Nations Photo (UN159152)

UN Security Council *Resolution 687* called for the disposal of Iraq's weapons of mass destruction. Saddam Hussain had stockpiled enormous quantities of chemical munitions, and while many of these insidious weapons were indeed destroyed, large quantities remained hidden. Iraq's refusal to cooperate with international inspection teams has caused sanctions to remain in place.

Iraq's economic recovery after the war with Iran, despite heavy external debts, suggested rapid growth in the 1990s. Gross domestic product was expected to rise by 5 percent a year due to increasing oil revenues. Even in 1988, Iraq's GDP of $50 billion was the highest in the Arab world, after Saudi Arabia's. With a well-developed infrastructure and a highly trained workforce, Iraq appeared ready to move upward into the ranks of the developed nations.

THE UN EMBARGO

Iraq's invasion and occupation of Kuwait and the resulting Gulf War drew a red line through these optimistic prospects. Bombing raids destroyed much of Iraq's infrastructure, knocking out electricity grids, bridges, and sewage and water-purification systems and refurbishing of industries, oil refineries and installations in particular. UN resolutions imposing sanctions on the country after the Gulf War added to its economic distress. The on-going insurgency with its attacks on oil installations and water and sewer systems have seriously delayed the restoration of this essential infrastructure.

The UN embargo that was imposed on Iraq after the Gulf War to force compliance with resolutions ordering the country to dismantle its weapons program has not only brought development to a halt but has

also caused untold suffering for the Iraqi population. The resolutions in question were *Resolution 687,* which required the destruction of all missile, chemical, and nuclear facilities; *Resolution 713,* which established a permanent UN monitoring system for all missile test sites and nuclear installations; and *Resolution 986,* which allowed Iraq to sell 700,000 barrels of oil per day for six months, in return for its compliance with the first two resolutions. Of the $1.6 billion raised through oil sales, $300 million would be paid into a UN reparations fund for Kuwait. Another $300 million would be put aside to finance the UN monitoring system as well as providing aid for the Kurdish population. The remainder would revert to Iraq to be used for purchases of food and medical supplies.

Saddam Hussain initially refused to be bound by *Resolution 986,* calling it an infringement of Iraq's national sovereignty. But in 1996, he agreed to its terms. By then, the Iraqi people were nearly destitute, suffering from extreme shortages of food and medicines. The United Nations estimated that 750,000 Iraqi children were "severely malnourished." Half a million had died, and the monthly death toll from malnutrition-related illnesses was averaging 5,750, the majority of them children under age five, due to lack of basic medicines and hospital equipment.[12] In 1998, the Security

Council increased approved Iraqi oil revenues to $5.26 billion every six months. Higher world oil prices and exemptions to make up for earlier shortfalls in its export quota due to equipment breakdowns brought total revenues to $7 billion in 1999. In all, Iraq has received $40 billion from oil sales since 1996. However, the UN Compensation Commission, which is responsible for reparations to companies and individuals for losses sustained during the occupation of Kuwait, disbursed $365 million in payment of claims in September 2001. The largest single payment, $176.3 million, went to Botas Petroleum Pipeline Corporation to cover losses caused by the shutdown in the pipeline from Iraq to the Turkish port of Iskenderun. Since its establishment, the commission has paid $35.4 billion to claimants.

Despite disagreement within the Security Council over the scope, effectiveness and moral legitimacy of the sanctions, they were kept in force in 6-month increments up until the U.S. invasion and overthrow of the Iraqi regime, the last renewal being in December 2002. However concern in many countries about their devastating effect on the population and the economy led the Arab states and others who had been Iraq's trading partners to bypass or simply ignore them. By 2001, 20 countries had resumed regular air service to Baghdad International Airport,

ACHIEVEMENTS

Despite the chronic political instability and violence, in a few areas national life is slowly returning to normal. The first-ever Iraqi entry in the Cannes Film Festival received honorable mention in 2005. A film festival in Baghdad in the summer featured a children's theater venue of 58 documentaries on subjects ranging from sheep farming to whimsical animations, all done by young filmmakers. The Iraqi National Junior Baseball League, founded in 2003 with U.S. help, now has 26 teams in 18 provinces, and in July 2004 it was admitted as a member by the National Olympic Committee. Despite the violence, ordinary Iraqis surf the Web, attend school, frequent outdoor markets, go to movies and as far as possible lead normal lives.

much of it in humanitarian supplies. They included Turkey, Egypt and Syria, all former members of the Desert Storm coalition. A free-trade agreement with Syria in January 2001 would triple the annual trade volume of the two countries, to $1 billion, and additional contracts with Jordan, Lebanon and the UAE would generate $4.7 billion in Iraqi exports. Unfortunately the "oil-for-food" program, rather than bringing material benefits to the people, resulted in huge profits for the companies involved. In what one writer called "the many streams that fed the river of graft to Hussein," some 4,758 companies in the program paid over $1.8 billion in illegal contributions, most of it in kickbacks on aid contracts or surcharges on oil purchases. An independent un-sponsored report on late 2005 detailed a range of abuses, involving companies or individuals from 66 countries. As examples it cited a $7,000 surcharge on a $70,000 Daimler-Chrysler contract and kickbacks by 3 Siemens companies of $1.6 million. Companies benefiting from the arrangement were given oil barrel allocations based on their level of public opposition to the sanctions.

GLORIOUS LEADER, VANQUISHED SURVIVOR

During his two-plus decades in power Saddam Hussain held complete sway over this people, his authority buttressed by a loyal army, his specially-trained Revolutionary Guards, an efficient security service and a huge corps of informers covering practically every street corner in Iraq. Internally his control over the Ba'th Party was enhanced by the Tikritis, his relatives and other Sunni loyalists from Tikrit. The title of a book by Kanan Makiya (pseud. Samir al-Khalil) *Republic of Fear*, accurately describes his country albeit written in exile. Under Saddam Iraq had become a huge prison; its people in a very real sense had

bartered their freedom for his protection. After the UN imposed sanctions their jailers became even more isolated. The impact of sanctions in fact fell heaviest on the middle class. Prior to the 1990s This sector of society had profited from oil-based development to become the best-educated and most productive in all the Arab states. But Saddam Hussain's excesses and particularly his ill-advised foreign policy ventures effectively ruined this class.

Saddam Hussain's central role as ruler of Iraq was underscored during his years in power by his extreme visibility. In addition to his numerous palaces, gigantic statues and posters of him in cities, on highway billboards, and before banks and other public buildings served as constant reminders of the Glorious Leader. In Baghdad victory arches and a statue of him were erected on the first anniversary of the U.S. air strikes. Saddam called it a "great victory," similar to the "defence" of the homeland after the 1991 Gulf War.

In spite of his absolute power, Saddam's rule was not entirely free from attempted coups. An army coup by officers from the Dalaimi clan, traditional rivals of his Tikriti clan, was thwarted with considerable difficulty in June 1995. A much greater threat was the defection to Jordan of two of his sons-in-law with their wives and some Sunni officers. One son-in-law, Hussein Kamel, had headed the Iraqi secret-weapons program and was a member of the inner circle around the president. Kamel was given asylum in Jordan and talked openly of leading a coup to overthrow Saddam. But in time, the Iraqi exiles became an embarrassment to their hosts. Seven months after their arrival, the party left Jordan to return to Iraq, having received an offer from Saddam to pardon them as "repentant sinners."

Accepting their repentance was the last thing on Saddam's mind. After the exiles had returned to Tikrit they were surrounded in the family compound and killed in a fierce fire fight by rival clan members loyal to the regime, who declared that they had stained the family honor. Saddam's daughters severed all contact with their father, and in March 2003, following his overthrow, they returned to Jordan and asked for asylum. In approving the asylum offer, a Jordanian official commented that "they are Arab women who have run out of options."

Saddam's skill in evading the direct impact of sanctions on his lifestyle and playing off more powerful countries against each other for Iraq's benefit was more than equaled by his internal actions. Those who survived arrest, torture and incarceration in his infamous prisons, many of them skilled professionals, usually fled into exile. The

assassination of Grand Ayatollah Sadiq al-Badr, spiritual head of the Shia community in Iraq, was a grim warning that anyone who spoke out against the Iraqi leader or questioned his decisions would suffer the same fate.

Until early 2003 it seemed that the Glorious Leader would survive UN sanctions, U.S. air attacks and international isolation and maintain his grip on power. Internal opposition was ruthlessly suppressed. The murder of the spiritual leader of the Shia community served as a warning of what future dissidents might expect. As a result the only organized opposition to the regime operated outside the country. It was the Iraqi National Congress, headed by Ahmad Chalabi. In 1999 Congress appropriated $97 million for its activities. Little of this money was actually spent. However, under the new Bush administration the Pentagon began paying the organization $340,000 per month. The payments were described as part of an "intelligence collection program" authorized by Congress under the Iraq Liberation Act. Unfortunately most INC reports were either falsified or incorrect. As a "favorite Iraqi" of the U.S. Defense Department, Chalabi returned to Iraq after many years in exile, most of it in Jordan. After the January 2005 elections he was considered briefly as a possible prime minister. As a secular Shiite Arab he enjoys some support within the Shia community, but his association with the U.S. has proven more handicap than asset.

Another of Chalabi's pet projects was a satellite TV station called Liberty TV. It went on the air in August 2001. The $1 million start-up costs and $1.3 million operating fees would be paid from new Congressional appropriations. Being broadcast by satellite and therefore not subject by jamming from within Iraq, it would theoretically reach that portion of the Iraqi elite with access to satellite dishes. However, Liberty TV was quickly eclipsed by the Qatar-based satellite TV station al-Jazeera, which captivated the Arab world as well as Iraq with its freewheeling American-style news and information programs. In December 2001 the State Department suspended funding, citing a lack of proper accounting procedures for its U.S. funds.

The other organized opposition group outside the country, the Supreme Council for Islamic Revolution in Iraq (SCIRI), was formed in 1982 in Iran. It was essentially an umbrella group for various Saddam opponents, most of them Shia. SCIRI's original goal was to establish an Islamic regime in Iraq similar to Khomeini's in Iran. Its military wing, the Badr Brigade, fought with Iranian forces against Iraq in the later stages of that war.

The Brigade entered Iraq after Saddam's fall and has cooperated with U.S. forces, even to the extent of laying down its weapons on request. SCIRI's spiritual leader, the Ayatollah Baqr al-Hakim, served as liaison between the two forces before his untimely assassination, and his brother Abdelaziz, also a Shia cleric, was appointed as a member of the Governing Council when it was formed in August 2003.

Saddam Hussain's ruthless elimination of political opponents, along with his autocratic rule, have identified him with the Iraqi nation to a greater degree than that of any of his predecessors. For more than two decades he has been the arbiter of power, the ultimate dispenser of justice, the sole formulator of national policy.

Following the September 11, 2001, terrorist bombings in the United States, Saddam denounced the action and the killing of innocent civilians. However, there were huge public demonstrations in Baghdad, presumably government-sponsored, to protest continued U.S. and British air raids on Iraqi territory and the U.S. military campaign in Afghanistan. Protesters carried banners that read "Down With American Terrorism Against Islam." Subsequently in January 2002 President George Bush included Iraq in his "axis of evil" of countries sponsoring terrorism and accused it of violating the 1972 treaty banning bacteriological, chemical, and other weapons of mass destruction. (Iraq is a signatory to the treaty.)

THE U.S. INVASION AND ITS IMPACT

The invasion of March 2003 and overthrow of Saddam Hussein by U.S. and British forces is dealt with as a conflict issue elsewhere in this book (see Theater of Conflict essay). But its impact on the Iraqi people as they struggle to put years of authoritarian rule behind them and construct a viable system of government, one based on law and human rights and buttressed by constitutional protections has been, continues to be extremely difficult especially given the widespread insurgency.

From its beginnings as an artificial nation–state patched together by outsiders, Iraq has always lacked the essential ingredients for successful nationhood. It was traditionally fragmented into many different groups with different and often opposed identities. Iraqi society was and to a great extent still is tribal, ethnic, religious, linguistic, urban and rural but not "national." Saddam Hussain was able to override these differences by sheer force or personality and absolute power. Amid a host of negative contributions, the Butcher of Baghdad must be credited for forcing these disparate elements into an Iraqi unitary state.

What lies ahead for this battered nation? In retrospect it is clear that the fall of Saddam Hussain could not be accomplished in any other way than by external invasion. But as should have been expected, the abrupt removal of an absolute ruler and the collapse of his regime left a huge political vacuum in Iraq.

What seems to have escaped awareness by U.S. policymakers was the deplorable condition of the Iraqi economy. This fact alone militated against any possibility of an on-going weapons program there. The years of UN sanctions and U.S. bombings had destroyed most of its infrastructure. Roads, electricity, the water system and health care had dropped to a primitive level, comparable with that of Bangladesh. The sudden collapse of the regime also set off an orgy of looting, revenge killings and destruction of the remaining essential services and facilities. A tragic loss was the looting of the National Museum in Baghdad with its priceless collection of artifacts. (Fortunately many were returned and thousands of others had been hidden by museum curators, but thousands more are still unaccounted for.)

Following President Bush's April 2003 announcement that the war had ended, Coalition forces set out to eliminate the top Ba'th leadership. A pack of 55 large cards with their faces displayed was circulated, in an eerie repetition of Nuri al-Said's "pack of cards" which he used to characterize the government during the monarchy. The cards brought results, abetted by large bounties. Those captured or killed included Saddam's two sons, Uday and Qusay, killed in a gunfight after an anonymous tip, and "Chemical Ali," first reported killed in an air strike but arrested subsequently. In December 2003 the Glorious Leader himself, bearded, disheveled and anything but "glorious," was trapped in an underground hideout in Adwar, near Tikrit.

Subsequently the fallen dictator was brought to trial after inordinate delays; CIA investigators had been the only persons permitted access to him. However he frequently interrupted the proceedings by walking out or insisting that as president he could not be tried in an Iraqi court of law. The trial was resumed in February 2006 under a different judge, Raouf Abdul-Rahman (a Kurd). While the overall intent was to try Saddam for war crimes, the particular charge was that of his direct linkage with the massacre of 148 Shia men and boys in the village of Dujail.

The "war that ended" with a presidential announcement months ago unfortunately continues and has changed significantly in scope. Although U.S. casualties continued to mount, the brunt of the violence is directed at ordinary Iraqis, the police, security services, school children, women, and all who seem in any way connected with or supported by the U.S. forces. It is a vicious cycle—the more American troops attempt to curb the violence, the more they alienate the Iraqi people. As an Iraqi policeman told a reporter, "they (the soldiers) treat us like Palestinians. They treat us like dirt. Our chief of police is in jail right now." [13]

Timeline: PAST

1520–1920
Border province of the Ottoman Empire

1920–1932
British mandate

1932
Independent kingdom under Faisal I

1958
The monarchy is overthrown by military officers

1968
The Ba'th Party seizes power

1975
The Algiers Agreement between the shah of Iran and Hussain ends Kurdish insurrection

1980s
Iran–Iraq War; diplomatic relations are restored with the United States after a 17-year break

1990s
Iraq invades and occupies Kuwait, leading to the brief but intense Gulf War; Saddam Hussain retains power

PRESENT

2000s
Despite continuing UN sanctions, Saddam remains firmly in control

Meanwhile Iraq's children are trickling back to their reopened schools, business improves, the sanctions are lifted and a sense of normalcy is slowly returning. Some have argued that the Iraq of the future may already have a role model in the Kurdish region, with its institutions well in place. In September 2003 a 25-member Governing Council took office with the U.S. Coalition administrator's approval. Its members are drawn from the Shia, Sunni and Kurdish populations, the Turkomans, the Communist Party, clerics and women. On September 10 Hoshyar Zebari, a Kurdish Council member, took his seat in the Council of Ministers of the Arab League as Iraq's representative. His opening statement may well portend the Iraqi future: "the new Iraq will be based on diversity, democracy, constitutional law and respect for human rights. It will stand firm against terrorism, from which it is now suffering." In the light of this statement, the land between the rivers may well become the role model for Arab Middle Eastern democracy that George Bush insists it will.

THREE ELECTIONS

The year 2005 has been marked by three sets of elections, in which the Iraqi people for the first time in their history have begun to sort out what they can become as a sovereign state. The first election, January, elected a 275-member Transitional National Assembly. While the Sunni population largely boycotted the proceedings, some 8.4 million valid votes were cast, half from the 4 provinces of Baghdad, Sulaimaniya, Basra and Erbil. The United Iraqi Alliance, a Shia coalition, won 48 percent of the popular vote to 26 percent for the Kurdistan Alliance and 14 percent for the Iraqi List, a Sunni-Shia coalition headed by interim prime minister Ayad Allawi. The successful Alliance candidates included the ever-resourceful Ahmed Chalabi.

Iraq's second "election" was the national referendum to approve the new national constitution. To win their support, Sunnis were promised the right to propose constitutional amendments during the first 4 months of the Transitional National Assembly, which took office in January 2005.

On December 15, 2005, Iraq held its first free multi-party parliamentary election in half a century. Voters would choose from some 6,650 candidates belonging to 307 parties, most of them belonging to larger political alliances. Despite threats of violence 70 percent of the country's 15 million eligible voters cast their ballots. The election was organized differently from the Transitional one, as voters chose candidates ac-

cording to provinces and districts. The election pattern also guaranteed 15 percent of sets to women and 13 percent to Sunnis.

The election results were somewhat expected given Iraq's ethnic/religious makeup. The United Iraqi Alliance, a Shia coalition, won 128 seats, with the two Sunni Alliance blocs taking 55 and the Kurdish Alliance List 53. Because no group held a clear majority, much hard bargaining remained before a government could be put in place. The one positive step taken as of early March was the choice of prime minister, and Shia lawmakers exercised their constitutional rights in re-electing Jaafari to the post by a one-vote margin.

NOTES

1. K. S. Husry, "The Assyrian Affair of 1933," *International Journal of Middle East Studies* (1974), p. 166. The Assyrians are also called Chaldeans.
2. Muhammad A. Tarbush, *The Role of the Military in Politics: A Case Study of Iraq to 1941* (London: Kegan Paul, 1982), p. 50.
3. Richard F. Nyrop, *Iraq: A Country Study.* Washington D.C.: American University, Foreign Area Studies (1979), p. 38. Faisal I had noted sadly just before his death: "There is no Iraqi people but unimaginable masses of human beings, devoid of any patriotic feeling, connected by no common tie, perpetually ready to rise against any government." Quoted in Hanna Batatu, *The Old Social Classes and the Revolutionary Movements in Iraq* (Princeton, NJ: Princeton University Press, 1978), pp. 25–26.
4. Joe Stork, "State Power and Economic Structure…" in Tim Niblock, *Iraq: The Contemporary State* (London: Croom Helm, 1982), p. 44.
5. Milton Viorst, "Letter From Baghdad," *The New Yorker* (June 24, 1991), p. 61.
6. Ofra Bengio, *Saddam's Word: Political Discourse in Iraq* (London: Oxford University Press, 1988), p. 24.
7. Khidr Hamza, with Jeff Stein, *Saddam's Bombmaker* (New York: Scribner's, 2001). Hamza was head of the team that designed the bomb.

8. "Anafal" is the Arabic name of the 8th sura (chapter) of the Koran and appeared as a revelation to Muhammad after the battle of Badr, the first victory of the Muslims over their Meccan enemies. It was viewed by them (and by Saddam) as proof that God and right were on their side. See Human Rights Watch, *Iraq's Crime of Genocide: The Anfal Campaign Against the Kurds* (New Haven, CT: Yale University Press, 1995), p. 4. The campaign would have never come to light but it was fully documented when 18 tons of Iraqi government documents were captured by Kurdish pesh mergas during the uprising that followed the Gulf War.
9. Henri J. Barkey, "Kurdish Geopolitics," *Current History* (January 1997), p. 2.
10. The INA was "managed" from Jordan by a special CIA team. After the coup had been thwarted—the regime had advance warning through penetration of the CIS's satellite-technology communications system—the team received a message: "We have arrested all your people. You might as well pack up and go home." The CIA team did just that. Andrew Cockburn and Peter Cockburn, *Out of the Ashes: The Resurrection of Saddam Hussain* (New York: HarperCollins, 1999), p. 229.
11. In June 2003 heavy spring rains and snow melt in the Turkish highlands where the rivers begin led occupation Forces to join with the remaining marsh Arabs (the Ma'adan) to open floodgates and levees to allow water to flow back into the marshes. An international scientific team has been formed to resurrect the marshes and it may be that in this case nature can heal herself with human help!
12. A judge who served on the Iraq Court of Appeals and had dared to rule one of Saddam's edicts unconstitutional was thrown in jail at the notorious Abu Ghraib prison outside Baghdad. A visitor who knew him visited the prison after the occupation and noted, in addition to chambers for torture and cells where prisoners were packed like vermin, there was a long bar with a deep pit with room for several nooses to be used at the same time.
13. Christian Parenti, "Two Sides: Scenes from a Nasty Brutish War," *The Nation*, February 25, 2004, p. 14.

Israel (State of Israel)

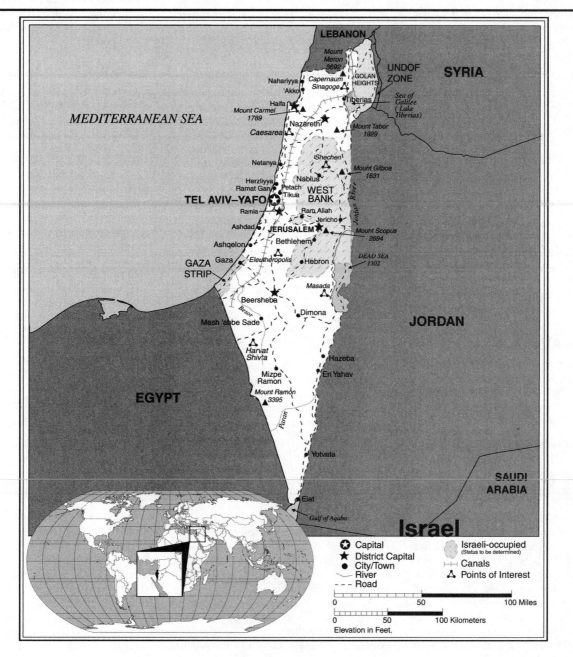

Israel Statistics

GEOGRAPHY

Area in Square Miles (Kilometers): 8,017 (20,770) (about the size of New Jersey)

Capital (Population): Tel Aviv (356,000) recognized by most countries; Jerusalem (591,000) claimed as the capital but not internationally recognized

Environmental Concerns: limited arable land and fresh water; desertification; air and groundwater pollution; fertilizers; pesticides

Geographical Features: desert in south; low coastal plain; central mountains; Jordan Rift Valley

Climate: hot and dry in southern Negev Desert area

PEOPLE

Population

Total: 6,276,883 (includes 187,000 Jewish settlers in the occupied West Bank, 20,000 in the Golan Heights, 5,000 in the occupied Gaza Strip, and about 177,000 in East Jerusalem)
Annual Growth Rate: 1.52%
Rural/Urban Population Ratio: 9/91
Major Languages: Hebrew; English; Arabic
Ethnic Makeup: 80% Jewish; 20% non-Jewish (mostly Arab)
Religions: 76.5% Jewish; 15.5% Muslim including Druze; 2% Christian; 6% others

Health

Life Expectancy at Birth: 78.86 years (male); 81 years (female)
Infant Mortality Rate: 7.7/1,000
Physicians Available (Ratio): 1/206 people

Education

Adult Literacy Rate: 95.4%
Compulsory (Ages): 5–16; free

COMMUNICATION

Telephones: 2,656,000 main lines
Daily Newspaper Circulation: 271 per 1,000 people
Televisions: 290 per 1,000 people
Users: 21 (2000)

TRANSPORTATION

Highways in Miles (Kilometers): 9,603 (15,464)
Railroads in Miles (Kilometers): 379 (610)
Usable Airfields: 54
Motor Vehicles in Use: 1,544,000

GOVERNMENT

Type: republic
Independence Date: May 14, 1948 (from a League of Nations mandate under British administration)
Head of State/Government: President Moshe Katsav (mainly ceremonial); Acting Prime Minister Ehud Olmert
Political Parties: Pending March 28, 2006 elections coalition government continues with Likud Bloc and Labor Alignment as the majority parties. Other parties include Shinui, Shas, National Religious Party and 2 small Arab parties. A new centrist party, Kadima, was formed by Prime Minister Ariel Sharon before his incapacitation.
Suffrage: universal at 18

MILITARY

Military Expenditures (% of GDP): 8.75%
Current Disputes: Evacuation of settlers and military forces from the Gaza Strip in November 2005 ended one phase of the Israeli-Palestinian conflict. Dispute with Syria over Israeli occupation and annexation of the Golan Heights is unresolved but inactive. Israel has officially relinquished sovereignty over 4 West Bank towns but remains in occupation

ECONOMY

Currency ($U.S. Equivalent): 48 shekels = $1
Per Capita Income/GDP: $20,800/$129 billion
GDP Growth Rate: 3.9%
Inflation Rate: 0%
Unemployment Rate: 10.7%
Labor Force: 2,400,000
Natural Resources: copper; phosphates; bromide; potash; clay; sand; timber; manganese; natural gas; oil
Agriculture: citrus fruits; vegetables; cotton; beef; poultry; dairy products
Industry: food processing; diamond cutting; textiles and apparel; chemicals; high-technology projects; wood and paper; others
Exports: $34.4 billion (primary partners United States, United Kingdom, Benelux)
Imports: $36.8 billion (primary partners United States, Benelux, Germany)

SUGGESTED WEBSITE

http://lcweb2.loc.gov/frd/cs/iltoc.html

Israel Country Report

Israel, the Holy Land of Judeo–Christian tradition, is a very small state about the size of New Jersey. Its population is also smaller than those of most of its neighbors', with low birth and immigration rates. Population growth within these limits would be manageable. But until recently, the country's very existence was not accepted by its neighbors (even now, it is not recognized by all), and its borders remained temporary ones under the 1949 armistice agreements that ended the first Arab–Israeli War.

The country occupies a larger land area than it held at the time of its independence in 1948, due to expansion wars with its neighbors. The border with Egypt was defined by a 1979 peace treaty; subsequently, Israel retroceded the Sinai Peninsula, which had been occupied by Israeli forces in 1973. In 1994, Israel signed a peace treaty with Jordan. Among its provisions, their common border was demarcated, and areas in the Galilee were returned to Jordan. Peace treaties with Israel's other two Arab neighbors have yet to be signed. Israeli forces occupied a nine-mile-wide, self-declared "security zone" along the Lebanese border from 1982 until 2000, when they were withdrawn.

Although it is small, Israel has a complex geography, with a number of distinct regions. The northern region, Galilee, is a continuation of the Lebanese mountains, but at a lower altitude. The Galilee uplands drop steeply on three sides: to the Jordan Valley on the east; to a narrow coastal plain on the west; and southward to the Valley of Esdraelon, a broad inland valley from the Mediterranean to the Jordan River, which is fertile and well watered and has become important to Israeli agriculture.

Another upland plateau extends south from Esdraelon for about 90 miles. This area contains the ancient Jewish heartland—Judea and Samaria to Israelis, the West Bank to Palestinians—which is supposed to serve as the core of the self-governing Palestinian state as defined in the 1993 Oslo Agreement. This plateau gradu-ally levels off into semidesert, the barren wilderness of Judea. The wilderness merges imperceptibly into the Negev, a desert region that comprises 60 percent of the land area but has only about 12 percent of the population.

TERRITORIAL CHANGES

The return of the Gaza Strip, captured from Egypt in the 1967 Six-Day War, to Palestinian control in 2005 leaves three territories held by Israel which were not part of the Jewish state approved in the 1947 UN partition plan for Palestine. They are the West Bank (of the Jordan River,) East Jerusalem (captured from Jordan at the same time) and the Golan Heights, captured from Syria in 1973 and annexed unilaterally in 1981. Since the 1973 Yom Kippur War the United Nations has maintained a small observer force in the demilitarized zone between Syrian and Israeli territory on the Golan. Syria's nonrecognition of Israel and Israel's insistence that the

Israelis regard Jerusalem as the political and spiritual capital of a Israel. East Jerusalem was annexed from Jordan after the 1967 Six-Day War, and returning this part of the city to Jordan has never been seriously considered by Israel.

Golan Heights are essential to its security have made resolution of the dispute all but impossible. In January 2000, Syrian and Israeli negotiators met in Shepherdstown, West Virginia, under U.S. sponsorship, but they were unable to reconcile their conflicting claims to the territory.

The 1993 Oslo Agreement and the 1998 Wye Agreement, signed by then-prime minister Binyamin Netanyahu and the late Yassir Arafat, leader of the Palestine Liberation Organization (PLO), set guidelines for the gradual transfer of these territories to Palestinian self-rule (excluding the Golan Heights, which is slated to be returned to Syria eventually). The Wye Agreement specified that 15 percent of the West Bank would be transferred by February 2000. In September 1999, Israel transferred 7 percent (160 square miles) of West Bank land to Palestinian control. An additional 5 percent was transferred in January 2000. The following summer, the new Israeli prime minister, Ehud Barak, met with Arafat at the presidential retreat in Camp David, Maryland, under U.S. sponsorship, and offered to turn over 94 percent of the West Bank for inclusion in the new Palestinian state. The Palestine National Authority (PNA) would be given administrative rule over the Dome of the Rock, excluding the

Wailing Wall, as well as the non-Jewish quarters of East Jerusalem. It was the most generous offer yet made to the Palestinians. Unfortunately Arafat's rejection of the offer on the grounds that it did not recognize the "right of return" of Palestinian refugees to their former homes and villages with appropriate compensation effectively blocked what might have been a settlement of the protracted conflict.

The issue of East Jerusalem—the "Old City" sacred to three faiths but revered especially by Jews as their spiritual, emotional, and political capital—remains a difficult one to resolve. From 1949 to 1967, it was under Jordanian control, with Jews prohibited from visiting the Wailing Wall and other sites important in the history of Judaism. Almost the only contact between the divided sectors of east and west was at the Mandelbaum Gate. In the 1967 Six-Day War, Israeli forces captured the Old City in fierce fighting. Since then its holy sites, sacred to Muslims, Christians, and Jews, have been opened to denominational and religious use. The Israeli government and its people are united in regarding Jerusalem as their eternal and political capital, although it is recognized as such by very few countries. Until very recently this status was considered nonnego-

tiable by Israelis. Acting Prime Minister Olmert, in his first major policy address, stated that in any final settlement Jerusalem would remain a unified city under Israeli control.

THE POPULATION

The great majority of the Israeli population are Jewish. Judaism is the state religion; Hebrew, the ancient liturgical language revived and modernized in the twentieth century, is the official language, although English is widely used. Language and religion, along with shared historical traditions, a rich ancient culture, and a commitment to the survival of the Jewish state, have fostered a strong sense of national unity among the Israeli people. They are extremely nationalistic, and these feelings are increased because of hostile neighbors. Most Israelis believe that their neighbors are determined to destroy their state, and this belief has helped to develop a "siege mentality" among them. This siege mentality has deep roots. Nobel Peace Prize–winner Elie Wiesel has defined it as follows: "Jewish history, flooded by suffering but anchored in defiance, describes a permanent conflict between *us* [Jews] and the others. Ever since

Abraham [the father of Judaism], we have been on one side and the rest of the world on the other."[2]

Except for a small population of Jews that remained in the region (the village of Peki'in in Galilee is said to be the only one with an unbroken Jewish presence over the past 2,000 years), Jews dispersed throughout the world after Jerusalem's conquest by Roman legions in A.D. 70. Those who settled in Europe are called Ashkenazis (from Ashkenaz, Genesis 10:3). Most of them settled in France and Germany, and later in Central/Eastern Europe. Other Jewish communities found refuge in Spain, where they later prospered under Islamic rule for seven centuries (A.D. 711–1492). Such Jewish intellectual leaders as Moses Ben Maimon (Maimonides) rendered important services to Muslim rulers, and their role in trans-Mediterranean trade was so valuable that their Hispano–Moorish commercial language, Ladino, became the lingua franca of Christian–Islamic economic relations.

The Reconquista (Reconquest) of all Spain by the crusading army of Christian rulers Ferdinand of Leon and Isabella of Castile brought misfortune and tragedy to both Jews and Muslims. Both groups were ordered to convert or face expulsion. Jews who chose to convert were called Conversos, or less flatteringly *marranos* (lit. "pigs"), and many of them came under suspicion by the Spanish Inquisition for preserving and practicing their Jewish faith in secret. In 1511 A.D. the Spanish king, Philip II, issued an "Edict of Expulsion" ordering all Muslims and Jews to leave Spain. As a result the remaining Jews left their Spanish homeland, settling either in Ottoman lands or Morocco and other areas of Muslim North Africa, where they were assured of safety and protection. As a result of this second Diaspora they became known collectively as Sephardim.[3]

The diversity among incoming Jews, particularly the Sephardic communities, was so great during the early years of independence that the government developed a special orientation program of Hebrew language and culture, called Ulpan (which is still in use), to help with their assimilation. Some Sephardic groups have prospered and gained economic and political equality with Ashkenazis. The election in 2000 of Moshe Katsav, originally from Yazd (Iran) as president and of Moroccan-born Amir Peretz in 2005 as head of the Labor Alignment, has generated new-found pride in the Sephardic community.

Another difference among Israelis has to do with religious practice. The Hasidim, or Orthodox Jews, strictly observe the rules and social practices of Judaism and live in their own separate neighborhoods within

cities. Reform Jews, by far the majority, are Jewish in their traditions, history, and faith, but they modify their religious practices to conform to the demands of modern life and thought. Both the Orthodox and Reform Jews have chief rabbis who sit on the Supreme Rabbinical Council, the principal interpretive body for Judaism.

Small groups of Jews of ancient origins, the Samaritans, have lived for centuries in Palestine in two locations, Holon (near Tel Aviv) and Nablus (in the West Bank). They are descended from one of the ancient Jewish tribes, one that broke away from the mainstream Jerusalem-based community over the location of Abraham's putative sacrifice of Isaac to God. (They believe that it took place on Mt. Gerizim rather than Mt. Moriah.) After the 1948 war for Israeli independence, they were separated; the Holon community was now part of Israel, while the Nablus Samaritans came under Jordanian rule. The 1967 Six-Day War reunited them, and since then they have served as intermediaries between their Palestinian neighbors and the Israeli authorities, being acceptable to both. The Samaritan community currently numbers just over 600. Most speak Arabic as well as Hebrew and are at home in both cultures.

Relations between the majority Reform Jews in Israel and the much smaller Orthodox community were marked by occasional incidents of friction but overall coexistence until the rise to power of Menachem Begin's Likud Bloc. Begin's own political party, Herut, always emphasized the country's biblical heritage in its platform. But during Begin's period in office, the small religious parties that represent the Orthodox acquired political power because they were essential to the coalition government. The higher birth rate among haredim and other factors, such as exemption from military service for students enrolled in *yeshivas* (religious schools), gives them political clout to supplement the respect accorded them by Reform Jews due to their strict observance of Judaic law and custom. Their importance in Israel's fractious politics was amply demonstrated in the January 2003 Knesset elections. The Likud Party won the majority of seats with 38, but in order to govern the country it had to accept a coalition with Shinui and Shas, the ultra-Orthodox parties, which won 30.

Differences in historical experiences have also divided the Ashkenazis. Most lived in Central/Eastern Europe, sometimes isolated from other Jews as well as from their Christian neighbors. At one time, "They were closed off in a gigantic ghetto called the Pale of Settlement, destitute, deprived of all political rights, living in the twilight of a slowly disintegrating medieval

world."[4] However, by the nineteenth century, Jews in Western Europe had become politically tolerated and relatively well-off, and, due to the Enlightenment, found most occupations and professions open to them. These "emancipated" Jews played a crucial role in the Zionist movement, but the actual return to Palestine and settlement was largely the work of Ashkenazis from Central/Eastern Europe. The former Soviet Union, which had a Jewish population of 3.5 million, nearly all in Russia itself, had not supported the creation of the State of Israel and did not establish diplomatic relations until 1990, when a consular office was opened in Tel Aviv. However, U.S. restrictions on the entry of Soviet Jews caused the majority of them to emigrate to Israel. By 1992, some 350,000 had arrived in Israel. The majority were highly educated and professionally trained, but they were often unable to find suitable jobs and placed an added strain on housing and social services in Israel. One reason for the Israeli request to the United States for $10 billion in loan guarantees, which was held up by the George Bush administration and partially released by its successor, the Clinton administration, was to obtain funds for housing Soviet immigrants. Disillusionment with their experiences and lack of professional opportunities in Israel led the immigrants to form their own political party, Yisrael Ba'Aliya, to press for better conditions. The party joined the Binjamin Netanyahu coalition after the Knesset (Parliament) elections and shifted to support for Ehud Barak after he became prime minister. Its leader, Natan Sharansky, was named minister of interior in return for his party's support. His appointment put him in direct conflict with Shas, which had previously controlled that ministry. Yisrael Ba' Aliya was also promised $65 million for jobs and housing for Soviet emigrants.

The *Aliyah* ("going up" in Hebrew, i.e., to Israel) policy has resulted in the return of many formerly isolated Jewish communities to the new homeland. The first group to arrive were from Yemen, where they had lived in scattered villages, working as craftsmen and using Aramaic in their liturgy. Some 90 percent of them were airlifted directly to Israel. Although Yemen closed its borders in the 1960s, most of the remaining Jews were allowed to emigrate, leaving about 300 members of this oldest of diaspora communities.

Since the 1980s nearly 100,000 Ethiopian Jews have migrated to Israel. They have been reasonably well absorbed into the larger Israeli community with one exception, that of the Falash Mura. The ancestors of this group supposedly converted to Christianity centuries ago, probably

© Neil Beer/Getty Images (DIL60250)

The Wailing Wall, a focal point of Jewish worship, is all that remains of the ancient temple destroyed by the Roman legions led by Titus in A.D. 70. The Wailing Wall stands as a place of pilgrimage for devout Jews throughout the world.

through coercion as was the case with the Conversos of Spain. Like the Conversos, however, they continued to practice their Jewish faith privately and more recently in public. Fears that this rather impoverished community would impose additional burdens on Israeli social services led the government in 2003 to require that only those Falashas who could document Jewish ancestry on the maternal side would be allowed into Israel. Some 15–20,000 of them were believed to meet the requirement.

Up until now Israel has limited Falasha immigration to 300 per month. But in January 2005 as the result of a civil action suit an accelerated immigration process for them was approved. Beginning in June 2005, 600 per month will be admitted, with all Falash scheduled to be moved to Israel by the end of 2007. In the interim they have been essentially uprooted from their villages and are encamped in hovels near the Israeli embassy in Addis Ababa. Those Falashas now in Israel, some 17,000, will be granted citizenship. It had been withheld pending verification by a rabbinical scholar of their Jewish lineage.

Israel has two important non-Jewish minorities, totaling about 1 million—about 20 percent of the population. The larger group consists of Muslim and some Christian Arabs who stayed after Israel achieved statehood. This Arab population was ruled under military administration from 1948 until 1966, when restrictions were lifted and the Arabs ostensibly became full citizens. However they are still prohibited from serving in the armed forces and do not have priority with Israeli Jews in education, housing and jobs. The Christian Arab population, originally much larger but now forming at most 3 percent, has experienced similar treatment, and events such as the siege of the Bethlehem Church of the Nativity by Israeli forces due to the presence there of Palestinian gunmen has encouraged many to migrate.

Efforts by Israeli Arabs to gain economic and social equality with Jews has concentrated on improving their political representation in recent year. The first Arab political party was formed in 1988 and won one seat in the Knesset. In 1996, the party won four seats. In 1999, 95 percent of the Arab electorate voted for Barak; they were a major factor in his victory over Netanyahu.

Thus far, political representation in the Knesset has not been accompanied by social and economic equality for Israel's Arabs. The Barak government took several small steps in that direction in the new century. El Al, the Israeli airline, hired its first Arab flight attendant and appointed its first Arab ambassador in the foreign service.

The resumption of the Palestinian Intifada in 2000, after Ariel Sharon's visit to the Dome of the Rock and its Islamic holy places to emphasize Israel's sovereignty over its entire territory, led to a demonstration by Israeli Arabs to show their sympathy with the Palestinian cause. It was broken up by police, with 13 Arabs killed. In September 2003, after a 3-year investigation, a committee formed by ex-Prime Minister Barak found police commanders guilty of using excessive force. The committee's report, the fifth such in Israel's independent history, underscored the failure of successive governments to deal fairly with its Arab minority.

In August 2003, the Knesset passed the Citizenship and Entry Law, which prohibits Palestinian spouses of Israeli Arabs from citizenship and residence in Israel. In effect it bans family reunification. The law is in effect for one year but may be renewed indefinitely by the Knesset. The Association for Civil Rights in Israel, a watchdog human rights group, has appealed to the Supreme Court to over-turn it on grounds that it "violates the right to a family life enshrined in Israeli and international law." [5]

A second non-Jewish minority, the Druze, live not only in Israel but also in mountain enclaves in Lebanon and Syria. They form a majority of the population in the Golan Heights, occupied by Israeli forces in the 1967 Six-Day War and annexed in 1981. They practice a form of Islam that split off from the main body of the religion in the tenth century. Most Druze have remained loyal to the Israeli state. In return, they have been given full citizenship, are guaranteed freedom to practice their faith under their religious leaders, and may serve in the armed forces. Some 233 Druze have died in Israel's wars.

At present about 70,000 Druze live in Israel in 16 large villages in the Galilee and near Haifa. Another 16,000 Druze live in the Israeli-annexed Golan Heights, where they are physically separated from their families on the Syrian side by the UN demilitarized zone. These Druze have rejected Israeli citizenship. When Israel annexed the Golan Heights unilaterally in 1981, they reacted with a six-month-long general strike; it ended only when the government agreed not to force citizenship on the Druze. In 1997, on the 15th anniversary of the strike, the villagers showed their continued defiance of Israeli rule by flying the Syrian flag over their schools.

Despite their relative freedom, the Druze in Israel, like the Israeli Arabs, experience some discrimination in educational and work opportunities. The discrepancy is particularly noticeable for Druze women. In 1988, the first national movement for the advancement of Druze women was formed. Named the Council of Druze Women, it works to help them to reach the educational and social levels of Israeli women and to protect them from the abuses of what is still a patriarchal society, one in which there is even a religious ban on women driving.

There are two other small minority groups in Israel, both Sunni Muslims. The Circassians, descendants of warriors from the Caucasus brought in centuries ago to help Muslim armies drive the Christian Crusaders from Palestine, have been completely integrated into Israeli society. The second minority group, the Bedouins, formerly roamed the barren uplands of Judea and the Negev Desert. Until recently they followed their traditional nomadic lifestyle unhindered. But economic hardship, along with lack of water supplies, has led increasing numbers to move onto state lands, which cover much of the Negev. Like rural migrants in many countries they settle down, building shanty villages (which are considered illegal under Israeli housing laws). By 2003 there were 46 such "unrecognized villages," with a population of 70,000.

Other Bedouin clans, who had been removed from the Sinai Peninsula when it was returned to Egypt in 1979 and resettled in Gaza, were again evacuated in 2005 prior to its return to Palestinian control. As Israeli citizens the government said it was obligated to do so for them. They will join other semi-nomadic Bedouin in the Negev, near the Dead Sea. Commenting prophetically on their second removal from their homes, one Bedouin said: "This is a hard situation for us. But I see my future as a Bedouin and an Israeli citizen."[6]

HISTORY

The land occupied by the modern State of Israel has been a crossroads for conquering armies and repository for many civilizations during its long history. As such, it is part of biblical history and an important component of the Judeo-Christian heritage. The ruins of Megiddo, near Haifa, are believed by scholars to be the site of Armageddon, described in *Revelations* as the final clash between good and evil resulting in the end of the world.

Jews believe that Israel is the modern fulfillment of the biblical covenant between God and a wanderer named Abraham (Abram) that granted him a homeland in a particular place. Jews have held fast to this covenant during the centuries of their exile and captivity in foreign lands. Each of these periods of exile is called a *diaspora* ("dispersion"). The most important one, in terms of modern Israel, took place in the first century A.D. Abraham's descendants through Isaac had called the land given to them in covenant Judea and Samaria; and when it became part of the expanding Roman Empire, it was known as the province of Judea. Initially the Romans preferred to rule the unruly Jews indirectly through appointed governors. The best-known of them, Herod, "bought" his appointment from Rome and then elevated his status to that of King. Aside from ruling Judea and Samaria at the time of Jesus, he made noteworthy contributions to its people, including renovation of the Temple.

After Herod's death Rome imposed direct rule over the province by appointed governors called *procurators*, the best-known being Pontius Pilate. In 69–70 A.D., the Jews rebelled. The forces of Roman general (later emperor) Titus then besieged Jerusalem. The city fell in A.D. 70, and Roman legions sacked and destroyed much of it. (A portion of the Temple Wall remains standing to this day. It is called the Wailing Wall because Jews come there to pray and mourn the loss of the original central shrine to their faith.)

From then until the twentieth century, most Jews were dispersed, living among alien peoples and subject to foreign rulers. Periodically persecuted and often mistrusted, they coexisted with the populations around them, in part by preserving their ancient rituals and customs but also due to the restrictions imposed on them by their non-Jewish rulers. Thus, in the Islamic world Jews were required to wear distinctive dress with armbands and pay a separate poll tax in order to observe their religious ceremonies. Their places of worship as well as their homes could not be built higher than Islamic ones, and in other ways they were made to feel inferior to their Muslim neighbors.

Zionism

The organized movement to reestablish a national home for dispersed Jews in biblical Judea and Samaria in accordance with God's promise is known as Zionism. It became, however, more of a political movement formed for a particular purpose: to establish by Jewish settlement a homeland where dispersed Jews may gather, escape persecution, and knit together the strands of traditional Jewish faith and culture. As a political movement, it differs sharply from spiritual Zionism, the age-old dream of "the return." Most Orthodox Jews and traditionalists opposed *any* movement to reclaim Palestine; they believed that it is blasphemy to do so, for only God can perform the miracle of restoring the Promised Land. The reality of the establishment of the Jewish state by force of arms, with a secular political system backed by strong Jewish nationalism, has created what one author calls "an unprecedented Jewish dialogue with power, an attempt to historicize the Jewish experience as a narrative of liberation by armed Jews."[7] On May 15, the anniversary of Israel's independence, many Haredim display a black flag rather than the Israeli one, marking their belief that Zionism has usurped the role of the Messiah and prevented His return.

Zionism as a political movement began in the late nineteenth century. Its founder was Theodore Herzl, a Jewish journalist from Vienna, Austria. Herzl had grown up in the Jewish Enlightenment period. Like other Western European Jews, he came to believe that a new age of full acceptance of the Jewish community into European life had begun. He was bitterly disillusioned by the wave of Jewish persecution that swept over Central/Eastern Europe after the murder of the liberal Russian czar, Alexander II, in 1881. He was even more disillusioned by the trial of a French Army officer, Alfred Dreyfus, for treason. Dreyfus, who was Jewish, was convicted after a trumped-up trial brought public protests that he was part of an antigovernment Jewish conspiracy.

Herzl concluded from these events that the only hope for the long-suffering Jews, especially those from Central/Eastern Europe, was to live together, separate from non-Jews. In his book *The Jewish State,* he wrote: "We have sincerely tried everywhere to merge with the national communities in which we live, seeking only to preserve the faith of our fathers. It is not permitted to us."[8]

Herzl had attended the Dreyfus trial as a journalist. Concerned about growing anti-Semitism in Western Europe, he organized a conference of European Jewish leaders in 1897 in Basel, Switzerland. The conference ended with the ringing declaration that "the aim of Zionism is to create a Jewish homeland in Palestine secured by public law." Herzl wrote in an appendage: "In Basel I have founded the Jewish state."[9]

The Zionists hoped to be allowed to buy land in Palestine for Jewish settlements. But the Ottoman government in Palestine would not allow them to do so. Small groups of Eastern European Jews escaping persecution made their way to Palestine and established communal agricultural settlements called *kibbutzim*. Those immigrants believed that hard work was essential to the Jewish return to the homeland. Work was sacred, and the only thing that gave the Jews the right to the soil of Palestine was the "betrothal of toil." This belief became a founding principle of the Jewish state.

The Balfour Declaration

Although the Zionist movement attracted many Jewish supporters, it had no influence with European governments, nor with the Ottoman government. The Zionists had difficulty raising money to finance land purchases in Palestine and to pay for travel of emigrants.

DEVELOPMENT

By international standards, Israel is a "developed" country with a per capita income, GDP and other economic levels comparable to those of Europe. But neither prosperity nor economic recession in the long fun will enable the country to integrate itself into the larger regional economy and establish true peace. A 2-year recession caused by the Palestinian intifada and collapse of the Internet affected tourism and the high-tech industry for a time, both of these being critical to economic growth. However the change from an economy based on agriculture and light industry to one of telecommunications, electronics and life-science industries made possible a strong economic revival in 2004–2005. While their Palestinian neighbors have fallen back to 3rd world status, Israel's GDP rose to 4.2 percent in 2004 and 3.9 percent in 2005, with tourism up by 40 percent in the same period.

It appeared in the early 1900s that the Zionists would never reach their goal. But World War I gave them a new opportunity. The Ottoman Empire was defeated, and British troops occupied Palestine. During the war, a British Zionist named Chaim Weizmann, a chemist, had developed a new type of explosive that was valuable in the British war effort against Germany. Weizmann and his associates pressed the British government for a commitment to support a home for Jews in Palestine after the war. Many British officials favored the Zionist cause, among them Winston Churchill. During his term as colonial secretary after World War I, he organized the

1921 Cairo Conference, which among other issues confirmed the League of Nations' assignment of Palestine to Britain as a mandate. Churchill then planted a tree on Mt. Scopus next to the new Hebrew University as a symbol of the British commitment to "some sort of national home there" for the Zionists.

The British Mandate

The peace settlement arranged after World War I by the new League of Nations gave Palestine to Britain as a mandate. As a result, the name Palestine came into common usage for the territory. It is probably derived from Philistine, from the original tribal inhabitants (who are also called Canaanites), but this covenanted land had been ruled by many other peoples and their rulers for centuries, due to its location as a strategic corridor between Asia and Africa. After it became part of the Ottoman Empire, along with Lebanon it was divided into *vilayets* (provinces), those of Beirut and Acre respectively; Jerusalem was administered separately as a *sanjak* (subprovince). The majority of the population were small farmers living in compact villages and rarely traveling elsewhere. Most were Muslims, but there was a substantial minority of Christians. Leadership, such as it was, was held by a small urban elite, the principal families being the Husseinis and Nashashibis of Jerusalem.

After World War I and the peace settlement, the Zionists assumed that the Balfour Declaration authorized them to begin building a national home for dispersed Jews in Palestine. Weizmann and his colleagues established the Jewish Agency to organize the return. Great Britain's obligation under the mandate was to prepare Palestine's inhabitants for eventual self-government. British officials assigned to Palestine tended to favor Jewish interests over those of the native population. This was due in part to their Judeo–Christian heritage, but also to the active support of Jews in Britain during the war. In addition to Weizmann's contribution, the Jewish Legion, a volunteer group, had fought with British forces against the Turks.

Britain's "view with favor" toward Zionism weighed heavily in the application of mandate requirements to Palestine. Jews were allowed to emigrate, buy land, develop agriculture, and establish banks, schools, and small industries. The Jewish Agency established a school system, while former members of the dispersed Jewish Legion regrouped into what became Haganah, the defense force for the Jewish community.

Compounding the difficulties of adjustment of two different peoples to the same land was the fact that most Zionist

leaders had never been to Palestine. They envisaged it as an empty land waiting for development by industrious Jews. David Ben-Gurion, for example, once claimed that one could walk for days there without meeting a soul; Palestine, he told his compatriots, was a land without a people for a people without a land. Palestine was indeed underpopulated, but it did have a substantial population, many of its members living in villages settled by their ancestors centuries earlier. Referring to the Balfour Declaration, Tom Segev observed that "the Promised Land had, by the stroke of a pen, become twice-promised."[10]

Palestinian Arabs were opposed to the mandate, to the Balfour Declaration, and to Jewish immigration. They turned to violence on several occasions, against the British and the growing Jewish population. In 1936, Arab leaders called a general strike to protest Jewish immigration, which led to a full-scale Arab rebellion. The British tried to steer a middle ground between the two communities. But they were unwilling (or unable) either to accept Arab demands for restrictions on Jewish immigration and land purchases or Zionist demands for a Jewish majority in Palestine. British policy reports and White Papers during the mandate wavered back and forth. In 1937, the Peel Commission, set up after the Arab revolt, recommended a halt to further Jewish immigration, and subsequently the 1939 "White Paper" stated that the mandate should be replaced by a self-governing Arab state with rights assured for the Jewish minority.

One important difference between the Palestinian Arab and Jewish communities was in their organization. The Jews were organized under the Jewish Agency, which operated as a "state within a state" in Palestine. Jews in Europe and the United States also contributed substantially to the agency's finances and made arrangements for immigration. The Palestinian Arabs, in contrast, were led by heads of urban families who often quarreled with one another. The Palestinian Arab cause also did not have outside Arab support; leaders of neighboring Arab states were weak and were still under British or French control.

A unique feature of Zionism that helped strengthen the Jewish pioneers in Palestine in their struggle to establish their claim to the land were the kibbutzim. Originally, there were two types, *moshavim* and *kibbutzim*. The moshavim, cooperative landholders' associations whose members worked the land under cooperative management and lived in nearby villages, have largely disappeared with urbanization. The kibbutzim, in contrast, were collective-ownership communities with self-contained, communal-living arrangements; members shared labor, income, and expenses. Over the years, kibbutzim have played a role that is disproportionate to their size and numbers, not only in building an integrated Jewish community in Palestine but also in the formation of the Israeli state. David Ben-Gurion, Israel's first prime minister, lived in and retired to Kibbutz Sde Boker, in the Negev. Shimon Peres, twice prime minister and longtime public official, wrote of his youth on a kibbutz in these moving terms: "We saw it as the solution to the evils of urban industrialized society. I dreamed of my future as a brawny, sunburned kibbutz farmer, plowing the fields by day, guarding the perimeter by night on a fleet-footed horse. The kibbutz would break new ground, literally; it would make the parched earth bloom and beat back the attacks of marauders who sought to destroy our pioneering lives."[11]

FREEDOM

Israel is a multiparty democracy with a Parliament (Knesset) and representative political and judicial institutions. Its Basic Laws guarantee free speech, among other human rights. However the full range of human rights has never been extended to Israeli Arabs. They are discriminated against in access to higher education, adequate schools, jobs and other "guaranteed" rights. The Israeli Arab political position has improved in recent years with the formation of "Arab" political parties and election of Arab members to the Knesset. However, in January 2003 two Arab members were stripped of their parliamentary immunity on the grounds that they had questioned the "Jewish character" of the state, arguing that it should be a "state for all citizens."[21]

Adolf Hitler's policy of genocide (total extermination) of Jews in Europe, developed during World War II, gave a special urgency to Jewish settlement in Palestine. American Zionist leaders condemned the 1939 British White Paper and called for unrestricted Jewish immigration into Palestine and the establishment of an independent, democratic Jewish state. After World War II, the British, still committed to the White Paper, blocked Palestine harbors and turned back the crowded, leaking ships carrying desperate Jewish refugees from Europe. World opinion turned against the British. Supplies of smuggled weapons enabled Haganah to fend off attacks by Palestinian Arabs, while Jewish terrorist groups such as the Stern Gang and Irgun Zvai Leumi carried out acts of murder and sabotage against British troops and installations, the most sensational being the bombing of the King David Hotel in Jerusalem, headquarters of the British military command and administration.

PARTITION AND INDEPENDENCE

In 1947, the British decided that the Palestine mandate was unworkable. They asked the United Nations to come up with a solution to the problem of "one land, two peoples." A UN Special Commission on Palestine recommended partition of Palestine into two states—one Arab, one Jewish—with an economic union between them. A minority of UNSCOP members recommended a federated Arab–Jewish state, with an elected legislature and minority rights for Jews. The majority report was approved by the UN General Assembly on November 29, 1947, by a 33–13 vote, after intensive lobbying by the Zionists. The partition would establish a Jewish state, with 56 percent of the land, and a Palestinian Arab state, with 44 percent. The population at that time was 60 percent Arab and 40 percent Jewish. Due to its special associations for Jews, Muslims, and Christians, Jerusalem would become an international city administered by the United Nations.

Abba Eban, who had been "present at the creation" and served the Israeli state as a senior statesman and troubleshooter in many capacities until his recent death, recalled in his memoirs, "President Truman told me: 'Quite simply, you got your state because you made feasible proposals and your adversaries did not. If Israel had asked for a Jewish state in the whole of the land of Israel it would have come away with nothing. An Arab and a Jewish state side by side with integrated economies was something that American ethics and logic could absorb.'"[12]

The Jewish delegation, led by Eban and David Ben-Gurion, accepted the partition plan approved by the UN General Assembly. But Palestinian Arab leaders, backed strongly by the newly independent Arab states, rejected the plan outright. On May 14, 1948, in keeping with Britain's commitment to end its mandate, the last British soldiers left Palestine. Ben-Gurion promptly announced the "birth of the new Jewish State of Israel." On May 15, the United States and the Soviet Union recognized the new state, even as the armies of five Arab states converged on it to "push the Jews into the sea."

INDEPENDENT ISRAEL

Long before the establishment of Israel, the nation's first prime minister, David Ben-Gurion, had come to Palestine as a youth. After a clash between Arab nomads and Jews from the kibbutz where he lived had injured several people, Ben-Gurion wrote prophetically, "It was then I realized ... that sooner or later Jews and Arabs would fight over this land, a tragedy since intelligence and good will could have avoided all bloodshed."[13] In the five decades of independence, Ben-Gurion's prophecy has been borne out in five Arab–Israeli wars. In between those wars, conflict between Israel and the Palestinians has gone on more or less constantly, like a running sore.

Some 700,000 to 800,000 Palestinians fled Israel during the War for Independence. After the 1967 Six-Day War, an additional 380,000 Palestinians became refugees in Jordan. Israeli occupation of the West Bank brought a million Palestinians under military control.

The unifying factor among all Palestinians is the same as that which had united the dispersed Jews for 20 centuries: the recovery of the sacred homeland. Abu Iyad, a top Palestine Liberation Organization leader, once said, "... our dream....[is] the reunification of Palestine in a secular and democratic state shared by Jews, Christians and Muslims rooted in this common land.... There is no doubting the irrepressible will of the Palestinian people to pursue their struggle ... and one day, we will have a country."[14] The land vacated by the Palestinians has been transformed in the decades of Israeli development. Those Israelis actually born in Palestine—now in their third generation—call themselves *Sabras,* after the prickly pear cactus of the Negev. The work of Sabras and of a generation of immigrants has created a highly urbanized society, sophisticated industries, and a productive agriculture. Much of the success of Israel's development has resulted from large contributions from Jews abroad, from U.S. aid, from reparations from West Germany for Nazi war crimes against Jews, and from bond issues. Yet the efforts of Israelis themselves should not be understated. Ben-Gurion once wrote, "Pioneering is the lifeblood of our people....We had to create a new life consonant with our oldest traditions as a people. This was our struggle."[15]

ISRAELI POLITICS: DEMOCRACY BY COALITION

Israel is unique among Middle Eastern states in having been a multiparty democracy from its establishment as a state. It does not have a written constitution, mainly because secular and Orthodox communities cannot agree on its provisions. The Orthodox community, for example, argues that it already has its constitution, in the Bible and Torah. For the Orthodox, a constitution would be something imported from a foreign country such as a Canada or Sweden, not a document that truly belonged to Israel.

In place of a constitution, the Israeli state is governed by a series of Basic Laws. They include the Law of Return, by which diaspora Jews may return and are automatically granted Israeli citizenship. In addition, a series of seven Basic Laws established the Knesset, the national army (Israel Defense Forces), the office of president, the legal system, and so on. Two new Basic Laws issued in 1992 provide for direct election of the prime minister and for recognition of human dignity and rights before the power of the state. The Law for Direct Election of the prime minister was invoked in the last elections but has been criticized, and will probably be revoked in the near future.

Power in the Israeli political system rests in the unicameral Knesset. It has 120 members who are elected for four-year terms under a system of proportional representation from party lists. The prime minister and cabinet are responsible to the Knesset, which must approve all policy actions. The new Direct Elections law also has a provision for the removal from office of a prime minister, either through a 61-member no-confidence vote or through impeachment for "crimes of moral turpitude." The possibility of this type of removal loomed large for a time in 1997 during the Bar-On affair, "Israel's Watergate."[16]

HEALTH/WELFARE

The 2002–2003 recession has punched large holes in the safety net provided for Israelis by the government and its adjunct organization, Histadrut. The increases in both unemployment and families below the poverty line suggest that the country can no longer function as a welfare state. One result of the recession is that more Russian Jews are emigrating to Germany than to Israel due to the more stable German economy.

Ben-Gurion's Labor Party controlled the government for the first three decades of independence. However, the party seldom had a clear majority in the Knesset. As a result, it was forced to join in coalitions with various small parties. Israeli political parties are numerous. Many of them have merged with other parties over the years or have broken away to form separate parties. The Labor Party itself is a merger of three Socialist labor organizations. The two oldest parties are Agudath Israel World Organization (founded in 1912), which is concerned with issues facing Jews outside of Israel as well as within,

and the Israeli Communist Party (Rakah, founded in 1919).

The Labor Party's control of Israeli politics began to weaken seriously after the October 1973 War. Public confidence was shaken by the initial Israeli defeat, heavy casualties, and evidence of Israel's unpreparedness. Austerity measures imposed to deal with inflation increased Labor's unpopularity. In the 1977 elections, the opposition, the Likud bloc, won more seats than Labor but fell short of a majority in the Knesset. The new prime minister, Menachem Begin, was forced to make concessions to smaller parties in order to form a governing coalition.

However, the Israeli invasion of Lebanon in 1982 weakened the coalition. It seemed to many Israelis that for the first time in its existence, the state had violated its own precept that wars should be defensive and waged only to protect Israeli land. The ethical and moral implications of Israel's occupation, and in particular the massacre by Lebanese Christian militiamen of Palestinians in refugee camps in Beirut who were supposedly under Israeli military protection, led to the formation in 1982 of Peace Now, an organization of Israelis who mounted large-scale demonstrations against the war and are committed to peace between Israel and its Arab neighbors.

Begin resigned in 1983, giving no reason but clearly distressed not only by the difficulties in Lebanon but also by the death of his wife. He remained in seclusion for the rest of his life. He died in 1992.

RECENT INTERNAL POLITICS

The Labor Party won the majority of seats in the Knesset in the 1984 elections—but not a clear majority. As a result, the two major blocs reached agreement on a "government of national unity," the first in Israel's history. The arrangement established alternating two-year terms for each party's leader as prime minister. Shimon Peres (Labor) held the office from 1984 to 1986 and Yitzhak Shamir (Likud) from 1986 to 1988.

In the 1988 elections, certain fundamental differences between Labor and Likud emerged. By this time the first Palestinian *intifada* ("uprising") was in full swing. It would not only change the relationship between Israelis and Palestinians forever but would also alter the norms of Israeli politics. Labor and Likud differed over methods of handling the uprising, but they differed even more strongly in their views of long-term settlement policies toward the Palestinians. Labor's policy was to "trade land for peace," with some sort of self-governing status for the occupied territories and peace treaties with its Arab neighbors

guaranteeing Israel's "right to exist." Likud would give away none of the sacred land; it could not be bartered for peace.

The election results underscored equally deep divisions in the population. Neither party won a clear majority of seats in the Knesset; Likud took 40 seats, Labor 39. Four minority ultra-religious parties gained the balance of power, with 15 percent of the popular vote and 18 seats. Their new-found political power encouraged the religious parties to press for greater control on the part of Orthodox Jewry over Israeli life. However, a proposed bill to amend the Law of Return to allow only Orthodox rabbis to determine "who is a Jew" for citizenship purposes aroused a storm of protest among diaspora Jews, who are mostly Reform or Conservative and would be barred from citizenship. In February 2002, the Israeli Supreme Court ruled that both movements would be listed as Jews in the official census registry. For Reform and Conservative Jews living in Israel, it was a major step toward official recognition.

The 1992 Knesset election ended 15 years of Likud dominance. Labor returned to power, winning a majority of seats. However, the splintered, multiparty Israeli electoral system denied it an absolute majority. Concessions to minority parties enabled Labor to establish a functioning government, and party leader Yitzhak Rabin was named prime minister. With the support of these minority parties, notably Shas, an ultra-religious, non-Zionist party of mostly Sephardic Jews, and the left-wing Meretz Party, the Rabin government could count on 63 votes in the Knesset. This majority ensured support for the government's policies, including the 1993 Oslo agreements with the Palestinians.

JEWISH EXTREMISTS, ARAB EXTREMISTS

The deep divisions in Israeli society regarding future relations with the Palestinian population in the occupied territories (for many Israelis, these lands are Judea and Samaria, part of the ancestral Jewish homeland) were underscored by the uncovering in the 1980s of a Jewish underground organization that had attacked Palestinian leaders in violent attempts to keep the territories forever Jewish. The group had plotted secretly to blow up the sacred Islamic shrines atop the Dome of the Rock. A number of the plotters were given life sentences by a military court. But such was the outcry of support from right-wing groups in the population that their sentences were later commuted by then-president Chaim Herzog.

A more virulent form of anti-Arab, anti-Palestinian violence emerged in 1984 with the founding of Kach, a political party that advocated expulsion of all Arabs from Israel. Its founder, Brooklyn, New York–born Rabbi Meir Kahane, was elected to the Knesset in 1984, giving him parliamentary immunity, and he began organizing anti-Arab demonstrations. The Knesset subsequently passed a law prohibiting any political party advocating racism in any form from participation in national elections. On that basis, the Israeli Supreme Court barred Kach and its founder from participating in the 1988 elections. Kahane was murdered by an Egyptian-American while in New York for a speaking engagement. His son Binyamin formed a successor party, *Kahane Chai* ("Kahane Lives"), based in the West Bank Jewish settlement of Tapuah. Both Kach and Kahane Chai were labeled terrorist groups by the U.S. Department of State. They were also outlawed in Israel after a member, Baruch Goldstein, murdered 29 Muslims in a mosque in Hebron. In September 2000, Binyamin Kahane and his wife were killed in an ambush while driving their children to school, in another blow to the struggling Palestinian–Israeli peace negotiations.

Arab extremism, or, more accurately, Palestinian extremism, evolved in the 1990s largely as a result of Palestinian anger and disillusionment over the peace agreements with Israel, which were seen as accommodation on the part of Palestinian leaders, notably Yassir Arafat, to Israel rather than negotiations to establish a Palestinian state. The main Palestinian extremist group is *Hamas* (the Arabic acronym for the Islamic Resistance Movement, or IRM). Hamas developed originally as a Palestinian chapter of the Muslim Brotherhood, which has chapters in various Islamic Arab countries where it seeks to replace their secular regimes by a government ruled under Islamic law. However, Hamas broke with its parent organization over the use of violence, due largely to the lack of success of the intifada in achieving Palestinian self-rule.

A number of violent attacks on Israelis, including the murder of a border policeman in 1992, led Israel to deport 415 Hamas activists to southern Lebanon. However, the Lebanese government refused to admit them. Lebanon and other Arab countries filed a complaint with the UN Security Council. The Council passed *Resolution 799,* calling for the return of the deportees. Although Israel seldom responds to UN resolutions, in this case the 1993 Oslo Agreement provided additional motivation, and eventually the deportees were allowed to return to their homes.

Israel's Greatest Friend

Israel's relationship with the United States has been close since the establishment of the Jewish state, in large part due to the large American Jewish population and its unstinting support. This friendship has been severely tested on two occasions. During the 1967 Six-Day War, the *Liberty,* a slow-moving, lightly armed U.S. Navy ship working with the U.S. National Security Agency to monitor the conflict, was attacked by Israeli aircraft and torpedo boats off the Mediterranean coast. Thirty-four American sailors were killed, and 171 wounded. Israeli statements that the attack was a "mistake" were accepted at face value by the Lyndon Johnson administration, which was concerned with maintaining good relations with Israel at that time of wartime crisis. However, more recent information from the National Security Agency and other archives suggest that it was deliberate.

A second major strain on U.S.–Israeli relations was the Pollard affair. It involved an American Jew, Jonathan Jay Pollard, who was convicted of spying on his own country for Israel. Pollard's reports on U.S. National Security Agency intelligence-collecting methods and his duplication of military satellite photographs seriously compromised U.S. security. They also damaged Israel's image in the American Jewish community. One U.S. Jewish leader asked, "Is Israel becoming an ugly little Spartan state instead of the light of the world?"[17]

In the years following Pollard's incarceration, many efforts have been made by Israeli and American Jewish organizations to have his sentence commuted. As yet, successive U.S. presidents have not done so. But these two negative events have not altered the "special relationship" between Israel and its major ally. The country remains the recipient of the largest amount of U.S. aid worldwide. With very few interruptions, the notable one being President Eisenhower's intervention in the 1956 Suez War, the United States has consistently supported Israel's foreign policy. The acclaimed "road map to peace" between the Israelis and the Palestinians issued by President George Bush in 2003 does require that Israel take certain steps toward the establishment of a Palestinian state. While the "road map" has largely disappeared or been overtaken by recent events, pressure from the Bush administration was to some extent responsible for Israel's withdrawal from Gaza and its cession of certain West Bank towns to Palestinian authority.

The puzzling question of "Who is a Jew?" became a major issue separating Orthodox and Reform and Conservative Je-

wry in 1998 and 1999. A bill introduced in the Knesset in 1997 by the three religious parties would ban Reform and Conservative Jews from serving on religious councils. The action was rejected by the Israeli Supreme Court, and in January 1998, the first non-Orthodox representatives took their seats on the Haifa Council.

In December 1998, the Court ruled that the system of exemptions from military service for yeshiva students (those enrolled in religious-study programs) was illegal because it was not anchored in law. The court ordered the Knesset to pass legislation dealing with the issue. Ehud Barak had pledged the end of the draft-deferral system during his campaign for the prime ministership. But after his election, he approved changing the system rather than eliminating it. Under the changes, yeshiva students would be exempt from service from ages 18 to 23. They would be allowed a "year of deliberation" thereafter; they could go to trade school or get jobs. At age 24, if they decided to leave the yeshiva, they would be required to do four months' service, either national or military, in a special unit.

Despite Barak's backtracking on the draft-deferral issue, other issues continued to divide Orthodox from secular Jews. Shas, flexing its political muscle after joining Barak's coalition, gained control of the Labor Ministry and promptly enforced a 1953 law that banned employment of Jewish teenagers on the Sabbath. As a result, a McDonald's franchise in Jerusalem was hit with $20,000 in fines.

The *haredim* (Orthodox) community holds itself apart from Reform and Conservative Parties on a number of issues, notably use of roads on the Sabbath, attendance at movies, and other activities. Until recently only Orthodox rabbis could perform marriages and approve divorces. In effect this allowed them alone to decide who was, and who was not, a Jew. Under a government ruling in 2001 the Interior Ministry would be responsible for recognizing and registering civil marriages involving Jews performed at consulates abroad. And in a further broadening of the Jewish identity issue, the Israeli Supreme Court ruled in March 2005, by a 7–4 vote, that non-Orthodox conversions to Judaism must be recognized as legitimate by the state regardless of where they were, and are, performed. The case that triggered this decision was a civil action suit by some 14 converts in this category, one of them a Colombian who had emigrated to Israel in 1991 but was not able to convert there since he was non-Orthodox. He then went to Buenos Aires, where he converted to Reform Judaism in 1997. Upon returning to Israel, however, his conversion was not recognized, leading to the civil suit. The court's decision would set a precedent for some 250,000 Israelis whose religious status is currently "undefined" and will further limit Orthodox control of religious affairs.

Other issues separating the two communities include closing of roads near haredim neighborhoods on the Sabbath, and the government's refusal to recognize non-Orthodox Conservative, Reform, and Reconstruction Judaism as legitimately Jewish, for fear of offending the Orthodox community.

The very nature of the Jewish state, founded by secular Zionists but balanced uneasily between secularism and orthodoxy, was challenged in December 2004 in a civil suit brought before the High Court of Justice. The suit, filed by both Jews and Arabs, enjoins the Court to order the Interior Ministry to inscribe them as "Israeli" in the Registry of Population. While this may seem a frivolous matter, it is critical in that the state is asked to recognize an inclusive earned form of nationality coterminous with citizenship. Thus far the Court has not ruled definitively on the suit, but a favorable verdict, if so rendered, would go far toward establishing Israel as "Hebrew republic", with religion a matter of private conscience and voluntary assembly as it is in most countries.[18]

In September 2000 then-Prime Minister Barak proposed adding 3 basic laws to the canon as part of his "civil-social agenda." They covered due process of law, freedom of expression, and the public's right to education and housing. For various reasons they were never passed into law, and the Sharon government, dependent upon Orthodox support in the ruling coalition, has made no effort to re-introduce them for consideration.

In recent years, a group of academic scholars and political leaders in the Israel Democracy Institute have worked to lay the groundwork for the country's first written constitution, 50-odd years after one was supposed to be adopted under the terms of the declaration of independence of the state. In September 2000, the effort snowballed when Ehud Barak proposed its adoption as part of his "civil-social agenda." Three new Basic Laws were presented to the Knesset. They concerned due process of law, freedom of expression, and the right to education and housing for all citizens. If adopted, they would add important rights planks to the constitutional structure. However, political developments halted the process in midstream—the collapse of Barak's coalition; the election of a right-wing government under Ariel Sharon, dependent on Orthodox support; and the renewal of the Palestinian intifada.

A similar sense of incompleteness marks the Israeli justice system. Under the maxim that everything is justiciable, the Supreme Court, headed by long-time Chief Justice Aharon Barak, has taken the lack of a written constitution to use the Basic Laws to justify legal decisions in a number of areas. In 2003, for example, it granted same-sex partners of company employees the same benefits provided for opposite-sex ones. It intervened in the case of an Arab family who had been refused to build on property it owned in a Jewish village, and in 2005 it ordered the wall being built between Israel and Palestinian territory to be moved closer to the original "Green Line" of the occupied West Bank, seized in 1967.

THE INTIFADA

The Palestinian intifada in the West Bank and Gaza Strip, which began in December 1987, came as a rude shock to Israel. Coming as it did barely 2½ years after the trauma of the Lebanon War, the uprising found the Israeli public as well as its citizen army unprepared. The military recall of middle-aged reservists and dispatch of new draftees to face stone-throwing Palestinian children created severe moral and psychological problems for many soldiers.

Military authorities devised a number of methods to deal with the uprising. They included deportation of suspected terrorists, demolition of houses, wholesale arrests, and detention of Palestinians without charges for indefinite periods. In February 2006 the government officially halted the practice of demolishing homes suspected of harboring Palestinian militants, after military leaders determined that demolitions were ineffective as deterrents. According to B'Tselem, the Israeli human rights organization, some 2,500 houses have been demolished since 1967. However, growing international criticism of the policy of "breaking the bones" of demonstrators (particularly children) developed by then-defense minister Yitzhak Rabin brought a change in tactics, with the use of rubber or plastic dum-dum bullets, whose effect is less lethal except at close range.

The government also tried to break the Palestinian resistance through arbitrary higher taxes, arguing that this was necessary to compensate for revenues lost due to refusal of Palestinians to pay taxes, a slowdown in business, and lowered exports to the territories. A value added tax (VAT) imposed on olive presses just prior to the processing of the West Bank's major crop was a particular hardship. Along with the brutality of its troops, the tax-collection methods drove Palestinians and Israelis further apart, making the prospect of any amicable relationship questionable.

The opening of emigration to Israel for Soviet Jews added an economic dimension to the intifada. Increased expropriation of land on the West Bank for new immigrant families, along with the expansion of Jewish settlements there, added to Palestinian resentment. Many Palestinians felt that because the new immigrants were unable to find professional employment, they were taking menial jobs ordinarily reserved for Palestinian workers in Israel.

In October 1990, the most serious incident since the start of the intifada occurred in Jerusalem. Palestinians stoned a Jewish group, the Temple Mount Faithful, who had come to lay a symbolic cornerstone for a new Jewish Temple near the Dome of the Rock. Israeli security forces then opened fire, killing some 20 Palestinians and injuring more than 100. The UN Security Council approved a resolution condemning Israel for excessive response (one of many that the Israeli state has ignored over the years). Israel appointed an official commission to investigate the killings. The commission exonerated the security forces, saying that they had acted in self-defense.

Shamir's Election Plan

In May 1989, responding to threats by Labor to withdraw from the coalition and precipitate new elections, the government approved a plan drafted by Defense Minister Rabin for elections in the West Bank and Gaza as a prelude to "self-government." Under the plan, the Palestinians would elect one representative from each of 10 electoral districts to an Interim Council. The Council would then negotiate with Israeli representatives for autonomy for the West Bank and Gaza, as defined in the 1979 Camp David treaty. Negotiations on the final status of the territories would begin within three years of the signing of the autonomy agreement.

The implementation of the Rabin plan would have forced Israelis to decide between the Zionist ideal of "Eretz Israel" (the entire West Bank, along with the coast from Lebanon to Egypt, as the Jewish homeland) and the trading of "land for peace" with another nation struggling for its independence. The success of the intifada lay in demonstrating for Israelis the limits to the use of force against a population under occupation. It also served as a pointed reminder to Israelis that "incorporating the occupied territories would commit Israel to the perpetual use of its military to control and repress, not 'Arab refugees' but the whole Palestinian population living in these lands."[19]

THE PEACE AGREEMENT

Prior to September 1993, there were few indications that a momentous breakthrough in Palestinian–Israeli relations was about to take place. The new Labor government had cracked down on the Palestinians in the occupied territories harder than had its predecessor in six years of the intifada. In addition to mass arrests and deportations of persons allegedly associated with Hamas, the government sealed off the territories, not only from Israel itself but also from one another. With 120,000 Palestinians barred from their jobs in Israel, poverty, hunger, and unemployment became visible facts of life in the West Bank and the Gaza Strip.

However, what 11 rounds of peace talks, five wars, and 40 years of friction had failed to achieve was accomplished swiftly that September, with the signing of a peace and mutual-recognition accord between Israel and its long-time enemy. The accord was worked out in secret by Israeli and PLO negotiators in Norway and under Norwegian Foreign Ministry sponsorship. It provided for mutual recognition, transfer of authority over the Gaza Strip and the West Bank city of Jericho (but not the entire West Bank) to an elected Palestinian council that would supervise the establishment of Palestinian rule, withdrawal of Israeli forces and their replacement by a Palestinian police force, and a Palestinian state to be formed after a transitional period.

ACHIEVEMENTS

Israeli scientist Robert Aumann, emeritus professor at the Hebrew University of Jerusalem, shared the Nobel Prize in Economics with U.S. Professor Thomas Schilling in 2005 for their work in game theory, which explains the choices competitors make in situations which require strategic thinking. Aumann, a mathematician, was honored for his development of formal techniques for analysis of competitive behavior, a good example being the Oslo accords between Israeli and Palestinian negotiators. A special component of his contribution was that of making "irrational" behavior understandable and rational. In his acceptance speech Aumann was asked how his theory applied to the Palestinian-Israeli conflict. His reply was that "this conflict has been going on for 80 years. It is an on-going conflict; we study on-going reactions. But game theory can help people better understand the patterns of the conflict and be part of the solution.

Opposition to the accord from within both societies was to be expected, given the intractable nature of Palestinian–Israeli differences—two peoples claiming the same land. Implementation of the Oslo

agreements has been hampered from the start by groups opposed to any form of Palestinian–Israeli accommodation. On the Israeli side, some settler groups formed vigilante posses for defense, even setting up a "tent city" in Jerusalem to protest any give-away of sacred Jewish land. Palestinian gunmen and suicide bombers responded with attacks on Jews, sometimes in alleyways or on lonely stretches of road outside the cities, but also in public places. One of the bloodiest incidents in this tragic vendetta was the killing of 29 Muslim worshippers in a mosque in Hebron by an American emigrant, an Orthodox Jew, during their Friday service.

Labor's return to power in 1992 suggested that, despite this virulent opposition, the peace process would go forward under its own momentum. The new government, headed by Rabin as prime minister and Peres as foreign minister, began to implement the disengagement of Israeli forces and the transfer of power over the territories to Yassir Arafat's Palestine National Authority (PNA). The Gaza Strip, Jericho, and several West Bank towns were turned over to PNA control. Israel's seeming commitment to the peace process at that time helped to improve relations with its Arab neighbors. The Arab boycott of Israeli goods and companies was lifted. In 1994, the country signed a formal peace treaty with Jordan and opened trade offices in several Gulf states.

RABIN'S DEATH AND ITS CONSEQUENCES

The second stage in transfer of authority over the West Bank had barely begun when Rabin was assassinated by an Orthodox Jew while speaking at a Peace Now rally in Jerusalem on November 4, 1995. The assassination climaxed months of increasingly ugly anti-Rabin rhetoric orchestrated by Orthodox rabbis, settlers, and right-wing groups who charged the prime minister with giving away sacred Jewish land while gaining little in return. The assassin, Yigal Amir, stated in court that God had made him act, since the agreement with the Palestinians contradicted sacred Jewish religious principles. He stated: "According to *halacha* [Judaic tradition] a Jew who like Rabin gives over his people and his land to the enemy must be killed. My whole life I have been studying the *halacha* and I have all the data."[20]

Rabin's assassination marked a watershed in Israeli political life; for the first time an Israeli leader had been struck down by one of his own people. Amir's action was condemned abroad. Most secular Israelis regard it as a terrible and useless tragedy, and Israeli society as a whole has

yet to come to terms with the murder of its most respected statesman. Unfortunately, many Orthodox Jews subscribed to Amir's stated belief that he had acted rightly to preserve the sacred homeland. Rabin's widow, Leah, continued to call Amir a traitor to his country until her own death in 2001, and she traveled throughout the world to promote the cause of Israeli–Palestinian coexistence. In August 2001, however, President Katsav pardoned Amir's co-conspirator Margalit Has-Sheft, who had been serving a nine-month jail term for complicity in the assassination. Amir's brother and others associated with his plot remain in prison, but the pardon re-exposed the deep fault lines in Israeli society that had emerged with Rabin's death.

After Rabin's death, Peres was confirmed as his successor by a 111–9 Knesset vote. However, he was defeated in the 1996 elections by the new Likud leader, Benjamin Netanyahu. It was Israel's first direct election for prime minister. Peres had hurt his cause by undertaking the ill-advised "Operation Grapes of Wrath" into Lebanon, which resulted in the deaths of many Lebanese civilians. The vote was close, a razor-thin margin of 50.3 percent for Netanyahu to 49.6 percent for Peres. However, Likud failed to win a majority of seats in the Knesset and was forced into another coalition with the religious parties. Labor actually won more seats than Likud (34 to 32), but support from Shas and the National Religious Party (20 seats) and the Russian-immigrant Yisrael Ba'Aliya party gave Netanyahu a narrow majority in the Knesset.

Netanyahu's victory, by a scant 16,000 votes, foreshadowed what would prove to be a near-total deadlock in the peace process. The deadlock was also marked by a deep and angry division within the Israeli society. Divisions have always been present in Jewish society, in the Diaspora as well as in modern Israel. God Himself described them as a "stiff-necked People" (in Exodus) and this cultural trait, honed over centuries, has certainly helped them to survive and preserve their separate identity as a people. But in the modern world it has set Haredim against Reform and Conservative Jewry, culminating in the murder of Yitzhak Rabin.

Israel's 50th birthday in 1998 underscored the divisions. One reporter noted: "It was not just the crime but the popular support behind it that raised the question of whether the Jews, having created the state, possessed the civility they needed to preserve it."[22]

As one observer noted, "As withdrawal from the occupied territories approached, mutual respect and moderation recede. Religious Jews emitted a defiant self-righteousness, to which seculars seemed unable to respond. It was not just the crime but the popular support behind it that raised the question of whether the Jews, having created the state, possessed the civility they needed to preserve it." Israel's 50th birthday in May 1998 underscored the divisions. Even the reelection of Ezer Weizman to the largely ceremonial office of president in March for a second five-year term generated controversy. The Likud, angered at Weizman's public criticism of its policies, nominated an unknown to run against the popular president. However, Weizman was easily reelected by 63–49 vote in the Knesset.

In contrast to the vision of Israel's founders of Jewish immigrants from many countries and cultures unified into a model democracy, Israeli society as it exists today would be better described as a land of contentious tribes, traceable back to Abraham's expulsion of Hagar and her son Ishmael (ancestor of the Arabs) in favor of Isaac and his descendants. In addition to not honoring the Oslo agreements, Netanyahu encouraged increased Jewish settlements in the West Bank and Gaza. He used his office to cultivate various groups and parties at the expense of other groups, changing sides when it suited him. A Labor Party official observed before the 1999 election that Israeli elections are more anthropology than politics; the nation does not have a democratic "pendulum" that swings voters away from a leader en masse when his policies seem to be failing.

Elections for the office of prime minister had been scheduled for 2000, but Netanyahu advanced them by a year to preempt increasing opposition to his policies. However, his principal rival, retired general Ehud Barak, defeated him handily in the election, and Netanyahu gave up his seat in the Knesset and retired to private life. Unfortunately for Israel, Barak's large popular majority was negated by the country's multiparty system, which makes a majority difficult to establish in the Knesset. One Israel (formerly Labor), which was now the majority party, held 27 seats to 17 for Shas, 10 for the left-of-center Mereta Party, six each for Center (a new party) and Yisrael Ba' Aliya, and five each for the ultra-Orthodox United Torah Judaism and National Religious Party. Barak was forced to form a coalition of several parties in order to retain support for his peace initiatives with the Palestinians.

Effective coalition government requires concessions to minority parties in Israel's fractious political system. In return for Shas's support, the party was granted four ministerial portfolios, including that of the powerful Interior Ministry. Shas leaders then demanded more concessions; they included $36 million to bail out its bankrupt religious-schools system, as well as debt relief and tax exemptions for the party's social services. When these demands were rejected, Shas pulled out of the coalition. (Earlier, its leader, Rabbi Aryeh Deri, had been convicted of fraud, bribery, and misuse of public power while he was a cabinet minister. He was given a three-year jail sentence.) A similar misuse of public power blocked a bid by supporters of President Ezer Weizman to have him serve a third term. The office, although it is largely ceremonial, requires Knesset confirmation. Weizman had been charged with receiving cash gifts from business people and keeping them rather than turning them over to the state. The court found insufficient evidence for more than a reprimand, but Weizman resigned rather than continue in office under a cloud.

Ariel Sharon's well-publicized visit on September 28, 2000, to the Dome of the Rock not only set off a new Palestinian intifada; it also had a direct impact on the muddled Israeli political system. The collapse of Barak's coalition left him in the awkward position of negotiating a settlement with the Palestinians without the support of the Knesset or of his own people, particularly the latter, since the negotiations involved territorial concessions. As former Likud leader, Netanyahu seemed poised to take back the office of prime minister. However, Israel's election laws require that a candidate must be a Knesset member. With Netanyahu unable to run, the election narrowed to a choice for the voters between Barak and Sharon, with the latter winning by another huge margin. One major difference from the previous election was that voter turnout was 60 percent. Of those who voted, 62 percent either voted for Barak or turned in blank ballots in protest. Also, only 18 percent of Israeli Arabs cast their ballots; they had been a significant factor in Barak's 1999 election victory, supporting him by 95 percent.

The new prime minister took office amid foreboding on the part of many Israelis, as well as by the country's Arab neighbors and the world at large. Noting Sharon's record, which includes masterminding the 1982 invasion of Lebanon and ultimate responsibility for the slaughter of Palestinians by Israel's Christian allies in Lebanese refugee camps, one author suggests that his election "looked like appointing the village pyromaniac to head its fire brigade." In 2001, survivors of the families killed at the camps filed suit against him in a Belgian court, under the new international statute allowing for prosecution of national leaders for war crimes such as genocide.

Sharon's first year in office coincided with the renewal of the Palestinian intifada, which has brought Israel and the Palestinians into a head-on conflict verging on total war. The accelerated cycle of violence has been marked by new tactics on both sides—relentless suicide bombings by Palestinians, mostly against Israeli civilians; and use of massive retaliation by Israeli tanks, missiles, and helicopter gunships. The first Israeli government official was killed in November, and since then Israel has embarked on a policy of targeting Palestinians suspected of leading the violence. The death toll is modest compared with that in other areas of conflict, in Africa and elsewhere. But the impact of the conflict on the two societies has been traumatic.

By 2004 the conflict had taken on many aspects of a civil-war. On the Israeli side there is increasing divergence between the population at large, 39 percent of whom favor accommodation with the Palestinians, and the Sharon government. In January a military court sentenced five "refuseniks" to one-year jail terms for refusing to serve in the West Bank. Earlier, 27 Air Force pilots had refused to attack civilian areas there; they were joined by reservists from the elite Sayeret Matkal unit of the IDF. On the official side, several former heads of Shin Bet, the Israeli security service, publicly denounced the government's repressive measures, and the military Chief of Staff, General Ayalon, warned that the measures would only "generate hatred that will explode in our face." Several officials spoke openly of a unilateral withdrawal from the occupied territories. As a leading Israeli intellectual told a reporter, "Israel behaves like a society in a coma, completely opaque and closed. But there are cracks in that opacity, symptoms that something momentous is happening." [23]

The personal hostility between Sharon and the late Yassir Arafat did little to move the peace process forward. In December 2001 Sharon declared that Arafat was "irrelevant" and Israeli forces blockaded his Ramallah headquarters. For the rest of his life Arafat remained essentially under house arrest. In September 2004, with Israel" permission (under U.S. and European pressure) he was airlifted to a French military hospital. He died there on November 11, probably of a stroke brought on by a blood disorder, disseminated intravascular coagulation (D.I.C.) which had never been controlled.

A similar extremism marks the relationship between Sharon and Yassir Arafat. In

December 2001 Sharon declared Arafat "irrelevant" to the peace process. Israeli security forces blockaded his Ramallah headquarters, confining him to house arrest. Relations improve temporarily with the election of a prime minister to expedite the Palestinian side of the "road map to peace". However Arafat retains final decision-making powers in the Palestinian hierarchy and remains a symbol of hope for generation of Palestinians in their struggle to establish a state of their own.

To some extent, the Palestinian–Israeli conflict has become one of a personal vendetta between Ariel Sharon and Yassir Arafat. Bolstered by the hawks in his government and party, Sharon declared Arafat "irrelevant" to the peace process in December 2001, and Israeli tanks blockaded the Palestinian leader in his Ramallah headquarters. The blockade was lifted in May 2002 under heavy U.S. pressure. But given the supercharged atmosphere of Palestinian–Israeli hatred and the Sharon–Arafat personal vendetta, only strong and unremitting external mediation could create the conditions for a truce and peace negotiations.[23]

THE ECONOMY

In terms of national income and economic and industrial development, Israel is ahead of a number of Middle Eastern states that have greater natural resources. Agriculture is highly developed; Israeli engineers and hydrologists have been very successful in developing new irrigation and planting methods that not only "make the desert bloom" but also are exported to other nations under the country's technical-aid program. Agriculture contributes 7.5 percent of GDP annually.

Water in Israel

In Israel water usage is as much a political problem as it is economic. The 187,000 Jewish settlers in the West Bank use 76 gallons of water per day per capita, leaving the much larger Palestinian population 18.5 gallons daily, below WHO minimum health and sanitation standards. The Dead Sea, a vital water resource not only for Israel but also for Jordan, has shrunk from 50 miles wide in 1950 30 miles today, with its water level dropping 3 feet yearly. Israel's diversion of the Jordan River, its main recharge source, into a national water carrier pipeline in 2004, has lowered Dead Sea water input from 160 billion gallons annually to 16 billion, insufficient to maintain it at its present size. Hydrologists agree that given the influence of the agricultural lobby in Israel and demands of the settler population the Dead Sea's prospects for recovery are bleak. Even the tourist industry

has been affected. Elegant resorts like Ein Gedi, formerly located at waterside, are now a mile inland.

It is an crisis compounded by urban congestion and high pollution levels. Unfortunately, the higher priority placed on security and national defense by the Palestinian intifada has left these long-term economic issues unresolved.

The cutting and export of diamonds and other gemstones is a small but important industry. Since 1996, exports of rough diamonds from the Ramat Gan diamond exchange have increased 80 percent, from $5.2 billion to $9.8 billion. However, in June 2001, the DeBeers Consolidated Mines of South Africa, colossus of the diamond world, announced that it would privatize and sell its own diamond jewelry rather than go through Israeli merchants in the exchange. The action poses a serious threat to the Israeli share of the industry. The aircraft industry is the largest single industrial enterprise, but it has fallen on hard times. The national airline, El Al, which began its existence with the founding of the state, was restructured as a public enterprise in 1982 by the Begin government. It continued to lose money, in part due to high operating costs and unique security arrangements. A $600 million upgrade in equipment enabled the company to expand its flights, and in 1998, it showed a modest $25 million profit. In 2001, the high-speed wireless companies BreezeCom and Floware merged to form a single company capitalized at $330 million.

In January 2001, under the rubric "business is business" and despite strained diplomatic relations, Israel signed an agreement with Egypt to import $3 billion in natural gas annually until 2012. Despite angry Egyptian criticism of Israel's policies toward the Palestinians as "fellow-Arabs," EMG, a partnership of the Egyptian Natural Gas Corporation and other business groups, remained committed to the sale. The initial distribution took place in June 2003.

An added boost for the Israeli economy was the discovery recently of large natural-gas deposits offshore near the Gaza Strip, by the British energy company BP. Development of these resources will help meet Israel's chronic freshwater shortages by providing energy for the new desalination plants under construction near Tel Aviv. These plants, scheduled for completion in 2004, will provide 500 cubic meters of desalinated water toward meeting national needs.

Providing further evidence of the high level of Israeli technology, the Israel-based Solel Solar Systems currently produces 50 kilowatts annually of solar power, meeting 80 percent of the country's power needs in

hotels, hospitals, and other public buildings. Thus far, it is the only solar power system in the world that generates electricity from conventional turbines.

Israel's military superiority over that of its Arab neighbors was reemphasized in 2001 by successful tests of its Arrow II ABM missile. The Arrow II intercepted and destroyed a simulated missile similar to the Scuds fired by Iraq during the 1991 Gulf War. Although Israel's nuclear capability remains a closely-guarded secret, its stockpile is estimated to contain between 200 and 500 neutron bombs and small nuclear bombs, capable of killing humans but not damaging the environment. Also, the Israeli Navy has 3 missile-equipped submarines.

The Israeli economy has been strengthened through the success of its high-technology companies, many of which are subsidiaries of U.S. firms like Net Manage and Geotek Communications. Its cadre of army-trained computer experts and skilled Russian emigrés made the country second to the United States in the number of start-up companies in the 1990s.

Israel has 34 producing oil wells. After the 1967 War and the occupation of the Sinai Peninsula, Israel was able to exploit Sinai petroleum resources as well as the Alma oil fields in the Gulf of Suez. Twenty-five percent of domestic oil needs came from the Alma fields. In accordance with the Egyptian–Israeli peace treaty, these fields were returned to Egypt, with the stipulation that Israel be able to purchase Sinai oil at less than prices set by the Organization of Petroleum Exporting Countries (OPEC) for Egyptian oil on the world market.

Prior to 1977, when the Likud Bloc came to power, the economy was managed largely by the state in tandem with Histadrut, the Israeli labor confederation. Histadrut functions partly as a union but also as an employer, and it is the controlling factor in many industries. It negotiates for most labor contracts through bargaining with the government. The cost-of-living increases built into these contracts and the country's high level of defense expenditures sent the inflation rate out of sight in the 1980s

Inflation was also fueled by Likud's fiscal policies. Foreign-exchange controls were abolished, along with the travel tax and import licenses. The currency was allowed to float, and Israelis could open bank accounts in foreign currency. The results were balance-of-payments deficits and a drop in foreign-exchange reserves due to luxury imports paid for in foreign currencies. By 1985, the inflation rate had reached 800 percent.

The Labor government took office amid warnings from U.S. and other economists,

Timeline: PAST

1897
The Zionist movement is organized by Theodor Herzel

1917
The Balfour Declaration

1922–1948
British mandate over Palestine

1947–1948
A UN partition plan is accepted by the Jewish community; following British withdrawal, the State of Israel is proclaimed

1949
Armistices are signed with certain Arab states, through U.S. mediation

1967
The Six-Day War; Israeli occupation of East Jerusalem, the Gaza Strip, and the Sinai Peninsula

1973
Yom Kippur War

1979
Peace treaty with Egypt

1980s
The Israeli invasion of Lebanon

1990s
Israeli–Palestinian efforts toward peace; violence escalates on both sides in response; Prime Minister Yitzhak Rabin is assassinated

PRESENT

2000s
Ariel Sharon becomes prime minister, supervises withdrawal from Gaza, then is felled by strokes and succeeded by his deputy, Ehud Olmert

---

plus its own advisers, that the economy was on the point of collapse. Peres introduced a package of draconian reforms. Wages and prices were frozen, and imports of all but essential goods were banned. The package included devaluation of the shekel, replacing it with the Israeli new shekel, pegged to the dollar. Cost-of-living increases were eliminated from labor contracts, and 10,000 jobs were cut from the government bureaucracy.

These measures brought about a brief economic rebound. But the intifada, with its consequent loss of confidence on the part of foreign investors, led to a major recession in 2002. Exports dropped by 10 percent and both inflation and unemployment rates rose dramatically. In April 2003 the new Finance Minister, former Prime Minister Binyamin Netanyahu, introduced a program of drastic measures to reduce the budget deficit and stabilize the economy. They included a wage freeze for public employees, higher pension premiums and dismissal of several thousand workers.

In 2004, under Netanyahu's economic reform program, the Knesset approved a major reform of capital markets. I will require banks to sell their provident and mutual funds and reduce their insurance company holdings by 10 percent. The reform package will also privatize Bank Leumi, the country's largest.

Politics Trumps Economics

Netanyahu's defeat by Sharon as both head of government and of Likud removed him temporarily from politics, but in 2005 he was able to win back the party chairmanship after Sharon left to form his own centrist party, Kadima. Equally surprising was the defeat of long-time Labor Alignment leader Shimon Peres by a relative newcomer to politics, Amir Peretz. A Sephardic Jew (born in Morocco), Peretz had risen to become chairman of Histadrut

Sharon's decision to withdraw military forces from the Gaza Strip and evacuate all the Jewish settlers there, while popular with the majority of the population, generated fierce opposition not only from Likud members but also from the powerful settler movement. But despite Netanyahu's success in recapturing its chairmanship, the party itself was split between "pragmatists" and "purists." The former had come to the conviction that a Palestinian state in the still-occupied West Bank and now-free Gaza was inevitable. The latter held fast to the original Likud dream of a "greater Israel," including these territories, as part of God's covenant with the Jewish people. The extent of public support for Sharon's new party and its presumed goal of a settlement with the Palestinians was unclear as of the year's end, but opinion polls in late November indicated that the "old warrior" would be re-elected in the March 2006 election.

NOTES

1. Menachem Froman, "A Modest Proposal," *Jerusalem Report* (October 25, 1999).
2. John Noble Wilford, "A New Armageddon Erupts Over Ancient Battlefields," *The New York Times* (January 6, 2000).
3. Ashkenazi, derived from Ashkenaz (Genesis 10:3), is the name given to Jews who lived in Europe, particularly Germany, and followed particular traditions of Judaism handed down from biblical days. Sephardim (from Sepharah, Obadiah 1:20) refers to Jews originally from Spain who were expelled and emigrated to the Middle East–North Africa. R. J. Zwi Werblowsky and Geoffrey Wigoder, eds., *The Encyclopedia of the Jewish Religion* (New York: Holt, Rinehart & Winston, 1965).
4. Dan V. Segre, *A Crisis of Identity: Israel and Zionism* (Oxford, England: Oxford University Press, 1980), p. 25.
5. Ben Lynfield, in *The Christian Science Monitor*, August 8, 2003.
6. Herod's palaces, adorned with copious mosaics and furnished with mineral baths and saunas, have been excavated and reveal not only his opulent lifestyles but also his very real contributions to the architecture of his time, including renovation of the Temple. Other links with Jesus' time are less definite. An ossuary (burial box) discovered in 2002 and thought to contain the bones of his elder brother James has been proven to be a forgery.
7. Yaron Ezrahi, *Rubber Bullets: Power and Conscience in Modern Israel* (New York: Farrar, Straus & Giroux, 1997), p. 269.
8. Abraham Shulman, *Coming Home to Zion* (Garden City, NY: Doubleday, 1979), p. 14.
9. The text is in "Documents on Palestine," *The Middle East and North Africa* (London: Europa Publications, 1984), p. 58.
10. See Roy Jenkins, *Churchill: A Biography* (London: Macmillan, 2001) p. 360. The author also notes that Churchill had spoken out favorably about the Arabs at the Cairo Conference: "The British government is well-disposed to them and cherishes Arab friendship." A more specific commitment came in the form of a letter from Arthur James Balfour to Lord Rothschild, prominent statesman and a Zionist sympathizer, in 1917. Although the latter's qualified endorsement favoring a Jewish national home in Palestine included a commitment to guarantee the rights of the indigenous population, its approval by the cabinet was taken by the Zionists as official British support for their cause. In this way the letter became "The Balfour Declaration."
11. Shimon Peres, *Battling for Peace: A Memoir,* David Landau, ed. (New York: Random House, 1995), pp. 20–21.
12. Abba Eban, "Rebirth of a Nation," *Jerusalem Post* Supplement 5757 (May 1997), p. 2.
13. David Ben-Gurion, *Memoirs* (Cleveland: World Publishing, 1970), p. 58.
14. Abu Iyad with Eric Rouleau, *My Home, My Land: A Narrative of the Palestinian Struggle*, Linda Butler Koseoglu, tr. (New York: New York Times Books, 1981), pp. 225–226.
15. Ben-Gurion, *op. cit.*, p. 57.
16. The Bar-On affair involved Netanyahu's appointment of a "mediocre lawyer," Roni Bar-On, as attorney-general, supposedly as a favor to the leader of Shas, in return for his party's support for the coalition government. The Israeli Supreme Court ruled in June 1997 that there was insufficient evidence to prosecute Netanyahu, but the nearscandal led to a no-confidence resolution in the Knesset that the prime minister survived by a 55–50 vote. See *The New York Times* (June 24, 1997).
17. Robert Schreiber, in *New Outlook* (May/June 1987), p. 42.
18. Bernard Avishai, "Saving Israel from Itself", *Harper's Magazine*, January 2005, pp. 38–46.
19. Ezrahi, *op. cit.*, pp. 274–275.

20. Quoted in Amos Elon, *A Blood-Dimmed Tale: Despatches from the Middle East* (New York: Columbia University Press, 1997) (originally appeared in *The New York Review of Books,* December 21, 1995).

21. Milton Viorst, What Shall I Do With These People; Jews and the Fractious Politics of Judaism (New York: Free Press, 2002, p. 215).

22. *The Christian Science Monitor*, October 13, 2005.

23. Milton Viorst, *What Shall We Do With These People; Jews and the Fractious politics of Judaism* (New York: Free Press, 2002, p. 215).

Jordan (Hashimite Kingdom of Jordan)

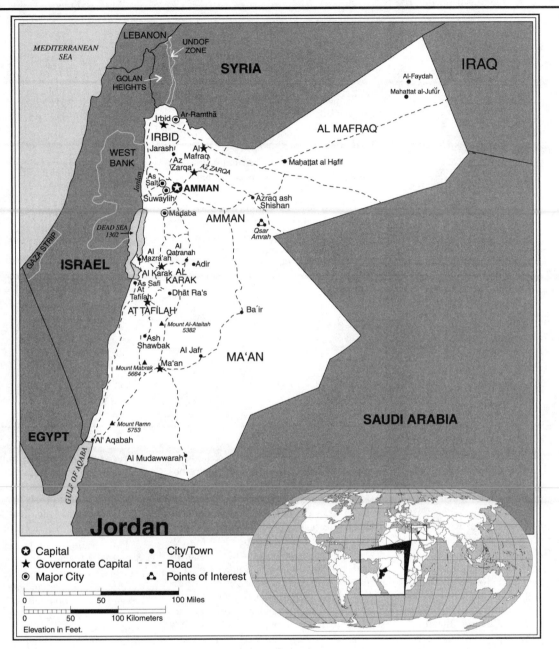

Jordan

- ✪ Capital
- ★ Governorate Capital
- ◉ Major City
- ● City/Town
- - - - Road
- △ Points of Interest

0 — 50 — 100 Miles
0 — 50 — 100 Kilometers
Elevation in Feet.

Jordan Statistics

GEOGRAPHY

Area in Square Miles (Kilometers): 35,000 (92,300) about the size of Indiana

Capital (Population): Amman (483,000)

Environmental Concerns: limited natural freshwater reserves; deforestation; overgrazing; soil erosion; desertification

Geographical Features: mostly desert plateau in the east; a highland area in the west; the Great Rift Valley separates the east and west banks of the Jordan River

Climate: mostly arid desert; a rainy season in the west

PEOPLE

Population

Total: 5,759,732

Annual Growth Rate: 2.56%

Rural/Urban Population Ratio: 28/72

Major Languages: Arabic; English widely understood

Ethnic Makeup: 98% Arab; 1% Circassian; 1% Armenian

Religions: 92% Sunni Muslim; 8% Christian and others

Health

Life Expectancy at Birth: 75.7 years (male); 80.8 years (female)

Infant Mortality Rate: 17.35/1,000 live births

Education

Adult Literacy Rate: 91.3%

Compulsory (Ages): 6–16; free

COMMUNICATION

Telephones: 622,600 main lines

Daily Newspaper Circulation: 62 per 1,000 people

Televisions: 176 per 1,000

Internet Users: 5 (2000)

TRANSPORTATION

Highways in Miles (Kilometers): 4,968 (8,000)

Railroads in Miles (Kilometers): 420 (677)

Usable Airfields: 18

Motor Vehicles in Use: 265,000

GOVERNMENT

Type: constitutional monarchy

Independence Date: May 25, 1946 (from League of Nations mandate)

Head of State/Government: King Abdullah II; Prime Minister Adnan Badran

Political Parties: Al-Umma (Nation) Party; Jordanian Democratic Popular Unity Party; Islamic Action Front; National Constitutional Party; many others

Suffrage: universal at 18

MILITARY

Military Expenditures (% of GDP): 14.6%

Current Disputes: none

ECONOMY

Currency ($U.S. Equivalent): 0.709 dinar = $1 (fixed rate)

Per Capita Income/GDP: $4,500/$25.5 billion

GDP Growth Rate: 5.1%

Inflation Rate: 3.2%

Unemployment Rate: officially 15%; more likely 25%–30%

Labor Force: 1,410,000

Natural Resources: phosphates; potash; shale oil

Agriculture: wheat; barley; fruits; tomatoes; olives; livestock

Industry: phosphate mining; petroleum refining; cement; potash; light manufacturing

Exports: $3.2 billion (primary partners India, United States, Iraq)

Imports: $7.6 billion (primary partners Germany, Saudi Arabia, China)

SUGGESTED WEBSITES

http://lcweb2.loc.gov/frd/cs/ jotoc.html

http://www.odci.gov/cia/ publications/factbook/index.html

Jordan Country Report

The Hashimite Kingdom of Jordan (previously called Transjordan; usually abbreviated to Jordan) is one of the smaller Middle Eastern nations. The country formerly consisted of two regions: the East Bank (lying east of the Jordan River) and the West Bank of the Jordan. Israel occupied the West Bank in June 1967, although the region continued to be legally and administratively attached to Jordan and salaries of civil servants and others were paid by the Jordanian government. In 1988, King Hussein formally severed the relationship, leaving the West Bank under Israeli occupation de facto as well as de jure. Between 1948 and 1967, Jordanian-occupied territory also included the old city of Jerusalem (East Jerusalem), which was annexed during the 1948 Arab-Israeli War.

Modern Jordan is an "artificial" nation, the result of historical forces and events that shaped the Middle East in the twentieth century. It had no prior history as a nation and was known simply as the land east of the Jordan River, a region of diverse peoples, some nomadic, others sedentary farmers and herders. Jordan's current neighbors are Iraq, Syria, Saudi Arabia, and Israel. Their joint borders were all established by the British after World War I,

when Britain and France divided the territories of the defeated Ottoman Empire between them.

Jordan's borders with Iraq, Syria, and Saudi Arabia do not follow natural geographical features. They were established mainly to keep nomadic peoples from raiding; over time, these borders have been accepted by the countries concerned. The boundary with Israel, which formerly divided the city of Jerusalem between Jordanian and Israeli control, became an artificial barrier after the 1967 Six-Day War and Israel's occupation of Jerusalem and the West Bank (of the Jordan River). The Jordan-Israel Peace Treaty of 1994 has resulted in a redrafting of borders. Israel returned 340 square miles captured in 1967 in the Arava Valley and south of the Galilee to Jordanian control. However, Israeli *kibbutzim* (communal farm settlements) will be allowed to continue cultivating some 750 acres in the territory under a 25-year lease.

HISTORY

The territory of modern Jordan was ruled by outside powers until it became an independent nation in the twentieth century. Under the Ottoman Empire, it was part of

the province of Syria. The Ottoman authorities in Syria occasionally sent military patrols across the Jordan River to "show the flag" and collect taxes, but otherwise they left the people of the area to manage their own affairs.[1]

This tranquil existence ended with World War I. The Ottomans were defeated and their provinces were divided into protectorates, called mandates, set up by the League of Nations and assigned to Britain and/or France to administer and prepare for eventual self-government. The British received a mandate over Palestine and extended its territory to include Transjordan east of the River Jordan. Due to their commitment to help Jews dispersed throughout the world to establish a national home in Palestine, the British decided to govern Transjordan as a separate mandate.

The terms of the mandate system required the protecting power (in this case, Britain) to appoint a native ruler. During the war, the British had worked with Sharif Husayn to organize an Arab revolt against the Ottomans. Husayn was a prominent Arab leader in Mecca who held the honorary position of "Protector of the Holy Shrines of Islam." Two of the sharif's sons, Faisal and Abdullah, had led the revolt, and

Jordan Tourism Board (JTB001)

Amman is a city built on seven hills. Against the slope of one hill is this large Roman amphitheater. The amphitheater, restored to its original seating capacity of 6,000, is now being used once again for grand performances, as it was almost 2,000 years ago.

the British felt that they owed them something. When Iraq was set up as a mandate, the British made Faisal its king. Abdullah was offered the Transjordan territory. Because the population was primarily pastoral, he chose the traditional title of emir, rather than king, considering it more appropriate.

EMIR ABDULLAH

Through his father, Abdullah traced his lineage back to the Hashim family of Mecca, the clan to which the Prophet Muhammad belonged. This ancestry gave him a great deal of prestige in the Arab world, particularly among the nomads of Transjordan, who had much respect for a person's genealogy. Abdullah used the connection assiduously to build a solid base of support among his kinspeople. When the country became fully independent in 1946, Abdullah named the new state the Hashimite Kingdom of Jordan.

Abdullah's new country had little to recommend it to outsiders except some fine Roman ruins and a great deal of empty

land. It was a peaceful, quiet place, consisting entirely of what is today the East Bank of the Jordan River, with vaguely defined borders across the desert. The population was about 400,000, mostly rural peasants and nomads; the capital, Amman, was little more than a large village spread over some of those Roman ruins.

During the period of the mandate (1921–1946), Abdullah was advised by resident British officials. The British helped him draft a constitution in 1928, and Transjordan became independent in everything except financial policy and foreign relations. But Emir Abdullah and his advisers ran the country like a private club. In traditional Arab desert fashion, Abdullah held a public meeting outside his palace every Friday; anyone who wished could come and present a complaint or petition to the emir.

Abdullah did not trust political parties or institutions such as a parliament, but he agreed to issue the 1928 Constitution as a step toward eventual self-government. He also laid the basis for a regular army. A British Army officer, John Bagot Glubb,

was appointed in 1930 to train the Transjordanian Frontier Force to curb Bedouin raiding across the country's borders. Under Glubb's command, this frontier force eventually became the Arab Legion; during Emir Abdullah's last years, it played a vital role not only in defending the kingdom against the forces of the new State of Israel but also in enlarging Jordanian territory by the capture of the West Bank and East Jerusalem.[2]

When Britain gave Jordan its independence in 1946, the country was not vastly different from the tranquil emirate of the 1920s. But events beyond its borders soon overwhelmed it, like the dust storm rolling in from the desert that sweeps everything before it. The conflict between the Arab and Jewish communities in neighboring Palestine had become so intense and unmanageable that the British decided to terminate their mandate. They turned the problem over to the United Nations. In November 1947, the UN General Assembly voted to partition Palestine into separate Arab and Jewish states, with Jerusalem to be an international city under UN administration.

The partition plan was not accepted by the Palestine Arabs, and as British forces evacuated Palestine in 1947–1948, they prepared to fight the Jews for possession of all Palestine. The State of Israel was proclaimed in 1948. Armies of the neighboring Arab states, including Jordan, immediately invaded Palestine. But they were poorly armed and untrained. Only the Jordanian Arab Legion gave a good account of itself. The Legion's forces seized the West Bank, originally part of the territory allotted to a projected Palestinian Arab state by the United Nations. The Legion also occupied the Old City of Jerusalem (East Jerusalem). Subsequently, Abdullah annexed both territories, despite howls of protest from other Arab leaders, who accused him of land grabbing from his "Palestine brothers" and harboring ambitions to rule the entire Arab world.

DEVELOPMENT

The Israel-Jordan peace treaty has enabled Jordan to save $893 million in debts to the U.S. and European countries. But continued budgetary deficits and the cutoff in trade with Iraq, the country's main trading partner, have hampered development. U.S. aid in setting up free trade Zones in 2000–2001 have generated 13,000 new jobs, and in July 2003 a thermal gas facility linked to Egypt via pipeline went into operation. It will supply Jordan with 2.7 billion cubic centimeters (cm^3) of Egyptian natural gas.

Jordan now became a vastly different state. Its population tripled with the addition of half a million West Bank Arabs and half a million Arab refugees from Israel. Abdullah still did not trust the democratic process, but he realized that he would have to take firm action to strengthen Jordan and to help the dispossessed Palestinians who now found themselves reluctantly included in his kingdom. He approved a new Constitution, one that provided for a bicameral Legislature (similar to the U.S. Congress), with an appointed Senate and an elected House of Representatives. He appointed prominent Palestinians to his cabinet. A number of Palestinians were appointed to the Senate; others were elected to the House of Representatives.

On July 20, 1951, King Abdullah was assassinated as he entered the Al Aqsa Mosque in East Jerusalem for Friday prayers. His grandson Hussein was at his side and narrowly escaped death. Abdullah's murderer, who was killed immediately by royal guards, was a Palestinian. Many Palestinians felt that Abdullah had betrayed them by annexing the West Bank and because he was thought to have carried on secret peace negotiations with the Israelis (recent evidence suggests that he did so). In his *Memoirs,* King Abdullah wrote, "The paralysis of the Arabs lies in their present moral character. They are obsessed with tradition and concerned only with profit and the display of oratorical patriotism."[3]

Abdullah dealt with the Israelis because he despaired of Arab leadership. Ironically, Abdullah's proposal to Britain in 1938 for a unified Arab-Jewish Palestine linked with Jordan, if it had been accepted, would have avoided five wars and hundreds of thousands of casualties. Yet this same proposal forms the basis for discussion of the Palestinian-Israeli settlement in recent years.[4]

KING HUSSEIN

Abdullah's son Crown Prince Talal succeeded to the throne. He suffered from mental illness (probably schizophrenia) and had spent most of his life in mental hospitals. When his condition worsened, advisers convinced him to abdicate in favor of his eldest son, Hussein.

At the time of his death from cancer in February 1999, Hussein had ruled Jordan for 46 years, since 1953—the longest reign to date of any Middle Eastern monarch and one of the longest in the world in the twentieth century. To a great extent he *was* Jordan, developing a small desert territory with no previous national identity into a modern state. A popular Jordanian saying was, "Hussein is Jordan and Jordan is Hussein," underlining the fusion of ruler and nation.

Yet during his long reign, Hussein faced and overcame many crises and challenges to his rule. These crises stemmed mainly from Jordan's involvement in the larger Arab-Israeli conflict. Elections for the Jordanian Parliament in 1956 resulted in a majority of West Bank Palestinian candidates. Controversy developed as these representatives pressed the king to declare his all-out support for the Palestinian cause. At one point, the rumor spread that he had been killed. Hussein immediately jumped into a jeep and rode out to the main Arab Legion base at Zerqa, near Amman. He then presented himself to the troops to prove that he was still alive and in command.[5]

The "Zerqa incident" illustrated Hussein's fine sense of timing, undertaking bold actions designed to throw real or potential enemies off balance. It also emphasized the importance of army support for the monarchy. The majority of the Arab Legion's soldiers came from Bedouin tribes, and for them, loyalty to the crown has always been automatic and unfailing.

Other challenges followed. Syrian fighters tried to shoot down King Hussein's plane in 1958, and Communists and Palestinian leaders plotted his overthrow.

The June 1967 Six-Day War produced another crisis in Jordan, this one not entirely of its own making. Israeli forces occupied 10 percent of Jordanian territory, including half of its best agricultural lands. The Jordanian Army suffered 6,000 casualties, most of them in a desperate struggle to hold the Old City of Jerusalem against Israeli attack. Nearly 300,000 more Palestinian refugees from the West Bank fled into Jordan. To complicate things further, guerrillas from the Palestine Liberation Organization, formerly based in the West Bank, made Jordan their new headquarters. The PLO considered Jordan its base for the continued struggle against Israel. Its leaders talked openly of removing the monarchy and making Jordan an armed Palestinian state.

By 1970, Hussein and the PLO were headed toward open confrontation. The guerrillas had the sympathy of the population, and successes in one or two minor clashes with Israeli troops had made them arrogant. They swaggered through the streets of Amman, directing traffic at intersections and stopping pedestrians to examine their identity papers. Army officers complained to King Hussein that the PLO was really running the country. The king became convinced that unless he moved against the guerrillas, his throne would be in danger. He declared martial law and ordered the army to move against them.

The ensuing Civil War lasted until July 1971, but in the PLO annals, it is usually referred to as "Black September," because of its starting date and because it ended in disaster for the guerrillas. Their bases were dismantled, and most of the guerrillas were driven from Jordan. The majority went to Lebanon, where they reorganized. In time they became as powerful there as they had been in Jordan.

For the remainder of his reign, there were no serious internal threats to King Hussein's rule. Jordan shared in the general economic boom in the Arab world that developed as a result of the enormous price increases in oil after the 1973 Arab-Israeli War. As a consequence, Hussein was able to turn his attention to the development of a more democratic political system. Like his grandfather, he did not entirely trust political parties or elected legislatures, and he was leery of the Palestinians' intentions toward him. He was also convinced that Jordan rather than the PLO should be the natural representative of the Palestinians. But he realized that in order to represent them effectively and to build the kind of Jordanian state that he could safely hand over to his successors, he would need to develop popular support in addition to that of the army. Accordingly, Hussein set up a National Consultative Council in 1978, as what he called an interim step toward democracy. The Council had a majority of Palestinians (those living on the East Bank) as members.

Hussein's arbitrary separation of Jordan from the West Bank has had important implications for internal politics in the kingdom. It enabled the king to proceed with political reforms without the need to involve the Palestinian population there. The timetable was accelerated by nationwide protests in 1989 over price increases for basic commodities. The protests turned swiftly to violence, resulting in the most serious riots in national history. Prime Minister Zaid Rifai was dismissed; he was held personally responsible for the increases and for the country's severe financial problems, although these were due equally to external factors. King Hussein appointed a caretaker government, headed by his cousin, to oversee the transitional period before national elections for the long-promised lower chamber of the Legislature.

In 1990, the king and leaders of the major opposition organization, the Jordanian National Democratic Alliance (JANDA), signed a historic National Charter, which provides for a multiparty political system. Elections were set under the Charter for an 80-member House of Representatives. Nine seats would be reserved for Christians and three for Circassians, an ethnic Muslim minority originally from the Caucasus.

In 1992, Hussein abolished martial law, which had been in effect since 1970. Henceforth, security crimes such as espionage would be dealt with by state civilian-security courts. New laws also undergirded constitutional rights such as a free press, free speech, and the right of public assembly.

With political parties now legalized, 20 were licensed by the Interior Ministry to take part in Jordan's first national parliamentary election since 1956. The election was scheduled for November 1993. However, the September 13 accord between Is-

rael and the PLO raised questions about the process. Many people in Jordan committed to democratization feared that the election would become a battle between supporters and opponents of the accord, since half the Jordanian population are of Palestinian descent. As a result, the government placed strict limits on campaigning. Political rallies were banned; the ban was rescinded by the courts several weeks before the election. Hussein also suspended the Parliament elected in 1990; and an amendment was added to the election law stipulating that voting would be "one person, one vote" rather than by party lists.

Despite these forebodings, the election went off on schedule, with few hitches. The results were an affirmation of Hussein's policy of gradual democratization. Pro-monarchy candidates won 54 of the 80 seats in the House of Representatives to 16 for the Islamic Action Front, the political arm of the Muslim Brotherhood. The remaining seats were spread among minor parties and independents. Voter turnout was 68 percent, far higher than in the 1990 election. The electorate also surprised by choosing the first woman member, Toujan Faisal, a feminist and television personality who, in the earlier campaign had been charged with apostasy by the Muslim Brotherhood.

Elections for the House of Representatives took place in November 1997, with pro-government candidates winning 62 of the 80 seats. The remainder were won by Islamic Action Front candidates and independents. However the restrictive press law, along with widespread disillusionment over the lack of positive benefits from the peace treaty with Israel and the increasing gap between rich and poor Jordanians, resulted in a turnout of barely 50 percent of Jordan's registered voters.

The 1997 parliamentary elections, which took place in November, continued the trend toward Hussein's version of representative government under firm monarchical rule. Government supporters won 62 of 80 seats in the lower house; the remainder went to independent candidates and Islamic Action Front candidates. However, a restrictive press law imposed in May led to

the suspension of 13 weekly newspapers. Opposition groups then called for a boycott of the election. The law was thrown out by the Higher Court of Justice (Jordan's Supreme Court) after the election. But the damage to Jordan's fledgling democracy had been done. Barely 50 percent of the country's 800,000 registered voters cast their ballots. The poor turnout was not a criticism of the king as much as it was due to public disappointment with the lack of visible benefits from the peace treaty with Israel and the widening gap between rich and poor in Jordan.

FOREIGN POLICY

During the 40-year cycle of hostilities between Israel and its Arab neighbors, there were periodic secret negotiations involving Jordanian and Israeli negotiators, including at times King Hussein himself, as Jordan sought to mend fences with its next-door neighbor. But in 1991 and 1992, Jordan became actively involved in the "peace process" initiated by the United States to resolve the vital issue of Palestinian self-government. As these negotiations proceeded, opponents of the process in Jordan did their best to derail them. Islamic Action Front members of the lower house of Parliament called for a vote of no-confidence in the government (but not in Hussein's leadership) for "treachery to Jordan and the Arab nation." However, the motion failed; most Jordanians supported the peace talks, and a majority of members of the House disagreed with the motion and voted against it.

Peace with Israel became a reality in October 1994, with Jordan the second Arab nation to sign a formal treaty with the Israeli state. Subsequently, the normalization of relations moved ahead with lightning speed. In July 1995, the Senate voted to annul the last anti-Israel laws still on the books. Embassies opened in Amman and Tel Aviv under duly accredited ambassadors. As Israel's first ambassador to Jordan observed, "I don't think there have ever been two countries at war for such a long time that have moved so quickly into peaceful cooperation."[6]

The treaty produced some positive results. Direct mail service between the two countries went into effect in February 1995. The maritime boundary in the Gulf of Aqaba was formally demarcated at midchannel, and Israeli tourists flocked to visit Petra and other historic Jordanian sites.

However, the benefits anticipated from the peace agreement and its spin-off in terms of U.S. aid and renewed close relations have been negated largely by the breakdown of the Oslo and subsequent agreements for Palestinian statehood. The late king Hussein worked tirelessly to mediate the conflict. King Abdullah II has been less involved. Other than closing down the Amman office of Hamas—the militant Islamic anti-Israeli organization in the West Bank and Gaza Strip—Abdullah's government observes the letter but not the spirit of the peace treaty. Early in 2001, Abdullah and Egyptian president Hosni Mubarak submitted a joint proposal to end the violence and to reinstate the peace talks on the basis of equality. Israel would halt settlement building in the West Bank in return for an end to the Palestinian intifada. But thus far, the proposal has achieved no results.

The Jordanian people have yet to follow up the peace treaty with cultural and social interchanges. Journalists who attended a workshop with their Israeli counterparts in Haifa in September 1999 were even expelled from the Jordan Press Association. They were reinstated only after issuing a public statement that "we still view Israel to be a conquest state and hold that it is impossible to conduct policies of normalization with her."[7]

The peace agreement has to some extent isolated Jordan from other Arab countries. Iraq, its major trading partner, broke relations after the Jordanian government allowed the Iraq National Accord, an umbrella opposition group working to overthrow Saddam Hussein, to set up an office in Amman. Inter-Arab solidarity was restored after the breakdown of the Palestinian-Israeli peace negotiations in 2000–2001. Saudi Arabia resumed oil shipments to Jordan after an eight-year break, caused by Jordan's alignment with Iraq, its principal supplier and trade partner. Kuwait also agreed to provide Jordan with 30 million barrels per day of oil; and in 2001, with the border open, regular deliveries of Iraqi oil were resumed.

The positive relationship established with successive U.S. administrations by the late king Hussein has been seriously weakened by the renewed cycle of Israeli-Palestinian conflict. In October 2001, the U.S. Senate approved the free-trade agreement with Jordan negotiated by the Bill Clinton administration and supported by that of George W. Bush. As a result, Jordan became the fourth country (after Canada, Mexico, and Israel) to enjoy a tariff-free association with the United States. Although it had been in the works long before the September 11, 2001, terrorist attacks on the United States, it offered an added incentive for Jordan to join the international coalition against terrorism. In November, King Abdullah offered troops in support of the offensive against Osama bin Laden and

his al Qaeda network. But he insisted that, in return, the United States would work for a "speedy solution" to the Israeli-Palestinian conflict.

ACHIEVEMENTS

Following the 2003 parliamentary election and cabinet selection, King Abdullah appointed Taghreed Hikmat, a lawyer and activist for women's rights, as the country's first female judge. Another of ex-Queen Noor's Quality of Life projects, the Village Business Incubator project in Umm Qasr, is the first of its kind in a rural area targeted particularly to help village women. Over a 15-year period the QLF has trained 1,500 persons from 22 countries to launch similar self-help projects.

The killing of U.S. diplomat Francis Foley in 2003 emphasized the challenge of Islamic extremists to King Abdullah's government. Foley's killers, who have yet to be apprehended, were said to be members of Takfit wal-Hijra, an affiliate of Osama bin Laden's al Qaeda. Several other bomb plots were uncovered and foiled by security police, but in November 2005 a coordinated attack on three Amman hotels frequented by foreigners killed 67 persons and injured some 150. The attacks were attributed to al Qaeda, whose leader in the Iraqi insurgency, Abu Musab al-Zarqawi, is a Jordanian. (The connection was exposed when one of the four bombers, a female, was captured when her explosives belt failed to ignite).

Subsequently Zarqawi's tribe in Jordan renounced him, thus removing the protection of his tribe, under a long-standing Arab tradition. Some 100,000 Jordanians marched in a demonstration in support of the king, shouting their opposition to al Qaeda. The government also issued strict anti-terrorism regulations. Among others, the new regulations specified that any non-Jordanian renting property would be reported to security forces within 48 hours.

THE ECONOMY

Jordan is rich in phosphates. Reserves are estimated at 2 billion tons, and new deposits are constantly being reported. Phosphate rock is one of the country's main exports, along with potash, which is mined on the Jordanian side of the Dead Sea.

The mainstay of the economy is agriculture. The most productive agricultural area is the Jordan Valley. A series of dams and canals from the Jordan and Yarmuk Rivers has increased arable land in the valley by 264,000 acres and has made possible production of high-value vegetable crops for export to nearby countries.

During the years of Israeli occupation of the West Bank, Jordan was estimated to have been deprived of 80 percent of its citrus crops and 45 percent of its vegetable croplands. It also lost access to an area that had provided 30 percent of its export market, as Israeli goods replaced Jordanian ones and the shekel became the medium of exchange there. The peace treaty guaranteed Jordan 7.5 billion cubic feet of water annually from the Jordan and Yarmuk Rivers, but to date the country has received less than half the agreed-on amount.

Jordan's economy traditionally has depended on outside aid and remittances from its large expatriate skilled labor force to make ends meet. A consequence of the Gulf War was the mass departure from Kuwait of some 350,000 Jordanian and Palestinian workers. Despite the loss in remittances and the added burden on its economy, Jordan welcomed them. But their return to Jordan complicated the nation's efforts to meet the requirements of a 1989 agreement with the International Monetary Fund for austerity measures as a prerequisite for further aid. The government reduced subsidies, but the resulting increase in bread prices led to riots throughout the country. The subsidies were restored; but they were again reduced in 1991, this time with basic commodities (including bread) sold at fixed low prices under a rationing system. Later price increases that were required to meet budgetary deficits in 1996 and 1998 met with little public protest, as the population settled down stoically to face a stagnating economy. One opposition leader remarked, "I don't want us to end up as an economic colony of the United States and Israel, enslaved as cheap labor."[8]

Jordan has backed strongly the development of a Palestinian state governed by the Palestine National Authority. Initially, the Jordanian government set up a preferential tax and customs exemption system for 25 Palestinian export products. A transit agreement reached in 1996 allows Palestinian exporters direct access to Aqaba port.

King Hussein's death and the accession to the throne of Abdullah rather than Crown Prince Hassan, the late king's brother, has caused concern over Jordan's political future. But a greater concern is that of the country's economic progress. As the result of the high birth rate—75 percent of the population are under age 29, with those age 15 to 29 accounting for 34.4 percent—there are not enough jobs. Unemployment for this age group is about 30 percent. A UN-financed Jordan Human Development Report 2000 found that Jordan's youth "are not well-equipped to meet the challenges of a globalizing world."

Timeline: PAST

1921
Establishment of the British mandate of Transjordan

1928
The first Constitution is approved by the British-sponsored Legislative Council

1946
Treaty of London; the British give Jordan independence and Abdullah assumes the title of emir

1948
The Arab Legion occupies the Old City of Jerusalem and the West Bank during the first Arab-Israeli War

1967
Jordanian forces are defeated by Israel in the Six-Day War; Israelis occupy the West Bank and Old Jerusalem

1970–1971
"Black September"; war between army and PLO guerrillas ends with expulsion of the PLO from Jordan

1990s
Politically, economically, and socially, Jordan is one of the primary losers in the Gulf crisis; Jordan signs a peace treaty with Israel; King Hussein dies and is succeeded by his son Abdullah

PRESENT

2000s
The Aqaba Special Economic Zone opens
Jordan supports the antiterrorism coalition

In addition to the free-trade pact, the United States has helped Jordan to set up a number of free-trade zones. A $300 million supplemental aid package for Jordan was approved by the U.S. Congress in 1999, and the country currently ranks fourth in the Middle East in U.S. aid appropriations (after Israel, Egypt, and Turkey).

King Hussein died in February 1999, and in accordance with his purported dying wish Abdullah, his son by an earlier marriage, succeeded him. It had been expected that Prince Hamza, his son by his fourth wife Queen Noor, would succeed him, and upon his accession Abdullah named Hamza, his half-brother, as crown prince. But in late November 2005 the king relieved him of his duties, saying that his job as heir to the throne impeded his ability to undertake more responsibilities.

Despite his youth and political inexperience, Abdullah's early actions indicated a much greater awareness of and concern for Jordan's economic problems than his late father, at least during Hussein's last years on the throne. Like the caliphs of old, King Abdullah went out in public in disguise, accompanied only by his driver, to learn about these problems first-hand. "There are sightings all over the place," he told an interviewer. "The bureaucrats are terrified. It's great"—as he changes into a wig, plastic glasses, false beard, and cane, before visiting the Finance Ministry.[9] He has inspected the free-trade zones incognito to hear complaints about fiscal mismanagement. Abdullah envisions Jordan as a "Middle Eastern Singapore," a model for development. "We can be symbols," he says, "for someone in Yemen who might say 'I don't want my country to be like it is today. I want it to be like the Jordanian or the Bahraini model, modern and progressive.'"[10]

In addition to a liberalized economy, the king launched in 2005 his new "national agenda." "It would," he said, "lead the country into a new age where there is more press freedom, health insurance for all, an independent judiciary, a more politically active public, political pluralism and … empowered women and youth."[11] One component of the agenda would be an electoral law giving Palestinians resident in Jordan greater representation in Parliament. Earlier, the number of seats in the legislature was increased from 80 to 120 in 2003. In the parliamentary elections for that year, the parties supporting the monarchy won half the seats. They included six women, resulting from Abdullah's new quota system intended to provide greater population balance and diversity.

The king also took an active role internationally in promoting for both Muslims and non-Muslims his moderate vision of Islam. In February 2006 he spoke to a gathering of U.S. clergy leaders emphasizing the linkages between Christianity, Judaism and Islam, and rejected the idea of a "clash of civilizations" which preoccupies many Westerners as well as some jihadists. "Our greatest challenge," he said, comes from violent extremists…. Extremism is a political movement under religious cover. Its adherents want nothing more than to pit us against each other, denying all that we have in common."[12]

NOTES

1. The Ottomans paid subsidies to nomadic tribes to guard the route of pilgrims headed south for Mecca. Peter Gubser, *Jordan: Crossroads of Middle Eastern Events* (Boulder, CO: Westview Press, 1983).

2. Years later, Glubb wrote, "In its twenty-eight years of life it had never been contemplated that the Arab Legion would fight an independent war." Quoted in Harold D. Nelson, ed., *Jordan, A Country Study* (Washington, D.C.: American University, Foreign Area Studies, 1979), p. 201.

3. King Abdullah of Jordan, *My Memoirs Completed,* Harold W. Glidden, trans. (London: Longman, 1951, 1978), preface, xxvi.

4. The text of the proposal is in Abdullah's *Memoirs, Ibid.,* pp. 89–90.

5. Naseer Aruri, *Jordan: A Study in Political Development (1925–1965)* (The Hague, Netherlands: Martinus Nijhoff, 1967), p. 159.

6. Quoted in *Middle East Economic Digest* (June 16, 1995).

7. Helen Schary Motro, "Israel the Invisible," *The Christian Science Monitor,* (January 12, 2000).

8. Laith Shbeilat, quoted in William A. Orme, "Neighbors Rally to Jordan," *The New York Times* (February 18, 1999).

9. Jeffrey Goldberg, "Learning to Be a King," *The New Times Magazine* (February 8, 2000).

10. *Ibid.*

11. Reported in *The Economist,* November 12, 2005.

12. Quoted in *The New York Times,* February 2, 2006.

Kuwait (State of Kuwait)

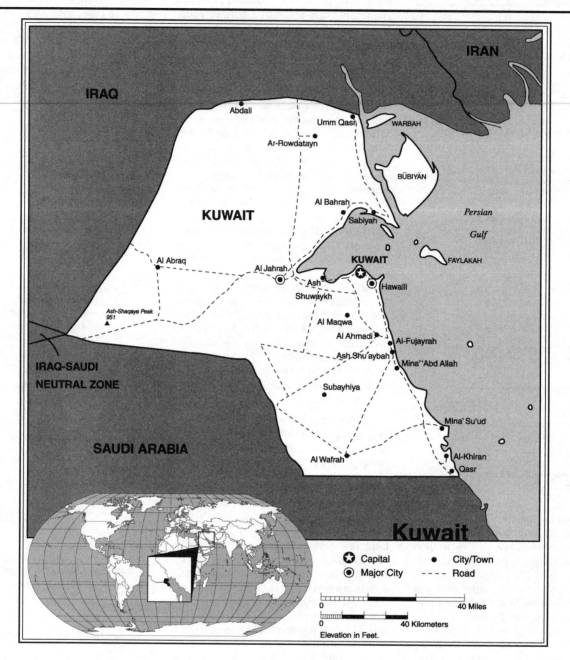

KUWAIT

IRAN

IRAQ

Abdali

Umm Qasr

WARBAH

Ar-Rowdatayn

BŪBIYĀN

Al Bahrah

Sabiyah

Persian

Gulf

Al Abraq

Al Jahrah

KUWAIT

FAYLAKAH

Ash Shuwaykh

Hawalli

Ash-Shaqaya Peak
951

Al Maqwa

Al Ahmadi

Al-Fujayrah

Ash Shu'aybah

Mina' 'Abd Allah

IRAQ-SAUDI
NEUTRAL ZONE

Subayhiya

Mina' Su'ud

Al-Khiran

SAUDI ARABIA

Al Wafrah

Qasr

Kuwait

⭐ Capital   • City/Town
◉ Major City   --- Road

0   40 Miles
0   40 Kilometers
Elevation in Feet.

Kuwait Statistics

GEOGRAPHY

Area in Square Miles (Kilometers): 6,880 (17,818) (about the size of New Jersey)

Capital (Population): Kuwait (277,000)

Environmental Concerns: limited natural freshwater reserves; air and water pollution; desertification

Geographical Features: flat to slightly undulating desert plain

Climate: intensely hot and dry summers; short, cool winters

PEOPLE

Population

Total: 2,335,648 (includes 1,291,354 non-nationals)

Annual Growth Rate: 3.3% (reflects increased immigration of expatriates, mostly Arabs, after the 1991 Gulf War.)

Rural/Urban Population Ratio: 3/97

Major Languages: Arabic; English

Ethnic Makeup: 45% Kuwaiti; 35% other Arab; 9% South Asian; 4% Iranian; 7% others

Religions: 85% Muslim; 15% Christian, Hindu, Parsi, and others

Health

Life Expectancy at Birth: 76 years (male); 78 years (female)
Infant Mortality Rate: 9.95/1,000 live births

Education

Adult Literacy Rate: 83.5%
Compulsory (Ages): 6–14; free

COMMUNICATION

Telephones: 412,000 main lines
Daily Newspaper Circulation: 401 per 1,000 people
Televisions: 390 per 1,000 people
Internet Service Providers: 2 (2000)

TRANSPORTATION

Highways in Miles (Kilometers): 2,763 (4,450)
Railroads in Miles (Kilometers): none
Usable Airfields: 7
Motor Vehicles in Use: 700,000

GOVERNMENT

Type: nominal constitutional monarchy
Independence Date: June 19, 1961 (from the United Kingdom)
Head of State/Government: Emir Shaykh Sabah Ahmad al-Sabah; Prime Minister Shaykh Nasir Muhammad Ahmad al-Sabah
Political Parties: none legal
Suffrage: formerly limited to male citizens over 21, women included as of 2005

MILITARY

Military Expenditures (% of GDP): 5.3%
Current Disputes: none. In 1994 Iraq formally accepted the UN-demarcated border between the two-countries and ended its claims to Kuwait and the offshore islands of Bubiyan and Warbah

ECONOMY

Currency ($U.S. Equivalent): 0.305 dinars = $1 (fixed rate)

Per Capita Income/GDP: $21,300/$48 billion
GDP Growth Rate: 6.8%
Inflation Rate: 2.3%
Unemployment Rate: 1.8% (official rate)
Labor Force: 1,420,000
Natural Resources: petroleum; fish; shrimp; natural gas
Agriculture: fish
Industry: petroleum; petrochemicals; desalination; food processing; construction materials; salt; construction
Exports: $27.42 billion (primary partners Japan, United States, Korea)
Imports: $11.12 billion (primary partners United States, Japan, Germany)

SUGGESTED WEBSITES

http://lcweb2.loc.gov/frd/cs/ kwtoc.html
http://kuwait-info.org

Kuwait Country Report

The State of Kuwait consists of a wedge-shaped, largely desert territory located near the head of the Persian Gulf and just southwest of the Shatt al-Arab. Kuwaiti territory includes the islands of Bubiyan and Failaka in the Gulf, both of them periodically claimed by Iraq. Kuwait also shares a Neutral Zone, consisting mainly of oil fields, which it administers jointly with Iraq and Saudi Arabia; oil production is supposedly divided equally among them. The Iraqi accusation that Kuwait was taking more than its share was one of the points of contention that led to Iraq's invasion of Kuwait in 1990.

Kuwait's location has given the country great strategic importance in the modern rivalries of regional powers and their outside supporters. The country played a major role in the Iran-Iraq War, supporting Iraq financially and serving as a conduit for U.S. naval intervention through the reflagging of Kuwaiti tankers. The Iraqi invasion reversed roles, with Iraq the aggressor and Kuwait both the victim and the target of UN/U.S.-led military action during the brief Gulf War in 1991.

HISTORY

Kuwait was inhabited entirely by nomadic peoples until the early 1700s. Then a number of clans of the large Anaiza tribal confederation settled along the Gulf in the current area of Kuwait. They built a fort for protection from raids—*Kuwait* means "lit-

tle fort" in Arabic—and elected a chief to represent them in dealings with the Ottoman Empire, the major power in the Middle East at that time. The ruling family of modern Kuwait, the al-Sabahs, traces its power back to this period.

DEVELOPMENT

Declining oil revenues resulted in budgetary deficits, which reached $6 billion in 1998 and 1999. With the Assembly dissolved, the emir issued some 60 decrees, intended to begin to privatize the economy and reduce expenditures by 20%. In 2001, 35% of ownership in the Kuwait Cement Company was turned over to private management.

Kuwait prospered under the al-Sabahs. Its well-protected natural harbor became headquarters for a pearl-fishing fleet of 800 dhows (boats). The town (also called Kuwait) became a port of call for British ships bound for India.

In the late 1700s and early 1800s, Kuwait was threatened by the Wahhabis, fundamentalist Muslims from central Arabia. Arab piracy also adversely affected Kuwait's prosperity. Kuwait's ruling shaykhs paid tribute to the Ottoman sultan in return for protection against the Wahhabis. However, the shaykhs began to fear that the Turks would occupy Kuwait, so they

turned to the British. In 1899, Shaykh Mubarak, who reigned from 1896 to 1915, signed an agreement with Britain for protection. In return, he agreed to accept British political advisers and not to have dealings with other foreign governments. In this way, Kuwait became a self-governing state under British protection.

During the 1890s, Kuwait had given refuge to Ibn Saud, a leader from central Arabia whose family had been defeated by its rivals. Ibn Saud left Kuwait in 1902, traveled in secret to Riyadh, the rivals' headquarters, and seized the city in a surprise raid. Kuwait thus indirectly had a hand in the founding of its neighbor state, Saudi Arabia.

INDEPENDENCE

Kuwait continued its peaceful ways under the paternalistic rule of the al-Sabahs until the 1950s. Then oil production increased rapidly. The small pearl-fishing port became a booming modern city. In 1961, Britain and Kuwait jointly terminated the 1899 agreement, and Kuwait became fully independent under the al-Sabahs.

A threat to the country's independence developed almost immediately, as Iraq refused to recognize Kuwait's new status and claimed the territory on the grounds that it had once been part of the Iraqi Ottoman province of Basra. Iraq was also interested in controlling Kuwaiti oil resources. The

Saudi Aramco World/PADIA (SA3670041)

Asian refugees at camp near Iraq/Jordan border who left Kuwait after Iraqi invasion.

ruling shaykh, now called emir, asked Britain for help, and British troops rushed back to Kuwait. Eventually, the Arab League agreed that several of its members would send troops to defend Kuwait—and, incidentally, to ensure that the country would not revert to its previous protectorate status. The Arab contingents were withdrawn in 1963. A revolution had overthrown the Iraqi government earlier in the year, and the new government recognized Kuwait's independence. However, the Ba'thist Party's concentration of power in Saddam Hussein's hands in the 1970s led to periodic Iraqi pressure on Kuwait, culminating in the 1990 invasion and occupation. After the expulsion of Iraqi forces, Kuwait requested a realignment of its northern border; and in 1992, the United Nations Boundary Commission approved the request, moving the border approximately 1,880 feet northward. The change gave Kuwait full possession of the Rumaila oil fields and a portion of the Iraqi Umm Qasr naval base. Kuwait had argued that the existing border deprived it of its own resources and access to its territorial waters as specified in the 1963 agreement. Some 3,600 UN observers were assigned to patrol the new border; and Kuwaiti workers dug a 130-mile trench, paid for by private donations, as a further protection for the emirate. Iraq's 1994 acceptance of the UN border demarcated between the two states has officially resolved the issue, although

before his overthrow by U.S. forces Iraqi president Saddam Hussein had his army conduct training exercises and maneuvers near the border.

FREEDOM

The National Assembly in May 2005 extended the right to vote and run for office in local elections to women. An earlier (2001) decree would allow them to join the police force.

REPRESENTATIVE GOVERNMENT

Kuwait differs from other patriarchally ruled Arabian Peninsula states in having a Constitution that provides for an elected National Assembly. Its 50 members are elected for four-year terms.

Friction developed between the Assembly and the ruling family soon after independence. Assembly members criticized Shaykh Abdullah and his relatives, as well as the cabinet, for corruption, press censorship, refusal to allow political parties, and insufficient attention to public services. Since all members of the ruling family were on the government payroll, there was some justification for the criticism.

Abdullah died in 1965, but his successor, Shaykh al-Sabah, accepted the criticism as valid. Elections were held in 1971 for a new Assembly.

Unfortunately for democracy in Kuwait, the new Assembly paid more attention to criticism of the government than to law making. In 1976, it was suspended by Shaykh al-Sabah. He died the following year, but his successor, Shaykh Jabir, reaffirmed the ruling family's commitment to the democratic process. A new Assembly was formed in 1981, with different members. The majority were traditional patriarchs loyal to the rulers, along with technical experts in various fields, such as industry, agriculture, and engineering. But the new Assembly fared little better than its predecessor in balancing freedom of expression with responsible leadership. The ruler suspended it, along with the Constitution, in 1986.

Pressures to reinstate the Assembly have increased in recent years. Just prior to the Iraqi invasion, the ruler had convened a 75-member National Council "to appraise our parliamentary experiment." The process was halted during the Iraqi occupation; but, after the Iraqi withdrawal and the return to Kuwait of the ruling family, the emir pledged to hold elections for a new Assembly in October 1992. The vote was limited to males over age 21 who could trace their residence in Kuwait back to 1920 or earlier. Under these rules, some 82,000 voters would elect two candidates from each of 25 constituencies to the 50-member Assembly.

The emir kept his pledge, and on October 5, 1992, the election took place as scheduled. More than half of the seats were won by critics of the government. They had campaigned on a platform of demands for government accountability and broadening of the franchise to include women. In the next Assembly election, in 1996, pro-government candidates won the majority of seats. But to appease his critics, the emir appointed a new cabinet with nine non-Sabah family members as ministers.

HEALTH/WELFARE

The 1.4 million foreign workers in Kuwait have had little protection under law until very recently. In November 2000, a large number of unemployed or underemployed Egyptians, the largest component of the expatriate labor force, rioted over bad working conditions and exploitation by sponsors who charge up to $3,000 for residency and work permits. The government agreed to review labor laws to limit payments to sponsors by mutual agreement.

The emir suspended the Assembly in May 1999, after opposition deputies had paralyzed government action by endless criticism of government ministers who were presenting their programs for legislative concurrence. However, he approved elections for a new Assembly to take place in July. During the suspension period he issued a decree giving women the right to vote and run for public office in 2003. (Under Kuwaiti law only native citizens or those naturalized for 30 years or more have the franchise; women are excluded along with the police and members of the armed services.) When the new Assembly (all-male) took office its members refused to approve the decree as required by Kuwaiti law.

Subsequently women undertook organized efforts to gain voting rights, appealing both to the Assembly and the courts. During the July 2003 Assembly election they even set up mock polling booths and cast ballots. Several of their appeals were rejected by the conservative-dominated legislature in close votes. But in May 2005 the Assembly approved the measure by a 35–23 vote. In part to mollify conservative legislators an addendum was attached to the right-to-vote bill that would require women voters and candidates for public office to abide by Shari'a Islamic law.[1]

The Assembly's improvement as a responsible legislative body was illustrated in the leadership crisis of early 2006. With the death of ruling emir Shaykh Jaber al-Sabah after 29 years of rule, and his successor, Shaykh Saad, ailing, the legislators exercised their constitutional right by deposing

him and appointing his brother Shaykh Ahmad al-Sabah as the new emir. Ahmad then named another brother as Crown Prince and a nephew as prime minister, breaking a long tradition that had alternated the top posts in the government between the Sabah family and the Salems, the other branch of the ruling family in Kuwait.

VULNERABILITY

Kuwait's location and its relatively open society make the country vulnerable to external subversion. In the early 1970s, the rulers were the target of criticism and threats from other Arab states because they did not publicly support the Palestinian cause. For years afterward, Kuwait provided large-scale financial aid not only to the Palestine Liberation Organization, but also to Arab states such as Syria and Jordan that were directly involved in the struggle with Israel because of their common borders.

A new vulnerability surfaced with the Iranian Revolution of 1979, which overthrew the shah. Kuwait has a large Shia Muslim population, while its rulers are Sunni. Kuwait's support for Iraq and the development of closer links with Saudi Arabia (and indirectly the United States) angered Iran's new fundamentalist rulers. Kuwaiti oil installations were bombed by Iranian jets in 1981, and in 1983 truck bombs severely damaged the U.S. and French Embassies in Kuwait City. The underground organization Islamic Jihad claimed responsibility for the attacks and threatened more if Kuwait did not stop its support of Iraq. Kuwaiti police arrested 17 persons; they were later jailed for complicity in the bombings. Since Islamic Jihad claimed links to Iran, the Kuwaiti government suspected an Iranian hand behind the violence and deported 600 Iranian workers.

Tensions with Iran intensified in the mid-1980s, as Iranian jets and missile-powered patrol boats attacked Kuwaiti tankers in the Gulf and pro-Iranian terrorists carried out a series of hijackings. A 1988 hijacking caused international concern when a Kuwait Airways 747 jet with several members of the royal family aboard was seized and its passengers held for 16 days while being shuttled from airport to airport. The hijackers demanded the release of the 17 truck bombers as the price for the hostages' freedom. The Kuwaiti government refused to negotiate; after the hostages were released through mediation by other Arab states, it passed a law making hijacking punishable by death.

Fear of Iran led Kuwait to join the newly formed Gulf Cooperation Council in 1981. The country also began making large purchases of weapons for defense, balancing

ACHIEVEMENTS

In addition to their recent success in gaining the right to vote, women have done well in the non-political arena. The first all-female national soccer tournament was held in April 2000 and there are regular leagues throughout the emirate. A woman heads the Kuwaiti Economists Association, and others have enjoyed success as journalists, newspaper columnists, university lecturers and doctors.

U.S. with Soviet equipment. Its arms buildup made it the world's third-highest defense spender, at $3.1 billion, an average of $2,901 per capita.

During the Iran–Iraq War, Kuwait loaned 13 tankers to the United States. They were reflagged and given U.S. naval escort protection to transit the Gulf. After the United States assumed a major role in the region due to the Iraqi invasion and the resulting Gulf War, Kuwait signed a 10-year mutual-defense pact, the first formal agreement of its kind for the Gulf states.

THE IRAQI OCCUPATION AND AFTERMATH

The seven months of Iraqi occupation (August 1990–February 1991) had a devastating effect on Kuwait. Some 5,000 Kuwaitis were killed, and the entire population was held hostage to Iraqi demands. Oil production stopped entirely. Iraqi forces opened hundreds of oil storage tanks as a defense measure, pouring millions of gallons of oil into the sea, thus creating a serious environmental hazard. (As they retreated, the Iraqis also set 800 oil wells on fire, destroying production capabilities and posing enormous technical and environmental problems. These conflagrations were not extinguished for nearly a year.) In Kuwait City, basic water, electricity, and other services were cut off; public buildings were damaged; shops and homes were vandalized; and more than 3,000 gold bars, the backing for the Kuwaiti currency, were taken to Iraq.

Some 605 Kuwaitis, out of thousands taken to Iraq, are still unaccounted for. Most of them are civilians, including 120 students, whose "crime" was noncooperation with the Iraqi occupation forces. The U.S. invasion and occupation of Iraq in 2003 offered new hope that some of them were still alive, and the Kuwait government offered $1 million for information on their whereabouts. However, in May an informant led U.S. forces to an abandoned military camp near Habbaniya where some 600 Kuwaiti prisoners are supposedly buried. According to the informant, they were brought to Baghdad in October 1991, blindfolded, lined up in horseshoe formation, killed by machine gun fire, hauled to the site in trucks and buried there.

In 2001, the UN Repatriations Commission, set up to recompense Kuwaitis and others for losses sustained during the occupation, approved payment of more than $360 million for this purpose. As a result, 230 Kuwait-based companies are to share $174 million in payment for their claims against Iraq.

Iraqi maneuvers near the Kuwait border and the incursion of Iraqi forces into UN-protected Kurdistan led the United States in 1996 to invoke the mutual-defense pact by sending an additional 5,000 ground troops to Kuwait to take part in a new attack on Iraq. The Kuwaiti government, which had not been consulted, agreed to accept only 3,500. But the contest of wills between the Clinton administration and Saddam Hussein over Iraq's secret weapons program, and the resulting fears of Kuwaitis of an Iraqi missile attack on the emirate led it to allow the stationing of 10,000 U.S. troops on its territory, along with Patriot missiles and fighter aircraft. After the September 11, 2001, terrorist bombings in the United States, Kuwait joined with other Gulf Cooperation Council member states in supporting the international coalition against terrorism.

After decades of hostility toward Israel and strong support for the Palestinian cause, Kuwait's leaders recently have begun to rethink national policy toward the Israeli state in the light of Israel's evacuation of the Gaza Strip. Thus the Kuwaiti newspaper *Al-Seyassa*, a voice of government opinion, published an op-ed column in October 2005 urging the termination of the country's long-standing trade embargo on Israel. Whether this would lead to further action remained doubtful, but the mere prospect of future normalization of relations with the Zionist state marked a milestone in itself.

THE PEOPLE

Until the economic recession in the region, the country had a high rate of immigration. As a result, there are more non-Kuwaitis than Kuwaitis in the population, though dislocation resulting from the Iraqi occupation has changed the balance. Today, approximately 45 percent of the population are native Kuwaitis.

About one third of the total population, both citizens and noncitizens, are Shi'a Muslims. After the 1979 Revolution in Iran, they were blamed for much of the unrest in the country; Shi'a terrorists were charged with the 1983 truck bombings of embassies in Kuwait City. The improve-

ment in Kuwait-Iran relations that followed the end of the Iran-Iraq War lessened this Shia anti-government activity, and Shi'a and Sunni residents suffered equally under the Iraqi occupation.

The insurgency against U.S. forces occupying Iraq and attempting to overthrow the U.S.-installed temporary government there spilled over into Kuwait early in 2005. Gun battles between Islamic militants and security forces in January killed 12 of the militants, who were said to be members of three cells of al Qaeda according to the Interior Ministry. The government also allocated 5.5 million dinars toward a campaign to promote moderate Islam in its opposition to Islamic extremists, and closed down two websites belonging to a radical preacher, one Amer Khleif al-Enezi, who was killed subsequently in a gun battle with police.

Before the Gulf War, the largest non-native population group was Palestinian. Although denied citizenship, the Palestinians were generally better educated and more industrious than the native Kuwaitis. Palestinians formed the nucleus of opposition to the ruling family, and a number of them collaborated with the Iraqi occupation forces. After the war, more than 600 Palestinians were tried and sentenced to prison terms for collaboration. Some 300,000 Palestinians abruptly left for Jordan. However, their management skills were not entirely missed, as their places were filled by Lebanese or Western expatriates. Low-level jobs also filled rapidly with the arrival of workers from less-developed countries. But their low pay and lack of fringe benefits triggered riots in April 2000 over a government plan to levy annual payments, up to $155, on foreign workers in return for their inclusion in the national health service. A Filipino doughnut-shop employee commented: "It's not a steep price, but it hurts when there's been no increase in my $300 a month salary."[1]

THE ECONOMY

Kuwait's only abundant resource is petroleum. Less than 1 percent of the land can be cultivated, and there is almost no fresh water. Drinking water comes from sea water converted to fresh water by huge desalination plants.

Kuwait's oil reserves of 94 billion barrels are the world's third largest, comprising 10 percent of global reserves. According to a 1996 study by the International Monetary Fund, the oil industry—and, with it, the economy—has recovered "impressively" from the effects of the Iraqi occupation. Oil production in 1997

reached 2 million barrels per day. The 1995–2000 Five-Year Plan approved by the Assembly projects a balanced budget by the end of the plan, largely through privatization of state enterprises, increased oil and non-oil revenues, and expansion of petrochemical industries.

Kuwait's economic recovery following the Iraqi occupation was hampered not only by the damage to its oil industry but also by an overstaffed public sector, huge welfare subsidies, and other factors. Despite some opposition in the Assembly, mainly on procedural grounds, the government went ahead with the reforms.

As a result, the Kuwaiti economy has rebounded to such an extent that, in 1996, Kuwait became the first Gulf state to receive an "A" rating from the International Banking Credit Association, an organization that evaluates countries on the basis of short- and long-term risks. In 2005 Kuwait's 950,000 citizens earned $45 billion in revenues.

In addition to restoring its oil fields to full production, by 2001 nearly all the land mines left by the retreating Iraqis had been cleared from Kuwait's vast stretches of desert. An enormous oil slick from oil-well destruction that had threatened to pollute the water supply (which comes from desalinization) had also been cleared up.

NOTE

1. Cameron Barr, in *The Christian Science Monitor* (March 2001).

Lebanon (Lebanese Republic)

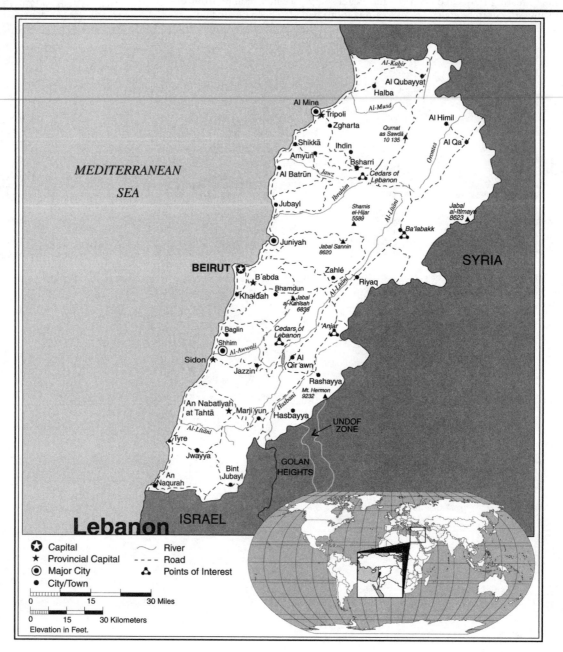

Lebanon Statistics

GEOGRAPHY

Area in Square Miles (Kilometers): 4,015 (10,452) (smaller than Connecticut)

Capital (Population): Beirut (1,826,000)

Environmental Concerns: deforestation; soil erosion; air and water pollution

Geographical Features: a narrow coastal plain; the Biqa' Valley separates Lebanon and the Anti-Lebanon Mountains

Climate: Mediterranean (hot, humid summers; cool, damp winters); heavy winter snows in mountains

PEOPLE

Population

Total: 3,826,018
Annual Growth Rate: 1.26%
Rural/Urban Population Ratio: 12/88
Major Languages: Arabic; French; English

Ethnic Makeup: 90% Arab; 10% Armenian and other

Religions: 17 sects recognized—Muslims 59% (Shia, Sunni, Ismaili, Alawite, also Druze), Christian 39% (Maronite, Melkite Catholic, Armenian Orthodox, Syrian Catholic, Armenian Catholic, Greek Orthodox, Greek Catholic, 2% Protestant

Health

Life Expectancy at Birth: 72.6 years (male); 75 years (female)
Infant Mortality Rate: 24.5/1,000 live births

Education

Adult Literacy Rate: 87.4%

COMMUNICATION

Telephones: 620,000 main lines
Daily Newspaper Circulation: 172 per 1,000 people
Televisions: 291 per 1,000 people
Users: 22 (2000)

TRANSPORTATION

Highways in Miles (Kilometers): 4,380 (7,300)

Railroads in Miles (Kilometers): 138 (222)
Usable Airfields: 8
Motor Vehicles in Use: 1,183,000

GOVERNMENT

Type: republic
Independence Date: November 22, 1943 (from League of Nations mandate under French administration)
Head of State/Government: President Emile Lahoud; Prime Minister Fouad Siniora
Political Parties: normally based on religious confessions as above; the new National Assembly elected in 2005 is divided into larger power blocs, two-thirds Christian, Sunni Muslim and Druze, one-third Shia Muslim (including Hizbullah and Amal)
Suffrage: compulsory for males at 21; authorized for women at 21 with elementary-school education

MILITARY

Military Expenditures (% of GDP): 4.8%
Current Disputes: Syrian troops continue de facto occupation in northeast and central areas; Hezbollah dispute with Israel over Shab'a Farms area in Israeli-occupied Golan Heights

ECONOMY

Currency ($U.S. Equivalent): 1,507 pounds = $1
Per Capita Income/GDP: $5,100/$18.2 billion
GDP Growth Rate: 4%
Inflation Rate: 2%
Unemployment Rate: 18%
Labor Force: 1,500,000 (plus an estimated 1,000,000 foreign workers)
Natural Resources: limestone; iron ore; salt; water; arable land
Agriculture: fruits; vegetables; olives; tobacco; hemp (hashish); sheep; goats
Industry: banking; food processing; jewelry; cement; textiles; mineral and chemical products; wood and furniture products; oil refining; metal fabricating
Exports: $27.42 billion (primary partners Japan, U.S., South Korea)
Imports: $11.12 billion (primary partners United States, Germany, Japan)

Lebanon Country Report

The Lebanese Republic is located at the eastern end of the Mediterranean Sea. The coastal plain, which contains the capital, Beirut, and all the other important cities, is narrow, rising just a few miles east of Beirut to a rugged mountain range, Mount Lebanon. Beyond Mount Lebanon is the Biqa', a broad, fertile valley that is the country's main wheat-growing region. At the eastern edge of the Biqa', the land rises again abruptly to the snow-capped Anti-Lebanon Range, which separates Lebanon from Syria.

Lebanon's geography has always been important strategically. Many invaders passed through it over the centuries on their conquests—Egyptians, Assyrians, Persians, Crusaders, Arabs, and Turks. However, they were seldom able to gain control of Mount Lebanon. For this reason, the mountain served as a refuge for ethnic and religious minorities, and it became in time the nucleus of the modern Lebanese state.

Lebanon's Mediterranean ports have traditionally served as an outlet for goods from the region's interior, notably Syria. Lebanese merchants have profited for centuries by being middlemen for this trade.

However, its strategic location and its role as a commercial entrepôt have hampered Lebanon's unification as a nation in the twentieth century. Unification and the establishment of a national identity have also been blocked by religious divisions and territorial rivalries by various clans. A Lebanese scholar described the country's political system as "a feudal hierarchy with fluctuating political influence, as powerful families asserted themselves to acquire power and prominence."[1]

HISTORY

In ancient times, Lebanon was known as Phoenicia. The Phoenicians were great traders who traveled throughout the Mediterranean and probably out into the Atlantic Ocean as far north as Cornwall in England, in search of tin, copper, and iron ore, which were valued in the ancient world for their many uses. Phoenician merchants established trading posts, some of which eventually grew into great cities.

No central government was ever established in Phoenicia itself. Phoenician towns like Byblos, Tyre, Sidon, and Tripoli were independent states, often in conflict or rivalry over trade with one another. This city-state rivalry has always been a feature of Lebanese life and is another reason for today's lack of a national Lebanese sense of unity.

Lebanon began to develop a definite identity much later, in the seventh century A.D. when a Christian group, the Maronites, took refuge in Mount Lebanon after they were threatened with persecution by the government of the East Roman or Byzantine Empire because of theological disagreements over the nature of Christ. The Muslim Arabs brought Islam to coastal Lebanon at about the same time, but they were unable to dislodge or convert the Maronites. Mount Lebanon's sanctuary tradition attracted other minority groups, Muslim as well as Christian. Shia Muslim communities moved there in the ninth and tenth centuries to escape persecution from Sunni Muslims, the Islamic majority. In the eleventh century, the Druze, adherents of an offshoot of Islam who followed the teachings of an Egyptian mystic and also faced persecution from Sunni Muslims, established themselves in the southern part of Mount Lebanon. These communities were originally quite separate.

In the modern period of Lebanese history, however, they have tended to overlap, a fact, David Gordon says, "that makes both for unity and in troubled times for a dangerous struggle for turf."[2]

Lebanon acquired a distinct political identity under certain powerful families in the sixteenth and seventeenth centuries. The Ottoman Turks conquered it along with the rest of the Middle East, but they were content to leave local governance in the hands of these families in return for tribute. The most prominent was the Ma'an family, who were Druze. Their greatest leader, Fakhr al-Din (1586–1635), established an independent principality that included all of present-day Lebanon, Israel, and part of Syria. It was during al-Din's rule that French religious orders were allowed to establish missions in the country, which facilitated later European intervention in Lebanon.

The Ma'ans were succeeded by the Shihabs, who were Maronites. Their descendants continue to hold important positions in the country, underscoring the durability of the extended-family system, which still dominates Lebanese politics. They also allied the Maronite Church with the Roman Catholic Church, an action that had great consequences in the twentieth century, when the Maronites came to view Lebanon as a "Christian island in a Muslim sea," preserving its unique Lebanese identity only through Western support.

European countries began to intervene directly in Lebanon in the nineteenth century, due to conflict between the Maronite and Druze communities. Mount Lebanon was occupied by Egyptian armies of the Ottoman khedive (viceroy) of Egypt, Muhammad Ali, in the 1830s. Egyptian development of Beirut and other coastal ports for trade purposes, particularly exports of Lebanese silk (still an important cash crop) at the expense of Mount Lebanon, and heavy taxes imposed by the khedive's overseers led to peasant uprisings in 1840 and 1857. By then the Ottomans had reestablished their authority, with European help. However, the European powers refused to allow the sultan to change Mount Lebanon's special status as an autonomous province. Ottoman governors resorted to intrigues with Maronite and Druze leaders, playing one against the other. The result was a Maronite-Druze civil war, which broke out in 1860. The cause was insignificant— "an affray between two boys, the shooting of a partridge or the collision of two pack animals," asserts one author; but whatever the spark, the two communities were ready to go for each other's throats.[3]

Although the Maronite fighters greatly outnumbered the Druze, the latter had better leadership. The Druze massacred 12,000 Christians and drove 100,000 from their homes during a four-week period. At that point, the European powers intervened to protect their coreligionists. French troops landed in Beirut and moved on to occupy Damascus. France and England forced the Ottoman sultan to establish Mount Lebanon as a self-governing province headed by a Christian governor. The province did not include Beirut. Although many Lebanese emigrated during this period because Mount Lebanon was small, rather poor, and provided few job opportunities, those who stayed (particularly the Maronites) prospered. Self-government under their own leader enabled them to develop a system of small, individually owned farms and to break their former dependence on absentee landowners. A popular saying among Lebanese at the time was, "Happy is he who has a shed for one goat in Mount Lebanon."[4]

The French Mandate

After the defeat of Ottoman Turkey in World War I, Lebanon became a French mandate. The French had originally intended the country to be included in their mandate over Syria; but in 1920, due to pressure from Maronite leaders, they separated the two mandates. "New" Lebanon was much larger than the old Maronite-Druze territory up on Mount Lebanon. The new "Greater Lebanon" included the coast—in short, the area of the current Lebanese state. The Maronites found themselves linked not only with the Druze but also with both Sunni and Shia Muslims. The Maronites already distrusted the Druze, out of bitter experience. Their distrust was caused by fear of a Muslim majority and fear that Muslims, being mostly Arabs, would work to incorporate Lebanon into Syria after independence.

DEVELOPMENT

The end of the civil war and Hariri's energetic reform program have brought significant improvements to the Lebanese infrastructure. The GDP growth rate reached a high of 8 percent in 1994, although dropping steadily thereafter to average 1 percent annually. An austerity program introduced in 2000 emphasized privatization of state-owned enterprises, reduction in the size of the bureaucracy and tax reform.

France gave Lebanon its independence in 1943, but French troops stayed on until 1946, when they were withdrawn due to British and American pressure on France. The French made some contributions to Lebanese development during the mandate, such as the nucleus of a modern army, development of ports, roads, and airports, and an excellent educational system dominated by the Université de St. Joseph, training ground for many Lebanese leaders. The French language and culture served until recently as one of the few things unifying the various sects and providing them with a sense of national identity.

THE LEBANESE REPUBLIC

The major shortcoming of the mandate was the French failure to develop a broad-based political system with representatives from the major religious groups. The French very pointedly favored the Maronites. The Constitution, originally issued in 1926, established a republican system under an elected president and a Legislature. Members would be elected on the basis of six Christians to five Muslims. The president would be elected for a six-year term and could not serve concurrently. (The one exception was Bishara al-Khuri [1943–1952], who served during and after the transition period to independence. The Constitution was amended to allow him to do so.) By private French-Maronite agreement, the custom was established whereby the Lebanese president would always be chosen from the Maronite community.

In the long term, perhaps more important to Lebanese politics than the Constitution is the National Pact, an oral agreement made in 1943 between Bishara al-Khuri, as head of the Maronite community, and Riad al-Sulh, his Sunni counterpart. The two leaders agreed that, first, Lebanese Christians would not enter into alliances with foreign (i.e., Christian) nations and Muslims would not attempt to merge Lebanon with the Muslim Arab world; and second, that the six-to-five formula for representation in the Assembly would apply to all public offices. The pact has never been put in writing, but in view of the delicate balance of sects in Lebanon, it has been considered by Lebanese leaders, particularly the Maronites, as the only alternative to anarchy.

Despite periodic political crises and frequent changes of government due to shifting alliances of leaders, Lebanon functioned quite well during its first two decades of independence. The large extended family, although an obstacle to broad nation building, served as an essential support base for its members, providing services that would otherwise have to have been drawn from government sources. These services included education, employment, bank loans, investment capital, and old-age security. Powerful families of different religious groups competed for power and influence but also coexisted, having had "the long experience with each other and with the rules and practices that make coexistence possible."[5] The freewheeling Lebanese economy was another important factor in Lebanon's relative stability. Per capita annual income rose from $235 in 1950 to $1,070 in 1974, putting Lebanon on a level with some of the oil-producing Arab states, although the country does not have oil.

The private sector was largely responsible for national prosperity. A real-estate boom developed, and many fortunes were made in land speculation and construction. Tourism was another important source of revenues. Many banks and foreign business firms established their headquarters in Beirut because of its excellent communications with the outside world, its educated, multilingual labor force, and the absence of government restrictions.

THE 1975–1976 CIVIL WAR

The titles of books on Lebanon in recent years have often contained adjectives such as "fractured," "fragmented," and "precarious." These provide a generally accurate description of the country's situation as a result of the Civil War of 1975–1976. The main destabilizing element, and the one that precipitated the conflict, was the presence and activities of Palestinians.

In some ways, Palestinians have contributed significantly to Lebanese national life. The first group, who fled there after the 1948 Arab-Israeli War, consisted mostly of cultured, educated, highly urbanized people who gravitated to Beirut and were absorbed quickly into the population. Many of them became extremely successful in banking, commerce, journalism, or as faculty members at the American University of Beirut. A second Palestinian group arrived as destitute refugees after the 1967 Six-Day War. They have been housed ever since in refugee camps run by the United Nations Relief and Works Agency. The Lebanese government provides them with identity cards but no passports. For all practical purposes, they are stateless persons.

Neither group was a threat to Lebanese internal stability until 1970, although Lebanon backed the Palestine Liberation Organization cause and did not interfere with guerrilla raids from its territory into Israel. After the PLO was expelled from Jordan, the organization made its headquarters in Beirut. This new militant Palestinian presence in Lebanon created a double set of problems for the Lebanese. Palestinian raids into Israel brought Israeli retaliation, which caused more Lebanese than Palestinian casualties. Yet the Lebanese government could not control the Palestinians. To many Lebanese, especially the Maronites, their government seemed to be a prisoner in its own land.

In April 1975, a bus carrying Palestinians returning from a political rally was ambushed near Beirut by the Kata'ib, members of the Maronite Phalange Party. The incident triggered the Lebanese Civil War of 1975–1976. The war officially ended with a peace agreement arranged by the Arab League.[6] But the bus incident also brought to a head conflicts derived from the opposing goals of various Lebanese power groups. The Palestinians' goal was to use Lebanon as a springboard for the liberation of Palestine. The Maronites' goal was to drive the Palestinians out of Lebanon and preserve their privileged status. Sunni Muslim leaders sought to reshape the National Pact to allow for equal political participation with the Christians. Shia leaders were determined to get a better break for the Shia community, generally the poorest and least represented in the Lebanese government.[7] The Druze, also interested in greater representation in the system and traditionally hostile to the Maronites, disliked and distrusted all of the other groups.

Like most civil wars, the Lebanese Civil War was fought by Lebanon's own people. It was a war that made no sense, where sides changed frequently and battles raged from street to street in Beirut. One of the casualties was the National Museum, a superb example of 1930s Mediterranean architecture, which held a priceless collection of antiquities. The building was literally in the line of fire between rival militias and lay in ruins after 1975. Its collection had mysteriously disappeared, but fortunately most items had been hidden by curators, and the restored museum reopened in 1997.

Eventually Lebanon's importance as a regional trade, banking, and transit center ensured that outside powers would intervene. Syrian troops were ordered by the Arab League to occupy the country. Their purpose was not only to end the conflict but also to block Palestinian aspirations to use Lebanon as a launching pad for the recovery of their lands in Israel. The Israelis encouraged a renegade Lebanese officer to set up an "independent free Lebanon" adjoining the Israeli border. The complexity of the situation was described in graphic terms by a Christian religious leader:

The battle is between the Palestinians and the Lebanese. No! It is between the Palestinians and the Christians. No! It is between Christians and Muslims. No! It is between Leftists and Rightists. No! It is between Israel and the Palestinians on Lebanese soil. No! It is between international imperialism and Zionism on the one hand, and Lebanon and neighboring states on the other.[8]

THE ISRAELI INVASION

The immediate result of the Civil War was to divide Lebanon into separate territories, each controlled by a different faction. The Lebanese government, for all practical purposes, could not control its own territory. Israeli forces, in an effort to protect northern Israeli settlements from constant shelling by the Palestinians, established control over southern Lebanon. The Lebanese-Israeli border, ironically, became a sort of "good fence" open to Lebanese civilians for medical treatment in Israeli hospitals.

In March 1978, PLO guerrillas landed on the Israeli coast near Haifa, hijacked a bus, and drove it toward Tel Aviv. The hijackers were overpowered in a shoot out with Israeli troops, but 35 passengers were killed along with the guerrillas. Israeli forces invaded southern Lebanon in retaliation and occupied the region for two months, eventually withdrawing after the United Nations, in an effort to separate Palestinians from Israelis, set up a 6,000-member "Interim Force" in Lebanon, made up of units from various countries, in the south. Although many Lebanese referred to them sarcastically as the "United Nothings," following the withdrawal of Israeli forces from Lebanon the peacekeeping force, now reduced to 2,000 men, moved up to monitor and patrol the UN-drawn boundary between Lebanon and Israel. Its mandate has been renewed a number of times, most recently for six months in 2005. However a Security Council resolution sponsored by France at the time called on Lebanon to extend its authority to the Israeli border and to disarm Hizbullah.

The Lebanese factions themselves continued to tear the nation apart. Political assassinations of rival leaders were frequent. Many Lebanese settlements became ghost towns; they were fought over so much that their residents abandoned them. Some 300,000 Lebanese from the Israeli-occupied south fled to northern cities as refugees. In addition to the thousands of casualties, a psychological trauma settled over Lebanese youth, the "Kalashnikov generation" that knew little more than violence, crime, and the blind hatred of religious feuds. (The Kalashnikov, a Soviet-made submachine gun, became the standard toy of Lebanese children.)[9]

HEALTH/WELFARE

The withdrawal of Israeli forces from southern Lebanon left Hizbullah as the sole on-site agency for reconstruction of that war-torn region. In late 2000, teams of fighters-turned-humanitarian-workers cleaned village streets, set up potable water dispensers, sent mosquito-spraying trucks into the villages, and established and equipped mobile health clinics. Schools were reopened, and the former Israeli hospital at Bint Jbail, the regional capital, is now managed completely by Hizbullah doctors and nurses.

The Israeli invasion of Lebanon in June 1982 was intended as a final solution to the Palestinian problem. It didn't quite work out that way. The Israeli Army surrounded Beirut and succeeded with U.S. intervention in forcing the evacuation of PLO guerrillas from Lebanon. Some of the Lebanese factions were happy to see them go, particularly the Maronites and the Shia community in the south. But they soon discovered that they had exchanged one foreign domination for another. The burden of war, as always, fell heaviest on the civilian population. A Beirut newspaper estimated almost 50,000 civilian casualties in the first two months of the invasion. Also, the Lebanese discovered that they were not entirely free of the Palestinian presence. The largest number of PLO guerrillas either went to Syria and then returned secretly to Lebanon or retreated into the Biqa' Valley to take up new positions under Syrian Army protection.

Israeli control over Beirut enabled the Christians to take savage revenge against the remaining Palestinians. In September 1983, Christian Phalange militiamen entered the refugee camps of Sabra and Shatila in West Beirut and massacred hundreds of people, mostly women and children. The massacre led to an official Israeli inquiry and censure of Israeli government and military leaders for indirect responsibility. But the Christian-dominated Lebanese government's own inquiry failed to fix responsibility on the Phalange.

The Lebanese Civil War supposedly ended in 1976, but it was not until 1990 that the central government began to show results in disarming militias and establishing its authority over the fragmented nation. Until then, hostage taking and clan rivalries underlined the absence in Lebanon of a viable national identity.

The 1982 Israeli invasion brought a change in government; the Phalange leader, Bashir Gemayel, was elected to head a "government of national salvation." Unfortunately for Bashir, his ruthlessness in his career had enabled him to compile an impressive list of enemies. He was killed by a bomb explosion at Phalange headquarters before he could take office. Gemayel was succeeded by his older brother, Amin. The new president was persuaded by U.S. negotiators to sign a troop-withdrawal agreement with Israel. However, the agreement was not supported by leaders of the other Lebanese communities, and in March 1984, Gemayel unilaterally repudiated it. The Israelis then began working their way out of the "Lebanese quagmire" on their own, and in June 1985, the last Israeli units left Lebanon. (However, the Israelis did reserve a "security zone" along the border for necessary reprisals for attacks by PLO or Shia guerrillas.)

The Israelis left behind a country that had become almost ungovernable. Gemayel's effort to restructure the national army along nonsectarian lines came to nothing, since the army was not strong enough to disband the various militias. The growing power of the Shia Muslims, particularly the Shia organization Amal, presented a new challenge to the Christian leadership, while the return of the Palestinians brought bloody battles between Shia and PLO guerrillas. As the battles raged, cease-fire followed cease-fire and conference followed conference, but without noticeable success.

The Israeli withdrawal left the Syrians as the major power brokers in Lebanon. In 1985, Syrian president Hafez al-Assad masterminded a comprehensive peace and reform agreement with Elie Hobeika, the commander of the Christian Falange militia that carried out the 1983 massacres of Palestinians at Sabra and Shatila refugee camps. The agreement would expand the National Pact to provide equal Christian–Muslim representation in the Chamber of Deputies. Hobeika was later ousted by one of his rivals and went into exile in Syria; his departure made the agreement worthless. He returned to Lebanon at the end of the Civil War and held several ministerial portfolios. In January 2002, however, he was killed in a car-bomb blast.

SYRIA INTERVENES

The collapse of peace efforts led Syria to send 7,000 heavily armed commandos into west Beirut in 1987 to restore law and order. They did restore a semblance of order to that part of the capital and opened checkpoints into east Beirut. But the Syrians were unable, or perhaps unwilling, to challenge the powerful Hizbullah faction (reputed to have held most Western hostages), which controlled the rabbit warren of narrow streets and tenements in the city's southern suburbs.

Aside from Hizbullah, Syria's major problem in knitting Lebanon together under its tutelage was with the Maronite community. With President Gemayel's six-year term scheduled to end in September 1988, the Syrians lobbied hard for a candidate of their choice. (Under the Lebanese parliamentary system, the president is elected by the Chamber of Deputies.) Unfortunately, due to the Civil War, only 72 of the 99 deputies elected in 1972, when the last elections had been held, were still in office. They rejected Syria's candidate, former president Suleiman Franjieh (1970–1976), because of his identification with the conflict and his ties with the Assad regime. When the Chamber failed to agree on an acceptable candidate, the office became vacant. Gemayel's last act before leaving office was to appoint General Michel Aoun, the commander of Christian troops in the Lebanese Army, to head an interim government. But the Muslim-dominated civilian government of Prime Minister Salim al-Hoss contested the appointment, declaring that it remained the legitimate government of the country.

BREAKDOWN OF A SOCIETY

The assassination in 1987 of Prime Minister Rachid Karami (a bomb hidden in the army helicopter in which he was traveling blew up) graphically underlined the mindless rejection of law and order of the various Lebanese factions. The only show of Lebanese unity in many years occurred at the funeral of former president Camille Chamoun, dead of a heart attack at age 87. Chamoun's last public statement, made the day before his death, was particularly fitting to this fractured land. "The nation is headed toward total bankruptcy and famine," he warned. The statement brought to mind the prophetic observations of a historian, written in 1966: "Lebanon is too conspicuous and successful an example of political democracy and economic liberalism to be tolerated in a region that has turned its back on both systems."[10]

The death of the Mufti (the chief religious leader of the Sunni Muslim community) in a car-bomb attack in 1989 confirmed Chamoun's gloomy prediction. The Mufti had consistently called for reconciliation and nonviolent coexistence between Christian and Muslim communities. The political situation remained equally chaotic. Rene Moawwad, a respected Christian lawyer, was elected by the Chamber to fill the presidential vacancy. However, he was murdered after barely 17 days in office. The Chamber then elected Elias Hrawi, a Christian politician from the Maronite stronghold of Zahle, as president. General Aoun contested the election, declaring himself the legitimate president of Lebanon, and holed up in the presidential palace in east Beirut, defended strongly by his Maronite militiamen.

But the Maronite community was as fragmented as the larger Lebanese community. Aoun's chief Christian rival, Samir Geagea, rejected his authority, and early in 1990, a renewed outbreak of fighting between their militias left east Beirut in shambles, with more than 3,000 casualties. After another shaky cease-fire had been reached, Syrian Army units supporting the regular Lebanese Army surrounded the Christian section. Aoun's palace became an embattled enclave, with supplies available only by running the Syrian blockade or from humanitarian relief organizations.

Aoun's support base eroded significantly in the spring, when his rival recognized the Hrawi government as legitimate and endorsed the Taif Accord.[11] In October, Hrawi formally requested Syrian military aid for the Lebanese Army. After an all-out assault on the presidential palace by joint Syrian-Lebanese forces, the general surrendered, taking refuge in the French Embassy and then going into exile.

Aoun's departure enabled the Hrawi government to begin taking the next step toward rebuilding a united Lebanon. This involved disarming the militias. The continued presence of Syrian forces was a major asset to the reconstituted Lebanese Army as it undertook this delicate process. A newspaper publisher observed that "the Syrian presence is a very natural fact for the Lebanese," echoing the Syrian president's statement to an interviewer: "Lebanon and Syria are one nation and one people, but they are two distinct states."[12] It was the first clear statement from any Syrian leader that Lebanon had a legitimate existence as a state.

Following the reestablishment of central-government authority, a new transitional Council of Ministers (cabinet) was appointed by President Hrawi in 1992. Its responsibilities were to stabilize the economy and prepare for elections for a new Chamber of Deputies. In preparation for the elections the Chamber was enlarged from 108 to 128 seats.

The first national elections since the start of the Civil War were held in 1992. Due in part to a boycott by Christian parties, which had demanded Syrian withdrawal as their price for participation, Shia candidates won 30 seats. Shia Amal leader Nabih Berri was elected speaker; Rafiq Hariri, a Sunni Muslim and millionaire (who had made his fortune as a contractor in Saudi Arabia) was named prime minister.

In any case, the growing demographic imbalance of Muslims and Christians indicated that, in the not too distant future, Lebanon would no longer be "a Christian island in a Muslim sea." By 1997, Christians numbered at most 30 percent of the population (composed of 800,000 Maronites, 400,000 Greek Orthodox, 300,000 Greek Catholics or Melkites, and 75,000 Armenians). Half a million Christians had left the country during the Civil War, along with top leaders such as Michel Aoun, Amin Gemayel, and Raymond Edde, in exile in Paris.

Changes in the election laws in the 1990s reshaped the Lebanese political configuration. The former governorates were incorporated into 14 constituencies, and the number of seats in the Chamber of Deputies (Parliament) per constituency increased from six to sixteen. Another change allowed government officials to run for president while still in office. As a result, Gen. Emile Lahoud, the army chief of staff, ran and was elected in October 1998 for a six-year term. Lahoud came under fire in 2005 during the "Independence Intifada" that led to Syria's withdrawal from Lebanon. However a constitutional amendment narrowly approved by the legislature would extend his term for an additional three years.

As army chief of staff, Lahoud, a Maronite, had been responsible for disarming the country's numerous militias. By 1993, the one remaining armed organization was Hizbullah. Its members were allowed to keep their weapons in order to deal with Israeli forces in the "security zone" along the border and with their allies, the South Lebanon Army (SLA). With weapons and training supplied by Iran and Syria, Hizbullah developed into a formidable fighting force. In 1998, its fighters succeeded in overrunning the main SLA base at Jezzin. The impending defeat of its Lebanese ally and the endless "war of attrition" with Hizbullah led to a complete withdrawal of Israeli forces from the self-proclaimed "security zone," inside the Lebanese border, in May 2000. Israeli prime minister Ehud Barak had originally set July 2000 for the withdrawal, but he accelerated the process for policy reasons. On May 24, the last Israeli soldiers pulled out of the zone, ending a 22-year occupation. Commenting on the departure, an Israeli tank officer observed: "You can't win a guerrilla war. We withdrew with dignity. I don't think we ran away."[13] Hizbullah leaders held a different view: To them, it was proof that the invincible Israeli Army had its weaknesses and could be defeated by unorthodox tactics.

The Israeli withdrawal resulted in jubilant celebrations throughout Lebanon. The government declared May 24 a national holiday, National Resistance Day, as crowds danced in the streets. Some 6,000 SLA militiamen fled into Israel with their families, fearing retribution by Hizbullah. Although the Israeli government provided housing and other benefits for them, the exiles felt uncomfortable in a "foreign" setting. They also faced unremitting hostility from Israeli Arabs. In 2000, they began returning to their homes in south Lebanon; by the end of 2001, some 2,344 had returned. The anticipated sectarian bloodbath did not materialize, and as a result the area remained at peace, ending 30 years of conflict.

Under the terms of UN *Resolution 425,* which had called for Israel's withdrawal as long ago as 1978, units of the United Nations Interim Force in Lebanon (UNFIL) moved to the border as peacekeepers. The border has been quiet since then—with one exception: Shebaa Farms. This mountain region of 15.6 square miles was originally Lebanese territory but had been awarded to Syria by the League of Nations in 1920, after the establishment of the Syrian and Lebanese mandates. For that reason, the United Nations had excluded it from Israel's "security zone." In May 2001, Hizbullah guerrillas abducted several Israeli soldiers stationed there, on the grounds that they were actually on Lebanese territory. Israel's response was an air attack on a nearby Syrian radar installation. It was Israel's first attack on Syria in 22 years.

Along with guerrilla warfare, Hizbullah engaged in a "propaganda war" with Israel through its satellite-television station, *Al-Manar* ("The Beacon"). Its programs were greatly expanded after the Israeli withdrawal from Lebanon and the revived Palestinian intifada. Its broadcasts in Hebrew to Israelis and in Arabic to the Palestinians, with video images of clashes and casualties, helped to strengthen the resolve of the Palestinians against the Israelis.

LEBANON AND THE WORLD

Aside from its vulnerability to international and inter-Arab rivalries because of internal conflicts, Lebanon drew world attention in the 1980s for its involvement in hostage taking. Lebanese militias such as Hizbullah, a Shia group backed by Iran as a means of exporting the Islamic Revolution, and shadowy organizations like the Islamic Jihad, Revolutionary Justice, and Islamic Jihad for the Liberation of Palestine kidnapped foreigners in Beirut. The conditions set for their release were rarely specific, and the refusal of the U.S. and other Western governments to "deal with terrorists" left them languishing in unknown prisons for years, seemingly forgotten by the outside world.

The changing Middle East situation and Lebanon's slow return to normalcy in the 1990s began to move the hostage-release process forward. Release negotiations were pursued by then-UN general-secretary Javier Pérez de Cuéllar. The UN team worked on two levels: Pérez de Cuéllar ran a high-profile diplomatic campaign by repeatedly visiting Iran, Syria, and Israel, while his long-time associate Giandomenico Picco conducted behind-the-scenes talks with Shia operatives in the Biqa' Valley located in the eastern part of Lebanon. Their efforts began to bear fruit: within a few months, the hostages, mostly British and American, were freed individually or in small groups. They included Terry Waite, an envoy of the archbishop of Canterbury originally sent to negotiate the hostages' release (he was charged mistakenly with espionage). Several had been professors at the American University of Beirut or Beirut College for Women, and the last to be released was Terry Anderson, a well-known *New York Times* correspondent.

Since then, Lebanese–American relations have remained stable. However, the September 11 terrorist bombings in the United States and President George W. Bush's effort to form an international antiterrorism coalition that would include Muslim states placed the Lebanese government in an awkward position. The U.S. ambassador to Lebanon commented in October 2001 that the country continued to shelter "terrorist organizations," including Hizbullah, since it had been responsible for the 1983 destruction of the American Embassy in Beirut and a truck-bomb onslaught on a U.S. Marine barracks that killed 241 Americans. Despite Hizbullah's newfound respectability as a social-service organization and political party represented in the Chamber of Deputies, the government feared that its past actions might motivate the United States to seek retribution, including Lebanon in its antiterrorism campaign. A Bush administration request to the Lebanese government to freeze Hizbullah assets as a "terrorist organization" was rejected.

SYRIA LEAVES LEBANON

The assassination of ex-prime minister Rafik Hariri by a remote-control bomb that blew up his motorcade in Beirut, killing him and 18 associates, was the catalyst in increasing Lebanese resentment to the point of bringing about a withdrawal of Syrian forces from the country. Syria disclaimed any connection with the murder, but UN investigators determined quite conclusively in 2005 that both pro-Syrian Lebanese police and Syrian intelligence agents were directly involved. Hariri had resigned in October, 2004, under Syrian pressure, and had begun to form an opposition coalition to win control of the Chamber of Deputies in the 2005 elections and form a new government independent of Syrian control.

Hariri's assassination triggered widespread demonstrations demanding Syrian troop withdrawal. The resulting "Cedar Revolution" (an indirect reference to the Lebanese national flag) briefly united all Lebanese fractions in a nonviolent campaign demanding an end to the Syrian occupation. With the help of external pressure on Syria and Security Council *Resolution 1559* requiring Syrian withdrawal it proved to be a successful revolution. Subsequently a new Chamber of Deputies was elected with Hariri's son Saad as prime minister.

THE ECONOMY

In the mid-1970s, the Lebanese economy began going steadily downhill. The Civil War and resulting instability caused most banks and financial institutions to move out of Beirut to more secure locations, notably Jordan, Bahrain, and Kuwait. Aside from the cost in human lives, Israeli raids and the 1982 invasion severely damaged the economy. The cost of the invasion in terms of damages was estimated at $1.9 billion. Remittances from Lebanese emigrants abroad dropped significantly. The Lebanese currency, once valued at 4.74 pounds to $1, had dropped in value to 3,000 to $1 by 1992, although rebounding to the present LL 1,507 to $1 eleven years later.

Yet by a strange irony of fate, some elements of the economy continued to display robust health. Most middle-class Lebanese had funds invested abroad, largely in U.S. dollar accounts, and thus were protected from economic disaster.

The expansion of the Civil War in 1989–1990 and the intervention of Syrian troops tested the survival techniques of the Lebanese people as never before. But they adjusted to the new "Battle of Beirut" with great inventiveness. A newspaper advertisement announced: "Civilian fortifications, 24-hour delivery service. Sandbags and barrels, full or empty." With the Syrian-Christian artillery exchanges concentrated at night, most residents fled the city then, returning after the muezzin's first call for morning prayers had in effect silenced the guns, to shop, to stock up on fuel smuggled ashore from small tankers, or to sample the luxury goods that in some mysterious way had appeared on store shelves.

The long, drawn-out civil conflict badly affected Lebanese agriculture, the mainstay of the economy. Both the coastal strip and the Biqa' Valley are extremely fertile, and in normal times produce crop surpluses for export. Lebanese fruit, particularly apples (the most important cash crop) and grapes, is in great demand throughout the Arab world. But these crops are no longer exported in quantity. Israeli destruction of crops, the flight of most of the farm labor force, and the blockade by Israeli troops of truck traffic from rural areas into Beirut had a devastating effect on production.

Lebanon produces no oil of its own, but before the Civil War and the Israeli invasion, the country derived important revenues from transit fees for oil shipments through pipelines across its territory. The periodic closing of these pipelines and damage to the country's two refineries sharply reduced revenues. The well-developed manufacturing industry, particularly textiles, was equally hard hit.

ACHIEVEMENTS

The cultivation of cannabis (opium poppy) in the Biqa' region, formerly one of the world's major sources, has been drastically reduced due to U.S. and Iranian aid, development of alternate crops, and cattle raising. Currently only 2,500 hectares are under cannabis cultivation.

The Syrian withdrawal resulted in significant changes in Lebanon's moribund political system, as the new election laws went into effect in time for the 2005 parliamentary elections. Former army chief of staff Michel Aoun, a Maronite, returned from exile and was elected along with 21 of his supporters. Hariri's son Saad, who had become head of the Sunni community after his father's murder, was also elected as the Sunni bloc won all 19 seats from the Beirut constituency. Hizbullah and its ally Amal won all 50 seats allotted to south Lebanon under the new configuration.

Timeline: PAST

9th–11th centuries
Establishment of Mount Lebanon as a sanctuary for religious communities

1860–1864
The first Civil War, between Maronites and Druze, ending in foreign military intervention

1920–1946
French mandate

1958
Internal crisis and the first U.S. military intervention

1975–1976
Civil war, ended (temporarily) by an Arab League-sponsored cease-fire and peacekeeping force of Syrian troops

1980s
Israeli occupation of Beirut; Syrian troops reoccupy Beirut; foreigners are seized in a new outbreak of hostage taking; the economy nears collapse

1990s
The withdrawal of Israeli forces from Lebanon; all foreign hostages are released; Lebanon begins rebuilding

PRESENT

2000s
Hizbullah's presence in Lebanon causes tension with the United States

Armed with $458 million in aid from the World Bank, the European Union, and the Paris-based Mediterranean Development Agency, the senior Hariri had launched a major economic reform drive in February 2001. He laid off 500 employees from the bloated public sector, and privatized the state-owned electricity company as a start toward further privatization. The cabinet also agreed to shut down the state-owned TeleLiban, saving $33 million a year. Elimination of the sugar subsidy will save another $40 million annually.

As is the case with most complex issues, Syria's departure from Lebanon had mixed consequences. Aside from the new-found Lebanese pride in political independence, the large number of Syrian construction workers involved in Hariri's reconstruction of Beirut were no longer available; they had left the country with the troops. New foreign-owned enterprises such as a Heineken brewery were put on hold, while the exodus of Syrian workers, who had been paid $6 per day in contrast to the $18 Lebanese minimum wage, left hundreds of low-paying jobs such as garbage collection unfilled. Beirut itself remained a half-finished vision, with its skyline dominated by idly swinging cranes.[14] Unfortunately with the withdrawal of Syrian forces and authority the confessional polarization of Lebanese sects and groups revived in full force. A series of political murders, the most important being that of anti-Syrian parliamentarian and critic Ghassan Tueni in December 2005, threatened to unravel the country's hard-won unity.

NOTES

1. Abdo Baaklini, *Legislative and Political Development: Lebanon 1842–1972* (Durham, NC: Duke University Press, 1976), pp. 32–34.

2. David C. Gordon, *The Republic of Lebanon: Nation in Jeopardy* (Boulder, CO: Westview Press, 1983), p. 4.

3. Samir Khlaf, *Lebanon's Predicament* (New York: Columbia University Press, 1987), p. 69.

4. Gordon, *op. cit.,* p. 19.

5. *Ibid.,* p. 25. See also Baaklini, *op. cit.,* pp. 200–202, for a description of the co-existence process as used by Sabri Hamadeh, for many years head of the assembly.

6. Whether the Civil War ever really ended is open to question. A cartoon in a U.S. newspaper in August 1982 shows a hooded skeleton on a television screen captioned "Lebanon" saying, "And now we return to our regularly scheduled civil war." Gordon, *op. cit.,* p. 113.

7. Shia religious leader Imam Musa al-Sadr's political organization was named Harakat al-Mahrumin ("Movement of the Disinherited") when it was founded in 1969–1970. See Marius Deeb, *The Lebanese Civil War* (New York: Praeger, 1980), pp. 69–70.

8. Gordon, *op. cit.,* p. 110.

9. *Ibid.,* p. 125.

10. Charles Issawi, "Economic Development and Political Liberalism in Lebanon," in Leonard Binder, ed., *Politics in Lebanon* (New York: John Wiley, 1966), pp. 80–81.

11. The Taif Accord, signed under Arab League auspices in Taif, Saudi Arabia, changes the power-sharing arrangement in the Lebanese government from a 6:5 Christian-Muslim ratio to one of equal representation in the government. The powers of the president are also reduced.

12. *Middle East Economic Digest* (October 10, 1990).

13. Joel Greenberg, in *The New York Times* (May 24, 2000).

14. To balance these negative results, a young literature student from the Lebanese University told a reporter, "Today, for the first time, I felt that Lebanon was free, sovereign and independent." Sam Ghattas, Associated Press report, April 10, 2005.

Libya (Socialist People's Libyan Arab Jamahiriyya)

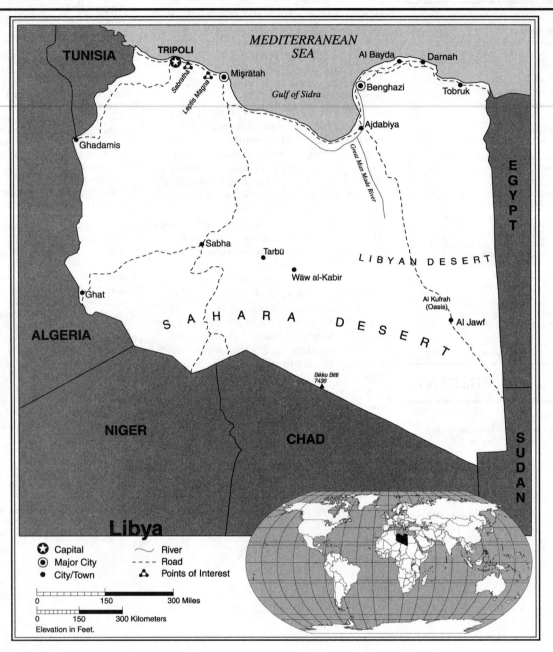

Libya

- ★ Capital
- ◉ Major City
- ● City/Town
- ⌇ River
- --- Road
- △ Points of Interest

0   150   300 Miles
0   150   300 Kilometers
Elevation in Feet.

Libya Statistics

GEOGRAPHY

Area in Square Miles (Kilometers):

679,147 (1,759,450) (about the size of Alaska)

Capital (Population): Tripoli (1,681,000)

Environmental Concerns: desertification; very limited freshwater resources

Geographical Features: mostly barren, flat to undulating plains, plateaus, depressions

Climate: Mediterranean along the coast; dry, extreme desert in the interior

PEOPLE

Population

Total: 5,765,563 (includes 166,510 non-nationals)

Annual Growth Rate: 2.33%

Rural/Urban Population Ratio: 14/86

Major Languages: Arabic; English; Italian
Ethnic Makeup: 97% Berber and Arab; 3% others
Religions: 97% Sunni Muslim; 3% others

Health

Life Expectancy at Birth: 74 years (male); 78 years (female)
Infant Mortality Rate: 24.6/1,000 live births

Education

Adult Literacy Rate: 82.6%
Compulsory (Ages): 6–15

COMMUNICATION

Telephones: 380,000 main lines
Daily Newspaper Circulation: 15 per 1,000 people
Televisions: 105 per 1,000 people
Internet Users: 1 (2000)

TRANSPORTATION

Highways in Miles (Kilometers): 15,180 (24,484)
Railroads in Miles (Kilometers): none

Usable Airfields: 139
Motor Vehicles in Use: 904,000

GOVERNMENT

Type: officially a *Jamahiriyya* ("state belonging to the people") with government authority exercised by a General Peoples' Congress
Independence Date: December 24, 1951 (from Italy)
Head of State/Government: Revolutionary Leader Mahammad Au Minyar al-Qadhafi holds no official title but serves as de facto head of state; Shukri Ghanem, Secretary of the GPC, is the equivalent to Prime Minister
Political Parties: none
Suffrage: universal and compulsory at 18

MILITARY

Military Expenditures (% of GDP): 3.9%
Current Disputes: Libya claims about 19,400 square kilometers of land in northern Niger and part of southeastern Algeria; Both disputes currently dormant

ECONOMY

Currency ($U.S. Equivalent): 1.305 dinars = $1
Per Capita Income/GDP: $6,700/$37.4 billion
GDP Growth Rate: 4.9%
Inflation Rate: 18.5%
Unemployment Rate: 2.9%
Labor Force: 1,500,000
Natural Resources: petroleum; natural gas; gypsum
Agriculture: wheat; barley; olives; dates; citrus fruits; vegetables; peanuts; beef; eggs
Industry: petroleum; food processing; textiles; handicrafts; cement
Exports: $18.65 billion (primary partners Italy, Germany, Spain)
Imports: $7.2 billion (primary partners Italy, Germany, Tunisia)

SUGGESTED WEBSITES

http://lcweb2.loc.gov/frd/cs/lytoc.html
http://home.earthlink.net/~dribrahim/

Libya Country Report

The Socialist People's Libyan Arab Jamahiriyya (Republic), commonly known as Libya, is the fourth largest of the Arab countries. Since it became a republic in 1969, it has played a role in regional and international affairs more appropriate to the size of its huge territory than to its small population.

Libya consists of three geographical regions: Tripolitania, Cyrenaica, and the Fezzan. Most of the population live in Tripolitania, the northwestern part of the country, where Tripoli, the capital and major port, is located. Cyrenaica, in the east along the Egyptian border, has a narrow coastline backed by a high plateau (2,400-feet elevation) called the Jabal al-Akhdar ("Green Mountain"). It contains Libya's other principal city, Benghazi. The two regions are separated by the Sirte, an extension of the Sahara Desert that reaches almost to the Mediterranean Sea. Most of Libya's oil fields are in the Sirte.

The Fezzan occupies the central part of the country. It is entirely desert, except for a string of widely scattered oases. Its borders are with Chad, Algeria, Niger, and Sudan. The border with Chad, established during French colonial rule in sub-Saharan Africa, was once disputed by Libya. The matter was settled through international mediation, with the border formally demarcated in 1994. Libya also claims areas in northern Niger and southeastern Algeria left over from the colonial period, when they formed part of the French West African empire. In the Libyan view, these areas should have been transferred to its control under the peace treaty that established the Libyan state and relinquishment of French control.

HISTORY

Until modern times, Libya did not have a separate identity, either national or territorial. It always formed a part of some other territorial unit and in most cases was controlled by outsiders. However, control was usually limited to the coastal areas. The Berbers of the interior were little affected by the passing of conquerors and the rise and fall of civilizations.

Libya's culture and social structure have been influenced more by the Islamic Arabs than by any other invaders. The Arabs brought Islam to Libya in the early seventh century. Arab groups settled in the region and intermarried with the Berber population to such an extent that the Libyans became one of the most thoroughly Arabized peoples in the Islamic world.

DEVELOPMENT

Although continued U.S. sanctions prohibit American firms from operating in Libya, improved relations with other countries have begun to generate diversification of the Libyan economy. An agreement with Ireland to import 50,000 live Irish cattle was concluded in March 2001, and Italy's export credit agency wrote off $230 million in Libyan debts to encourage investment by Italian firms. New oil discoveries in the Murzuq field have aided economic recovery.

Coastal Libya, around Tripoli, was an outlying province of the Ottoman Empire for several centuries. Like its urban neighbors Tunis and Algiers, Tripoli had a fleet of corsairs that made life dangerous for European merchant ships in the Mediterranean. When the United States became a Mediterranean trading nation, the corsairs of Tripoli included American ships among their targets. The USS *Philadelphia* was sent to Tripoli to "teach the corsairs a lesson" in 1804, but it got stuck on a sandbar and was captured. Navy Lieutenant Stephen Decatur led a commando raid into Tripoli harbor and blew up the ship, inspiring the words to what would become the official U.S. Marine hymn: "From the halls

United Nations Photo (UN43392)

The Fezzan occupies the central part of Libya and is entirely desert, except for some widely scattered oases. This oasis is called Bu Gheilan.

of Montezuma to the shores of Tripoli...." Ironically the Marines never reached Tripoli, although a land campaign led by the itinerant American soldier of fortune William Eaton, who had been hired by a dispossessed member of the ruling dynasty to recover his throne, got only as far as Darna, well east of the capital.

This ruling dynasty, founded by an ex-Ottoman officer named Ahmad Karamanli, controlled Tripoli and its coastline from the 18th century on. Its corsair fleet was allied with the Ottoman navy. In 1815, at the end of the Napoleonic Wars, British and French naval forces defeated the Tripolitan corsairs and forced the Karamanli ruler of that time to end payment of tribute. Lacking its main source of revenue, his population rebelled. By 1835 France had seized control of Algiers and Tunis, the other main corsair states. To forestall further territorial losses the Ottoman sultan, Tripoli's nominal ruler, sent forces to occupy the city and for the first time place it under direct Ottoman rule.

The Sanusiya Movement

At various stages in Islam's long history, new groups or movements have appeared committed to purifying or reforming Islamic society and taking it back to its original form of a simple community of believers led by just rulers. Several of these movements, such as the Wahhabis of Saudi Arabia, were important in the founding of modern Islamic states. The movement called the Sanusiya was formed in the nineteenth century. In later years, it became an important factor in the formation of modern Libya.

The founder, the Grand Sanusi, was a religious teacher from Algeria. He left Algeria after the French conquest and settled in northern Cyrenaica. The Grand Sanusi's teachings attracted many followers. He also attracted the attention of the Ottoman authorities, who distrusted his advocacy of a strong united Islamic world in which Ottomans and Arabs would be partners. In 1895, to escape from the Ottomans, the Grand Sanusi's son and successor moved Sanusiya headquarters to Kufra, a remote oasis in the Sahara.

The Sanusiya began as a peaceful movement interested only in bringing new converts to Islam and founding a network of *zawiyas* ("lodges") for contemplation and monastic life throughout the desert. But when European countries began to seize territories in North and West Africa, the Sanusi became warrior-monks and fought the invaders.

Italy Conquers Libya

The Italian conquest of Libya began in 1911. The Italians needed colonies, not only for prestige but also for the resettlement of poor and landless peasants from Italy's crowded southern provinces. The Italians expected an easy victory against a weak Ottoman garrison; Libya would become the "Fourth Shore" of a new Roman Empire from shore to shore along the Mediterranean. But the Italians found Libya a tougher land to subdue than they had expected. Italian forces were pinned to Tripoli and a few other points on the coast by the Ottoman garrison and the fierce Sanusi warrior-monks.

The Italians were given a second chance after World War I. The Ottoman Empire had been defeated, and Libya was ripe for the plucking. The new Italian government of swaggering dictator Benito Mussolini sent an army to occupy Tripolitania. When the Italians moved on Cyrenaica, the Grand Sanusi crossed the Egyptian border into exile under British protection. The Italians found Cyrenaica much more difficult to control than Tripolitania. It is ideal guerrilla country, from the caves of Jabal al-Akhdar to the stony plains and dry, hidden *wadis* (river beds) of the south. It took nine years (1923–1932) for Italy to overcome all of Libya, despite Italy's vast superiority in troops and weapons. Sanusi guerrilla bands harried the Italians, cutting supply lines, ambushing patrols, and attacking convoys. Their leader, Shaykh Omar Mukhtar, became Libya's first national hero.

After the 1969 Revolution, the government strove to develop many aspects of the country. These local chiefs are meeting to plan community development.

The Italians finally overcame the Sanusi by the use of methods that do not shock us today but seemed unbelievably brutal at the time. Cyrenaica was made into a huge concentration camp, with a barbed-wire fence along the Egyptian border. Nomadic peoples were herded into these camps, guarded by soldiers to prevent them from aiding the Sanusi. Sanusi prisoners were pushed out of airplanes, wells were plugged to deny water to the people, and flocks were slaughtered. In 1931, Omar Mukhtar was captured, court-martialed, and hanged in public. The resistance ended with his death.

The Italians did not have long to cultivate their Fourth Shore. During the 1930s, they poured millions of lire into the colony. A paved highway from the Egyptian to the Tunisian border along the coast was completed in 1937; in World War II, it became a handy invasion route for the British. A system of state-subsidized farms was set up for immigrant Italian peasants. Each was given free transportation, a house, seed, fertilizers, a mule, and a pair of shoes as inducements to come to Libya. By 1940, the Italian population had reached 110,000, and about 495,000 acres of land had been converted into productive farms, vineyards, and olive groves.[1]

Independent Libya

Libya was a major battleground during World War II, as British, German, and Italian armies rolled back and forth across the desert. The British defeated the Germans and occupied northern Libya, while a French army occupied the Fezzan. The United States later built an important air base, Wheelus Field, near Tripoli. Thus the three major Allied powers all had an interest in Libya's future. But they could not agree on what to do with occupied Libya.

Italy wanted Libya back. France wished to keep the Fezzan as a buffer for its African colonies, while Britain preferred self-government for Cyrenaica under the Grand Sanusi, who had become staunchly pro-British during his exile in Egypt. The Soviet Union favored a Soviet trusteeship over Libya, which would provide the Soviet Union with a convenient outlet in the Mediterranean. The United States waffled but finally settled on independence, which would at least keep the Soviet tentacles from enveloping Libya.

Due to lack of agreement, the Libyan "problem" was referred to the United Nations General Assembly. Popular demonstrations of support for independence in Libya impressed a number of the newer UN members; in 1951, the General Assembly approved a resolution for an independent Libyan state, a kingdom under the Grand Sanusi, Idris.

THE KINGDOM OF LIBYA

Libya has been governed under two political systems since independence: a constitutional monarchy (1951–1969); and a Socialist republic (1969–), which has no constitution because all power "belongs" to the people. Monarchy and republic have had almost equal time in power. But Libya's sensational economic growth and aggressive foreign policy under the republic need to be understood in relation to the solid, if unspectacular, accomplishments of the regime that preceded it.

At independence, Libya was an artificial union of the three provinces. The Libyan people had little sense of national identity or unity. Loyalty was to one's family, clan, village, and, in a general sense, to the higher authority represented by a tribal confederation. The only other loyalty linking Libyans was the Islamic religion. The tides of war and conquest that had washed over them for centuries had had little effect on their strong, traditional attachment to Islam.[2]

Political differences also divided the three provinces. Tripolitanians talked openly of abolishing the monarchy. Cyrenaica was the home and power base of King Idris; the king's principal supporters were the Sanusiya and certain important families. The distances and poor communication links between the provinces contributed to the impression that they should be separate countries. Leaders could not even agree on the choice between Tripoli and Benghazi for the capital. For his part, the king distrusted both cities as being corrupt and overly influenced by foreigners. He had his administrative capital at Baida, in the Jabal al-Akhdar.

The greatest problem facing Libya at independence was economics. Per capita income in 1951 was about $30 per year; in 1960, it was about $100 per year. Approximately 5 percent of the land was marginally usable for agriculture, and only 1 percent could be cultivated on a permanent basis. Most economists considered Libya to be a hopeless case, almost totally dependent on foreign aid for survival. (It is interesting to note that the Italians were seemingly able to force more out of the soil, but one must remember that the Italian government poured a great deal of money into the country to develop the plantations, and credit must also be given to the extremely hard-working Italian farmer.)

Despite its meager resources and lack of political experience, Libya was valuable to the United States and Britain in the 1950s and 1960s because of its strategic location. The United States negotiated a long-term lease on Wheelus Field in 1954, as a vital link in the chain of U.S. bases built around the southern perimeter of the Soviet Union due to the Cold War. In return, U.S. aid of $42 million sweetened the pot, and Wheelus became the single largest employer of Libyan labor. The British had two air bases and maintained a garrison in Tobruk.

Political development in the kingdom was minimal. King Idris knew little about parliamentary democracy, and he distrusted political parties. The 1951 Constitution provided for an elected Legislature, but a dispute between the king and the Tripolitanian National Congress, one of several Tripolitanian parties, led to the outlawing of all political parties. Elections were held every four years, but only property-owning adult males could vote (women were granted the vote in 1963). The same legislators were reelected regularly. In the absence of political activity, the king was the glue that held Libya together.

THE 1969 REVOLUTION

At dawn on September 1, 1969, a group of young, unknown army officers abruptly carried out a military coup in Libya. King Idris, who had gone to Turkey for medical treatment, was deposed, and a "Libyan Arab Republic" was proclaimed by the officers. These men, whose names were not known to the outside world until weeks after the coup, were led by Captain Muammar Muhammad al-Qadhafi. He went on Benghazi radio to announce to a startled Libyan population: "People of Libya… your armed forces have undertaken the overthrow of the reactionary and corrupt regime.… From now on Libya is a free, sovereign republic, ascending with God's help to exalted heights."[3]

Qadhafi's new regime made a sharp change in policy from that of its predecessor. Wheelus Field and the British air bases were evacuated and returned to Libyan control. Libya took an active part in Arab affairs and supported Arab unity, to the extent of working to undermine other Arab leaders whom Qadhafi considered undemocratic or unfriendly to his regime.[4]

REGIONAL POLICY

To date, Qadhafi's efforts to unite Libya with other Arab states have not been successful. A 1984 agreement for a federal union with Morocco, which provided for separate sovereignty but a federated Assembly and unified foreign policies, was abrogated unilaterally by the late King Hassan II, after Qadhafi had charged him with "Arab treason" for meeting with Israeli leader Shimon Peres. Undeterred, Qadhafi tried again in 1987 with neighboring Algeria, receiving a medal from President Chadli Bendjedid but no other encouragement.

United Nations Photo (UN81801)

Muammar al-Qadhafi led a group of army officers in the military coup of 1969 that deposed King Idris. In later years, Qadhafi gained world-wide notoriety for his apparent sanction of terrorism.

Although distrustful of the mercurial Libyan leader, other North African heads of state have continued to work with him on the basis that it is safer to have Qadhafi inside the circle than isolated outside. Tunisia restored diplomatic relations in 1987, and Qadhafi agreed to compensate the Tunisian government for lost wages of Tunisian workers expelled from Libya during the 1985 economic recession. Qadhafi also accepted International Court of Justice arbitration over Libya's dispute with Tunisia over oil rights in the Gulf of Gabes. In 1989, Libya joined with other North African states in the Arab Maghrib Union, which was formed to coordinate their respective economies. However, the AMU has yet to become a viable organization due to political differences among its members, in particular the Western Sahara dispute between Algeria and Morocco.

With little to show for his efforts to unite the Arab countries, Qadhafi turned his attention to sub-Saharan Africa. He had abolished the Secretariat for Arab Unity as a government ministry in 1997, and subsequently black African workers were invited to come and work in Libya. By 2000, nearly a million had arrived, most of them from Nigeria, Chad, and Ghana. Economic problems in sub-Saharan Africa caused thousands more to use Libya as an escape route for Europe, many of them also fleeing from civil war in Côte d' Ivoire and Sierra Leone. The flood of migrants generated tension between them and Libyan natives; the latter viewed the migrants as agents of social misbehavior ranging from prostitution to drug usage and AIDS. In August 2000, the Libyan government deported several thousand African workers. They were hauled to the Niger border in trucks and dumped across the border there. Qadhafi had announced earlier that a "United States of Africa" would come into existence in March 2001 under Libyan sponsorship. But for once the Libyan people did not agree with him; "We are native Arabs, not Africans," they told their leader.

SOCIAL REVOLUTION

Qadhafi's desert upbringing and Islamic education gave him a strong, puritanical moral code. In addition to closing foreign bases and expropriating properties of Italians and Jews, he moved forcefully against symbols of foreign influence. The Italian cathedral in Tripoli became a mosque, street signs were converted to Arabic, nightclubs were closed, and the production and sale of alcohol were prohibited.

But Qadhafi's revolution went far beyond changing names. In a three-volume work entitled *The Green Book,* he described his vision of the appropriate political system for Libya. Political parties would not be allowed, nor would constitutions, legislatures, even an organized court system. All of these institutions, according to Qadhafi, eventually become corrupt and unrepresentative. Instead, "people's committees" would run the government, business, industry, and even the universities. Libyan embassies abroad were renamed "people's bureaus" and were run by junior officers. (The takeover of the London bureau in 1984 led to counterdemonstrations by Libyan students and the killing of a British police officer by gunfire from inside the bureau. The Libyan bureau in Washington, D.C., was closed by the U.S. Federal Bureau of Investigation and the staff deported on charges of espionage and terrorism against Libyans in the United States.) The country was renamed the Socialist People's Libyan Arab Jamahiriyya, and titles of government officials were eliminated. Qadhafi became "Leader of the Revolution," and each government department was headed by the secretary of a particular people's committee.

Qadhafi then developed a so-called Third International Theory, based on the belief that neither capitalism nor communism could solve the world's problems. What was needed, he said, was a "middle way" that would harness the driving forces of human history—religion and nationalism—to interact with each other to revitalize mankind. Islam would be the source of that middle way, because "it provides for the realization of justice and equity, it does not allow the rich to exploit the poor."[5]

THE ECONOMY

Modern Libya's economy is based almost entirely on oil exports. Concessions were granted to various foreign companies to explore for oil in 1955, and the first oil strikes were made in 1957. Within a decade, Libya had become the world's fourth-largest exporter of crude oil. During the 1960s, pipelines were built from the oil fields to new export terminals on the Mediterranean coast. The lightness and low sulfur content of Libyan crude oil make it highly desirable to industrialized countries, and, with the exception of the United States, differences in political viewpoint have had little effect on Libyan oil sales abroad.

After the 1969 Revolution, Libya became a leader in the drive by oil-producing countries to gain control over their petroleum industries. The process began in 1971, when the new Libyan government took over the interests of British Petroleum in Libya. The Libyan method of nationalization was to proceed against individual companies rather than to take on the "oil giants" all at once. It took more than a decade before the last company, Exxon, capitulated. However, the companies' $2 billion in assets were left in limbo in 1986, when the administration of U.S. president Ronald Reagan imposed a ban on all trade with Libya to protest Libya's involvement in international terrorism. President George Bush extended the ban for an additional year in 1990, although he expressed satisfaction with reduced Libyan support for terroristic activities, one example being the expulsion from Tripoli of the Palestine Liberation Front, a radical opponent of Yassir Arafat's Palestine Liberation Organization.

Libya's oil reserves are estimated at 36 billion barrels, along with 52 trillion cubic feet of recoverable natural gas reserves. The low sulfur content of Libyan oil and its proximity to the Mediterranean coast, which keeps transport costs low, have made its oil highly marketable. With oil production reaching a record 1.4 million barrels per day, Libya has been able to build a strong petrochemical industry. The Marsa Brega petrochemical complex is one of the world's largest producers of urea, although a major contract with India was canceled in 1996 due to UN sanctions on trade with Libya.

Until recently, industrial-development successes based on oil revenues enabled Libyans to enjoy an ever-improving standard of living, and funding priorities were shifted from industry to agricultural development in the budget. But a combination of factors—mismanagement, lack of a cadre of skilled Libyan workers, absenteeism, low motivation of the workforce, and a significant drop in revenues (from $22 billion in 1980 to $7 billion in 1988)—cast doubts on the effectiveness of Qadhafi's *Green Book* socialist economic policies.

FREEDOM

The General People's Congress has the responsibility for passing laws and appointing a government. In 1994, the GPC approved legislation making Islamic law applicable in the country. They concerned retribution and blood money; rules governing wills, crimes of theft and violence, protection of society from things banned in the Koran, marriage, and divorce; and a ban on alcohol use.

In 1988, the leader began closing the book. As production incentives, controls on both imports and exports were eliminated, and profit sharing for employees of small businesses was encouraged. In 1990, the General People's Congress (GPC), Libya's equivalent of a parliament, began a restructuring of government, adding new secretariats (ministries) to help expand economic development and diversify the economy.

In January 2000, Qadhafi marched into a GPC meeting waving a copy of the annual budget. He tore up the copy and ordered most of the secretariats abolished. Their powers would be transferred to "provincial cells" outside of Tripoli. Only five government functions—finance, defense, foreign affairs, information, and African unity— would remain under central-government control. In October of that year, Qadhafi ordered further cuts, continuing his direct management of national affairs. For the first time he named a prime minister, Mubarak al-Shamekh, to head the stripped-down government. The secretariat for information was abolished, and the heads of the justice and finance secretariats summarily dismissed. The head of the National Oil Company (NOC), Libya's longest-serving government official, was transferred to a new post; Qadhafi had criticized the NOC for mismanagement of the oil industry and lack of vision.

Libya also started developing its considerable uranium resources. A 1985 agreement with the Soviet Union provided the components for an 880-megawatt nuclear-power station in the Sirte region. Libya has enough uranium to meet its foreseeable domestic needs. The German-built chemical-weapons plant at Rabta, described by Libyans as a pharmaceutical complex but confirmed as to its real function by visiting scientists, was destroyed in a mysterious fire in the 1980s. A Russian-built nuclear reactor at Tajoora, 30 miles from Tripoli, suffered a similar fate, not from fire but due to faulty ventilation and high levels of radiation. But the Libyans have pressed on. An underground complex at Mount Tarhuna, south of Tripoli, was completed in 1998 and closed subsequently to international inspection. Libya claims that it is part of the Great Man-Made River (GMR) project and thus not subject to such inspections.

The GMR, a vast $30 billion complex of pipelines to draw water from underground Saharan aquifers, was begun in 1983. A component of Qadhafi's vision of a self-sufficient sovereign Libya, its goals were to expand irrigation in the fertile coastal agricultural area and improve the potable water supply in Libyan cities. It was planned in five stages. Stage three, reached in 2005, has reached the second goal but not the first. Although work continues on expanding the pipeline network, progress has been slow, due to faulty engineering, collapsed pipe connections and other problems. Whether the GMR will ever bring economic benefits enough to offset its enormous cost remains to be seen. But in non-economic terms it has helped link Libya's disparate land areas and build national unity and a sense of shared national pride in its people.

In addition to its heavy dependence on oil revenues, another obstacle to economic development in Libya is derived from an unbalanced labor force. One author observed, "Foreigners do all the work. Moroccans clean houses, Sudanese grow vegetables, Egyptians fix cars and drive trucks. Iraqis run the power stations and American and European technicians keep the equipment and systems humming. All the Libyans do is show up for makework government jobs."[6] Difficult climatic conditions and little arable land severely limit agricultural production; the country must import 75 percent of its food.

AN UNCERTAIN FUTURE

The revolutionary regime has been more successful than the monarchy was in making the wealth from oil revenues available to ordinary Libyans. Per capita income, which was $2,170 the year after the revolution, had risen to $10,900 by 1980. U.S sanctions and the drop in global oil prices have resulted in sharp reductions; per capita income was $8,900 in 2001 and $6,700 in 2005.

This influx of wealth changed the lives of the people in a very short period of time. Semi-nomadic tribes such as the Qadadfas of the Sirte (Qadhafi's kin) have been provided with permanent homes, for example. Extensive social-welfare programs, such as free medical care, free education, and low-cost housing, have greatly enhanced the lives of many Libyans. However, this wealth has yet to be spread evenly across society. The economic downturn of the 1990s produced a thriving black market, along with price gouging and corruption in the public sector. In 1996, Libya organized "purification committees," mostly staffed by young army officers, to monitor and report instances of black-market and other illegal activities.

HEALTH/WELFARE

In addition to 1 million sub-Saharan African workers, Libya has made use of skilled workers as well as unskilled ones from many other Arab countries. Palestinian workers were expelled after the 1993 Oslo Agreement with Israel, which Qadhafi opposed vehemently. GPC regulations supposedly limit the number of foreign workers, who are mostly Egyptians and Tunisians. Altogether they make up about 3 percent of the total population.

Until recently, opposition to Qadhafi was confined almost entirely to exiles abroad, centered on former associates living in Cairo, Egypt, who had broken with the Libyan leader for reasons either personal or related to economic mismanagement. But economic downturns and dissatisfaction with the leader's wildly unsuccessful foreign-policy ventures increased popular discontent at home. In 1983, Qadhafi had introduced two domestic policies that also generated widespread resentment: He called for the drafting of women into the armed services, and he recommended that all children be educated at home until age 10. The 200 basic "people's congresses," set up in 1977 to recommend policy to the national General People's Congress (which in theory is responsible for national policy), objected strongly to

both proposals. Qadhafi then created 2,000 more people's congresses, presumably to dilute the opposition, but withdrew the proposals. In effect, suggested one observer, *The Green Book* theory had begun to work, and Qadhafi didn't like it.

Qadhafi's principal support base rests on the armed forces and the "revolutionary committees," formed of youths whose responsibility is to guard against infractions of *The Green Book* rules. "Brother Colonel" also relies upon a small group of collaborators from the early days of the Revolution, and his own relatives and members of the Qadadfa form part of the inner power structure. This structure is highly informal, and it may explain why Qadhafi is able to disappear from public view from time to time, as he did after the United States conducted an air raid on Tripoli in 1986, and emerge having lost none of his popularity and charismatic appeal.

In recent years, disaffection within the army has led to a number of attempts to overthrow Qadhafi. The most serious coup attempt took place in 1984, when army units allied with the opposition Islamic Front for the Salvation of Libya, based in Cairo and headed by several of Qadhafi's former associates, attacked the central barracks in Tripoli where he usually resides. The attackers were defeated in a bloody gun battle. A previously unknown opposition group based in Geneva, Switzerland, claimed in 1996 that its agents had poisoned the camel's milk that Qadhafi drinks while eating dates on his desert journeys, but proof of this claim is lacking.

However, the Libyan leader's elusiveness and penchant for secrecy make assessments of his continued leadership risky. According to the Tripoli rumor mill, someone attempts to assassinate Qadhafi every couple of months. But as yet no organized internal opposition has emerged, and the mercurial Libyan leader remains not only highly visible but also popular with his people.

INTERNAL CHANGES

Qadhafi has a talent for the unexpected that has made him an effective survivor. In 1988, he ordered the release of all political prisoners and personally drove a bulldozer through the main gate of Tripoli's prison to inaugurate "Freedom Day." Exiled opponents of the regime were invited to return under a promise of amnesty, and a number did so.

In June of that year, the GPC approved a "Charter of Human Rights" as an addendum to *The Green Book*. The charter outlaws the death penalty, bans mistreatment of prisoners, and guarantees every accused person the right to a fair trial. It also permits forma-

tion of labor unions, confirms the right to education and suitable employment for all Libyan citizens, and places Libya on record as prohibiting production of nuclear and chemical weapons. In March 1995, the country's last prison was destroyed and its inmates freed in application of the charter's guarantees of civil liberty.

THE WAR WITH CHAD

Libyan forces occupied the Aouzou Strip in northern Chad in 1973, claiming it as an integral part of the Libyan state. Occupation gave Libya access also to the reportedly rich uranium resources of the region. In subsequent years, Qadhafi played upon political rivalries in Chad to extend the occupation into a de facto one of annexation of most of its poverty-stricken neighbor.

But in late 1986 and early 1987, Chadian leaders patched up their differences and turned on the Libyans. In a series of spectacular raids on entrenched Libyan forces, the highly mobile Chadians, traveling mostly in Toyota trucks, routed the Libyans and drove them out of northern Chad. Chadian forces then moved into the Aouzou Strip and even attacked nearby air bases inside Libya. The defeats, with casualties of some 3,000 Libyans and loss of huge quantities of Soviet-supplied military equipment, exposed the weaknesses of the overequipped, undertrained, and poorly motivated Libyan Army.

ACHIEVEMENTS

The Great Man-Made River (GMR) called by Qadhafi the world's eighth wonder, is in the third stage of completion as noted above. In addition to the network of pipes, each one 13 feet in diameter, excess water pumped will be stored in the Kufra basin, which has a capacity for 5000 cubic meters. The water flow to Tripoli and Benghazi is presently 200 million cubic feet per day, sufficient to meet the needs of residents of both cities.

In 1989, after admitting his mistake, Qadhafi signed a cease-fire with then-Chadian leader Hissène Habré and agreed to submit the dispute over ownership of Aouzou to the International Court of Justice (ICJ). The ICJ affirmed Chadian sovereignty in 1994 on the basis of a 1955 agreement arranged by France as the occupying power there. Libyan forces withdrew from Aouzou in May, and since then the two countries have enjoyed a peaceful relationship. In 1998, the border was opened completely, in line with Qadhafi's policy of "strengthening neighborly relations."

FOREIGN POLICY

Libya's relations with the United States have remained hostile since the 1969 Revolution, which not only overthrew King Idris but also resulted in the closing of the important Wheelus Field air base. Despite Qadhafi's efforts in more recent years to portray himself and Libya as respectable members of the world of nations, the country remains on the U.S. Department of State's list as one of the main sponsors of international terrorism. In 1986, U.S. war planes bombed Tripoli and Benghazi in retaliation for the bombing of a disco in Berlin, Germany, which killed two U.S. servicemen and injured 238 others. The retaliatory U.S. air attack on Libya killed 55 Libyan civilians, including Qadhafi's adopted daughter. After numerous delays and conflicting evidence about Libya's role in the Berlin bombing, a trial began in 1998 for four persons implicated in the attack. Only one, a diplomat in the embassy in East Berlin (now closed), was a Libyan national. The trial ended in 2001 with the conviction of the four; they were given 12- to 14-year sentences.

Libya resumed its old role of "pariah state" in 1992 by refusing to extradite two officers of its intelligence service suspected of complicity in the 1988 bombing of a Pan American jumbo jet over Lockerbie, Scotland. The United States, France, and Britain had demanded the officers' extradition and introduced a resolution to that effect in the UN Security Council; in the event of noncompliance on Libya's part, sanctions would be imposed on the country. *Resolution 748* passed by a 10-to-zero vote, with five abstentions. A concurrent ruling by the ICJ ordered Libya to turn over the suspects or explain in writing why it was not obligated to do so.

Qadhafi, however, refused to comply with *Resolution 748.* He argued that the suspects should be tried (if at all) in a neutral country, since they could not be given a fair trial either in Britain or the United States.

The Security Council responded by imposing partial sanctions on Libya. Despite the partial embargo, Libya's leader continued to reject compliance with the resolution. As a result, the Security Council in 1993 passed *Resolution 883,* imposing much stiffer sanctions on the country. The new sanctions banned all shipments of spare parts and equipment sales and froze Libyan foreign bank deposits. International flights to Libya were prohibited. The only area of the economy not affected was that of oil exports, since Britain and other Western European countries are dependent on low-sulfur Libyan crude for their economies.

Despite the sanctions and Libya's isolation, Qadhafi continued to refuse to surrender the two Lockerbie suspects. He maintained that they were innocent and could not receive a fair trial except in a neutral country under international law.

The tug-of-war between the United Nations and its recalcitrant member went on for six years. In March 1998, the United Nations, set a 60-day deadline for compliance. Subsequently Qadhafi reversed his stance on the Lockerbie suspects. While he insisted that the Libyan government was not involved, he agreed to turn over the suspects to be tried in a neutral court under Scottish law. The two were then flown to the Netherlands, where they were tried in a court set up in an abandoned Dutch air base, Camp Zeist. The trial was marked by intricate legal maneuverings and some questionable evidence. In 2000, one of the suspects was acquitted. The other, former Libyan intelligence agent Abdel Basset al-Megrahi, was found guilty and sentenced to life imprisonment. In August 2003 there was a major step toward resolution of the Lockerbie issue when Libya formally accepted the responsibility for the bombing in a letter to the UN Security Council. The acceptance would expedite a $2.7 billion settlement with the families of the 270 victims, with each family to receive between $5 and $10 million. Libya agreed to deposit the funds in an international bank. The end of sanctions enabled U.S. companies to reinstate old concessions and compete favorably with other national companies in bidding for new ones. Among the successful bidders were Occidental Oil, Marathon and Amerada Hess.

PROSPECTS

The tide of fundamentalism sweeping across the Islamic world and challenging secular regimes has largely spared Libya thus far, although there were occasional clashes between fundamentalists and police in the 1980s, and in 1992, some 500 fundamentalists were jailed briefly. However, the bloody civil uprisings against the regimes in neighboring Algeria and Egypt caused Qadhafi in 1994 to reemphasize Libya's Islamic nature. New laws passed by the General People's Congress would apply Islamic law (Shari'a) and punishments in such areas as marriage and divorce, wills and inheritance, crimes of theft and violence (where the Islamic punishment is cutting off a hand), and for apostasy. Libya's tribal-based society and Qadhafi's own interpretation of Islamic law to support women's rights and to deal with other social issues continue to serve as obstacles to Islamic fundamentalism.

Timeline: PAST

1835
Tripoli becomes an Ottoman province with the Sanusiya controlling the interior

1932
Libya becomes an Italian colony, Italy's "Fourth Shore"

1951
An independent kingdom is set up by the UN under King Idris

1969
The Revolution overthrows Idris; the Libyan Arab Republic is established

1973–1976
Qadhafi decrees a cultural and social revolution with government by people's committees

1980s
A campaign to eliminate Libyan opponents abroad; the United States imposes economic sanctions in response to suspected Libya-terrorist ties; U.S. planes attack targets in Tripoli and Benghazi; Libyan troops are driven from Chad, including the Aouzou Strip

1990s
Libya's relations with its neighbors improve; the UN votes to impose sanctions on Libya for terrorist acts; Qadhafi comes to an agreement with the UN regarding the trial of the PanAm/Lockerbie bombing suspects

PRESENT

2000s
Resolution of the Lockerbie issue ends UN and U.S. sanctions. increased oil production and wider distribution of revenues should improve per capita income.

Though no rules for succession to Qadhafi are in place, his son Saif al-Islam increasingly speaks for him at public and international meetings such as the 2005 Davos economic summit

On September 7, 1999, the Libyan leader celebrated his 30th year in power with a parade of thousands of footsoldiers, along with long-range missiles and tanks, through the streets of Tripoli. Libyan jets, many of them piloted by women, flew overhead.

At 60-plus the charismatic Libyan leader shows no sign of relinquishing power and seems in excellent health. In the absence of a formal succession process (Qadhafi has no official title), speculation centers on his oldest son, Muhammad Sayf al-Islam. However, a younger son, El-Saadi, represented Libya on an official visit to Japan in 2001.

The lifting of UN sanctions on the country resulted from Qadhafi's acceptance of international jurisdiction in the

Lockerbie case. As a result, relations with Europe have been normalized. The moribund tourist industry is also beginning to show signs of life, offering desert oases, splendid beaches and well-preserved Roman ruins to prospective visitors. In December 2003 Qadhafi again confounded his critics by agreeing to discontinue Libya's nuclear weapons development program and open its facilities to international inspection. The country also signed the Nuclear Non-Proliferation Treaty. In March 2004 the Libyan leader ordered 3,300 chemical bombs destroyed and agreed to halt further production.

Although he remains hostile to the U.S. and equally to Israel, Qadhafi has cultivated an image of respectability in recent years and has vigorously promoted African unity. During the July 2003 meeting of the Organization for African Unity, he described HIV/AIDS as a "peaceful virus". Along with malaria and sleeping sickness, he said it was God's way of keeping white colonizers out of Africa. Salvos such as this one have earned him considerable support from other African nations. In 2003, with African backing, Libya was elected to the chairmanship of the UN Commission on Human Rights.

NOTES

1. "[I]rrigation, colonization and hard work have wrought marvels. Everywhere you see plantations forced out of the sandy, wretched soil." A. H. Broderick, *North Africa* (London: Oxford University Press, 1943), p. 27.

2. Religious leaders issued a *fatwa* ("binding legal decision") stating that a vote against independence would be a vote against religion. Omar el Fathaly, et al., *Political Development and Bureaucracy in Libya* (Lexington, KY: Lexington Books, 1977).

3. See *Middle East Journal,* vol. 24, no. 2 (Spring 1970), Documents Section.

4. John Wright, *Libya: A Modern History* (Baltimore, MD: Johns Hopkins University Press, 1982), pp. 124–126. Qadhafi's idol was former Egyptian president Nasser, a leader in the movement for unity and freedom among the Arabs. While he was at school in Sebha, in the Fezzan, he listened to Radio Cairo's Voice of the Arabs and was later expelled from school as a militant organizer of demonstrations.

5. *The London Times* (June 6, 1973).

6. Khidr Hamza, with Jeff Stein, *Saddam's Bombmaker* (New York: Scribner's, 2000), p. 289. The author was head of the Iraqi nuclear-weapons program before defecting to Libya and eventually the United States.

7. Donald G. McNeil Jr., in *The New York Times* (February 1, 2001).

Morocco (Kingdom of Morocco)

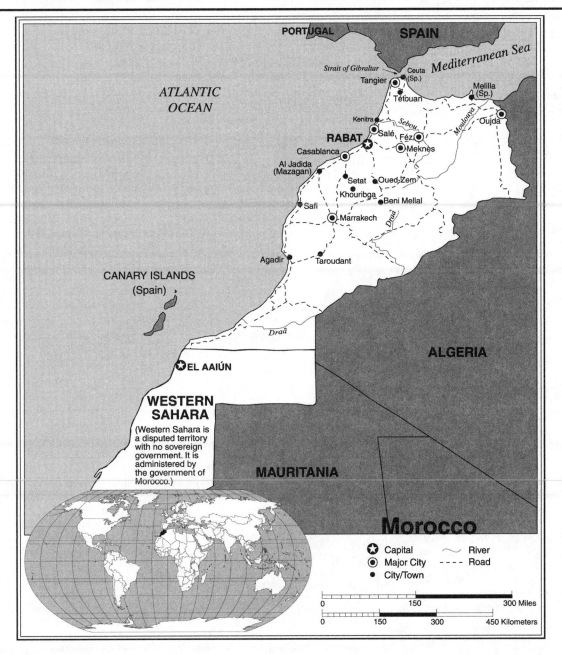

Morocco Statistics

GEOGRAPHY

Area in Square Miles (Kilometers):
274,400 (710,850) including the Western Sahara (102,675 square miles-266,000 sq.km.) about the size of California

Capital (Population): Rabat (1,293,000)

Environmental Concerns: land degradation; desertification; soil erosion; overgrazing; contamination of water supplies; oil pollution of coastal waters

Geographical Features: the northern coast and interior are mountainous, with large areas of bordering plateaux, intermontane valleys, and rich coastal plains; south, southeast and entire Western Sahara is desert

Climate: varies from Mediterranean to desert

PEOPLE

Population

Total: 32,725,847
Annual Growth Rate: 1.57%
Rural/Urban Population Ratio: 47/53

127

Major Languages: Arabic; Tama-zight; various Berber dialects; French
Ethnic Makeup: 64% Arab; 35% Berber; 1% non-Moroccan and Jewish
Religions: 99% Sunni Muslim; 1% Christian and Jewish

Health

Life Expectancy at Birth: 68 years (male); 73 years (female)
Infant Mortality Rate: 41.6/1,000 live births

Education

Adult Literacy Rate: 51.7%
Compulsory (Ages): 7–13

COMMUNICATION

Telephones: 1,515,000 main lines
Daily Newspaper Circulation: 13 per 1,000 people
Televisions: 93 per 1,000 people
Internet Service Providers: 8 (2000)

TRANSPORTATION

Highways in Miles (Kilometers): 37,649 (60,626)

Railroads in Miles (Kilometers): 1,184 (1,907)
Usable Airfields: 69
Motor Vehicles in Use: 1,278,000

GOVERNMENT

Type: constitutional monarchy
Independence Date: March 2, 1956 (from France)
Head of State/Government: King Muhammad VI; Prime Minister Driss Jettou
Political Parties: National Rally of Independents; Popular Movement; National Democratic Party; Constitutional Union; Socialist Union of Popular Forces; Istiqlal; Kutla Bloc; Party of Progress and Socialism; others
Suffrage: universal at 18

MILITARY

Military Expenditures (% of GDP): 5%
Current Disputes: final resolution on the status of Western Sahara remains to be worked out; from time to time Morocco demands the retrocession of Ceuta and

Melilla, cities located physically within its territory but considered extensions of mainland Spain (plazas de soberaniá by the Spanish government)

ECONOMY

Currency ($U.S. Equivalent): 8.868 dirhams = $1
Per Capita Income/GDP: $4,200/$134.6 billion
GDP Growth Rate: 4.4%
Inflation Rate: 2%
Unemployment Rate: 12%
Labor Force: 11,000,000
Natural Resources: phosphates; iron ore; manganese; lead; zinc; fish; salt
Agriculture: barley; wheat; citrus fruits; wine; vegetables; olives; livestock
Industry: phosphate mining and processing; food processing; leather goods; textiles; construction; tourism
Exports: $9.754 billion (primary partners France, Spain, United Kingdom)
Imports: $15.6 billion (primary partners France, Spain, Germany)

Morocco Country Report

The Kingdom of Morocco is the westernmost country in North Africa. Morocco's population is the second largest (after Egypt) of the Arab states. The country's territory includes the Western Sahara (a claim made under dispute), formerly two Spanish colonies, Rio de Oro and Saguia al-Hamra. Morocco annexed part in 1976 and the balance in 1978, after Mauritania's withdrawal from its share, as decided in an agreement with Spain. Since then, Morocco has incorporated the Western Sahara into the kingdom as its newest province.

Two other territories physically within Morocco remain outside Moroccan control. They are the cities of Ceuta and Melilla, both located on rocky peninsulas that jut out into the Mediterranean Sea. They have been held by Spain since the fifteenth century. (Spain also owns several small islands off the coast in Moroccan territorial waters in the Mediterranean.) Spain's support for Morocco's admission to the European Union (EU) as an associate member has eased tensions between them over the enclaves. An additional economic advantage to Morocco is that each day some 40,000 Moroccans cross legally into them for work.

In 1986 a Spanish law excluding Moroccan Muslim residents from Spanish citizenship led to protests among them. The

Moroccan government did not pursue the protests, and in 1988 the question of citizenship became moot when the Spanish Cortes (Parliament) passed a law formally incorporating Ceuta and Melilla into Spain as overseas territories.

DEVELOPMENT

Morocco has important reserves of phosphate rock, particularly in the Western Sahara. It also has exportable supplies of certain rare metal, such as antimony. Unfortunately it lacks oil resources. An oil strike in the Sahara in 2000 proved abortive. Abundant rainfall has improved agricultural production; GDP growth presently averages 5 percent annually.

In the last couple of years the two cities have been all but overwhelmed by migrants, most of them undocumented, from sub-Saharan Africa seeking to cross into Europe, where they hope to find jobs and a better life than that available in their own countries of origin. Some 15,000 attempted the hazardous crossing in 2004 as compared with 350 in all in the previous 6-year period. A mass breakthrough in October, which resulted in many casualties as the migrants scaled the razor-wire fences around the enclaves, led the Spanish gov-

ernment to send military units to guard the borders. On its side, the Moroccan government stepped up efforts to block illegal migration through its territory and break up the criminal networks bringing the migrants northward. Some 1,000 of them were captured and interned in the desert town of Bou Arfa, near Goulimine, prior to their deportation. The UN and human rights groups, notably Doctors Without Borders, criticized the deportations. In a response in October 2005, large-scale demonstrations took place in Rabat, with marchers carrying signs that read "we are all Africans; Morocco cannot become Europe's immigration policeman."[1]

Although Morocco is considerably better off than many of its African neighbors who provide the bulk of would-be immigrants, its growing youth population and lack of jobs continue to encourage Moroccans particularly in the lower age groups to attempt the often risky crossing into Europe in search of employment. Thus in December 2005 Italy reported a 15 percent increase in illegal immigrants; a third of them were Moroccans. This was due largely to the closure of Ceuta and Melilla. As an example of this economic pressure, a coal mine in the rural east near Jerada, operated by a Russian company, was closed

Hamilton Wright/Government of Morocco (GM001)
Morocco has a rich history. The Karawiyyin Mosque at Fez was founded in the ninth century A.D. and is the largest mosque in North Africa. It is also the seat of one of Africa's oldest universities.

abruptly in the late 1990s as the company shifted its operations to Poland and the Ukraine. As a result the community found itself with no jobs and no prospects.

Morocco is a rugged land, dominated by several massive mountain ranges. The Rif Range, averaging 7,000 feet in elevation, runs parallel to the Mediterranean, isolating the northern region from the rest of the country. The Atlas Mountains dominate the interior of Morocco. The Middle Atlas begins south of the Rif, separated by the Taza Gap (the traditional gateway for invaders from the east), and extends from northeast to southwest to join the High Atlas, a snow capped range containing North Africa's highest peak. A third range, the Anti-Atlas, walls off the desert from the rest of Morocco. These ranges and the general inaccessibility of the country have isolated Morocco throughout most of its history, not only from outside invaders but internally as well, because of the geographical separation of peoples.

Moroccan geography explains the country's dual population structure. About 35 percent of the population are Berbers, descendants of the original North Africans. The Berbers were, until recently, grouped into tribes, often taking the name of a common ancestor, such as the Ait ("Sons of") 'Atta of southern Morocco.[2] Invading Arabs converted them to Islam in the eighth century but made few other changes in Berber life. Unlike the Berbers, the majority of the Arabs who settled in Morocco were, and are, town-dwellers. The Berbers, more than the Arabs, derived unity and support from their extended families rather than from state control, whether real or putative.

The fact that the Arabs were invaders caused the majority of the Berbers to withdraw into mountain areas. They accepted Islam but held stubbornly to their basic independence. Much of Morocco's past history consisted of efforts by various rulers, both Berber and Arab, to control Berber territory. The result was a kind of balance-of-power political system. The rulers had their power bases in the cities, while the rural groups operated as independent units. Moroccan rulers made periodic military expeditions into Berber territory to collect tribute and if possible to secure full obedience from the Berbers. When the ruler was strong, the Berbers paid up and submitted; when he was weak, they ignored him. At times Berber leaders might invade "government territory," capturing cities and replacing one ruler with another more to their liking. When they were not fighting with urban rulers, different Berber groups fought among themselves, so the system did little to foster Moroccan national unity.

HISTORY

Morocco has a rich cultural history, with many of its ancient monuments more or less intact. It has been governed by some form of monarchy for over a thousand years, although royal authority was frequently limited or contested by rivals. The current ruling dynasty, the Alawis, assumed power in the 1600s. One reason for their long rule is the fact that they descend from the Prophet Muhammad. Thus, Moroccans have had a real sense of Islamic traditions and history through their rulers.

The first identifiable Moroccan "state" was established by a descendant of Muhammad named Idris, in the late eighth century. Idris had taken refuge in the far west of the Islamic world to escape civil war in the east. Because of his piety, learning, and descent from Muhammad, he was accepted by a number of Berber groups as their spiritual and political leader. His son and successor, Idris II, founded the first Moroccan capital, Fez. Father and son established the principle whereby descent from the Prophet was an important qualification for political power as well as social status in Morocco.

The Idrisids ruled over only a small portion of the current Moroccan territory, and, after the death of Idris II, their "nation" lapsed into decentralized family rule. In any case, the Berbers had no real idea of nationhood; each separate Berber group thought of itself as a nation. But in the eleventh and twelfth centuries, two Berber confederations developed that brought imperial grandeur to Morocco. These were the Almoravids and the Almohads. Under their rule, North Africa developed a political structure separate from that of the eastern Islamic world, one strongly influenced by Berber values.

The Almoravids began as camel-riding nomads from the Western Sahara who were inspired by a religious teacher to carry out a reform movement to revive the true faith of Islam. (The word *Almoravid* comes from the Arabic *al-Murabitun,* "men of the ribat," rather like the crusading religious orders of Christianity in the Middle Ages.) Fired by religious zeal, the Almoravids conquered all of Morocco and parts of western Algeria.

A second "imperial" dynasty, the Almohads, succeeded the Almoravids but improved on their performance. They were the first, and probably the last, to unite all of North Africa and Islamic Spain under one government. Almohad monuments, such as the Qutubiya tower, the best-known landmark of Marrakesh, and the Tower of Hassan in Rabat, still stand as reminders of their power and the high level of the Almohads' architectural achievements.

The same fragmentation, conflicts, and Berber/Arab rivalries that had undermined their predecessors brought down the Almohads in the late thirteenth century. From then on, dynasty succeeded dynasty in power. An interesting point about this cyclical pattern is that despite the lack of political unity, a distinctive Moroccan style and culture developed. Each dynasty contributed something to this culture, in architecture, crafts, literature, and music. The interchange between Morocco and Islamic Spain was constant and fruitful. Poets, musicians, artisans, architects, and others traveled regularly between Spanish and Moroccan cities. One can visit the city of Fez today and be instantly transported back into the Hispano-Moorish way of life of the Middle Ages.

Mulay Ismail

The Alawis came to power and established their rule partly by force, but also as a result of their descent from the Prophet Muhammad. This link enabled them to win the support of both Arab and Berber populations. The real founder of the dynasty was Mulay Ismail, one of the longest-reigning and most powerful monarchs in Morocco's history.

Mulay Ismail unified the Moroccan nation. The great majority of the Berber groups accepted him as their sovereign. The sultan built watchtowers and posted permanent garrisons in Berber territories to make sure they continued to do so. He brought public security to Morocco also; it was said that in his time, a Jew or an unveiled woman could travel safely anywhere in the land, which was not the case in most parts of North Africa, the Middle East, and Europe.

Mulay Ismail was a contemporary of Louis XIV, and the reports of his envoys to the French court at Versailles convinced him that he should build a capital like it. He chose Meknes, not far from Fez. The work was half finished when he died of old age. The slaves and prisoners working on this "Moroccan Versailles" threw down their shovels and ran away. The enormous unfinished walls and arched Bab al-Mansur ("Gate of the Victorious") still stand today as reminders of Mulay Ismail's dream.

Mulay Ismail had many wives and left behind 500 sons but no instructions as to which should succeed him. After years of conflict, one of his grandsons took the throne as Muhammad II. He is important for giving European merchants a monopoly on trade from Moroccan ports (in wool, wax, hides, carpets, and leather) and for being the first non-European monarch to recognize the United States as an independent nation, in 1787.[3]

The French Protectorate

In the 1800s and early 1900s, Morocco became increasingly vulnerable to outside pressures. The French, who were established in neighboring Algeria and Tunisia, wanted to complete their conquests. The nineteenth-century sultans were less and less able to control the mountain Berbers and were forced to make constant expeditions into the "land of dissidence," at great expense to the treasury. They began borrowing money from European bankers, not only to pay their bills but also to finance arms purchases and the development of ports, railroads, and industries to create a modern economy and prove to the European powers that Morocco could manage its own affairs. Nothing worked; by 1900, Morocco was so far in debt that the French took over the management of its finances. (One sultan, Abd al-Aziz, had bought one of everything he was told about by European salesmen, including a gold-plated toy train that carried food from the kitchen to the dining room of his palace.) Meanwhile, the European powers plotted the country's downfall.

FREEDOM

The "freest and fairest" elections in Moroccan history established a new 325-seat Chamber of Representatives equally balanced among the leading political parties. Some 35 women candidates were elected. King Muhammad VI's "National Action Plan" raises the legal marriageable age for women to 18 and gives other rights to them. In 2003 the king proposed revisions to the 1957 *Mudawanna* (Family Law) which were approved by the Chamber. They now have the right to file for divorce, share equally in family property, and travel without prior consent from male family members.

In 1904, France, Britain, Spain, and Germany signed secret agreements partitioning the country. The French would be given the largest part of the country, while Spain would receive the northern third as a protectorate plus some territory in the Western Sahara. In return, the French and Spanish agreed to respect Britain's claim to Egypt and Germany's claim to East African territory.

The ax fell on Morocco in 1912. French workers building the new port of Casablanca were killed by Berbers. Mobs attacked foreigners in Fez, and the sultan's troops could not control them. French troops marched to Fez from Algeria to restore order. The sultan, Mulay Hafidh (Hafiz), was forced to sign the Treaty of Fez, establishing a French protectorate over southern Morocco. The sultan believed that he had betrayed his country and died shortly thereafter, supposedly of a broken heart. Spain then occupied the northern third of the country, and Tangier, the traditional residence of foreign consuls, became an international city ruled by several European powers.

The French protectorate over Morocco covered barely 45 years (1912–1956). But in that brief period, the French introduced significant changes into Moroccan life. For the first time, southern Morocco was brought entirely under central government control, although the "pacification" of the Berbers was not complete until 1934. French troops also intervened in the Spanish Zone to help put down a rebellion in the Rif led by Abd al-Krim, a *Qadi* ("religious judge") and leader of the powerful Ait Waryaghar tribe.[4]

The organization of the protectorate was largely the work of the first French resident-general, Marshal Louis Lyautey. Lyautey had great respect for Morocco's past and its dignified people. His goal was to develop the country and modernize the sultan's government while preserving Moroccan traditions and culture. He preferred the Berbers to the Arabs and set up a separate administration under Berber-speaking French officers for Berber areas.[5]

Lyautey's successors were less respectful of Moroccan traditions. The sultan, supposedly an independent ruler, became a figurehead. French *colons* (settlers) flocked to Morocco to buy land at rock-bottom prices and develop vineyards, citrus groves, and orchards. Modern cities sprang up around the perimeters of Rabat, Fez, Marrakesh, and other cities. In rural areas, particularly in the Atlas Mountains, the French worked with powerful local chiefs (*qaids*). Certain qaids used the arrangement to become enormously wealthy. One qaid, al-Glawi, as he was called, strutted about like a rooster in his territory and often said that he was the real sultan of Morocco.[6]

Morocco's Independence Struggle

The movement for independence in Morocco developed slowly. The only symbol of national unity was the sultan, Muhammad ibn Yusuf. But he seemed ineffectual to most young Moroccans, particularly those educated in French schools, who began to question the right of France to rule a people against their will.

The hopes of these young Moroccans got a boost during World War II. The Western Allies, Great Britain and the United States, had gone on record in favor of the right of subject peoples to self-determination after the war. When U.S. president Franklin D. Roosevelt and British prime minister Winston Churchill came to Casablanca for an important wartime conference, the sultan was convinced to meet them privately and get a commitment for Morocco's independence. The leaders promised their support.

However, Roosevelt died before the end of the war, and Churchill was defeated for reelection. The French were not under any pressure after the war to end the protectorate. When a group of Moroccan nationalists formed the Istiqlal ("Independence") Party and demanded the end of French rule, most of them were arrested. A few leaders escaped to the Spanish Zone or to Tangier, where they could operate freely. For several years, Istiqlal headquarters was the home of the principal of the American School at Tangier, an ardent supporter of Moroccan nationalism.

With the Istiqlal dispersed, the sultan represented the last hope for national unity and resistance. Until then, he had gone along with the French; but in the early 1950s, he began to oppose them openly. The French began to look for a way to remove him from office and install a more cooperative ruler.

In 1953, the Glawi and his fellow qaids decided, along with the French, that the time was right to depose the sultan. The qaids demanded that he abdicate; they said that his presence was contributing to Moroccan instability. When he refused, he was bundled into a French plane and sent into exile. An elderly uncle was named to replace him.

The sultan's departure had the opposite effect from what was intended. In exile, he became a symbol for Moroccan resistance to the protectorate. Violence broke out, French settlers were murdered, and a Moroccan Army of Liberation began battling French troops in rural regions. Although the French could probably have contained the rebellion in Morocco, they were under great pressure in neighboring Algeria and Tunisia, where resistance movements were also under way. In 1955, the French abruptly capitulated. Sultan Muhammad ibn Yusuf returned to his palace in Rabat in triumph, and the elderly uncle retired to potter about his garden in Tangier.

INDEPENDENCE

Morocco became independent on March 2, 1956. (The Spanish protectorate ended in April, and Tangier came under Moroccan control in October, although it kept its free-port status and special banking and currency privileges for several more years.) It began its existence as a sovereign state with a number of assets—a popular ruler, an established government, and a well-developed system of roads, schools, hospitals, and industries inherited from the protectorate. Against these assets were the liabilities of age-old Arab-Berber and inter-Berber conflicts, little experience with political parties or democratic institutions, and an economy dominated by Europeans.

The sultan's goal was to establish a constitutional monarchy. His first action was to give himself a new title, King Muhammad V, symbolizing the end of the old autocratic rule of his predecessors. He also pardoned the Glawi, who crawled into his presence to kiss his feet and crawled out backwards as proof of penitence. (He died soon thereafter.) However, the power of the qaids and pashas ended; "they were compromised by their association with the French, and returned to the land to make way for nationalist cadres, many…not from the regions they were assigned to administer."[7]

Muhammad V did not live long enough to reach his goal. He died unexpectedly in 1961 and was succeeded by his eldest son, Crown Prince Hassan. Hassan II ruled until his death in 1999. While he fulfilled his father's promise immediately with a Constitution, in most other ways Hassan II set his own stamp on Morocco.

The Constitution provided for an elected Legislature and a multiparty political system. In addition to the Istiqlal, a number of other parties were organized, including one representing the monarchy. But the results of the French failure to develop a satisfactory party system soon became apparent. Berber-Arab friction, urban-rural distrust, city rivalries, and inter-Berber hostility all intensified. Elections failed to produce a clear majority for any party, not even the king's.

In 1965, riots broke out in Casablanca. The immediate cause was labor unrest, but the real reason lay in the lack of effective leadership by the parties. The king declared a state of emergency, dismissed the Legislature, and assumed full powers under the Constitution.

For the next dozen years, Hassan II ruled as an absolute monarch. He continued to insist that his goal was a parliamentary system, a "government of national union." But he depended on a small group of cronies, members of prominent merchant families, the large Alawi family, or powerful Berber leaders as a more reliable group than the fractious political parties. The dominance of "the king's men" led to growing dissatisfaction and the perception that the king had sold out to special interests. Gradually, unrest spread to the army, previously loyal to its commander-in-chief. In 1971, during a diplomatic reception, cadets from the main military academy invaded the royal palace near Rabat. A number of foreign diplomats were killed and the king held prisoner briefly before loyal troops could crush the rebellion. The next year, a plot by air-force pilots to shoot down the king's plane was narrowly averted. The two escapes helped confirm in Hassan's mind his invincibility under the protection of Allah.

But they also prompted him to reinstate the parliamentary system. A new Constitution issued in 1972 defined Morocco "as a democratic and social constitutional monarchy in which Islam is the established religion."[8] However, the king retained the constitutional powers that, along with those derived from his spiritual role as "Commander of the Faithful" and lineal descendant of Muhammad, undergirded his authority.

HEALTH/WELFARE

In October 2000, the International Labor Organization (ILO) ranked Morocco as the 3rd-highest country in the world, after China and India, in the exploitation of child labor. Moroccan children as young as 5, all girls, are employed in the carpet industry, working up to 10 hours per day weaving the carpets that are at present Morocco's major source of foreign currency. In recent years, more and more Moroccan minors have been leaving their families and migrating illegally to Europe through the Spanish port city of Ceuta. Some 3,500 did so in 2002.

INTERNAL POLITICS

Morocco's de facto annexation of the Western Sahara has important implications for future national development due to the territory's size, underpopulation, and mineral resources, particularly shale oil and phosphates. But the annexation has been equally important to national pride and political unity. The "Green March" of 350,000 unarmed Moroccans into Spanish territory in 1975 to dramatize Morocco's claim was organized by the king and supported by all segments of the population and the opposition parties. In 1977, opposition leaders agreed to serve under the king in a "government of national union." The first elections in 12 years were held for a new Legislature; several new parties took part.

The 1984 elections continued the national unity process. The pro-monarchist Constitutional Union (CU) party won a majority of seats in the Chamber of Representatives (Parliament). A new party, the National Rally of Independents (RNI),

Hamilton Wright/Government of Morocco (GM002)

Tangier was once a free city and port. Just across the Strait of Gibraltar from Spain, it now is Morocco's northern metropolis. Modernization and expansion of port facilities to accommodate large cruise ships and tankers got under way in 1999.

formed by members with no party affiliations, emerged as the chief rival to the CU.

New elections were scheduled for 1989 but were postponed three times; the king said that extra time was needed for the economic-stabilization program to show results and generate public confidence. The elections finally took place in two stages in 1993: the first for election of party candidates, and the second for trade-union and professional-association candidates. The final tally showed 195 seats for center-right (royalist) candidates, to 120 for the Democratic-bloc opposition. As a result, coalition government became necessary. The two leading opposition parties, however—the Socialist Union of Popular Forces (USFP) and the Istiqlal—refused to participate, claiming election irregularities. Opposition from members of these parties plus the Kutla Bloc, a new party formed from the merger of several minor parties, blocked legislative action until 1994. At that point, the entire opposition bloc walked out of the Legislature and announced a boycott of the government.

King Hassan resolved the crisis by appointing then-USFP leader Abdellatif Filali as the new prime minister, thus bringing the opposition into the government. The king continued with this method of political reconciliation by appointing the new head of the USFP, Abderrahmane Youssoufi, to the position after the latter's return from political exile in 1998.

A referendum in 1996 approved several amendments to the constitution. One in particular replaced the unicameral legislature by a bicameral one. The Chamber of Representatives (lower house) is to be elected directly, for 5-year terms. The chamber of Counselors (upper house) is to be two-thirds elected and one-third appointed. In September 2002 elections were held for the 325-seat lower house. Some 26 parties, a dozen of them brand new, presented candidates. Described by the government as the first free and fair election in national history, they resulted in much higher turnout than the 58 percent of the vote-buying, tainted 1997 election.

The election results underlined the broad spectrum of Moroccan politics. The Socialist Union of Popular Forces (USFP), headed by then Prime Minister Youssoufi, won 50 seats, the venerable Istiqlal Party 48, the National Rally of Independents 41, the National Popular Movement 27. The Party of Justice and Development, which had replaced the banned Islamist Justice and Development, won 42 seats; its predecessor had held only 18 in the outgoing Chamber. Also noteworthy was the election of 35 women; a quota of 30 had been reserved for them.

FOREIGN RELATIONS

During his long reign, King Hassan II served effectively in mediating the long-running Arab-Israeli conflict. He took an active part in the negotiations for the 1979 Egyptian-Israeli peace treaty and for the treaty between Israel and Jordan in 1994. For these services he came to be viewed by the United States and by European powers as an impartial mediator. However, his absolute rule and suppression of human rights at home caused difficulties with Europe. The European Union suspended $145 million in aid in 1992; it was restored only after Hassan had released long-time political prisoners and pardoned 150 alleged Islamic militants. In 1995, Morocco became the second African country, after Tunisia, to be granted associate status in the EU.

Thus far, Morocco's only venture in "imperial politics" has been in the Western Sahara. This California-size desert territory, formerly a Spanish protectorate and then a colony after 1912, was never a part of the modern Moroccan state. Its only connection is historical—it was the headquarters and starting point for the Al-moravid dynasty, camel-riding nomads who ruled western North Africa and southern Spain in the eleventh century. But the presence of so much empty land, along with millions of tons of phosphate rock and potential oil fields, encouraged the king to "play international politics" in order to secure the territory. In October 1975 the king led a "Green March" of half a million Moroccans armed only with Korans into the Spanish Sahara "to recover sacred Moroccan territory. As a result, Spain agreed in

1976 to cede it jointly to Morocco and Mauritania, in two zones, one-third to Mauritania and two-thirds to Morocco. After the overthrow of the Mauritanian government by a military coup in 1978, its new leaders turned over its zone to Morocco.

Since then Morocco's control has been challenged by Polisario, an acronym for the military wing of the Saharan nationalist movement. The latter's goal is a sovereign state, the Sahrawi Arab Democratic Republic (S.A.D.R.) It has been recognized as such by a number of African countries. However the territory remains under Moroccan control; it has been defined as the country's newest province. As a frontier province, over the past three decades thousands of Moroccan settlers have moved there, encouraged by free land, farm equipment, subsidized housing and other inducements.

Conflict between the Moroccan army and Polisario forces operating from Algerian bases with Algerian support for their cause continued throughout the 1980s. It eventually led the army to construct a 350-mile "Sand Wall" from the Atlantic Ocean around the province's land border. Meanwhile some 140,000 Sahrawis became refugees, crowded into four small camps in southern Algeria.

The dispute eventually came before the UN in 1991. In 1995 the Security Council, prodded by Algeria, approved *Resolution 995*. It called for voter registration for a referendum to determine the future of the territory. The two parties agreed to a cease-fire, and a UN Observer Force, MINURSO, was established to monitor the cease-fire and supervise voter registration. Due largely to Morocco's intransigence, the referendum has yet to be held, despite efforts at mediation by ex-U.S. Secretary of State James Baker and others.

The "glacial chill" between Morocco and the Polisario thawed a bit in 2005, when the latter released 404 Moroccan prisoners in a "humanitarian gesture." But with the demographic balance tipped in its favor by large-scale Moroccan settlement, it seemed that nothing short of strong international pressure would move the country toward acceptance of *Resolution 995*.

A decade later, the referendum seems less and less likely to be held. King Hassan II unilaterally named the territory Morocco's newest province, and by 2001 Moroccan settlers formed a majority in the population of 244,593. In December 2001, French president Jacques Chirac made an official visit to Morocco and saluted the country for the development of its "southern provinces." Earlier, the United Nations had appointed former U.S. secretary of state James Baker as mediator between Morocco and the Polisario. After several failed attempts at mediation, Baker submitted a plan for postponement of the referendum until 2006. In the interim, the Sahrawis would elect an autonomous governing body, with its powers limited to local and provincial affairs. The voting list would include all residents. The Security Council approved the Baker plan. But in view of the extensive Moroccanization of the territory, its self-government under Sahrawi leadership remained highly unlikely. King Muhammad made his first visit there in 2002, receiving a thunderous welcome from the Moroccan settler population. Emphasizing its integration into the kingdom as its newest province, he approved offshore oil-exploration concessions and released 56 Sahrawi political prisoners as a "sign of affection for the sons of the Sahara."

THE ECONOMY

Morocco has many of certain resources but too little of other, critical ones. It has two thirds of the world's known reserves of phosphate rock and is the top exporter of phosphates. The major thrust in industrial development is in phosphate-related industries. Access to deposits was one reason for Morocco's annexation of the Western Sahara, although to date there has been little extraction there due to the political conflict. The downturn in demand and falling prices in the global phosphates market brought on a debt crisis in the late 1980s. Increased phosphate demand globally and improved crop production following the end of several drought years have strengthened the economy. Privatization of the government-owned tobacco monopoly, the first industry to be so affected, generated a budgetary surplus in 2002.

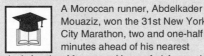

ACHIEVEMENTS

A Moroccan runner, Abdelkader Mouaziz, won the 31st New York City Marathon, two and one-half minutes ahead of his nearest rival. Another Moroccan, Youssef el Aynaoui, had become one of the world's premier tennis players and competed well in major tournaments before his retirement.

The country also has important but undeveloped iron-ore deposits and a small but significant production of rare metals such as mercury, antimony, nickel, and lead. In the past, a major obstacle to development was the lack of oil resources. Prospects for oil improved in 2000 when the U.S. oil company Skidmore Energy was thought to have struck oil near Talsinnt, in the eastern Sahara. But the find, which the king had declared to be God's gift to Morocco, turned out to be mud. In 2001, the French oil company TotalFinaElf and Kerr-McGee of Texas were granted parallel concessions of 44,000 square miles offshore in Western Saharan waters near Dakhla.

Although recurring droughts have hampered improvement of the agricultural sector, it still accounts for 20 percent of gross domestic product and employs 50 percent of the labor force. Production varies widely from year to year, due to fluctuating rainfall. Abundant rains in 2001 resulted in bumper crops and a 25 percent increase in agricultural output, with 8 percent growth in GDP for that year.

The fisheries sector is equally important to the economy, with 2,175 miles of coastline and half a million square miles of territorial waters to draw from. Fisheries account for 16 percent of exports; annual production is approximately 1 million tons. The agreement with the European Union for associate status has been very beneficial to the industry. Morocco received $500 million in 1999–2001 from European countries in return for fishing rights for their vessels in Moroccan territorial waters.

But the economic outlook and social prospects remain bleak for most people. Although the birth rate has been sharply reduced, job prospects are limited for the large number of young Moroccans entering the labor force each year. The "suicide bombers," who attacked a Jewish community center, a hotel, foreign consulates and other structures in Casablanca in May 2003, killing some 41 persons, were said to belong to the radical Islamist organization al-Sirat al-Mustakim (Righteous Path), believed to be linked with al Qaeda. However, the fact that they came mostly from the impoverished Thomasville slum area of the city suggests that they acted not out of a desire to overthrow the monarchy, but out of frustration with the problems that face Morocco's youth today, namely unemployment, poverty and lack of opportunities in the workplace. As one observer noted of those arrested (only 12 were suicide bombers), little distinguished them from the group of young men idling in the streets or hawking designer sunglasses at intersections around town.

PROSPECTS

King Hassan II died in July 1999. The monarch had ruled his country for 38 years—the second-longest reign in the Middle East. Like King Hussein of Jordan, Hassan became identified with his country to such a degree that "Hassan was Morocco, and Morocco was Hassan." But unlike Jordan's ruler, Hassan combined religious with secular authority. Among his many titles was that of "Commander of the Faithful," and the affection felt for him by most Moroccans, particularly women and youth, was amply visible during his state funeral. His frequent reminders to the nation in speeches and broadcasts that "I am the person entrusted by God to lead you" clearly identified him in the public mind not only as their religious leader but also as head of the family.

The king's eldest son, Crown Prince Muhammad, succeeded him without incident as Muhammad VI. Morocco's new ruler began his reign with public commitments to reform human-rights protections and an effort to atone for some aspects of Hassan's autocratic rule. One of his first actions was the dismissal of Interior Minister Driss Basri, the acknowledged "power behind the throne." He had been considered largely responsible for the "Years of Lead" during Hassan's rule, when human rights were routinely violated and opposition political leaders jailed on various pretexts by the police, army and security services.

Muhammad VI also publicly admitted the existence of the Tazmamat "death camp" and other camps in the Sahara, where rebel army officers and political prisoners were held, often for years and without trial or access to their families. (The family of General Oufkir, leader of the 1972 attempted coup who was later executed, were among those held, but they managed to escape.)[8] The new king also committed $3.8 million in compensation to the families of those who had been imprisoned.

The rise of Islamic fundamentalism as a political force is an obstacle to Muhammad VI's vision of Morocco. Hassan II kept fundamentalists on a tight rein. He suppressed the main Islamist movement, Adil wa Ihsan ("Justice and Charity"), and sent its leader to a mental institution. In 2004 the king established the Equity and Reconciliation Commission, headed by former Marxist leader Driss Benzekri, to investigate rights violations during the Years of Lead. By April 2005, when it was disbanded, the Commission had investigated 22,000 cases and compensated the families involved on behalf of the victims. Unfortunately because it had been established as only an investigative body, it could not prosecute defendants. Any such action would have to be taken through the judiciary.

Mohammed VI also issued in 2000 a "National Action Plan" which included women's rights, a free press, and other elements lacking in the country's social structure. But actions attributed to Islamic fundamentalists, such as the Casablanca bombings, provoked a heavy reaction par-

ticularly in the security services. Independent publications such as *Le Journal*, the most popular French-language magazine, and newspapers were closed. The head of the Moroccan Association for Human Rights was arrested and beaten, and riot police broke up demonstrations protesting the restrictions, arresting 800 persons. A police officer noted that "we don't want the chaos of a second intifada," a reference to the current Palestinian uprising against Israel.

The king's major social reform effort to date is the Family Code of Laws, approved by the Chamber of Representatives in 2003. Among other provisions it makes wives equal to their husbands in ownership of property and allows them to initiate divorce proceedings. In a separate action, Muhammad VI appointed the first female Royal Counselor, and reserved 30 seats for women in the Chamber.

Reform in Morocco continues to face numerous obstacles, nonetheless. A 2005 Report by Reporters Without Borders indicated that 80 percent of the country's journalists did not feel free to write about many issues, despite the protection presumably afforded by the Action Plan. One reporter noted in the Report that "while the practice of freedom is clear, the legal guarantee is not there. I would not necessarily say that we are in a process of democratization. I would say we are in a process."[9]

Partly to ward off the fundamentalist threat, but also to placate Berber leaders (who have long felt excluded from the political process and marginalized as a culture group), the king in his July 2001 Speech from the Throne announced the formation of a Royal Institute for the Preservation of Berber Culture. He also directed the Ministry of Education to incorporate Tamazight, the principal Berber language, into the national educational curriculum.

NOTES

1. Sue Miller, "Migration Station", *Christian Science Monitor*, June 26, 2003.
2. See David M. Hart, *Dadda 'Atta and His Forty Grandsons* (Cambridge, England: Menas Press, 1981), pp. 8–11. Dadda 'Atta was a historical figure, a minor saint or marabout.
3. Harold D. Nelson, ed., *Morocco, A Country Study* (Washington, D.C.: American University, Foreign Area Studies, 1978), p. 112.
4. The oldest property owned by the U.S. government abroad is the American Consulate in Tangier; a consul was assigned there in 1791.
5. See David Woolman, *Rebels in the Rif: Abd 'al Krim and the Rif Rebellion* (Palo Alto, CA: Stanford University Press, 1968). On the Ait Waryaghar, see David M. Hart, *The*

Ait Waryaghar of the Moroccan Rif: An Ethnography and a History (Tucson, AZ: University of Arizona Press, 1976). Abd 'al Krim had annihilated a Spanish army and set up a Republic of the Rif (1921–1926).
6. For a detailed description of protectorate tribal administration, see Robin Bidwell, *Morocco Under Colonial Rule* (London: Frank Cass, 1973).
7. Mark Tessler, "Morocco: Institutional Pluralism and Monarchical Dominance," in W. I. Zartman, ed., *Political Elites in North Africa* (New York: Longman, 1982), p. 44.
8. See Malika Oufkir, with Michele Fitoussi, *Stolen Lives: Twenty Years in a Desert Jail* (New York: Hyperion Books, 1999). Another prisoner, Ahmed Marzouki, recently published his memoir of life there. Entitled *Cell 10*, it has sold widely in Morocco.
9. Geoff Pingree and Lisa Abend, "Morocco moves gradually to address past repression." *The Christian Science Monitor*, September 23, 2005.

Timeline: PAST

788–790
The foundations of the Moroccan nation are established by Idris I and II, with the capital at Fez

1062–1147
The Almoravid and Almohad dynasties, Morocco's "imperial period"

1672
The current ruling dynasty, the Alawi, establishes its authority under Mulay Ismail

1912
Morocco is occupied and placed under French and Spanish protectorates

1956
Independence under King Muhammad V

1961
The accession of King Hassan II

1975
The Green March into the Western Sahara dramatizes Morocco's claim to the area

1980s
Bread riots; agreement with Libya for a federal union; the king unilaterally abrogates the 1984 treaty of union with Libya

1990s
Elections establish parliamentary government; King Hassan dies and is succeeded by King Muhammad VI

PRESENT

2000s
King Muhammad VI works to improve human rights
The economic picture brightens

Oman (Sultanate of Oman)

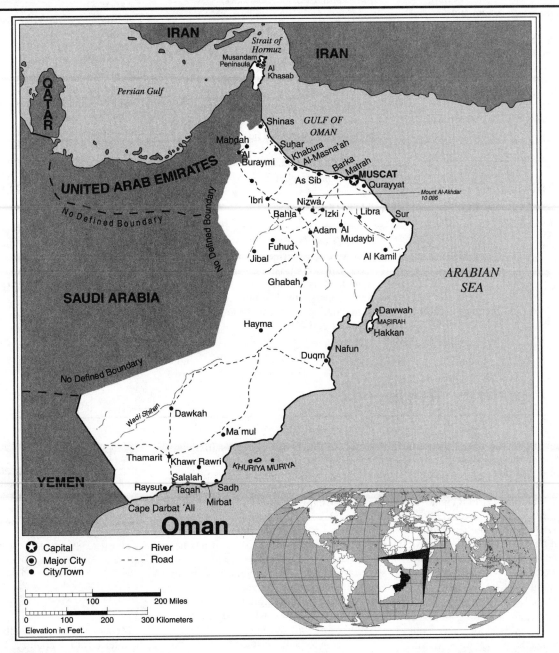

Oman Statistics

GEOGRAPHY

Area in Square Miles (Kilometers):

82,009 (212,460) (about the size of Kansas)

Capital (Population): Muscat (400,000)

Environmental Concerns: rising soil salinity; beach pollution from oil spills; very limited freshwater

Geographical Features: central desert plain; rugged mountains in the north and south

Climate: coast, hot and humid; interior, hot and dry

PEOPLE

Population

Total: 3,001,583 (includes 577,293 non-nationals)

Annual Growth Rate: 3.32%

Rural/Urban Population Ratio: 22/78

135

Major Languages: Arabic; English;
various South Asian languages
Ethnic Makeup: almost entirely Arab;
small Baluchi, South Asian, and African
groups
Religions: 75% Ibadi Muslim; remainder
Sunni Muslim, Shia Muslim, some Hindu

Health

Life Expectancy at Birth: 70 years (male);
75 years (female)
Infant Mortality Rate: 19.5/1,000 live births

Education

Adult Literacy Rate: 75.8%

COMMUNICATION

Telephones: 220,000 main lines
Daily Newspaper Circulation: 31 per
1,000 people
Televisions: 711 per 1,000 people
Internet Service Provider: 1 (2000)

TRANSPORTATION

Highways in Miles (Kilometers): 20,369
(32,800)

Railroads in Miles (Kilometers): none
Usable Airfields: 143
Motor Vehicles in Use: 347,000

GOVERNMENT

Type: monarchy; the monarch's absolute
power is limited by the 1996 Basic Law
Independence Date: 1650 (expulsion of
the Portuguese)
Head of State/Government: Sultan and
Prime Minister Qabus ibn Said Al Said is
both head of state and head of
government
Political Parties: none
Suffrage: universal at 21

MILITARY

Military Expenditures (% of GDP): 11.4%
Current Disputes: boundary with United
Arab Emirates defined bilaterally in
2002. Other boundaries, with UAE
Emirates Ras al-Khaymah and Sharjah,
which separate the Musandam Peninsula
from Oman proper, are administrative
and not treaty-defined

ECONOMY

Currency ($U.S. Equivalent): 0.385
rials = $1
Per Capita Income/GDP: $13,100/$38
billion
GDP Growth Rate: 1.2%
Inflation Rate: -0.8%
Labor Force: 920,000
Natural Resources: petroleum; copper;
asbestos; marble; limestone; chromium;
gypsum; natural gas
Agriculture: dates; limes; bananas; alfalfa;
vegetables; camels; cattle; fish
Industry: crude-oil production and
refining; natural-gas production;
construction; cement; copper
Exports: $13.14 billion (primary partners
Japan, China, Thailand)
Imports: $6.3 billion (primary partners
United Arab Emirates, Japan, United
Kingdom)

SUGGESTED WEBSITES

http://lcweb2.loc.gov/frd/cs/
omtoc.htm
http://www.oman.org/

Oman Country Report

The Sultanate of Oman was, at least until about 1970, one of the least-known countries in the world. Yet it is a very old country with a long history of contact with the outside world. Merchants from Oman had a near monopoly on the trade in frankincense and myrrh. Oman-built, shallow-draught, broad-beamed ships called dhows crisscrossed the Indian Ocean, trading with India and the Far East.

In the twentieth century, Oman became important to the outside world for two primary reasons: it began producing oil in the 1960s; and it has a strategic location on the Strait of Hormuz, the passageway for supertankers carrying Middle Eastern oil to the industrialized nations. Eighty percent of Japan's oil needs passes through Hormuz, as does 60 percent of Western Europe's. A Swiss journalist called the Omanis "sentinels of the Gulf" because they watch over this vital traffic.

GEOGRAPHY

Oman is the third-largest country in the Arabian Peninsula after Yemen and Saudi Arabia. However, the population is small, and large areas of land are uninhabited or sparsely populated. The geographical diversity—rugged mountains, vast gravelly plains, and

deserts—limits large-scale settlement. The bulk of the population is centered in the Batinah coastal plain, which stretches from the United Arab Emirates border south to the capital, Muscat. Formerly this area was devoted to fishing and agriculture; but with the rapid development of Oman under the current sultan, it has become heavily industrialized, with extensive commerce. The ancient system of *falaj*—underground irrigation channels that run for miles, bringing water downhill by gravity flow—has made farming possible, although the agricultural sector has been adversely affected in recent years by prolonged drought. Oman's southern Dhofar Province is more fertile and productive than the rest of the country, due to monsoon rains. In addition to citrus and other tropical fruits, Oman is the major world source of frankincense and gum from a small tree that grows wild and has been prized since ancient times.

Behind Oman's coast is the Jabal al-Akhdar ("Green Mountain"), a spine of rugged mountains with peaks over 10,000 feet. The mountains form several disconnected chains, interspersed with deep, narrow valleys where village houses hang like eagles' nests from the mountaintops, above terraced gardens and palm groves.

DEVELOPMENT

Vision Oman 2020, the sultan's blueprint for long-term growth, sets among its objectives increasing economic diversity to reduce dependence on oil, developing a competitive private sector producing manufactured goods for export, and Omanization of the labor force. With one-half million youths entering the job market by 2005, reducing dependence on foreign workers is critical.

Most of Oman's oil wells are located in the interior of the country. The interior is a broad, hilly plain dotted with oasis villages, each one a fortress with thick walls to keep out desert raiders. The stony plain eventually becomes the Rub al-Khali ("Empty Quarter"), the great uninhabited desert of southeastern Arabia.

Omani territory includes the Musandam Peninsula, at the northeastern tip of Arabia projecting into the Strait of Hormuz. The peninsula and the neighboring Midha oasis are physically separated from the rest of Oman by U.A.E. territory. In 1995, the Omani border with Yemen was formally demarcated in accordance with a UN-sponsored 1992 agreement. The oasis of Buraimi, on the Oman/Saudi Arabia/

S.M. Amin/Saudi Aramco World/PADIA (SA1977063)

Boys study the Koran at a village in Oman. When Qabus ibn Said came to power in 1970, replacing his father, he targeted education, health care, and transportation as prime development areas.

U.A.E. border, is currently under U.A.E. control, although it is claimed by both Saudi Arabia and Yemen. The surrounding desert hinterland is shared by the three states and remains undefined.

HISTORY

As was the case elsewhere in Arabia, the early social structure of Oman consisted of a number of tribal groups. Many of them were and still are nomadic (Bedouin), while others became settled farmers and herders centuries ago. The groups spent much of their time feuding with one another. Occasionally, several would join in an alliance against others, but none of them recognized any higher authority than their leaders.

In the seventh century A.D., the Omanis were converted to Islam. They developed their own form of Islam, however, called Ibadism, meaning "Community of the Just," a branch of Shia Islam. The Ibadi peoples elect their own leader, called an Imam. The Ibadi Imams do not have to be

descendants of the prophet Muhammad, as do the Imams in the main body of Shia Muslims. The Ibadi community believes that anyone, regardless of background, can be elected Imam, as long as the individual is pious, just, and capable. If no one is available who meets those requirements, the office may remain vacant.

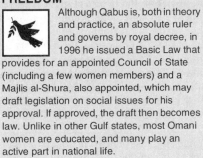

FREEDOM

Although Qabus is, both in theory and practice, an absolute ruler and governs by royal decree, in 1996 he issued a Basic Law that provides for an appointed Council of State (including a few women members) and a Majlis al-Shura, also appointed, which may draft legislation on social issues for his approval. If approved, the draft then becomes law. Unlike in other Gulf states, most Omani women are educated, and many play an active part in national life.

Ibadi Imams ruled interior Oman with the support of family shaykhs until the eighteenth century. Well before then, however,

coastal Oman was being opened up to foreign powers. The Portuguese captured Muscat in the 1500s for use as a stopping place for their ships on the trade route to India. (An Omani served as navigator to Portuguese admiral Vasco da Gama in his voyage across the Indian Ocean to India.) They built two great forts guarding the entrance to Muscat harbor, forts that still stand, giving the town its picturesque appearance. The Portuguese were finally driven out in 1650. Since that time, Oman has not been ruled directly by any foreign power.

The current ruling dynasty in Oman is the Al Bu Said Dynasty. It has been in power since 1749, when a chief named Ahmad ibn Said defeated an Iranian invasion and established his authority over most of Oman. But, for most of the period, Oman actually had two rulers—a sultan ruling in Muscat and an Imam ruling in the interior at the same time.

The most successful Omani sultan before the twentieth century was Said ibn Sultan (1804–1856). He added Dhofar

Province and Zanzibar, on the East African coast, to Omani territory. Sultan Said had good relations with Britain. He signed a treaty with the British that stated, "the friendship between our two states shall remain unshook to the end of time." The sultan also signed a friendship treaty with the United States in 1833; in 1836, to the surprise of the New York Port authorities, an Omani ship docked in New York harbor. Its captain said that the sultan had sent him to get to know the Americans whom he had heard so much about and to arrange trade contacts. Just under 170 years later, some 100 "cultural ambassadors" from Oman, including indigo dyers, shipwrights, bagpipers and sword dancers, arrived in Washington DC to participate in the 39th annual Smithsonian Folklife Festival, representing the first Arab nation to do so.[1]

HEALTH/WELFARE

Mobile health units that travel to remote areas have helped bring about a steep decline in infant mortality rates, from 34.3 per 1,000 in 1995 to 21.7 per 1,000 in 2003. Effective family planning programs using mobile health units that cover the rural countryside have helped lower the high birth rate from 3.41 percent in 1998 to 2.5 percent in 2003. Oman's health care system was declared to be 8th best in the world by the World Health Organization (WHO) in 2000.

After Said's death, a number of ethnic, tribal, and religious differences re-asserted themselves, and Oman lost its importance in regional affairs. Its territory was again restricted to its small corner of southeastern Arabia. The opening of the Suez Canal in 1869 diverted shipping to new Red Sea routes, and ships no longer called at Muscat harbor. Piracy and the slave trade, both of which had provided revenues for the sultan, were prohibited by international law. For the rest of the 1800s and most of the 1900s, Oman sank back into isolation, forgotten by the world. Only Britain paid the Omanis any attention, giving the sultan a small subsidy in the event that Oman might be of some future use to it.

In the early twentieth century, the Imams of inner Oman and the sultans ruling in Muscat came to a complete parting of the ways. In 1920, a treaty between the two leaders provided that the sultan would not interfere in the internal affairs of inner Oman. Relations were reasonably smooth until 1951, when Britain recognized the independence of the Sultanate of Muscat-Oman, as it was then called, and withdrew its advisers. Subsequently, the Imam declared inner Oman to be a separate state from the sultanate. A number of Arab states supported the Imam on the grounds that the sultan was a British puppet. Conflict between the Imam and the sultan dragged on until 1960, when the sultan finally reestablished his authority.

Oman's ruler for nearly four decades in the twentieth century was Sultan Said ibn Taimur (1932–1970). The most interesting aspect of his reign was the way in which he stopped the clock of modernization. Oil was discovered in 1964 in inland Oman; within a few years, wealth from oil royalties began pouring in. But the sultan was afraid that the new wealth would corrupt his people. He refused to spend money except for the purchase of arms and a few personal luxuries such as an automobile, which he liked to drive on the only paved road in Salalah. He would not allow the building of schools, houses, roads, or hospitals for his people. Before 1970, there were only 16 schools in all of Oman. The sole hospital was the American mission in Muscat, established in the 1800s by Baptist missionaries. All 10 of Oman's qualified doctors worked abroad, because the sultan did not trust modern medicine. The few roads were rough caravan tracks; many areas of the country, such as the Musandam Peninsula, were inaccessible.

ACHIEVEMENTS

A new Iranian-built power plant began operations in 1999 in Oman, meeting domestic needs for electricity. Oman is also self-sufficient in cement and textiles, most of the latter made in factories located in the Rusayl free-trade zone near Muscat. The zone now has more than 60 industries and produces $60 million in finished goods, generating $24 million in exports.

The sultan required the city gates of Muscat to be closed and locked three hours after sunset; no one could enter or leave the city after that. Flashlights were prohibited, since they were a modern invention; so were sunglasses and European shoes. Anyone found on the streets at night without a lighted kerosene lantern could be imprisoned. In the entire country, there were only about 1,000 automobiles; to import a car, one had to have the sultan's personal permission. On the darker side, slavery was still a common practice. Women were almost never seen in public and had to be veiled from head to foot if they so much as walked to a neighbor's house to visit. And on the slightest pretext, prisoners could be locked up in the old Portuguese fort at Muscat and left to rot.

As the 1960s came to an end, there was more and more unrest in Oman. The opposition centered around Qabus ibn Said, the sultan's son. Qabus had been educated in England. When he came home, his father shut him up in a house in Salalah, a town far from Muscat, and refused to give him any responsibilities. He was afraid of his son's "Western ideas."

On July 23, 1970, supporters of Crown Prince Qabus overthrew the sultan, and Qabus succeeded him. Sultan Qabus ibn Said brought Oman into the twentieth century in a hurry. The old policy of isolation was reversed.

Qabus also ended a long-running rebellion in Dhofar. His father had considered the province his personal estate, and did nothing to develop it. In 1962 Marxist-trained Omanis backed by the Peoples' Democratic Republic of Yemen (PDRY) next door led a rebellion against Said's rule. Relations between the two countries remained poor even after the sultan had crushed the rebellion in 1975, with the help of troops from Britain and Iran. The unification of the two Yemens in 1990 improved prospects for an Omani-Yemeni reconciliation, which was confirmed by the 1992 border agreement and the 1995 demarcation.

OMANI SOCIETY

Oman today is a land in flux, its society poised between the traditional past and a future governed increasingly by technology. An Omani business executive or industrial chief may wear a Western suit and tie to an appointment, but more than likely he will arrive for his meeting in a *dishdasha* (the traditional full-length robe worn by Gulf Arabs), with either a turban or an embroidered skullcap to complete the outfit. A ceremonial dagger called *kanjar* will certainly hang from his belt or sash. He will have a cellular phone pressed to his right ear and a digital watch on his wrist, courtesy of Oman's extensive trade with Japan. Older Omani women are also traditional in costume, covered head to toe with the enveloping *chador* and their faces (except for the eyes) hidden behind the black *batula,* the eagle-like mask common in the region. But increasingly their daughters and younger sisters opt for Western clothing, with only head scarves to distinguish them as Muslims.

In social, economic, and even political areas of Omani life, Qabus has brought about changes that have proceeded at a dizzying pace during his three-decade rule. Education, health care, and roads were his three top priorities when he took office. By 2003 Oman had 1,022 schools, 49 hospitals, 199 health clinics and over 20,000 miles of paved highways. The enrollment is 48.7 percent female. Sultan Qabus University, which opened in 1986 with a student body of 3,000, now has 6,000 students. These efforts, along with numerous adult-education programs, have increased Oman's literacy rate to 80 percent.

The sultan has also begun the process of replacing authoritarian rule by representative government. In 1996, his silver-anniversary year, he issued a Basic Law setting up a Majlis al-Shura (Council of State). Its 82 members are appointed by the ruler to represent Oman's provinces wilayats and cities. The Majlis has neither veto nor legislative powers, but it acts as an advisory body in the drafting of laws and the national budget. In 1998, an amendment to the Basic Law established local and municipal councils in order to exercise internal authority in these areas.

THE ECONOMY

Oman began producing and exporting oil in limited quantities in 1967. The industry was greatly expanded after the accession of Sultan Qabus. It is managed by a national corporation, Petroleum Development Oman (PDO). Oil production in the 1990s reached 900,000 barrels per day but was reduced to 860,000 bpd in 2000–2001, in accordance with OPEC production cuts. Oman's oil reserves are 5.7 billion barrels. Natural-gas reserves are 29.3 trillion cubic feet. The new liquefied natural gas (LNG) plant at Qalhat produced 6.6 million tons of LNG in 2001. Some 4 million tons were exported to South Korea in 2000; it was the largest single gas-export contract arranged with two state companies.

In its search for ways to supplement its oil income, Oman in 1996 formed a Caspian Sea Consortium with Russia, Kazakhstan, and several U.S. oil companies to link its refinery with Kazakh oil fields via new pipelines. Consortium ownership is divided between Russia (24 percent), Kazakhstan (19 percent), and Oman (7 percent), with the balance being held by the U.S. companies Chevron, Lukoil, and Arco. Oman also became the first Gulf country to establish a privately-owned electricity grid. In the industrial sector the Oman Oil Company has undertaken a number of new joint ventures, such as new refineries and a fertilizer plant being developed in conjunction with Indian industrialists. In 2005 the country's second oil refinery came on stream, along with an aluminum smelter at Sohar, on the Batinah coast.

Barely 2 percent of Oman's land is arable. Rainfall averages two to four inches annually except in monsoon-drenched Dhofar, and recent drought has largely dried up the long-established falaj system. The interior oases and Dhofar provide for intensive cultivation of dates. They also grow coconuts and various other fruits. Agriculture provides 35 percent of non-oil exports and employs 12 percent of the labor force.

The fishing industry employs 10 percent of the working population, but obsolete equipment and lack of canning and freezing plants have severely limited the catch in the past. Another problem is the unwillingness of Omani fishermen to move into commercial production; most of them catch just enough fish for their own use. The Oman Fish Company was formed in 1987 to develop fishery resources, financing the purchase by fishermen of aluminum boats powered by outboard motors to replace the seaworthy but slow traditional wooden dhows. A new fish-processing plant at Rusayl, built at a cost of $34 million, and the enlargement of the main fishing harbor at Raysut, generated a 5.6 percent increase in that sector of the economy in 2003.

FOREIGN RELATIONS

Oman joined with other Arab countries in opening links with Israel after its Oslo agreements with the Palestine Liberation Organization for Palestinian autonomy and the 1994 Jordan-Israel peace treaty. However, the freeze in Arab-Israeli relations ordered by the Arab League caused Oman to cancel the proposed Israeli trade mission in Muscat.

As a member of the Gulf Cooperation Council, Oman has become active in regional affairs, a role emphasized by its strategic location. Its long history of dealings with the United States—a relationship dating back to Andrew Jackson's presidency—has made Oman a natural partner in U.S. efforts to promote stability in the Gulf region. In 1980, Oman granted American military and naval personnel use of its Masirah Island and other military bases and the right to station troops and equipment there. In turn, the United States has provided new equipment to the Omani armed forces and built base housing for American personnel. During the Iran-Iraq War of 1980–1988, the country provided logistical support for U.S. warships escorting oil tankers in the Gulf to protect them from Iranian attacks, and U.S. jet fighters were based in Oman during the 1991 Gulf War.

After the September 11, 2001, terrorist attacks on the World Trade Center and the Pentagon, Sultan Qabus took the lead among the Gulf states in supporting the U.S.-led international coalition against terrorism. In October, 20,000 British troops arrived in Muscat to supplement the U.S. forces already there. The United States also reached agreement for a $1.1 billion arms sale to Oman. It included 12 F-16 fighter aircraft, Sidewinder air-to-air missiles, and Harpoon antiship missiles. However the Sultan's close alignment with Britain and the U.S. does not preclude his periodic adoption of an independent foreign policy. He maintained good relations with Iran during the Iran-Iraq war and with Iraq during its occupation of Kuwait and in the postwar UN sanctions period.

Timeline: PAST

1587–1588
The Portuguese seize Muscat and build massive fortresses to guard the harbor

1749
The Al Bu Said Dynasty is established; extends Omani territory

late 1800s
The British establish a de facto protectorate; the slave trade is supposedly ended

1951
Independence

1970
Sultan Said ibn Taimur is deposed by his son, Prince Qabus

1975
With British and Iranian help, Sultan Qabus ends the Dhofar rebellion

1980s
Oman joins the Gulf Cooperation Council Sultan Qabus sets up a Consultative Assembly as the first step toward democratization

1990s
The sultan focuses on expanding Oman's industrial base

PRESENT

2000s
Sultan continues modest democratization program by establishing universal suffrage and an elected Council of State with advisory powers and inclusion of women. The growing tourism sector welcomed 2 million visitors in 2002

Qatar (State of Qatar)

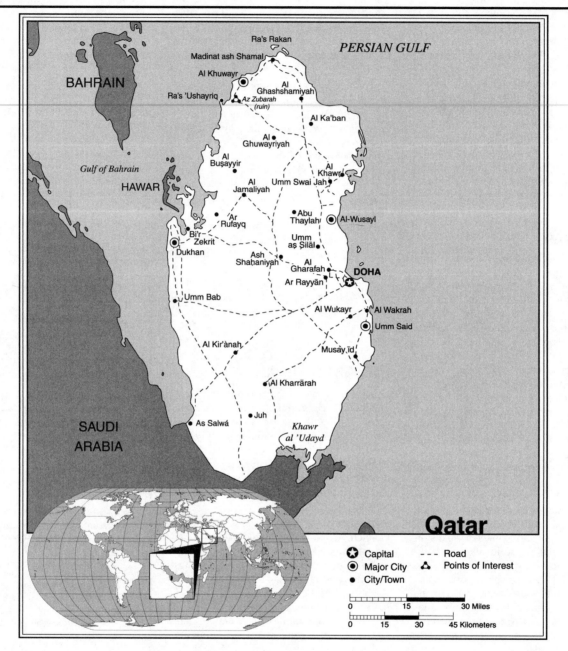

Qatar Statistics

GEOGRAPHY

Area in Square Miles (Kilometers): 4,400 (11,400) (about the size of Connecticut)

Capital (Population): Doha (340,000)

Environmental Concerns: limited natural freshwater supplies; increasing dependence on large-scale desalination facilities

Geographical Features: mostly flat and barren desert covered with loose sand and gravel

Climate: desert; hot and dry; humid and sultry summers

PEOPLE

Population

Total: 563,051

Annual Growth Rate: 2.6%

Rural/Urban Population Ratio: 8/92

Major Languages: Arabic; English widely used
Ethnic Makeup: 40% Arab; 18% Pakistani; 18% Indian; 10% Iranian; 14% others
Religions: 95% Muslim; 5% others

Health

Life Expectancy at Birth: 71 years (male); 76 years (female)
Infant Mortality Rate: 18.6/1,000 live births

Education

Adult Literacy Rate: 82.5%

COMMUNICATION

Telephones: 150,500 main lines
Daily Newspaper Circulation: 143 per 1,000 people
Televisions: 451 per 1,000 people
Internet Service Provider: 1 (2000)

TRANSPORTATION

Highways in Miles (Kilometers): 764 (1,230)
Railroads in Miles (Kilometers): none

Usable Airfields: 4
Motor Vehicles in Use: 183,000

GOVERNMENT

Type: traditional monarchy
Independence Date: September 3, 1971 (from the United Kingdom)
Head of State/Government: Emir Hamad bin Khalifa al-Thani; Prime Minister Abdallah bin Khalifa al-Thani
Political Parties: none. However the ruling emir approved a constitution in 2004 which was ratified in a popular referendum and went in effect in 2005. It provides for a popularly-elected 29-member Central Municipal Council with advisory powers to oversee municipal services.
Suffrage: universal at 21 under the constitution

MILITARY

Military Expenditures (% of GDP): 10%
Current Disputes: none; territorial dispute with Bahrain settled in 2001 by the International Court of Justice

ECONOMY

Currency ($U.S. Equivalent): 3.64 rials = $1 (fixed rate)
Per Capita Income/GDP: $23,200/$19 billion
GDP Growth Rate: 8.7%
Inflation Rate: 3%
Labor Force: 233,000
Unemployment Rate: 2.7%
Natural Resources: petroleum; natural gas; fish
Agriculture: fruits; vegetables; poultry; dairy products; beef; fish
Industry: crude-oil production and refining; fertilizers; petrochemicals; steel reinforcing bars; cement
Exports: $15 billion (primary partners Japan, Singapore, South Korea)
Imports: $6 billion (primary partners United Kingdom, France, Germany)

SUGGESTED WEBSITES

http://lcweb2.loc.gov/frd/cs/
qatoc.html
http://www.qatar-info.com/

Qatar Country Report

Qatar is a shaykhdom on the eastern (Gulf) coast of Arabia. It is the second-smallest Middle Eastern state, after Bahrain; but due to its oil wealth, it has an extremely high per capita annual income. Before 1949, when its oil exports began, there were about 20,000 Qataris, all descendants of peoples who had migrated to the coast centuries ago in search of a dependable water supply. Since then, rapid economic growth has attracted workers and residents from other Arab countries and distant Muslim states such as Pakistan. As a result, Qatar has a high number of immigrants and expatriates, which makes for some tension.

HISTORY

Although the peninsula has been inhabited since 4000 B.C., little is known of its history before the nineteenth century. At one time, it was ruled by the al-Khalifa family, the current rulers of Bahrain. It became part of the Ottoman Empire formally in 1872, but the Turkish garrison was evacuated during World War I. The Ottomans earlier had recognized Shaykh Qassim al-Thani, head of the important al-Thani family, as emir of Qatar, and the British followed suit when they established a protectorate after the war.

The British treaty with the al-Thanis was similar to ones made with other shaykhs in Arabia and the Persian Gulf in order to keep other European powers out of the area and to protect their trade and communications links with India. In 1916, the British recognized Shaykh Abdullah al-Thani, grandfather of the current ruler, as ruler of Qatar and promised to protect the territory from outside attack either by the Ottomans or overland by hostile Arabian groups. In return, Shaykhal-Thani agreed not to enter into any relationship with any other foreign government and to accept British political advisers.

DEVELOPMENT

Qatar's huge gas reserves are among the largest in the world, but are concentrated in a single field. They form the basis for ongoing economic development. New petrochemical and related fertilizer industries are beginning to diversify sources of revenue.

Qatar remained a tranquil British protectorate until the 1950s, when oil exports began. Since then, the country has developed rapidly, though not to the extent of

producing the dizzying change visible in other oil-producing Arab states.

INDEPENDENCE

Qatar became independent in 1971. The ruler, Shaykh Ahmad al-Thani, took the title of emir. Disagreements within the ruling family led the emir's cousin, Shaykh Khalifa, to seize power in 1972. Khalifa made himself prime minister as well as ruler and initiated a major program of social and economic development, which his cousin had opposed.

Shaykh Khalifa limited the privileges of the ruling family. There were more than 2,000 al-Thanis, and most of them had been paid several thousand dollars a month whether or not they worked. Khalifa reduced their allowances and appointed some nonmembers of the royal family to the Council of Ministers, the state's chief executive body. In 1992, he set up a Consultative Council of 30 members to advise the cabinet on proposed legislation and budgetary matters. Subsequently, the cabinet itself was enlarged, with new ministries of Islamic affairs, finance, economy, and industry and trade. While the majority of cabinet and Consultative Council members belonged to the royal family, the appointment of a number of nonfamily members to both these organizations heralded the "quiet revolution" toward power sharing to which Shaykh Khalifa was committed.

FOREIGN RELATIONS

Because of its small size, great wealth, and proximity to regional conflicts, Qatar is vulnerable to outside intervention. The government fears especially that the example of the Iranian Shia Revolution may bring unrest to its own Shia Muslim population. After the discovery of a Shia plot to overthrow the government of neighboring Bahrain in 1981, Qatari authorities deported several hundred Shia Qataris of Iranian origin. But thus far the government has avoided singling out the Shia community for heavy-handed repression, preferring to concentrate its efforts on economic and social progress. On the 10th anniversary of Qatar's independence, the emir said that "economic strength is the strongest guarantee that safeguards the independence of nations, their sovereignty, rights and dignity."[1]

Fears of a possible attack by Iran led the country to sign a bilateral defense agreement with Saudi Arabia in 1982. The Iraqi invasion of Kuwait exposed Saudi military weakness, and, as a result, Qatar turned to the United States for its defense. A Qatar official noted, "Saudi Arabia was the protector, but the war showed that the emperor had no clothes."[2] However, the close alliance has not brought full acceptance of U.S. Middle East policy. The emir joined other Arab leaders in criticizing Israel and its principal ally for repression of the renewed Palestinian intifada (uprising) after the collapse of the 1993 Oslo agreements for Palestinian self-determination. In 2000, the Israeli trade mission in Qatar, the last one active for the Gulf states, was ordered closed.

FREEDOM

Since he deposed his father, the ruling emir has abolished press censorship, established Al Jazeerah as a service of uncensored news to the Arab world, and appointed younger members of the ruling family to replace his father's advisers and ministers.

The continued U.S./UN sanctions imposed on Iraq after the Gulf War drew increased opposition among Qataris, as the extent of harm to the Iraqi civilian population became more evident. Qatar was the first Gulf state to criticize openly U.S. and British air attacks on Iraq.

However, the expanded U.S. military presence in the Gulf, and the Qatari government's fears of a threat to its territory by a remilitarized Iraq, resulted in 2000 in the establishment in the country of the largest American military base outside the continental United States. The base was placed on full military alert in July 2000, after the attack on the U.S. destroyer *Cole* in Aden (Yemen) harbor. The emir also completed the $1 billion air base at Al-Ubeid, which holds up to 100 fighter aircraft (Qatar's total air force consists of 10 aircraft). An attack in November 2001 by a lone gunman on guards at the base was a reminder of the unpopularity of the U.S. military presence in the Gulf region despite its low profile. But when he was criticized by Islamic opposition leaders with "Christianizing" the country by making the base available to the United States, the emir responded: "We intend to be essential to the American presence in the Gulf. What do we get in exchange? We don't need to spend a lot of money on defense, we'll be attractive to U.S. businessmen, and we'll get international status and prestige."[3]

HEALTH/WELFARE

Qatar's first private hospital opened in 1996 and is now fully staffed by Qatari doctors and nurses, who have replaced expatriate medical personnel. Among the Arab states, Qatar has an unusually high ratio of physicians to population.

Inasmuch as Iraq was apparently not involved or charged with complicity in the September 11, 2001, terrorist attacks on the United States, the Qatari government felt less constrained about actively participating in the international coalition being formed to combat global terrorism than it had before the attacks. The American military and air bases in Qatar went on full alert for the invasion of Afghanistan, and the troops were reinforced by U.S. special forces and fighter aircraft.

Qatar's main foreign-policy concern has involved the islands of Hawar and Fishat al-Duble, which lie off of its northwest coast. Ownership was disputed with Bahrain, which controlled them under arrangements made in the 1930s, when both countries were British protectorates. In 1992, Qatar unilaterally extended its territorial waters to 12 nautical miles to bring the islands and adjacent seabed under Qatari sovereignty. Bahrain filed a complaint with the International Court of Justice (the ICJ, or World Court). In 2001, the World Court confirmed Bahraini ownership of the islands and adjacent territorial waters. The

Qatari emir had said previously that he would not accept World Court arbitration, but following the issuance of the Court's decision, he agreed to accept Bahraini sovereignty. Under the terms of the Court's ruling, Qatar was awarded sovereignty over Zabarah and Janan Islands, and the elevation at low tide of Fasht and Dubal. Qatari ships were also guaranteed the right of unobstructed passage through the Bahrain territorial sea. Further evidence of the country's growing influence internationally was its election in 2005 to a 2-year term on the UN Security Council starting in January 2006.

THE ECONOMY

With oil reserves expected to be used up within 20-plus years at current extraction rates, the Qatari economy has not only diversified but more important, shifted to natural gas. The main source of development is the huge natural gas field at Ras Laffan, the world's third largest, which is being developed by Qatar Petroleum in cooperation with Western companies under the logo Oryx GTL. (The rare desert oryx is the country's national symbol.) Technical experts working for and with GTL are developing a process of conversion of liquefied natural gas (LNG) to clean-burning diesel fuel. Production to date has been small—34,000 barrels per day (bpd) as compared with 80 million bpd of global oil production. But in time the new process should help meet surging energy demands worldwide while reducing smog and other air pollutants.

Depletion of water supplies due to heavy demand and dependence on outdated desalination plants for its fresh water have prompted the country to undertake some innovative food-production projects. One such project, begun in 1988, uses solar energy and seawater to cultivate food crops in sand. As a result of such projects, Qatar produces sufficient food both to meet domestic needs and to export vegetables to neighboring states.

SOCIETAL CHANGES

Qatar was originally settled by nomadic peoples, and their influence is still strong. Traditional Bedouin values, such as honesty, hospitality, pride, and courage, have carried over into modern times.

Most Qataris belong to the strict puritanical Wahhabi sect of Islam, which is also dominant in Saudi Arabia. They are similar to Saudis in their conservative outlook, and Qatar generally defers to its larger neighbor in foreign policy. There are, however, significant social differences between Qataris and Saudis. Western movies may be shown in Qatar, for example, but not in Saudi Arabia. Furthermore, Qatar does not have religious police or "morals squads" to enforce Islamic conventions, and foreigners may purchase alcoholic beverages legally.

Qatar also differs from its Arab peninsular neighbors and the Arab world generally in permitting free discussion in the media of issues generally suppressed by Arab rulers. Following his accession to the throne, the new emir abolished press censorship and eliminated the Information Ministry from his cabinet. In 1997, his government licensed a new satellite TV network station called Al-Jazeera ("Peninsula," in Arabic), supported by an annual subsidy to meet its operating costs. Despite its freewheeling broadcasting style and frequent criticism of Arab rulers, including its own, the ruling emir does not attempt to censor the station or close it down. As one analyst noted, "its reporters … openly challenge the sycophantic tone of the state-media and the mainstream Arab press, both of which play down controversy and dissent."[4] Al-Jazeera has featured interviews with a variety of Islamic and other global leaders, ranging from ex-U.S. Secretary of State Colin Powell and European heads of state to Osama bin Laden. The station's frank approach to controversial issues and frequent criticism of U.S. foreign policy led President Bush in 2005 to declare he would shut it down, if he had the power to do so.

ACHIEVEMENTS

Qatar played host to the World Trade Conference annual meeting in November 2001. Despite threats of disruption by activists opposed to WTO policies, the meeting opened on schedule, albeit under tight security. It produced a compromise agreement among the 144 member states on global-market reforms that will expedite free trade while protecting the interests of the poorer nations.

The most significant societal change in Qatar involves the position of women. The first school for girls opened there in 1956.

But change in women's rights and roles has accelerated in recent years. In 1998, the new emir, who had deposed his father in 1995 while the latter was vacationing in Switzerland, granted women the right to vote and to run for and hold public office.

Roughly 30 percent of Qatari women are employed in the labor force, and many are not only educated but also well-qualified professionally. Unlike their sisters in some other Gulf states, they drive cars, work in offices, juggle careers and family responsibilities. As a Qatari female journalist told an interviewer: "It is not easy for a man to permit his wife to work, to appear on TV, to drive a car. We have special clubs for women, separate cinema areas, even women's banks, to reinforce each other."[5]

INTERNAL POLITICS

Crown Prince Shaykh Hamad bin Khalifa's "palace coup" was bloodless, although several attempts by supporters of the deposed ruler to overthrow his son have been thwarted. In July 1999, Shaykh Hamid al-Thani, the ruling emir's cousin and former chief of police, was arrested and brought back from his hiding place "somewhere abroad" for trial. He was charged with being the leader and organizer of the attempted coups. The ex-emir also agreed to return $2 billion in government funds that he had deposited abroad over the years in personal accounts.

Elections for a unicameral Central Municipal Council of 29 members for Doha were held first in 1999 and subsequently in 2003, two "firsts" for the emirate. Some 221 men and 3 women competed for the 29 seats. One of the women was elected, the first female to hold elective office not only in Qatar but in the entire Gulf region. As a first step toward constitutional government, the Council has consultative powers, although these are limited to improving municipal services.

The "quiet revolution" initiated by the new emir entered a new stage in March 1999, with elections for a Doha Central Mu-

nicipal Council, the country's first public elections. The Council does not have executive powers, but it is intended as a transitional body between patriarchal rule and the establishment of an elected parliament. All Qataris over age 18 were allowed to vote, including women (who make up 44 percent of registered voters). Six women ran with 221 men for the Council's 29 seats.

NOTES

1. Qatar News Agency (November 23, 1981).
2. Douglas Jehl, *The New York Times International* (July 20, 1997).
3. Mary Anne Weaver, "Democracy by Decree," *The New Yorker* (November 20, 2000), p. 57.
4. Fouad Ajami, "What the Muslim World Is Watching," *The New York Times Magazine* (November 18, 2001).
5. Weaver, *op cit.*

Timeline: PAST

1916
Britain recognizes Shaykh Abdullah al-Thani as emir

1949
The start of oil production in Qatar

1971
An abortive federation with Bahrain and the Trucial States (U.A.E.), followed by independence

1972
The ruler is deposed by Shaykh Khalifa

1990s
Qatar condemns the Iraqi invasion of Kuwait and expels resident Palestinians; Crown Prince Hamad al-Thani deposes his father and takes over as emir

PRESENT

2000s
World Court favors Bahrain against Qatar in territorial dispute

First Christian church in Qatar since Islam's arrival in 7th century A.D. established to serve the state's 70,000 Egyptian and Indian Christian workers

Saudi Arabia (Kingdom of Saudi Arabia)

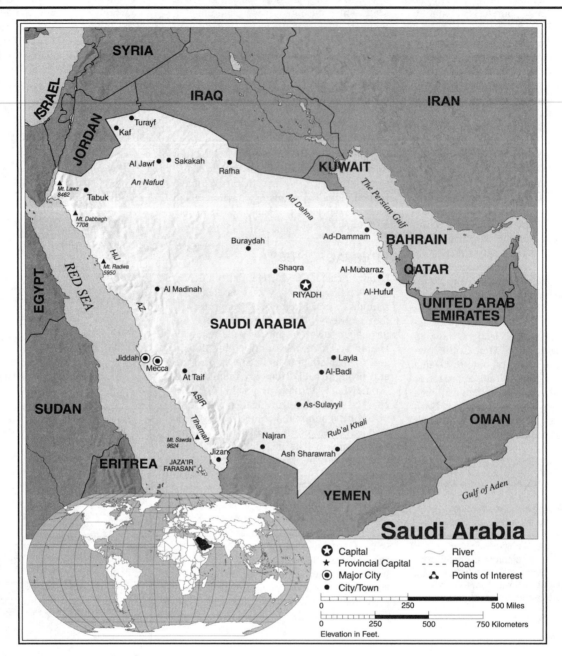

Saudi Arabia Statistics

GEOGRAPHY

Area in Square Miles (Kilometers):
756,785 (1,960,582) (about 1/5 the size of the United States)
Capital (Population): Riyadh (2,625,000)
Environmental Concerns: desertification; depletion of underground water resources; coastal pollution from oil spills

Geographical Features: mostly uninhabited sandy desert

Climate: harsh, dry desert, with great extremes of temperature

PEOPLE

Population

Total: 26,417,599 (includes 5,576,076 non-nationals)
Annual Growth Rate: 2.3%
Rural/Urban Population Ratio: 16/84

Major Languages: Arabic; English widely used

Ethnic Makeup: 90% Arab; 10% Afro-Asian

Religion: 100% Muslim

Health

Life Expectancy at Birth: 73 years (male); 77 years (female)

Infant Mortality Rate: 13.2/1,000 live births

Education

Adult Literacy Rate: 78.8%

COMMUNICATION

Telephones: 3,100,000, plus 1 million mobile cellular phones

Daily Newspaper Circulation: 54 per 1,000 people

Televisions: 257 per 1,000 people

Internet Service Providers: 42 (2001)

TRANSPORTATION

Highways in Miles (Kilometers): 87,914 (146,524)

Railroads in Miles (Kilometers): 863 (1,390)

Usable Airfields: 209

Motor Vehicles in Use: 2,800,000

GOVERNMENT

Type: hereditary monarchy

Independence Date: September 23, 1932 (unification)

Head of State/Government: King Abdullah ibn Aziz Al Saud, as of August 2005; he also serves as prime minister

Political Parties: none, but in 2005 the first elections in Saudi history were held for local and provincial councils and one-third of members in the Majlis al-Shura

Suffrage: none

MILITARY

Military Expenditures (% of GDP): 13%

Current Disputes: border with Yemen fixed by treaty but not demarcated due to frequent use by nomadic tribes. Boundary with UAE not formally demarcated but recognized de facto

ECONOMY

Currency ($U.S. Equivalent): 3.75 riyals = $1

Per Capita Income/GDP: $12,000/$310 billion

GDP Growth Rate: 5%

Inflation Rate: 0.8%

Labor Force: 6,620,000

Unemployment Rate: 25% for male Saudis

Natural Resources: petroleum; natural gas; iron ore; gold; copper

Agriculture: wheat; barley; tomatoes; melons; dates; citrus fruits; mutton; chickens; eggs; milk

Industry: crude-oil production; petroleum refining; basic petrochemicals; cement; construction; fertilizer; plastics

Exports: $113 billion (primary partners Japan, United States, South Korea)

Imports: $36 billion (primary partners United States, Japan, Germany)

SUGGESTED WEBSITES

http://lcweb2.loc.gov/frd/cs/
satoc.html
http://www.saudinf.com/main/
start.htm

Saudi Arabia Country Report

The Kingdom of Saudi Arabia is the geographical giant of the Arabian Peninsula. It is also a giant in the world economy because of its oil. To many people, the name Saudi Arabia is a synonym for oil wealth. Indeed, its huge oil reserves, large financial surpluses from oil production, and ability to use oil as a political weapon (as in the 1973 embargo) enable the country to play an important part in international as well as regional affairs.

Saudi Arabia's population is small in relation to the country's size and is heavily urbanized. Urban growth has been very rapid, considering that only 1 percent of the land can be used for agriculture and all employment opportunities are in the cities or in the oil-producing regions. The kingdom has relied strongly on expatriate workers, skilled as well as unskilled, in its development. The economic dislocation of the Gulf War, along with the support given Iraq by Palestinians and the government of Yemen, led to the expulsion of nearly 1 million foreign workers, most of them Palestinians and Yemenis. But due to the unwillingness of most Saudis to take on low-paying work that seems to be below them professionally, the government has had to continue its dependence on expatri-

ates. Some 67 percent of government jobs and 95 percent of those in private industry are still held by foreigners.

The country contains three main geographical regions: the Hejaz, along the Red Sea; the Nejd, a vast interior plateau that comprises the bulk of Saudi territory; and the Eastern Province. The kingdom's largest oases, al-Hasa and Safwa, are located in this third region, along with the major oil fields and industrial centers. The Empty Quarter (al-Rub' al-Khali), an uninhabited desert where rain may not fall for a decade or more, occupies the entire southeastern quadrant of the country.

THE WAHHABI MOVEMENT

In the eighteenth century, most of the area included in present-day Saudi Arabia was the home of nomads, as it had been for centuries. These peoples had no central government and owed allegiance to no one except their chiefs. They spent much of their time raiding one another's territories in the struggle for survival. Inland Arabia was a great blank area on the map—a vast, empty desert.

The only part of modern Saudi Arabia under any government control in the eigh-

teenth century was the Hejaz, which includes the Islamic holy cities of Mecca and Medina. It was a province of the Ottoman Empire, the major power in the Middle East at that time.

Saudi Arabia became a nation, in the modern sense of the word, in 1932. But the origins of the Saudi nation go back to the eighteenth century. One of the tribes that roamed the desert beyond Ottoman control was the tribe of Saud. Its leader, Muhammad Ibn Saud, wanted to gain an advantage over his rivals in the constant search for water and good grazing land for animals. He approached a famous religious scholar named Abd al-Wahhab, who lived in an oasis near the current Saudi capital, Riyadh (then a mud-walled village). Abd al-Wahhab promised Allah's blessing to Ibn Saud in his contests with his rivals. In return, the Saudi leader agreed to protect al-Wahhab from threats to his life by opponents of the strict doctrines he taught and preached, and he swore an oath of obedience to these doctrines. The partnership between these two men gave rise to a crusading religious movement called Wahhabism.

Wahhabism is basically a strict and puritanical form of Sunni Islam. The Wahhabi code of law, behavior, and conduct is mod-

S.M. Amin/Saudi Aramco World/PADIA (SA351024)

The Great Mosque at Mecca, in Saudi Arabia, is the holiest of shrines to Muslims. Historically, Mecca was the site at which Islam was founded, in the seventh century A.D., by the Prophet Muhammad. Pilgrims today flock to the Great Mosque to fulfill their Muslim duties as set down by the Five Pillars of Islam.

eled on that of the original Islamic community established in Mecca and Medina by the Prophet Muhammad. Although there has been some relaxation of the code due to the country's modernization, it remains the law of Saudi Arabia today. Interpretation of Islamic law is the responsibility of the *ulema* (a body of religious scholars and jurists, in Sunni Islam). As a result, Saudi society is more conservative and puritanical than many other Islamic societies, including those of its Persian Gulf neighbors. The Taliban, the Islamic fundamentalist movement that held power in Afghanistan from 1996 to 2001, is thus far the only movement in Islam to have embraced Wahhabism.

Although Wahhabi social and cultural restrictions are still very much in effect in the country and are enforced stringently by the religious police, modification of these restrictions has slowly become the norm in the new century. This is particularly evident regarding the rights and status of women. In 1990 a group of women from prominent families defied a government ban on female driving and drove their family cars into downtown Riyadh. The religious authorities promptly issued an edict emphasizing the ban, and the women were jailed briefly and deprived of their passports. As an indicator of changing conditions in the kingdom, the members of the group held a public celebration in 2005 to mark the 15th anniversary of their exploit. They were supported by a leading member of the Majlis al-Shura, who argued that

such a ban existed neither in the Qur'an nor in Islamic law. A poll in December 2005 indicated that 60 percent of Saudi males agreed that women should have the right to drive. However the poll results were qualified by Crown Prince Sultan bin Abdul-Aziz, who stated that the government had no objections as long as their husbands, fathers and brothers agreed.

Other recent changes in rigid Saudi Wahhabism included in 2005 the removal of a ban on forced marriages, prompted by the 50 percent divorce rate. Perhaps the greatest change was the country's first-ever election, held in April 2005 to elect members of municipal councils. In November 2005 a second election was held for members of chamber of commerce boards. Women were not only allowed to vote but to run for seats on the boards, and two were elected to the board in Jiddah. The late grand mufti, Shaykh Abd al-Aziz Bin Baz (who was famous for declaring that Earth was not round but flat), played such a role in the crisis that followed the Iraqi invasion of Kuwait.

The September 11, 2001, terrorist attacks in the United States and the resulting war on terrorism proclaimed by President George W. Bush against Osama bin Laden and his al Qaeda network placed the Saudi government in an awkward position. As a valued ally, it was expected to provide active support for the international antiterrorism coalition. But a large section of the Saudi population is opposed to U.S. policy

in the Middle East because of its support for Israel. A number of Wahhabi religious leaders ranged themselves in opposition to the monarchy due to its alliance with the "infidel West," some of them even urging its overthrow. In October 2001, a senior Wahhabi cleric, Shaykh Hamoud Ben Oqla, issued a *fatwa* (religious edict) to the effect that "it is a duty to wage jihad on anyone who supports the [American] attack on Afghanistan by hand, tongue or money; whoever helps the infidel against Muslims is to be considered an infidel." Although the Saudi government in October ended its recognition of the Taliban as the legitimate government of Afghanistan, the presence of 5,000 U.S. troops on "sacred Saudi Islamic soil" and bin Laden's popularity as a symbol of Muslim defiance against American "arrogance" in stationing them there forced the monarchy to walk a tightrope in balancing its international obligations with the views of its own people.

Fears that Saudi Arabia would be next on Saddam Hussein's invasion list after Kuwait led to the formation of the coalition of United Nations–sponsored forces that carried out Operation Desert Storm. This action involved stationing of American and other non-Muslim troops in the kingdom. The Saudi leadership was divided on the issue. But at this critical juncture, Bin Baz issued a fatwa. His edict said that in an extreme emergency, it was permissible for an Islamic state to seek help from non-Islamic ones. A later edict ruled that the campaign against Iraq was a *jihad*, further justifying the coalition and buildup of non-Muslim troops on Saudi soil.

In the late 1700s, the puritanical zeal of the Wahhabis led them to declare a "holy war" against the Ottoman Turks, who were then in control of Mecca and Medina, in order to restore these holy cities to the Arabs. In the 1800s, Wahhabis captured the cities. Soon the Wahhabis threatened to undermine Ottoman authority elsewhere. Wahhabi raiders seized Najaf and Karbala in Iraq, centers of Shia pilgrimage, and desecrated Shia shrines. In Mecca, they removed the headstones from the graves of members of the Prophet's family, because in their belief system, all Muslims are supposed to be buried unmarked.

The Ottoman sultan did not have sufficient forces at hand to deal with the Wahhabi threat, so he called upon his vassal, Muhammad Ali, the khedive (viceroy) of Egypt. Muhammad Ali organized an army equipped with European weapons and trained by European advisers. In a series of hard-fought campaigns, the Egyptian Army defeated the Wahhabis and drove them back into the desert.

Inland Arabia reverted to its old patterns of conflict. The Saudis and other rival tribes were Wahhabi in belief and practice, but this religious bond was countered by age-old disputes over water rights, territory, and control over trade routes. In the 1890s, the Saudis' major rivals, the Rashidis, seized Riyadh. The Saudi chief escaped across the desert to Kuwait, a town on the Persian Gulf that was under British protection. He took along his young son, Abd al-Aziz ibn Saud.

DEVELOPMENT

The new Saudi 5-Year Plan sets a growth rate of 3.16% annually, with increased diversification of the economy to reduce dependence on oil. With revenues dropping to half of the 1980 totals, an abnormally high birth rate and high unemployment, there are simply not enough jobs being created for the 100,000 Saudis entering the workforce each year.

IBN SAUD

Abd al-Aziz al-Rahman Al Sa'ud, usually referred to simply as Ibn Saud (son of Sa'ud), was the father of his country, in both a political and a literal sense.[1] He grew up in exile in Kuwait, where he brooded and schemed about how to regain the lands of the Saudis. When he reached age 21, in 1902, he decided on a bold stroke to reach his goal. On 5 Shawwal

1319 (January 1902), he led a force of 48 warriors across the desert from Kuwait to Riyadh. They scaled the city walls at night and seized the Rashidi governor's house, and then the fort in a daring dawn raid. The population seems to have accepted the change of masters without incident, while Bedouin tribes roaming in the vicinity came to town to pledge allegiance to Ibn Saud and applaud his exploit.

Over the next three decades, Ibn Saud steadily expanded his territory. He said that his goal was "to recover all the lands of our forefathers."[2] In World War I, he became an ally of the British, fighting the Ottoman Turks in Arabia. In return, the British provided arms for his followers and gave him a monthly allowance. The British continued to back Ibn Saud after the war, and in 1924, he entered Mecca in triumph. His major rival, Sharif Husayn, who had been appointed by the Ottoman government as the "Protector of the Holy Places," fled into exile. (Sharif Husayn was the great-grandfather of King Hussein I of Jordan.)

Ibn Saud's second goal, after recovering his ancestral lands, was to build a modern nation under a central government. He used as his motto the Koranic verse, "God changes not what is in a people until they change what is in themselves" *(Sura XIII, 2).* The first step was to gain recognition of Saudi Arabia as an independent state. Britain recognized the country in 1927, and other countries soon followed suit. In 1932, the country took its current name of Saudi Arabia, a union of the three provinces of Hejaz, Nejd, and al-Hasa.

INDEPENDENCE

Ibn Saud's second step in his "grand design" for the new country was to establish order under a central government. To do this, he began to build settlements and to encourage the nomads to settle down, live in permanent homes, and learn how to grow their own food rather than relying on the desert. Those who settled on the land were given seeds and tools, were enrolled in a sort of national guard, and were paid regular allowances. These former Bedouin warriors became in time the core of the Saudi armed forces.

Ibn Saud also established the country's basic political system. The basis for the system was the Wahhabi interpretation of Islamic law. Ibn Saud insisted that "the laws of the state shall always be in accordance with the Book of Allah and the Sunna (Conduct) of His Messenger and the ways of the Companions."[3] He saw no need for a written constitution, and as yet Saudi Arabia has none. Ibn Saud decreed that the country would be governed as an

Royal Embassy of Saudi Arabia, Washington, DC (RESA002)

King Faisal was instrumental in bringing Saudi Arabia into the world's international community and in establishing domestic plans that brought his country into the twentieth century.

absolute monarchy, with rulers always chosen from the Saud family. He was unfamiliar with political parties and distrusted them in principle; political organizations were therefore prohibited in the kingdom. Yet Ibn Saud was himself democratic, humble in manner, and spartan in his living habits. He remained all his life a man of the people and held every day a public assembly *(majlis)* in Riyadh at which any citizen had the right to ask favors or present petitions. (The custom of holding a daily majlis has been observed by Saudi rulers ever since.) More often than not, petitioners would address Ibn Saud not as Your Majesty but simply as Abd al-Aziz (his given name), a dramatic example of Saudi democracy in action.

Ibn Saud died in 1953. He had witnessed the beginning of rapid social and economic change in his country due to oil revenues. Yet his successors have presided over a transformation beyond the wildest imaginations of the warriors who had scaled the walls of Riyadh half a century earlier. Riyadh then had a population of 8,000. Today, it has 2.6 million inhabitants. It is one of the fastest-growing cities in the world.[4]

Ibn Saud was succeeded by his eldest surviving son, Crown Prince Saud. A number of royal princes felt that the second son, Faisal, should have become the new king because of his greater experience in foreign affairs and economic management. Saud's only experience was as governor of Nejd.

Although he was large and corpulent and lacked Ibn Saud's forceful personality, the new king was like his father in a number of ways. He was more comfortable in a desert tent than running a bureaucracy or meeting foreign dignitaries. Also, like his father, he had no idea of the value of money. Ibn Saud would carry a sackful of riyals (the Saudi currency) to the daily majlis and give them away to petitioners. His son, Saud, not only doled out money to petitioners but also gave millions to other members of the royal family. One of his greatest extravagances was a palace surrounded by a bright pink wall.[5]

By 1958, the country was almost bankrupt. The royal family was understandably nervous about a possible coup supported by other Arab states, such as Egypt and Syria, which were openly critical of Saudi Arabia because of its lack of political institutions. The senior princes issued an ultimatum to Saud: First he would put Faisal in charge of straightening out the kingdom's finances, and, when that had been done, he would abdicate. When the financial overhaul was complete, with the kingdom again on a sound footing, Saud abdicated in favor of Faisal.

The transfer of authority from Saud to Faisal illustrates the collective principle of government of the Saudi family monarchy. The sovereign rules in theory; but in practice, the inner circle of Saudi senior princes, along with ulema leaders, make all decisions concerning succession, foreign policy, the economy, and other issues. The reasons for a decision must always be guessed at; the Saudis never explain them. It is a system very different from the open, freewheeling one of Western democracies, yet it has given Saudi Arabia stability and leadership on occasions when crises have threatened the kingdom.

FAISAL AND HIS SUCCESSORS

In terms of state-building, the reign of King Faisal (1964–1975) is second in importance only to that of Ibn Saud. One author wrote of King Faisal during his reign, "He is leading the country with gentle insistence from medievalism into the jet age."[6] Faisal's gentle insistence showed itself in many different ways. Encouraged by his wife, Queen Iffat, he introduced education for girls into the kingdom. Before Faisal, the kingdom had had no systematic development plans. In introducing the first five-year development plan, the king said that "our religion requires us to progress and to bear the burden of the highest tradition and best manners."[7]

In foreign affairs, Faisal ended the Yemen Civil War on an honorable basis for both sides; took an active part in the Islamic world in keeping with his role as Protector of the Holy Places; and, in 1970, founded the Organization of the Islamic Conference, which has given the Islamic nations of the world a voice in international affairs. Faisal laid down the basic strategy that his successors have followed, namely, avoidance of direct conflict, mediation of disputes behind the scenes, and use of oil wealth as a political weapon when necessary. The king never understood the American commitment to Israel, any more than his father had. (Ibn Saud had met U.S. president Franklin D. Roosevelt in Egypt during World War II. Roosevelt, motivated by American Jewish leaders to help in the establishment of a Jewish homeland in Palestine, sought to convince Ibn Saud, as head of the only independent Arab state at that time, to moderate Arab opposition to the project.) But Faisal's distrust of communism was equally strong. This distrust led him to continue the ambivalent yet close Saudi alliance with the United States that has continued up to the present.

FREEDOM

Saudi Arabia's strict adherence to Islamic law not only imposes harsh punishments for many crimes, but also restricts human rights. The country ranks second in the world in executions per million population: 123 in 2000. Shari'a law applies equally to Saudis and non-Saudis; in 2001, 4 Britons were flogged publicly for dealing in alcohol. A new Code of Criminal Procedure took effect in July 2001. But although Saudi judges (qadis) in theory are bound to respect judicial procedure and legal rights (e.g. of lawyers to defend their clients), they often revert to arbitrary decisions and base these purely on Islamic law.

King Faisal was assassinated in 1975 by a deranged nephew while he was holding the daily majlis. The assassination was another test of the system of rule by consensus in the royal family, and the system held firm. Khalid, Faisal's eldest half-brother, his junior by six years, succeeded him without incident. He ruled until his death in 1982. The next-oldest half-brother, Fahd, succeeded him.

THE MECCA MOSQUE SIEGE

One of the most shocking events in Saudi Arabia since the founding of the kingdom was the seizure of the Great Mosque in Mecca, Islam's holiest shrine, by a group of fundamentalist Sunni Muslims in November 1979. The leader of the group declared that one of its members was the *Mahdi* (in Sunni Islam, the "Awaited One") who had come to announce the Day of Judgment. The group occupied the mosque for two weeks. The siege was finally overcome by army and national guard units, but with considerable loss of life on both sides. No one knows exactly what the group's purpose was, nor did it lead to any general expressions of dissatisfaction with the regime. But the incident reflects the very real fear of the Saudi rulers of a coup attempted by the ultra-religious right.

Although the Saudi government remains staunchly conservative, it has before it the example of Iran, where a similar Islamic fundamentalist movement overthrew a well-established monarchy. Furthermore, the Shia Muslim population of the country is concentrated in al-Hasa Province, where the oil fields are located. The government's immediate fear after the Great Mosque seizure was of an outside plot inspired by Iran. When this plot did not materialize, the Saudis feared Shia involvement. Outside of increased security measures, the principal result of the incident has been a large increase in funding for the Shia community to ease socioeconomic tensions.

THE ECONOMY

Oil was discovered in Saudi Arabia in 1938, but exports did not begin until after World War II. Reserves in 1997 were 261 billion barrels, 26 percent of the world's oil supply. The oil industry was controlled by Aramco (Arabian-American Oil Company), a consortium of four U.S. oil companies. In 1980, it came under Saudi government control, but Aramco continued to manage marketing and distribution services. The last American president of Aramco retired in 1989 and was succeeded by a Saudi. The company was renamed Saudi Aramco. But after a quarter-century of exclusion of foreign firms from the oil and gas industry, Saudi Arabia opened the gates in June 2001. A consortium of foreign oil companies was granted exploration rights in a desert area the size of Ireland. As a spin-off from the concession, the consortium will develop the existing South Ghawar gas field, and related power, desalination, and petrochemical plants.

The pressures of unemployment (14 percent for male Saudis), a population growth rate of 3.2 percent annually, and a stagnating economy have motivated the government to seek foreign capital investment. The new investment law passed in 2000 permits 100 percent foreign ownership of projects. Import duties were reduced from 12 to 5 percent in 2001. As a result, foreign investment doubled to $9 billion. Some 60 percent of this amount comes from two projects: a Japanese-built desalination plant, and a U.S. contract to build 3,000 schools with connections to the Internet.

King Faisal's reorganization of finances and development plans in the 1960s set the kingdom on an upward course of rapid development. The economy took off after 1973, when the Saudis, along with other Arab oil-producing states, reduced production and imposed an export embargo on Western countries as a gesture of support to Egypt in its war with Israel. After 1973, the price per barrel of Saudi oil continued to increase, to a peak of $34.00 per barrel in 1981. (Prior to the embargo, it was $3.00 per barrel; in 1979, it was $13.30 per barrel.) The outbreak of the Iran-Iraq War in 1980 caused a huge drop in world production. The Saudis took up the slack.

The huge revenues from oil made possible economic development on a scale undreamed of by Ibn Saud and his Bedouin warriors. The old fishing ports of Yanbu, on the Red Sea, and Jubail, on the Persian Gulf, were transformed into new industrial cities, with oil refineries, cement and petrochemical plants, steel mills, and dozens of related industries. Riyadh experienced a building boom; Cadillacs bumped into camels on the streets, and the shops filled up with imported luxury goods. Every Saudi, it seemed, profited from the boom through free education and health care, low-interest housing loans, and guaranteed jobs.

HEALTH/WELFARE

Although Saudi schools are administratively under the Ministry of Education, the curriculum is controlled by the religious authorities. After a disastrous fire at a girls' school in which a number of students died after religious police blocked their escape on grounds they were not fully covered, the government transferred responsibility for female education to the Ministry. Also a nonpartisan advisory group was formed in 2003 to revise the Saudi school curriculum to remove unfavorable references to other religions, and to strengthen instruction in higher education to prepare Saudi youth to better function in a world of globalization and high technology.

The economic boom also lured many workers from poor countries, attracted by the high wages and benefits available in Saudi Arabia. Most came from such countries as Pakistan, Korea, and the Philippines, but the largest single contingent was from Yemen, next door. However, the bottom dropped out of the Saudi economy in the late 1980s. Oil prices fell, and the kingdom was forced to draw heavily on its cash reserves. Yemen's support for Iraq during the Gulf War was the last straw; the Saudi government deported 850,000 Yemeni workers, seriously disrupting the Yemeni economy with the shutdown in remittances.

Continued low world oil prices in the 1990s had a very bad effect on what was formerly a freewheeling economy. As one reporter noted, "the days of oil and roses are over for ordinary Saudis."[8] In 1998, the country's oil income dropped 40 percent, to $20 billion; after a two-year surplus, the budget showed a $13 billion deficit. Lowered oil prices accounted only in part for the deficit. Monthly stipends ranging from $4,000 to $130,000 given to the 20,000-plus descendants of Ibn Saud continued to drain the treasury, while free education, health care, and other benefits guaranteed for all Saudis under the Basic Law of Government generated some $170 billion in internal debts. The steady downturn in the economy, along with significant population increase (it doubled in the past two decades) and high unemployment have called these social benefits into question. The age group most affected is that from 15 to 20. Although there is much truth to the observation by a Saudi prince that "ours is the most conservative country in the world," these mounting socio-economic problems clearly demand reforms in the nation's basic delivery system.

The Saudi economy experienced a 2-year surge in growth in 2003–2004, aided by the highest global oil price increases in two decades, with a $12 billion budget surplus after three years of deficits. The surplus is to be invested in infrastructure projects, particularly transportation and utilities. However, the country's heavy reliance on oil production continues to affect the non-oil sector. It grew barely 3.4 percent in 2003 compared with 14 percent for the oil sector. Unemployment remains high, with an estimated 13 percent of the work force idled. And the Saudi-ization of business and industry, with consequent reduction of expatriate labor, has been adversely affected by the lower job skills and expectations on young Saudis entering the labor force.

King Fahd died in August 2005, ending a decade in which he had been essentially wheelchair-bound and had ceded the responsibility for leadership to his half-brother Abdullah. As the longest-serving Saudi ruler after his father, he had presided over or initiated many significant developments. They included establishment of a more-or-less formal succession process, mediation to settle the civil war in Lebanon, and perhaps most important, the stationing of U.S. and other non-Muslim troops on sacred Islamic soil as a result of the Iraqi invasion of Kuwait. Ironically this action, although it came about logically out of fears that the Saudi regime was next on Saddam Hussein's list, has generated the organization of al Qaeda, headed by Osama bin Laden., whose goal is not only to destabilize the U.S but also to overthrow the Saudi government.

A CHANGING KINGDOM, INCH BY INCH

Its size, distance from major Middle Eastern urban centers, and oil wealth historically have insulated Saudi Arabia from the winds of political change. Domestic and foreign policy alike evolve from within the ruling family. Officials who undertake independent policy actions are quickly brought into line (an example being the freewheeling former oil minister Shaykh Zamani). The ruling family is also closely aligned with the ulema; Saudi rulers since Ibn Saud's time have held the title "Guardians of the Holy Mosques" (of Mecca and Medina), giving them a preeminent position in the Islamic world. The modern version of the Saudi-Wahhab partnership permits the ruler to appoint the Council of Senior Theologians, whose job it is to ensure Islamic cultural and social "rules" (such as women driving). In return, their presence and prescripts on Islamic behavior serve as an endorsement of the monarchy. But as one Saudi scholar told an interviewer, "the clerics can issue edicts to their hearts' content, answering weighty questions like whether or not a wife can wear jeans in front of her husband. Our clerics' religion has little to do with ethics, only lifestyle. They never do what they should do—denounce tyranny, injustice, corruption."[10]

Pressure to broaden participation in political decision making outside of the royals has increased markedly in recent years. This is due not only to greater contact by educated Saudis with more democratic political systems, but also to the vastly increased use of satellite dishes and the Internet. While agreeing in principle to changes, the House of Saud, strongly supported by the religious leaders, has held fast to its patriarchal system.

Given these strictures, it was somewhat surprising in 1991 when the ulema submitted a list of 11 "demands" to King Fahd. The most important one was the formation of a *Majlis al-Shura* (Consultative Council), which would have the power to initiate legislation and advise the government on foreign policy. The king's response, developed in deliberate stages with extensive behind-the-scenes consultation, in typical Saudi style, was to issue in February 1992 an 83-article "Organic Law," comparable in a number of respects to a Western constitution. The law sets out the basic rules for Saudi government.

A significant step in the country's glacial progress toward a more representative political system took place in early 2005. As mentioned earlier, the first public elections in history were held for members of newly-formed municipal councils. Voters were asked to choose half of some 178 council seats, the remainder being filled by government appointees with only men allowed to participate, voter turnout was small. Even so there were some surprising results. Islamist candidates in Riyadh won all the seats on its council, and in the primarily Shia Eastern Province Shia candidates also swept the ballot.

Saudi Arabia is defined in the Organic Law as an Arab Islamic sovereign state (Article 5), with Islam the state religion (Article 1), and as a monarchy under the rule of Ibn Saud's descendants. Other articles establish an independent judiciary under Islamic law (*shari'a*) and define the powers and responsibilities of the ruler.

Aside from some internal pressures, mainly from intellectuals, the main reason for Fahd's decision to broaden the political process was the Gulf War, which exposed the Saudi system to international scrutiny and pointed up the risks of patriarchal government. A major difference between the Saudi Organic Law and Western-style constitutions is the absence of references to political, civil, and social rights. Political parties as such remain illegal; but, in 1993, the first human-rights organization in the country, the Committee for the Defense of Legitimate Rights, was formed by a group of academics, tribal leaders, and government officials. Its members included the second-ranking religious scholar, Shaykh Abdullah al-Jubrien, and the former head of Diwan al-Mazalem, the Saudi equivalent of ombudsman. The Committee's goal was the elimination of oppression and injustice, which is considered an important part of its members' religious duty. But its emergence was perceived as a threat to the ulema. An edict condemned it, stating that there was no need for such an organization in a country ruled by Islamic law.

Satellite television, instant worldwide communications, the Internet, and other features of the contemporary interlinked world certainly threaten the self-imposed isolation of regimes such as Saudi Arabia's. In 2001, 42 Internet service providers were operational in the country. Although Internet usage is controlled through a single central-government authority that may block out websites considered pornographic or politically objectionable, Saudi Arabia's huge youth population (nearly 50 percent are under age 15) has easy access to satellite television or foreign websites, enabling viewers to circumvent such control. The opening of the U.S.-style Faisalah Mall in Riyadh in 2000 has at least provided Saudi young people with a public meeting-place and entertainment venue in the absence of movie theaters and other mass media centers.

Since the Gulf War, the country's purchases of large amounts of weaponry and the stationing of 5,000 American air and ground forces on Saudi soil have been strongly criticized by the Saudi public as well as by other Arab countries. Other Arab leaders have even accused Saudi Arabia (and Kuwait) of becoming U.S. satellites! In the past, the monarchy ignored such criticism; alignment with United States as its major ally and protector have been cornerstones of Saudi foreign policy since World War II. However, U.S. support for Israel and the presence of American forces on sacred Islamic soil have begun to fray the strands holding the alliance together. Further damage resulted from the revelation that the hijackers in the 9/11 attack on U.S. targets were Saudis. (Earlier, in 1996, a truck-bomb explosion at the U.S. Khobar base near Riyadh had killed 24 Americans and injured 400.).

Despite the American presence and its own vigilance, the country continues to be threatened by Islamic fundamentalists. After the seizure of the Great Mosque in Mecca by Sunni fundamentalists in 1979, several hundred of them were deported to Afghanistan, where they joined the resistance to the Soviet occupation. After the Soviet withdrawal in 1989, many of these "Arab Afghans" returned to Saudi Arabia and other Middle Eastern Islamic countries. Some even migrated to Europe or Canada, ultimately entering the United States. Most of them were Saudi nationals; they included Osama bin Laden. Although bin Laden was deprived of his Saudi citizenship and deported (to Sudan) in 1994, the nucleus of his terrorist organization remained in Saudi Arabia. It presents a serious widespread threat to the monarchy. In 2003 suicide bombers attacked residential districts in Riyadh, killing 9 Americans along with other non-Saudis. Since then Saudi security forces have been engaged in almost-weekly battles with al Qaeda militants organized into cells throughout the country. The government has offered bounties of $267,000 each for some 26 known militants. In September 2005 the U.S. consulate in Dhahran was closed due to extremist threats. Subsequently the Jiddah consulate was attacked, as al Qaeda continued to target Westerners in its effort to cripple the Saudi economy and overthrow the monarchy.

ACHIEVEMENTS

In 2000, Saudi doctors performed the world's first uterus transplant, from a 46-year-old woman to a 26-year-old who had a hemorrhage after childbirth. The transplant was successful at first, but it had to be removed after 90 days due to blood clotting in the patient.

The government also cooperated with U.S. and other intelligence organizations in closing down international banks that had served as fronts for al Qaeda funding. But due to the large number of mujahideen fighters who had returned from the wars in Afghanistan to infiltrate Suadi society, a full-scale crackdown remained difficult. Saudi officials estimated their number at between 10,000 and 60,000.

The presence of American, non-muslim forces in the country is certainly conductive to terrorism, as is the appeal of Osama bin Laden and his organization for young Saudis. But Saudi vulnerability stems more from internal weaknesses than external threats, as was the case earlier when Iraqi forces seized Kuwait. Not only has the economy not kept pace with the demands of a growing population, but the heavy hand of Wahhabism also limits social changes. To his credit, Prince Abdallah has made some changes.

The monarchy's gradualist approach to political reform took two small steps forward in 2004. The first was the formation of the Saudi Human Rights Association. Its mandate calls for investigation into reported violations of human rights. A second government-sponsored organization, the Council for the National Dialogue, is intended to bring together the various sectors of society for discussion of needed political reform issues. However, the detention in March of its key leaders on charges of anti-government criticism put a sharp brake on the reform movement, notably after religious leaders called them hostile to true Islam. The detainees were released subsequently but ordered to sign statements disavowing political action.

The Crown Prince has also taken some steps to revitalize the economy. Foreigners are now allowed to own property and new laws were approved in 2001–2002 to encourage foreign investment, particularly in the huge natural gas industry. Changes in the legal system, notably the Saudi-ization of banks and other industries, will in time reduce unemployment as more Saudis can be added to their staffs. However, the arrest in June 2003 of a prominent newspaper editor for publishing a critique of the religious establishment was a reminder that the Wahhabi side of government is still powerful. A Western diplomat provided a trenchant view of the situation: "The coasts of (Saudi Arabia) have been cosmopolitan for 5,000 years. In the middle of the country they have been goatherds for 5,000 years. If they could close the window on the world, they would. That's the problem."[11]

Saudi vulnerability stems from internal weaknesses rather than external threats, as was the case a decade ago when Iraqi forces had seized Kuwait and stood on the borders of the kingdom. King Fahd marked his 20th anniversary on the throne in 2001 in poor health, confined to a wheelchair, and reportedly suffering from Alzheimer's disease. Prior to his death and Abdullah's formal accession as king he had formed in 2000 a family council to manage royal-family affairs for the 7,000 or so princes; in 2001, he directed members to pay their own electricity and phone bills and placed a five-year moratorium on military contracts, another source of graft and kickbacks.

FOREIGN POLICY

The Iraqi invasion and occupation of Kuwait caused a major shift in Saudi policy, away from mediation in regional conflicts and bankrolling of popular causes (such as the Palestinian) to one of direct confrontation. For the first time in its history, the Saudi nation felt directly threatened by the actions of an aggressive neighbor. Diplomatic relations were broken with Iraq and subsequently with Jordan and Yemen, due to their support of the Iraqi occupation. Yemeni workers were rounded up and expelled, and harsh restrictions were imposed on Yemeni business owners in the kingdom. Profiting by the example of Israel's "security fence" built to block Palestinian gunmen and suicide attacks, Saudi Arabia closed its border with Yemen in January 2004. The purpose was to block entry of al Qaeda activists based in the latter country. Yemeni leaders protested the action as a violation of the Treaty of Taif. That treaty, approved in 2000, established a neutral zone between the two states which allows nomadic tribes to move about freely. After Yemeni president Saleh visited Riyadh to discuss the matter, the Saudis agreed to dismantle the barrier in return for increased joint border patrols to deal with smuggling and terrorist infiltration. Establishment of the UN/U.S.-led coalition against Iraq led to the stationing of foreign non-Muslim troops on Saudi soil, also a historic first.

The continued survival of Saddam Hussein's regime in Iraq resulted in huge Saudi purchases of U.S. military equipment, although a $1.7 billion arms deal was cancelled in 1999 due to the economic recession. In the last few years, the purchase of U.S. arms and the stationing of 5,000 American troops on Saudi soil has drawn criticism from other Arab countries. They have even accused the kingdom of becoming a U.S. satellite. In the past, Saudi rulers have ignored such criticism. Alignment with the United States as its major ally and arms supplier has been the cornerstone of Saudi policy since World War II. However, all-out U.S. support for Israel in its conflict with the Palestinians began to erode the alliance in 2000–2001. The government authorized public anti-American demonstrations in several Saudi cities, while the crown prince and other officials angrily criticized both the Clinton and George W. Bush administrations for their lack of even-handedness in the conflict. One observer noted: "The Islamists think the Saudis have sold out to the Americans, and the Americans think

they have sold out to the terrorists. Eventually this translates into an erosion of legitimacy—that if you are not satisfying the Arabs and Washington, you're on your own."[12]

The country's often difficult relationship with Iran underwent another change in 1991. The fall of the Iranian monarchy and establishment of the Islamic Republic had initially been welcomed by Saudi rulers because of the new regime's fidelity to Islamic principles. But, in 1987, Iranian pilgrims attending the pilgrimage to Mecca undertook anti-Saudi demonstrations that led to a violent confrontation with police, resulting in more than 400 casualties. The two countries broke diplomatic relations; and, in 1988, Saudi Arabia established a quota system for pilgrims on the basis of one pilgrim per 1,000 population. The quota system was described as necessary to reduce congestion on the annual pilgrimages, but in fact it would limit Iran to 50,000 pilgrims and limit Iranian-inspired political activism. Iran boycotted the pilgrimage in 1988 and 1989 as a result. Although it has remained free from politically-inspired violence since then, the hajj has been adversely affected in other ways. Some 1,426 pilgrims died on July 2, 1990, in a tunnel crush, and in the following year 340 were killed in fires in the overcrowded tent city of Mina, outside Mecca. In 2004 another stampede during the devil-stoning ritual at Mina caused nearly 500 casualties, and another in January 2006 killed 363. The difference between the hajj and other large-scale public events lies in its enormous logistical problems. As the director of public safety for the Saudi government observed: "When you get 300,000 people seeking to move all at once, accidents are bound to happen and they are quickly magnified."[13]

Relations with neighboring Gulf states have also improved. Long-time border disputes with Qatar and Yemen have been resolved amicably, with demarcation through largely featureless desert territory. In the Yemeni case, the border was demarcated by a joint arbitration commission to extend from Jebel Thar to the Omani border, on the basis of the 1934 Treaty of Taif.

Now that "terrorism" has become a household word in the West, and one particularly associated with Islam, many scholars and analysts have traced it to Wahhabism, Saudi Arabia's version of the faith. The partnership between the House of Sa'ud and the Wahhabi religious establishment has been a source of strength to the Saudi state. But it has also resulted in a state which supports and encourages militant Islam worldwide, providing significant funding thereto.

Timeline: PAST

1800
Wahhabis seize Mecca and Medina

1902
Ibn Saud captures Riyadh in a daring commando raid

1927
Ibn Saud is recognized by the British as the king of Saudi Arabia

1946
Oil exports get under way

1963
King Saud, the eldest son and successor of Ibn Saud, is deposed in favor of his brother Faisal

1975
Faisal is assassinated; succession passes by agreement to Khalid

1979
The Great Mosque in Mecca is seized by a fundamentalist Muslim group

1980s
King Khalid dies; succession passes to Crown Prince Fahd; Saudi jets shoot down an Iranian jet for violation of Saudi air space

1990s
Saudi Arabia hosts foreign troops and shares command in the Gulf War

PRESENT

2000s
The Saudi economy stabilizes
The Saudi-U.S. relationship is scrutinized in the wake of the September 11 terrorist attacks

NOTES

1. He had 24 sons by 16 different women during his lifetime (1880–1953). See William Quandt, *Saudi Arabia in the 1980's* (Washington, D.C.: Brookings Institution, 1981), Appendix E, for a genealogy.
2. George Rentz, "The Saudi Monarchy," in Willard A. Beling, ed., *King Faisal and the Modernization of Saudi Arabia* (Boulder, CO: Westview Press, 1980), pp. 26–27.
3. *Ibid.,* p. 29.
4. "Saudi Arabia's Centennial," *Aramco World,* Vol. 50, No. 1 (January–February 1999), pp. 21–22. The walls and gates were demolished in 1953 under the "relentless pressure" of modernization, but the Masmak and other structures dating from Ibn Saud's time have been preserved as museums to celebrate the nation's past.
5. The wall was torn down by his successor, King Faisal. Justin Coe, in *The Christian Science Monitor* (February 13, 1985).
6. Gordon Gaskill, "Saudi Arabia's Modern Monarch," *Reader's Digest* (January 1967), p. 118.
7. Ministry of Information, Kingdom of Saudi Arabia, *Faisal Speaks* (n.d.), p. 88.
8. Douglas Jehl, in *The New York Times* (March 20, 1999).

9. Susan Sachs, in *The New York Times* (December 4, 2000).
10. David Hirst, "Corruption, Hard Times Fuel Desert Discontent," *The Washington Times* (September 29, 1999), p. 150.
11. Quoted by Nicholas Blanford, "Reformist impulse in Saudi Arabia," *Christian Science Monitor,* June 5.
12. Stephen Schwartz, *The Two Faces of Islam the House of Sa'ud from Tradition to Terror* (New York: Doubleday, 2002 pp. 64–65. A man in Jiddah was given 4,750 lashes for adultery with his sister-in-law, although the Qur'anic limit is 100.
13. Hassan M. Fattah, "Why Mecca's Pilgrims Need Engineering, Not Just Prayer," *The New York Times*, January 17, 2006. An architect in Jidda noted that 'the three main variables in managing the hajj are density, space, and time. So far all they have been dealing with is space.' *Ibid.*

Sudan (Republic of the Sudan)

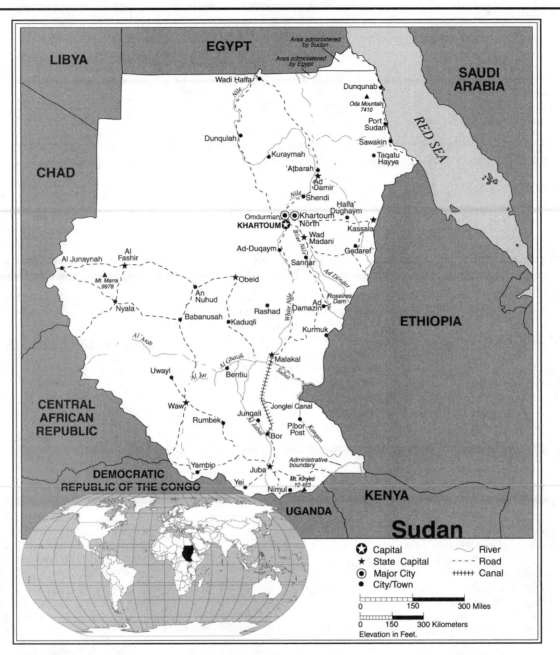

Sudan Statistics

GEOGRAPHY

Area in Square Miles (Kilometers):

892,068 (2,505,810) (about 1/4 the size of the United States)

Capital (Population): Khartoum (948,000)

Environmental Concerns: little potable water; threatened wildlife populations; soil erosion; desertification

Geographical Features: generally flat, featureless plain; mountains in the east and west

Climate: varies from arid desert in the north to tropical in the south

PEOPLE

Population

Total: 40,187,486

Annual Growth Rate: 2.6%

Rural/Urban Population Ratio: 68/32

Major Languages: Arabic (official); various Nubian, Nilotic and other

153

African languages in south; English widely used
Ethnic Makeup: 52% black; 39% Arab; 6% Beja; 3% others
Religions: 70% Sunni Muslim in north; 25% indigenous beliefs; 5% Christian, mostly in the south and Khartoum

Health

Life Expectancy at Birth: 57 years (male); 59.8 years (female)
Infant Mortality Rate: 62.5/1,000 live births

Education

Adult Literacy Rate: 61%

COMMUNICATION

Telephones: 162,000 main lines
Daily Newspaper Circulation: 21 per 1,000 people
Televisions: 8.2 per 1,000 people
Internet Users: 1 (2000)

TRANSPORTATION

Highways in Miles (Kilometers): 7,390 (11,900)
Railroads in Miles (Kilometers): 3,425 (5,516)
Usable Airfields: 61
Motor Vehicles in Use: 75,000

GOVERNMENT

Type: transitional
Independence Date: January 1, 1956 (from Egypt and the United Kingdom)
Head of State/Government: President Omar Hassan al-Bashir (continues in office under the 2005 peace agreement; no prime minister as such, but under the coalition government Gen. Kir Mayardit and Ali Osman Muhammad Taha serve as vice-presidents
Political Parties: under the peace accord former political "associations" approved as parties. They include the Popular National Congress (PNC), Democratic Unionist Party, Umma, National Democratic Alliance
Suffrage: universal at 17

MILITARY

Military Expenditures (% of GDP): 7.3% (est.)
Current Disputes: demarcation of entire boundary with Ethiopia underway as of 2002; dispute with Kenya over latter's claim to the "Ilmi Triangle"; dispute with Egypt over areas north and south of border demarcated by 1899 treaty along 22nd parallel of latitude

ECONOMY

Currency ($U.S. Equivalent): 257.9 dinars = $1
Per Capita Income/GDP: $1,900/$76 billion
GDP Growth Rate: 6.4%
Inflation Rate: 9%
Unemployment Rate: 18.7%
Labor Force: 11,000,000
Natural Resources: petroleum; small reserves of iron ore; copper; chromium ore; zinc; tungsten; mica; silver; gold
Agriculture: cotton; groundnuts; sorghum; millet; wheat; gum arabic; sesame; sheep
Industry: cotton ginning; textiles; cement; edible oils; sugar; soap distilling; shoes; petroleum refining; armaments
Exports: $3.39 billion (primary partners Japan, China, Saudi Arabia)
Imports: $3.496 billion (primary partners China, Saudi Arabia, Britain)

SUGGESTED WEBSITES

http://lcweb2.loc.gov/frd/cs/sdtoc.html

Sudan Country Report

Sudan is the largest nation on the African continent. It extends from its northern border with Egypt and the Libyan and Nubian Deserts southward deep into tropical Africa. Its territory includes the Blue and White Nile Rivers, which join at Khartoum to form the Nile, Egypt's lifeline.

The name of the country underscores its distinctive social structure. Centuries ago, Arab geographers named it Bilad al-Sudan, "Land of the Blacks." The northern half, including Khartoum, is Arabic in language, culture, and traditions, and Islamic in religion. However, the admixture of Arab and African peoples over 2,000 years has produced a largely black Arab population.

Southern Sudan is the home of a large number of black African tribes and tribal groups, the largest being the Dinka. Other important ones are the Shilluk, Nuer, and in western Sudan the Azande and Bor. They make up approximately 30–35 percent of the total population. About 5 percent are Christian.

The two halves of Sudan have little or nothing in common. The country's basic political problem is how to achieve unity between these two different societies, which were brought together under British rule to form an artificial nation.

HISTORY

The ancient history of Sudan, at least of the northern region, was always linked with that of Egypt. The pharaohs and later conquerors of Egypt—Persians, Greeks, Romans, and eventually the Arabs, Turks, and British—periodically attempted to extend their power farther south. The connection with Egypt became very close when the Egyptians were converted to Islam by invading armies from Arabia, in the seventh century A.D. As the invaders spread southward, they converted the northern Sudanese people to Islam, developing in time an Islamic Arab society in northern Sudan. Southern Sudan remained comparatively untouched, because it was isolated by the geographical barriers of mountain ranges and the great impassable swamps of the Nile.

The two regions were forcibly brought together by conquering Egyptian armies in the nineteenth century. The conquest became possible after the exploration of sub-Saharan Africa by Europeans. After the explorers and armies came slave traders and then European fortune hunters, interested in developing the gold, ivory, diamonds, timber, and other resources of sub-Saharan Africa.

The soldiers and slave traders were the most brutal of all these invaders, particularly in southern Sudan. In fact, many of the slave traders were Muslim Sudanese from the north. The Civil War between the Islamic north and the Christian/animist south began essentially in 1955 before independence and continued intermittently until 2005. However its roots in the nineteenth-century experiences of the southerners, as "memories of plunder, slave raiding and suffering" at the hands of slavers and their military allies were passed down from generation to generation.[1]

THE ORIGINS OF THE SUDANESE STATE

The first effort to establish a nation in Sudan began in the 1880s, when the country was ruled by the British as part of their protectorate over Egypt. The British were despised as foreign, non-Muslim rulers. The Egyptians, who made up the bulk of the security forces assigned to Sudan, were hated for their arrogance and mistreatment of the Sudanese.

In 1881, a religious leader in northern Sudan announced that he was the *Mahdi,* the "Awaited One," who, according to Sunni Islamic belief, would appear on Earth, sent by God to rid Sudan of its foreign rulers. The Mahdi called for a *jihad* (struggle or holy war) against the British and the Egyptians.

Sudanese by the thousands flocked to join the Mahdi. His warriors, fired by revolutionary zeal, defeated several British-led Egyptian armies. In 1885, they captured Khartoum, and, soon thereafter, the Mahdi's rule extended over the whole of present-day Sudan. For this reason, the Mahdi is remembered, at least in northern Sudan, as Abu al-Istiqlal, the "Father of Independence."[2]

The Mahdi's rule did not last long; he died in 1886. His chief lieutenant and successor, the Khalifa Abdallahi, continued in power until 1898, when a British force armed with guns mowed down his spear-carrying, club-wielding army. Sudan was ruled jointly by Britain and Egypt from then until 1955. Since the British already ruled Egypt as a protectorate, for all practical purposes joint rule meant British rule.

Under the British, Sudan was divided into a number of provinces, and British university graduates staffed the country's first civil service.[3] But the British followed two policies that have created problems for Sudan ever since it became independent. One was "indirect rule" in the north. Rather than developing a group of trained Sudanese administrators who could take over when they left, the British governed indirectly through local chiefs and religious leaders. The second policy was to separate southern from northern Sudan through "Closed Door" laws, which prohibited northerners from working in, or even visiting, the south.

Sudan became independent on New Year's Day 1956, as a republic headed by a civilian government. The first civilian government lasted until 1958, when a military group seized power "to save the country from the chaotic regime of the politicians."[4] But the military regime soon became as "chaotic" as its predecessor's. In 1964, it handed over power to another civilian group. The second civilian group was no more successful than the first had been, as the politicians continued to feud, and intermittent conflict between government forces and rebels in the southern region turned into all-out civil war.

In 1969, the Sudanese Army carried out another military coup, headed by Colonel Ja'far (or Gaafar) Nimeiri. Successive Sudanese governments since independence, including Nimeiri's, have faced the same basic problems: the unification of north and south, an economy hampered by inadequate transportation and few resources, and the building of a workable political system. Nimeiri's record in dealing with these difficult problems is one explanation for his longevity in power. A written Constitution was approved in 1973. Although political parties were outlawed, an umbrella political organization, the Sudan Socialist Union (SSU), provided an alternative to the fractious political jockeying that had divided the nation before Nimeiri.[5]

Nimeiri's firm control through the military and his effectiveness in carrying out political reforms were soon reflected at the ballot box. He was elected president in 1971 for a six-year term and was re-elected in 1977. Yet broad popular support did not generate political stability. There were a number of attempts to overthrow him, the most serious in 1971 and 1976, when he was actually captured and held for a time by rebels.

One reason for his survival may be his resourcefulness. After the 1976 coup attempt, for example, instead of having his opponents executed, he invited them and other opposition leaders to form a government of national unity. One of Nimeiri's major opponents, Sadiq al-Mahdi, a great-grandson of the Mahdi and himself an important religious leader, accepted the offer and returned from exile.

DEVELOPMENT

Renewal of oil production and the end of the civil war has enabled the country to move upward from the bottom of the Middle East economic ladder by a few steps. Per capita income in 2005 increased 9 percent over that of 2004 and oil output reached 500,000 barrels per day (bpd).

Nimeiri's major achievement was to end temporarily the Civil War between north and south. An agreement was signed in 1972 in Addis Ababa, Ethiopia, mediated by Ethiopian authorities, between his government and the southern Anya Anya resistance movement. The agreement provided for regional autonomy for the south's three provinces, greater representation of southerners in the National People's Assembly, and integration of Anya Anya units into the armed forces without restrictions.

THE COUP OF 1985

Nimeiri was reelected in 1983 for a third presidential term. Most of his political opponents had apparently been reconciled with him, and the army and state security forces were firmly under his control. It seemed that Sudan's most durable leader would round out another full term in office without too much difficulty. But storm clouds were brewing on the horizon. Nimeiri had survived for 16 years in power largely through his ability to keep opponents divided and off balance by his unpredictable moves. From 1983 on, however, his policies seemed designed to unite rather than divide them.

The first step in Nimeiri's undoing was his decision to impose Islamic law (*Shari'a*) over the entire country. The impact fell heaviest on the non-Muslim southern region. In a 1983 interview, Nimeiri explained his reasons for the action. His goal from the start of his regime, he said, was "to raise government by the book [i.e., the Koran] from the level of the individual to that of government." If the Sudanese, with their numerous ethnic and cultural differences and the country's vast size, were governed properly by God's Book, they would provide an example of peace and security to neighboring countries.[6]

In Nimeiri's view, the application of Islamic restrictions on alcohol, tobacco, and other prohibited forms of behavior was appropriate to Sudanese Muslims and non-Muslims alike, since "Islam was revealed to serve man and all its legislation has the goal of regulating family, social, and individual life and raising the level of the individual."[7]

The new draconian measures were widely resented, but particularly in the south, where cigarettes and home-brewed beer were popular palliatives for a harsh existence. When Nimeiri continued his "Islamic purification" process with a reorganization of Sudanese administration into several large regions in order to streamline the cumbersome bureaucracy inherited from the British, the southerners reacted strongly. Consolidation of three autonomous provinces into one directly under central-government control was seen by them as a violation of the commitment made to regional autonomy that had ended the Civil War. An organized guerrilla army, the Sudan People's Liberation Army (SPLA), resumed civil war under the expert leadership of U.S.-trained colonel John Garang. The rebels' new strategy was not only to oppose government troops but also to strike at development projects essential to the economy. Foreign workers in the newly developed oil fields in southwestern Sudan were kidnapped or killed; as a result, Chevron Oil Company halted all work on the project.

A crackdown on Islamic fundamentalist groups, particularly the Muslim Brotherhood,

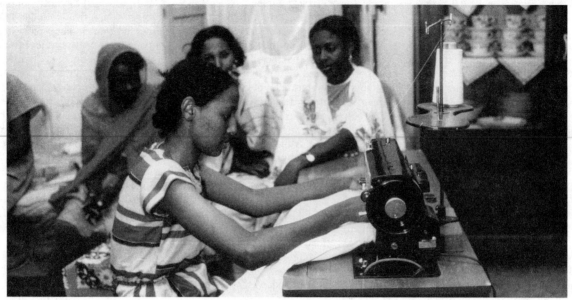

The attainment of political stability is important to Sudanese development, but will be a strong economy that makes for lasting peace. Job training is extremely important. This woman in a sewing class in Khartoum represents the need to create a skilled-labor pool.

added to Nimeiri's growing list of opponents. Members of the Brotherhood had been active in implementing Islamic law as the law of the land, but Nimeiri felt that they had gone too far. By late 1984, it appeared that the president had angered or alienated everybody in the country, all for different reasons.

In the end, it was the failure of his economic policies rather than anything else that brought about Nimeiri's fall. The International Monetary Fund imposed strict austerity requirements on Sudan in 1984 as a prerequisite to a $90 million standby loan to enable the country to pay its mounting food and fuel bills. The food bills were aggravated by famine, the fuel bills by the necessity to import almost all fuel requirements. The IMF insisted on drastic budget cuts, devaluation of currency, and an end to subsidies on basic commodities. If Nimeiri had been able to carry out these reforms, he would have stood a chance of restoring the country to solvency and his own rule to respectability. Protests turned to riots, mainly over the end of price subsidies and a consequent 33 percent increase in the prices of such necessities as bread, sugar, and cooking oil. Other protests erupted over the application of Islamic law, especially the ban on alcohol, which brought thousands of Sudanese into the streets shouting "We want beer! We want beer!"

Nimeiri's departure for the United States to seek further economic help triggered a general strike in 1985. A genuine national movement arose, uniting students and professionals with the urban poor, all demanding that Nimeiri resign. Fearing anarchy or an uprising by young army officers, the senior military leaders moved quickly, took over the government, and ordered Nimeiri deposed. Crowds in Khartoum shouted, "Nimeiri the butcher is finished; the country belongs to the people." "He's nothing, let him sell lemons," cried one demonstrator, and others tore Nimeiri's picture from their devalued bank notes.[8]

The new military government, headed by General Abd al-Rahman Swareddahab, a highly respected senior officer, promised to hold elections within a year to restore civilian rule and to revive political parties. That promise was kept: In 1986, elections were held for a new People's Assembly. Two revived pre-Nimeiri parties, the Umma and the Democratic Unionist Party (DUP), won the majority of seats, with the fundamentalist National Islamic Front emerging as a strong third party. Sadiq al-Mahdi, head of the Umma, automatically became prime minister; his principal rival, DUP leader Ahmed Ali al-Mirghani, was chosen as president. The new prime minister chose a coalition cabinet to begin the arduous process of restoring the democratic process to Sudan after 15 years of Nimeiri.

But the euphoria over the departure of "Nimeiri the Butcher" soon gave way to the realization that the problems that had daunted him remained unresolved. They included heavy foreign indebtedness, a weak economy, inefficient agricultural production, an inadequate transportation system, party and personal rivalries, and extreme distrust between north and south in the divided Sudanese nation.

INTERNAL PROBLEMS

The al-Mahdi government had no more success than its predecessors in resolving Sudan's endemic political disunity. Efforts to limit the application of Islamic law throughout the country were blocked by the National Islamic Front (NIF) in 1988. The Civil War then heated up. SPLA success in capturing the principal towns in the south led the DUP to sign a separate agreement with the rebels for a cease-fire. The People's Assembly rejected the agreement, and the DUP then withdrew from the government.

Faced with the imminent collapse of civilian authority, the armed forces again seized power, in Sudan's fourth military coup since independence. The army moved after food shortages and soaring inflation, fed by war costs of $1 million a day, led to riots in Khartoum and other cities. A Revolutionary Council, headed by Lieutenant General Omar Hassan al-Bashir, suspended the Constitution and arrested government leaders.

In 1992, Bashir appointed a 300-member National Transitional Assembly to lay the groundwork—at least in theory—for a return to civilian rule. Its members included military leaders (those who sat on the ruling Revolutionary Council), provincial (state) governors, and some former government leaders. Its primary function was to implement Council decrees; during the transitional period; however, it could develop legislation.

The regime also sought to broaden its popular base through the establishment of local elections. The elections were held in two stages, the first stage being the election of people's congresses (at the village and town level); in the second stage of the process, the congresses then elected provincial legislatures. Due to the Civil War, the southern region remained unrepresented.

The gradual return to representative government in the 1990s improved Sudan's image internationally. Elections were held in 1996 for a 400-member National People's Assembly and to choose a president. Not surprisingly, Bashir was elected president; he received 75.5 percent of the 5.5 million votes cast. (The south was excluded from the election process.) Subsequently, Hassan al-Turabi, leader of the NIF and the architect of Sudanese Islam, was elected Speaker of the new Assembly.

In June 1998, the regime enacted a number of constitutional reforms. They allow political parties to form, although they are to be registered officially as "associations." Freedoms of speech, assembly, and the press are guaranteed under the reforms, although political parties that receive foreign funding or "go beyond the bounds of religion" can be proscribed.

HEALTH/WELFARE

Development of the oil industry by an international consortium of oil companies has been widely criticized abroad because it excludes the non-Muslim population from the benefits of production and export. In March 2000, the Clinton administration barred U.S. companies from doing business with the consortium for this reason.

In 1998 and 1999, the regime called on political opponents to return and help build a "new democratic Sudan." Several did so. They included Nimeiri and Sadeq al-Mahdi. However, rivalry between Bashir and Turabi, which came to a head in 1999, halted the restoration of representative government. Turabi presented constitutional amendments to the Assembly in December. If approved, they would abolish the position of prime minister (held under the 1998 reorganization of his government) as well as his control over provincial governorships. The president could also be impeached by a two-thirds vote in the Assembly. Inasmuch as Turabi's party controlled the great majority of its 400 seats, approval of the amendments seemed inevitable.

On December 12, however, the embattled president struck back. He declared a three-month state of emergency, sus-

pended the Constitution and the Assembly, and dismissed his entire cabinet. In January 2000, he removed all state governors from office and appointed a new cabinet, with most of the ministries now held by his own supporters. "God willing with this team we will guide Sudan toward peace," he told the nation in a public address.

In March 2001, the conflict between the two leaders intensified when Turabi was arrested. He was charged with complicity with the southern rebels, because he had signed a "memorandum of understanding" with SPLA leaders. The memorandum called for an end to the two-year-old state of emergency and recognition of Sudan's religious and cultural pluralism.

Prior to his arrest, Turabi had formed a political "association" (so called because political parties remain banned) to replace the NIF. The new association, the Popular National Congress (PNC), had boycotted the December 2000 presidential election on grounds that the election had been "cooked and prearranged" by Bashir and his supporters. Bashir was then reelected by an 86 percent majority of voters, although only 66 percent of those eligible to vote cast their ballots. Former Sudanese dictator Jaafar Nimeiri, invited to return after years of exile in Egypt, ran against the president as an independent, receiving 9 percent of the popular vote.

Military rule eased in 2001. Bashir announced an amnesty for opponents of his regime, and two important civilian leaders accepted the offer. Former prime minister Sadeq al-Mahdi returned from voluntary exile in December 2000. In December 2001, Ahmed al-Mirghani, deputy head of the Democratic Unionist opposition party, returned to a welcome by tens of thousands after a 12-year exile in Egypt. He was greeted by Bashir in person as a "unifying symbol of the state," part of the president's newest effort at political reconciliation as a prelude to restoration of civilian government.

THE CIVIL WAR

The government gained some ground against rebels in the Civil War in late 1991, when the SPLA split into contending factions. One faction, led by Lieutenant Rick Machar, accused SPLA commander John Garang of a dictatorial reign of terror within the organization. The split became tribal when Nuer troops of Machar's faction invaded Dinka territory; the Dinkas are Garang's main supporters. Some 100,000 Dinkas fled their homeland during the fighting. Sudan government forces took advantage of internal SPLA rivalry to capture several important southern towns during an offensive in March 1992.

The ethnic killings of Nuers and Dinkas, along with famine (which has been intensified by the SPLA infighting), led other African states and, in late 1993, the United States to attempt to mediate and bring the two factions together as a prelude to ending the Civil War. But even the presence of former U.S. president Jimmy Carter as mediator failed to bridge the differences separating the two SPLA leaders. And the major differences between southerners and northerners—imposition of Islamic law on non-Muslims, revenue sharing among regions, states' rights and powers versus those of the national government—seemed insurmountable.

In the late 1990s, the National Islamic Front, renamed the National Congress Party in an effort to soften its fundamentalist image, won almost complete control over the Revolutionary Council. The security forces, the judiciary, and the universities were purged of moderate or liberal staff members and replaced by NIF fundamentalists. In 1991, the regime had bowed to NIF pressure and issued an edict making Islamic law the "law of the land" in both north and south Sudan. As a result, the Civil War intensified.

In 1997, the government signed peace agreements with several SPLA factions after rebel successes threatened to win the entire southern region. The agreements specified a four-year autonomy period for the south. At the end of the period, the population could choose between independence and integration on a basis of equality with the Muslim north.

The Bush administration appointed ex-U.S. Senator John Danforth as its special envoy to mediate in the conflict. Earlier, the two SPLA factions had united temporarily to present a common front against government forces. Danforth's first step in mediation was to negotiate a cease-fire in the Nuba region. Its population has been caught between government forces and the SPLA, and as a result is living under near-famine conditions. A cease-fire was arranged in January 2002 and continues to hold as of this writing. However, air attacks by government helicopter gun ships on UN food distribution centers elsewhere in February caused Danforth to suspend his mediation efforts.

Early in 2003 the peace talks were renewed, this time in neighboring Kenya and under joint Libyan-Egyptian sponsorship. But many questions remained to be resolved. They included allocation of a fair proportion of oil revenues to southerners (where the oil fields are located), the nature of the proposed transitional government, and restitution to families of southerners kidnapped and sold into slavery. As a "sweetener" the U.S. warned that unless the government negotiated in good faith, it would reimpose economic sanctions on the country and begin funding the SOLA. But as a former government official observed: "One amazing feature of the Sudanese conflict is its tenacity. Like the mythical phoenix, whenever the flames of war die down—it surprises friend and foe by emerging from the ashes reinvigorated." [9]

The appointment in October 2001 by the George W. Bush administration of former senator John Danforth as its special envoy to mediate the Civil War underscored expanded U.S. concern about Sudan's involvement as a state sponsor of terrorism. But of equal concern was the Sudanese regime's record of repression and denial of human rights to its southern non-Muslim population.

Earlier, the Garang and Machar factions of the SPLA had patched up their differences to present a united front against government forces. Although he had arrived with no detailed peace program, Danforth urged an end to attacks on civilians, abductions, and forced slavery of southerners, and asked for a permanent cease-fire in the Nuba Mountains, whose inhabitants have been caught between government and SPLA forces and deprived of food. Danforth negotiated a cease-fire in that region in January 2002.

After three years of off-and-on intense negotiations, effectively mediated by Kenya's Gen. Lazaro Sumbeiyo, SPLA chief John Garang and President Bashir signed a far-reaching agreement to end Africa's longest civil war. Altogether it had lasted half a century and caused 2 million casualties. Sumbeiyo's mediation was critical in moving the process forward. As U.S. Senator John Danforth, his predecessor as mediator, observed: "He should win the Nobel Peace Prize. His ability to be an honest broker was essential and very largely responsible for the result." [10]

Under the peace agreement (which was signed by ex-U.S. Secretary of State Colin Powell as a witness), Garang was sworn in as vice-president of the new government. He became the first southerner to hold a top cabinet post. Garang's tragic death in a plane crash in August raised doubts about the peace pact, although his second-in-command was immediately named to replace him.

The pact establishes a 6-year interim period; at the end of the period the southern population will vote on whether to secede or remain in the Sudanese union.

Although the civil war has officially ended, continued conflict in the northwestern province of Darfur, the country's poorest, threatens to undermine the peace pact. This conflict has pitted government-backed militias (*janjaweed*, in Arabic) against rebel groups separate from the SPLA. It began in 2003. The janjaweed, ignoring several cease-fires, carried on a systematic campaign against the population, burning villages, destroying crops and kidnapping or killing villagers there. By 2005 it had claimed over 200,000 lives and forced 2.4 million people from their homes, most of them fleeing as refugees into camps hastily set up in neighboring Chad. In March, the UN Security Council approved a resolution imposing an arms freeze and travel ban on Sudan. A second resolution established a UN peacekeeping mission to the devastated province. The Organization for African Unity (OAU) then agreed to provide a 2,000-man peacekeeping force drawn from its members, while a UN report documented the mass killings and "crimes against humanity" of the janjaweed but stopped short of charges of genocide.

Early in November the OAU force was increased to 7,000. But in a province as large as Darfur, these numbers proved totally inadequate. A number of peacekeepers were killed, and attacks by Sudan government aircraft on civilians sent an additional 10,000 refugees fleeing into Chad. To make matters worse, the refugees began kidnapping Sudanese aid workers to call international attention to their plight, while the janjaweed turned to attacks on vehicles belonging to the aid agencies.

Early in 2006 the Darfur conflict intensified as janjaweed began raiding across the Chad border, not only attacking refugee camps but also driving Chadians from their homes and villages and stealing their crops and cattle. An agreement between the two governments mediated by Libyan leader Muammar al-Qadhafi eased tensions temporarily; the two countries had accused each other of fostering anti-state insurgencies. But without a large-scale intervention by outside forces and a much larger UN peacekeeping force, the conflict seemed likely to drag on indefinitely.

THE ECONOMY

Although the attainment of political stability is important to Sudanese development, much depends on building the economy. The Sudanese economy is largely dependent on agriculture. The most important crop is cotton. Until recently, the only other Sudanese export crop of importance was gum arabic.

Because Sudan has great agricultural potential, due to its rivers, alluvial soils, and vast areas of unused arable land, Nimeiri had set out in the 1970s to develop the country into what experts told him could be the "breadbasket" of the Middle East. To reach this ambitious goal, some cotton plantations were converted to production of grain crops. The huge Kenana sugar-refinery complex was started with joint foreign and Sudanese management; the long-established Gezira cotton scheme was expanded; and work began on the Jonglei Canal, intended to drain a vast marshy area called the Sudd ("swamp") in the south, in order to bring hundreds of thousands of acres of marshlands under cultivation. But the breadbasket was never filled. Mismanagement and lack of skilled labor delayed some projects, while others languished because the roads and communications systems needed to implement them did not exist.

FREEDOM

Sudan's 1998 Constitution guarantees civil rights for all citizens and establishes separation of powers in government and a multiparty political system. Although the National Assembly has been reinstated and a presidential election was held in 2000, civil rights are limited by the state of emergency and the exclusion from the political process of the southern region.

It may be that oil, rather than agriculture, holds the key to Sudan's economic growth and indirectly its internal peace. Oil was discovered by Chevron in the southwestern region in the 1970s. The two oil fields there were being developed when the north-south war resumed. In 1984, three foreign oil workers were killed by rebels in an attack on the Bentiu facility, and Chevron withdrew and closed down its entire installation.

In the late 1990s, a temporary halt to the Civil War made resumption of oil exploration feasible. A consortium of three foreign oil companies (Talisman of Canada and the state oil companies of China and Malaysia), along with the Sudan National Oil Company, built a 936-mile, $1.2 billion pipeline from the former Chevron fields near Haglig northeast to Port Sudan. The pipeline was completed in less than a year. Exports from the Port Sudan refinery had reached 320,000 bpd by 2005. In April some 60 nations pledged $4.5 billion for reconstruction of the southern provinces, including $765 million from the European Commission. It would be used for food, refugee resettlement, schools, roads, and hospitals. The total cost of reconstruction was estimated to be $7.9 billion. The U.S. contribution would be $1.7 billion; however this was to be contingent on resolution of the Darfur conflict.

Although it was criticized for supporting a repressive government with oil revenues, enabling it to purchase new weaponry, Talisman officials insisted that they were helping the Sudanese people to meet urgent social needs.

But the message reached only deaf ears in the United States. In June 2001, Congress passed a series of measures that would prohibit foreign oil companies working in Sudan from raising capital in the United States, or trading their securities in American financial markets in order to finance their oil operations. The actions would force them to give up their seats on the New York Stock Exchange or Nasdaq.

FAMINE

The Sudanese people traditionally have lived in a barter economy, with little need for money. Huge budget deficits and high prices for basic commodities hardly affect the mass of the population. But the Civil War and a 12-year drought cycle in the sub-Saharan Sahel region, which includes Sudan, have changed their subsistence way of life into one of destitution.

The drought became critical in 1983, and millions of refugees from Ethiopia and Chad, the countries most affected, moved into temporary camps in Sudan. Then it was Sudan's turn to suffer. Desperate families fled from their villages as wells dried up, cattle died, and crops wilted. By 1985, an estimated 9 million people, half of them native Sudanese, were dying of starvation. Emergency food supplies from many countries poured into Sudan; but due to inadequate transportation, port delays, and diversion of shipments by incompetent or dishonest officials, much of this relief could not be delivered to those who most needed it. Bags of grain lay on the docks, waiting for trucks that did not come because they were immobilized somewhere else, stuck in the sand or mired in the mud of one of Sudan's few passable roads.

ACHIEVEMENTS

Although Sudan failed to win a seat on the UN Security Council in 2000, the country gained one on the Human Rights Council when the U.S. seat became vacant.

Heavy rains in the rainy season regularly washed out sections of track of Sudan's one railroad, the only link with remote provinces other than intermittent air drops. By 1987, it was estimated that 2,000 children a day were dying from malnutrition-related diseases.

Prodded into action in 1989, after a drought-related famine had caused 250,000 deaths, the United Nations organized "Operation Lifeline Sudan," a consortium of two of its agencies (Unicef and the World Food Program) and 40 humani-

tarian nongovernmental organizations (NGOs). Humanitarian aid averaging $1 million a day reduced the number of Sudanese requiring emergency relief in the 1990s. However, prolonged drought and the ongoing Civil War increased their numbers significantly in 2000. As of January 2001, the World Food Program (WFP) was feeding 1.7 million people in Sudan, most of them in the provinces of Darfur, Kordofan, Equatoria, and Jonglei. A ban on cattle imports due to outbreaks of foot-and-mouth disease elsewhere added to the misery of southerners, many of whom have only their cattle herds as their assets.

These difficulties have been compounded by the periodic ban on relief flights imposed by the government and requisitioning of food stocks, sometimes by the military, but also by the SPLA or local militias. The Sudanese military's policy forcibly removing villagers from oil-field areas and using food as a weapon, along with rape, forced labor, and abduction of southern youths to be taken to Khartoum as slaves or household servants for Muslim families, has turned a civil war into one of extermination of people.

FOREIGN POLICY

Aside from the internal devastation of the Civil War, Sudan's somewhat unwitting involvement in events outside its borders has affected its economic survival as well as its political stability. The country sided with Iraq during the Gulf War, and consequently 300,000 Sudanese expatriate workers were expelled from the Arab states in that area. Their return put further strain on the weak economy and eliminated $445 million annually in worker remittances, which had been an important source of revenue. The regime's reinstatement of Parliament and gradual steps toward representative government helped to improve Sudan's relations with its neighbors. Diplomatic relations were restored with Egypt and Eritrea in 2000. A treaty with Uganda withdrew Sudanese support for the Lord's Resistance Army, an opposition force to the Ugandan government.

Sudan's identification with international terrorism led the UN Security Council to approve *Resolution 1044* in 1996. It imposed economic sanctions on the country. The United States put into place its own sanctions after Sudanese nationals were implicated in the 1993 bombing of the World Trade Center building in New York.

In August 1998, the Clinton administration's firm belief that Sudan was a major sponsor of international terrorism led to the bombing of the Al-Shifa pharmaceutical plant near Khartoum. The plant, one of six

Timeline: PAST

1820
An Egyptian province under Muhammad Ali

1881
Mahdi rebellion against the British and Egyptians

1898
The British recapture Khartoum; establishment of joint Anglo-Egyptian control

1955
The Civil War begins

1956
Sudan becomes an independent republic

1969
Nimeiri seizes power

1980s
Nimeiri is overthrown in a bloodless coup; millions of people die of starvation; the Civil War resumes in the south

1990s
The regime institutes systematic slavery

PRESENT

2000s
Sudan remains ravaged by the Civil War Sudan cooperates in the international effort against terrorism

in the country, produced drugs, medicines, and veterinary medications. In January it had been granted a $199,000 contract to ship 100,000 cartons of Shifazole (an antibiotic used to treat parasites in animals) to Iraq. The shipment was to be made for humanitarian purposes; hence, it would be exempt from UN-imposed sanctions on that country. But the United States claimed that the shipment would include a chemical that could be used to manufacture the nerve gas UFX. Subsequent investigation proved that the Al-Shifa plant was involved exclusively with production of pharmaceuticals and that the United States had erred in bombing it. Since then the Sudanese government has preserved the ruined plant as a showcase symbolizing its mistreatment at the hands of the world's only superpower.

Ironically, the American missile attack united the Sudanese people behind their often-reviled regime. Mobs destroyed the unmanned U.S. embassy in Khartoum as angry demonstrators shouted "the tomb will await our enemies whatever the cost." A Sudanese scholar-professor and former member of the government, ousted for his arguments in favor of secular Islamic reform, noted that "The U.S. has shot itself in the foot.... This is not terrorism. You don't deal with a resurgent Muslim world through cruise missiles."

U.S. sanctions on Sudan for its support of international terrorism and harboring of terrorist mastermind Osama bin Laden led to his expulsion from Sudan in 1996. In its continuing effort to regain international respectability, the regime also offered the use of its military facilities to the United States after the September 11, 2001, terrorist bombings. It signed UN Security Council *Resolutions 1044, 1054,* and *1066,* which require the suspension of support for terrorist groups, and began rounding up the remaining members of bin Laden's network. As one observer noted, "Sudan is now effectively eliminated as one of the biggest bases of operation for bin Laden."[11]

The country's emerging respectability internationally encouraged the U.S. to remove its "pariah state" image and work for greater cooperation. U.S. Secretary of State Colin Powell attended the peace negotiations in Kenya and promised to bring both parties to the White House for formal signatures of the projected peace treaty.

NOTES

1. Dunstan Wai, *The African-Arab Conflict in the Sudan* (New York: Africana Publishing, 1981), p. 32.
2. Southerners are not so favorable; in the south, the Mahdi's government was as cruel as the Egyptian. *Ibid.,* p. 31.
3. Peter M. Holt, in *The History of the Sudan,* 3rd ed. (London: Weidenfeld and Nicolson, 1979), p. 123, quotes the British governor as saying that they were recruited on the basis of "good health, high character and fair abilities."
4. *Ibid.,* p. 171.
5. The SSU is defined as "a grand alliance of workers, farmers, intellectuals, business people and soldiers." Harold D. Nelson, ed., *Sudan, A Country Study* (Washington, D.C.: American University, Foreign Area Studies, 1982), p. 199.
6. Quoted in Tareq Y. Ismael and Jacqueline S. Ismael, *Government and Politics in Islam* (New York: St. Martin's Press, 1985), Appendix, pp. 148–149.
7. *Ibid.,* p. 150.
8. *The Christian Science Monitor* (April 16, 1985).
9. Mansour Khalid, quoted in *The Economist,* May 10, 2003.
10. Quoted in *The Christian Science Monitor,* September 12, 2005.
11. Robin Wright, in *The Los Angeles Times* (October 2001).

Syria (Syrian Arab Republic)

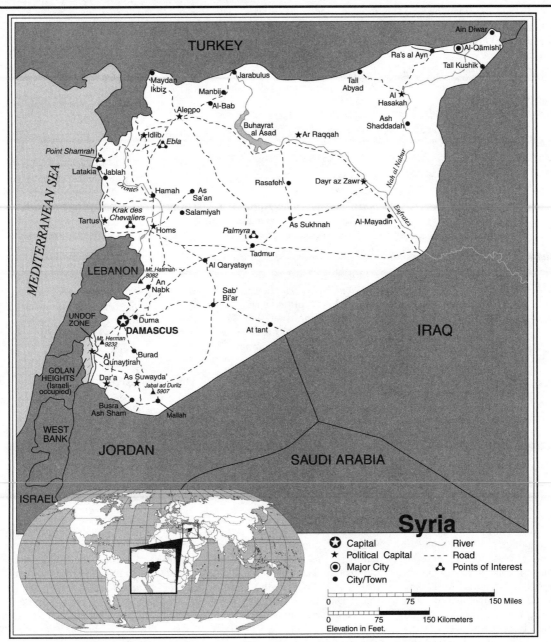

Syria Statistics

GEOGRAPHY

Area in Square Miles (Kilometers):

71,500 (185,170) (about the size of North Dakota)

Capital (Population): Damascus (1,549,000)
Environmental Concerns: deforestation; overgrazing; soil erosion; desertification; water pollution; insufficient potable water

Geographical Features: primarily semiarid and desert plateau; narrow coastal plain; mountains in the west
Climate: predominantly desert; considerable variation between the interior and coastal regions

161

PEOPLE

Population

Total: 18,881,361 (plus 20,000 in the Israeli-occupied Golan Heights, of which 18,000 are Druze, 2,000 Alawites)

Annual Growth Rate: 2.34%

Rural/Urban Population Ratio: 47/53

Major Languages: Arabic, Kurdish, various minority languages, e.g. Aramaic, Hebrew

Ethnic Makeup: 90% Arab; 10% Kurd, Armenian, and others

Religions: 74% Sunni Muslim; 16% Alawite, Druze, and other Muslim sects; 10% Christian and Jewish

Health

Life Expectancy at Birth: 68 years (male); 71 years (female)

Infant Mortality Rate: 29.53/1,000 live births

Physicians Available (Ratio): 1/953 people

Education

Adult Literacy Rate: 76.9%

Compulsory (Ages): 6–12

COMMUNICATION

Telephones: 1,313,000 main lines

Daily Newspaper Circulation: 19 per 1,000 people

Televisions: 49 per 1,000 people

Internet Users: 1 (2000)

TRANSPORTATION

Highways in Miles (Kilometers): 25,741 (41,451)

Railroads in Miles (Kilometers): 1,650 (2,750)

Usable Airfields: 100

Motor Vehicles in Use: 353,000

GOVERNMENT

Type: republic under a military regime since March 1963

Independence Date: April 17, 1946 (from a League of Nations mandate under French administration)

Head of State/Government: President Bashar al-Assad; Prime Minister Muhammad al-Utri

Political Party: The dominant Syrian Ba'th Party formed a coalition with several minor parties in 2003 prior to elections for a 250-member unicameral Peoples Council, winning 167–83

Suffrage: universal at 18

MILITARY

Military Expenditures (% of GDP): 5.9%

Current Disputes: Golan Heights is Israeli-occupied; dispute with Turkey over Turkish water-development plans

ECONOMY

Currency ($U.S. Equivalent): 46 Syrian pounds = $1

Per Capita Income/GDP: $3,400/$60 billion

GDP Growth Rate: 2.3%

Inflation Rate: 2.1%

Unemployment Rate: 30%

Labor Force: 5,012,000

Natural Resources: petroleum; phosphates; chrome and manganese ores; asphalt; iron ore; rock salt; marble; gypsum; hydropower

Agriculture: wheat; barley; cotton; lentils; chickpeas; olives; sugar beets; beef; mutton; eggs; poultry; milk

Industry: petroleum; textiles; food processing; beverages; tobacco; phosphate-rock mining

Exports: $6 billion (primary partners Germany, Italy, UAE)

Imports: $5 billion (primary partners Italy, Germany, China)

SUGGESTED WEBSITES

http://lcweb2.loc.gov/frd/cs/ sytoc.html

Syria Country Report

The modern Syrian state is a pale shadow of ancient Syria, which at various times in its history was a great kingdom, incorporating the lands of present-day Lebanon, Israel, Iraq, Jordan, and a part of Turkey within its boundaries. Ancient Syria was also a part of the great civilization centered in Mesopotamia. Recent discovery by archaeologists of a 6,000-year-old city at Hamonkar, in northeastern Syria near the Iraqi and Turkish borders, has pushed back the start of urban design centuries earlier than that of the Sumerians.

Syrian kings figure prominently in the Old Testament as rivals to those of Israel and Judah. One of these kings, Antiochus, divided the empire of Alexander the Great. Antiochus's kingdom dominated the Near East prior to the establishment of the Roman empire, with Syria as its center.

Syria also figured prominently in the expansion of Islam. After the death of the Prophet Muhammad, his successors, called *caliphs*, expanded Islamic rule over a territory greater than Rome. They moved their capital from Mecca to Damascus. The Umayyad Caliphate, so called because of its family origins in Muhammad's clan, spread the Arabic language and Islamic culture from North Africa to the western border of India. Due to its centrality, Arab geographers and cartographers termed Syria *Bilad ash-Sham,* literally "east," whence the sun rose over the lands of Islam.

Modern Syria is a nation of artificial boundaries. Its borders were determined by agreement between France and Britain after World War I. The country's current boundaries are with Turkey, Iraq, Jordan, Israel, and Lebanon. (The only one of these boundaries in dispute is the Golan Heights, which was seized and annexed by Israel in the 1970s.) The border with Turkey is defined by a single-track railroad, perhaps the only case in the world of a railroad put to that use. Syria's other borders are artificial lines established by outside powers for convenience.

Syria is artificial in another sense: Its political system was established by outside powers. Since becoming independent in 1946, the Syrians have struggled to find a political system that works for them. The large number of coups and frequent changes in government are evidence of this struggle. The most stable government in Syria's independent history is the current one, which has been in power since 1970.

Syrian political instability stems from the division of the population into separate ethnic and religious groups. The Syrians are an amalgamation of many different ethnoreligious groups that have settled the region over the centuries. The majority of the population are Sunni Muslim Arabs. The Alawis form the largest minority group. Although the Alawis are nominally Shia Muslims, the Sunni Muslims distrust them—not primarily because of religion, but because of the secret nature of their rituals and because as a minority they are very clannish. The next-largest minority, the Druze, live in Israel and Lebanon as well as Syria. They

Damascus Online/Ayman Haykal (DAMA001)

The Dome of the Eagle (Qubbat Al-Nisr) is considered one of the architectural highlights of the Omayyad Mosque.

are nominally Muslims, but their (secret) rituals include Christian liturgical elements such as the Eucharist, when they drink the blood and eat the body not of Christ but of Ali, Muhammad's closest relative and for Shias his legitimate successor as Caliph of Islam.

The largest non-Arab minority is the Kurds, some 1.7 million, forming 9 percent of the population. They are Sunni Muslims, forming part of the large Kurdish population spread over mountainous areas in eastern Turkey, northern Iraq and Iran. In 1962 some 120,000 Kurds were stripped of citizenship by the Syrian government, allegedly for advocating an independent Kurdish state. In 2005 the regime of President Bashar al-Assad, faced with a political crisis due to its involvement in the murder of Lebanon's ex-Prime Minister Rafik Hariri and the resulting international fallout, agreed formally to restore citizenship to its Syrian Kurds in the near future. (Currently some 300,000 Kurds are classified as "foreigners" and carry red ID cards in lieu of passports.)

Syria also has small but long-established Christian and Jewish communities. The Alawi regime, itself a minority (15 percent) of the population, allows them full exercise of their religious rights and services. Some Christian communities still use the ancient Aramaic language in their liturgy, an example being the village of Qamishli, near the Turkish border.

Although Syrian cities are slowly becoming more homogeneous in population, the different communities still constitute a majority in certain areas. Thus, Alawis make up 60 percent of the population of the northern coast. The Druze predominate in Jabal Druze, near the Lebanese border, and in that part of the Golan Heights still under Syrian control. Kurds are found mostly north of Aleppo and eastward toward the Turkish border.

HISTORY

Syria's greatest period was probably that of the Umayyad caliphs (A.D. 661–750). These caliphs were rulers of a vast Islamic empire. The first Umayyad caliph, Mu'awiya, is considered one of the political geniuses of Islam. He described his political philosophy to a visitor as follows:

> I apply not my lash where my tongue suffices, nor my sword where my whip is enough. If there be one hair binding me to my fellow men I let it not break. If they pull I loosen; if they loosen I pull.[1]

During this period of Umayyad rule, Damascus became a great center of learning and culture. The Caliph Abd al-Malik built not only the great mosque there but also the Dome of the Rock in Jerusalem, Islam's most sacred site after Mecca and Medina. But his successors were less capable, emphasizing Arabs as central to the faith and superior to non-Arab converts. The Umayyads even imposed a special tax on these new converts.

FREEDOM

The 1973 constitution defines Syria as a socialist, popular democracy. But this constitution remains superseded by martial law, which has been in effect since 1963. The late President Hafez al-Assad's Ba'thist regime severely limited press freedom and human rights. His successor has restored some of them, but the secret police continue to arrest opposition leaders or close down discussion groups deemed hostile.

The Umayyads were overthrown in 750 A.D. by a rival group with Persian backing, and the caliphate capital was moved from Damascus to Baghdad. From then on until Syria became an independent republic in the 20th century, its destiny was controlled by outside forces.

After the Ottoman Turks had established their empire and expanded their rule to the Arab lands of the Middle East, Syria became an Ottoman province governed by a pasha. His main responsibilities were to keep order and collect taxes, with a specified amount remitted to the Sultan's government in Constantinople. The balance was his to keep as compensation for his unpaid appointment. As was the case in Lebanon and Palestine, the Syrian province was named for its principal city, Damascus. The name "Syria" did not come into use until shortly before World War I and was adopted by France after the war when Syria became a League of Nations mandate under French sponsorship.

Syria was ruled by the Ottoman Turks for four centuries as a part of their empire. It was divided into provinces, each governed by a pasha. In mountain areas such as Lebanon, then part of Syria, the Ottomans delegated authority to the heads of powerful families or leaders of religious communities. The Ottomans recognized each of these communities as a *millet,* a Turkish word meaning "nation." The religious head of each millet represented the millet in dealings with Ottoman officials. The Ottomans, in turn, allowed each millet leader to manage the community's internal affairs. The result was that Syrian society became a series of sealed compartments. The millet system has disappeared, but its effects have lingered to the present, making national unity difficult.

The French Mandate

In the nineteenth century, as Ottoman rule weakened and conflict developed among Muslim, Christian, and Druze communities in Syria, the French began to intervene directly in Syria to help the Maronite Christians. French Jesuits founded schools for Christian children. In 1860, French troops landed in Lebanon to protect the Christian Maronites from massacres by the Druze. French forces were withdrawn after the Ottoman government agreed to establish a separate Maronite region in the Lebanese mountains. This arrangement brought about the development of Lebanon as a nation separate from Syria. The Christians in Syria were less fortunate. About 6,000 of them were slaughtered in Damascus before Ottoman troops restored order.[2]

In the years immediately preceding World War I, numbers of young Syrian Christians and some Muslims were exposed through mission schools to ideas of nationalism and human rights. A movement for Arab independence from Turkish rule gradually developed, centered in Damascus and Beirut. After the start of World War I, the British, with French backing, convinced Arab leaders to revolt against the Ottoman government. The Arab army recruited for the revolt was led by Emir Faisal, the second son of Sharif

Husayn of Mecca, leader of the powerful Arab Hashimite family, and the Arab official appointed by the Ottomans as "Protector of the Holy Shrines of Islam." Faisal's forces, along with a British army, drove the Ottomans out of Syria. In 1918, the emir entered Damascus as a conquering hero, and in 1920 he was proclaimed king of Syria.

Faisal's kingdom did not last long. The British had promised the Arabs independence in a state of their own, in return for their revolt. However, they had also made secret agreements with France to divide the Arab regions of the defeated Ottoman Empire into French and British protectorates. The French would govern Syria and Lebanon; the British would administer Palestine and Iraq. The French now moved to collect their pound of flesh. They sent an ultimatum to Faisal to accept French rule. When he refused, a French army marched to Damascus, bombarded the city, and forced him into exile. (Faisal was brought back by the British and later was installed as the king of Iraq under a British protectorate.)

What one author calls the "false dawn" of Arab independence was followed by the establishment of direct French control over Syria.[3] The Syrians reacted angrily to what they considered betrayal by their former allies. Resistance to French rule continued throughout the mandate period (1920–1946), and the legacy of bitterness over their betrayal affects Syrian attitudes toward outside powers, particularly Western powers, to this day.

The French did some positive things for Syria. They built schools, roads, and hospitals, developed a productive cotton industry, and established order and peaceful relations among the various communities. But the Syrians remained strongly attached to the goals of Arab unity and Arab independence, first in Syria, then in a future Arab nation.[4]

INDEPENDENT SYRIA

Syria became independent in 1946. The French had promised the Syrians independence during World War II but delayed their departure after the war, hoping to keep their privileged trade position and military bases. Eventually, pressure from the United States, the Soviet Union, and Britain forced the French to leave both Syria and Lebanon.

The new republic began under adverse circumstances. Syrian leaders had little experience in government; the French had not given them much responsibility and had encouraged personal rivalries in their divide-and-rule policy. The Druze and

UPI/Bettmann (BETT001)

In 1920, following the successful expulsion of the Ottoman government from Syria in 1918, Emir Faisal (pictured above) was named king of Syria.

Alawi communities feared that they would be under the thumb of the Sunni majority. In addition, the establishment in 1948 of the State of Israel next door caused great instability in Syria. The failure of Syrian armies to defeat the Israelis was blamed on weak and incompetent leaders.

For two decades after independence, Syria had the reputation of being the most unstable country in the Middle East. There were four military coups between 1949 and 1954 and several more between 1961 and 1966. There was also a brief union with Egypt (1958–1961), which ended in an army revolt.

One reason for Syria's chronic instability was that political parties were simply groups formed around individuals. At independence, the country had many such parties. Other parties were formed on the basis of ideology, such as the Syrian Communist Party. In 1963, one party, the Ba'th, acquired control of all political activities. Since then, Syria has been a single-party state.

THE BA'TH

The Ba'th Party (the Arabic word *ba'th* means "resurrection") began in the 1940s as a political party dedicated to Arab unity. It was founded by two Damascus schoolteachers, both French-educated: Michel Aflaq, a Greek Orthodox Christian, and Salah Bitar, a Sunni Muslim. In 1953, the Ba'th merged with another political party, the Arab Socialist Party. Since then, the formal name of the Ba'th has been the Arab Socialist Resurrection Party.

The Ba'th was the first Syrian political party to establish a mass popular base and to draw members from all social classes. Its program called for freedom, Arab unity, and socialism. The movement for Arab unity led to the establishment of the branches of the party in other Arab countries, notably Iraq and Lebanon. The party appealed particularly to young officers in the armed forces; and it attracted strong support from the Alawi community, because it called for social justice and the equality of all Syrians.

The Ba'th was instrumental in 1958 in arranging a merger between Syria and Egypt as the United Arab Republic (U.A.R.). The Ba'thists had hoped to undercut their chief rival, the Syrian Communist Party, by the merger. But they soon decided that they had made a mistake. The Egyptians did not treat the Syrians as equals but as junior partners. Syrian officers seized control and expelled the Egyptian advisers. It was the end of the U.A.R.

For the next decade, power shifted back and forth among military and civilian factions of the Ba'th Party. The process had little effect on the average Syrian, who liked to talk about politics but was wary, with good reason, of any involvement. Gradually, the military faction got the upper hand; and, in 1970, Lieutenant General Hafez al-Assad, the defense minister of one of the country's innumerable previous governments, seized power in a bloodless coup.[5]

THE HAFEZ AL-ASSAD REGIME

Syria can be called a presidential republic, in the sense that the head of state has extensive powers, which are confirmed in the Constitution approved in 1973. He decides and executes policies, appoints all government officials, and commands the armed forces. He is also head of the Ba'th Party. Under the Constitution, he has unlimited emergency powers "in case of grave danger threatening national unity or the security... of the national territory" (Article 113), which only the president can determine.

Hafez al-Assad ruled Syria for nearly three decades, becoming in the process the longest-serving elected leader of any Arab state. He was first elected in 1971 (as the only candidate), and thereafter for five consecutive seven-year terms, the last in 1999. Over the years he broadened the political process to some extent, establishing a People's Assembly with several small socialist parties as a token opposition body in the Legislature. In 1990, elections were held for an enlarged, 250-member Assembly. Ba'th members won 134 seats to 32 for the opposition; the remainder were won by independents. Assad then approved the formation of a National Progressive Front, which included the independents. But mindful of Syria's long history of political instability in the years before he took office, he decreed that its only function would be approval of laws issued by the Ba'th Central Committee.

Syria's Role in Lebanon

Assad's position was strengthened domestically in the 1970s due to his success (or perceived success) in certain foreign-policy actions. The Syrian Army fought well against Israel in the October 1973 War, and Syria subsequently received both military and financial aid from the Soviet Union as well as its Arab brothers. The invitation by the Arab League for Syria to intervene in Lebanon, beginning with the 1975–1976 Lebanese Civil War, was widely popular among Syrians. They never fully accepted the French action of separating Lebanon from Syria during the mandate period, and they continue to maintain a proprietary attitude toward Lebanon. Assad's determination to avoid conflict with Israel led him in past years to keep a tight rein on Syrian-based Palestine Liberation Organization (PLO) operations. The al-Saiqa Palestinian Brigade was integrated into the Syrian Army, for example. However, Assad's agreement to join a Middle East conference with other Arab states and Israel in 1991 resulted in the release of all PLO activists held in detention in Syria.

HEALTH/WELFARE

Syria's high birth rate has generated a young population. With insufficient jobs available, the unemployment rate is around 30%. A mandatory family-planning program will eventually lower the birth rate, and a grant from the European Union is being used to expand the public sector and provide salary increases for those already employed there.

When the Lebanese Civil War broke out, Assad pledged that he would control the Palestinians in Lebanon. He sent about 2,000 al-Saiqa guerrillas to Beirut in early 1976. The peacekeeping force approved by the Arab League for Lebanon included 30,000 regular Syrian troops. For all practical purposes, this force maintained a balance of power among Lebanese factions until the Israeli invasion of June 1982. It then withdrew to the eastern Biqa' Valley, avoiding conflict with Israeli forces and providing sanctuary to Palestinian guerrillas escaping from Beirut.

From this vantage point, Syria made a number of attempts to broker a peace agreement among the various Lebanese factions. However, all of them failed, owing in large measure to the intractable hostility separating Muslim from Christian communities and intercommunal rivalries among the militias. In 1987, faced with a near-total breakdown in public security, Assad ordered 7,000 elite Syrian commandos into West Beirut. Syrian forces maintained an uneasy peace in the Lebanese capital until 1989, when they were challenged directly by the Christian militia of General Michel Aoun, who refused to accept Syrian authority and declared himself president. Syrian forces surrounded the Christian enclave and, early in 1990, mounted a massive assault, backed by heavy artillery, that finally broke the Christian resistance. Aoun took refuge in the French Embassy and then went into exile.

Syrian troops in Lebanon were reduced in stages after the end of the Lebanese civil war. However the assassination of Lebanon's former Prime Minister, Rafik Hariri, in February 2005 and the purported involvement of Syrian intelligence agents and government officials in the murder triggered massive demonstrations in Lebanon demanding the withdrawal of the remaining Syrian troops. Essentially peaceful demonstrations, the "Cedars of Lebanon" movement which developed led to UN Security Council *Resolution 1559*, requiring troop withdrawal and an international investigation. While the Syrian government denied any involvement and questioned the reliability of witnesses appearing before the investigating committee, it withdrew its remaining troops and closed the offices of the much-feared Syrian intelligence service (the *mukhabarat*) who had ruled Lebanon for more than 20 years.

Internal Opposition

Opposition to the Hafez al-Assad regime was almost nonexistent in the 1990s. A major cause for resentment among rank-and-file Syrians, however, is the dominance of the Alawi minority over the government, armed forces, police, and intelligence services. The main opposition group was the Syrian branch of the Muslim Brotherhood (a Sunni organization spread throughout the Arab world). The Brotherhood opposed Assad because of his practice of advancing Alawi interests over those of the Sunni majority. Its main stronghold was the ancient city of Hama, famed for its Roman waterwheels. In 1982, Assad's regular army moved against Hama after an ambush of government officials there. The city was almost obliterated by tanks and artillery fire, with an estimated 120,000 casualties. Large areas were bulldozed as a warning to other potentially disloyal elements in the population.[6]

The "lessons of Hama" have not been forgotten. The calculated savagery of the attack was meant not only to inflict punishment but to provide a warning for future generations of Syrians. It did have a positive result. In ensuring the survival of his regime, Assad guaranteed political stability, along with prosperity for the largely Sunni merchant class. Thus other Arab states look toward the "Hama solution" with nostalgia.

Assad's control over the various levers of power, notably the intelligence services (*mukhabarat*), the security police, and the military, ensured his rule during his lifetime, despite his narrow support base as head of a minority group. After Hama, no organized opposition group remained to challenge his authority. As a result, he was able to give Syria the political stability that his predecessors had never provided.

ACHIEVEMENTS

Ghada Shonaa brought honor and glory to Syria when she became the first Syrian to win an Olympic gold medal. She won the women's heptathlon at the 1996 Olympic Games in Atlanta, Georgia. And although the United States and Syria remain estranged politically, the Host House Trio, an American jazz group, joined with Syrian musicians in 1999 in a concert in Damascus.

Syrian popular support for the aging president grew in the 1990s, as he continued to resist accommodation with Israel, while other Arab states were establishing relations or even recognizing the State of Israel. This broader support enabled Assad to loosen the reins of government. At the start of his fourth term he included several Sunni ministers in his cabinet. Political

prisoners were released, most of them Muslim Brotherhood members. In 1999, he ordered a general amnesty for 150,000 prisoners to "clean out the jails"; most of them had been jailed for smuggling, desertion from the armed forces, or various economic crimes.

DEVELOPMENT

Bashar's pledge of economic reform made after he took office in 2000 has brought some openness to the sluggish economic system. Internet service providers are omnipresent, although emails can be, and often are, monitored by the government. One industry that is booming is telecommunications, with cell phones in common use. The country has a new stock exchange, import taxes have been reduced drastically, and procedures for foreign investment made less draconian in order to attract outside support for the important oil industry.

THE ECONOMY

At independence, Syria was primarily an agricultural country, although it had a large merchant class and a free-enterprise system with considerable small-scale industrial development. When it came to power, the Ba'th Party was committed to state control of the economy. Agriculture was collectivized, with land expropriated from large landowners and converted into state-managed farms. Most industries were nationalized in the 1960s. The free-enterprise system all but disappeared.

Cotton was Syria's principal export crop and money earner until the mid-1970s. But with the development of oil fields, petroleum became the main export. Syria produced enough oil for its own needs until 1980. However, the changing global oil market and the reluctance of foreign companies to invest in Syrian oil exploration under the unfavorable concession terms set by the government have hampered development. Oil production, formerly 580,000 barrels per day, fell to 340,000 bpd in the mid-1990s. It increased to 450,000 bpd in 2000 and 550,000 bpd in 2001, due largely to imports of Iraqi oil for further export through the Kirkuk-Banias pipeline.

Unrest among the Kurdish population, in whose territory the oil fields are located, has deterred the search for new oil resources. Production from existing fields has leveled off at 500,000 bpd. Lacking new discoveries and major foreign investment the industry will certainly decline in the coming years.

Agriculture, which accounts for 30 percent of gross domestic product annually at present and employs 33 percent of the labor force, benefited in the early 1990s from expanded irrigation, which brought additional acreage under cultivation. Production of cotton, the major agricultural crop, reached a record 1.1 million tons in 2000, with 270,000 tons exported.

The end of Syria's special relationship with the Soviet Union due to the breakup of that country in 1991 encouraged a modest liberalization of the Ba'thist economic system. A prominent exiled businessman who had been one of Assad's bitterest critics returned in 1993 to set up a retail store chain similar to London's Marks & Spencer, taking advantage of new tax exemptions and other incentives. However economic growth in the early 1990s was blocked by a recession at the end of the decade. It was compounded by population growth at an explosive 3.4 percent level, foreign debts of $10–13 billion (mostly to the former Soviet Union) and an unfortunate mixture of too many government workers and landless peasants in the labor force. The unemployment level at that time was 30 percent. The association agreement with the European Union negotiated in 2004 has encouraged the regime to liberalize the foreign investment laws and simplify tax regulations.

Hafez al-Assad's death in June 2000 and the accession of his son Bashar to the presidency have been felt most strongly in the economic sector. In December 2000, the Ba'th Regional Command, the party's central committee, approved the establishment of private banks, ending 40 years of state monopoly over banking and foreign-exchange transactions. For the first time in the lives of most of them, Syrians no longer had to go across the Lebanese border to stash their illegal dollars in Lebanese banks or to use them to buy imported goods unavailable in their own country.

FOREIGN RELATIONS

Syria's often prickly relations with its neighbors and its rigid opposition to Israel have made the country the "odd man out" in the region at various times. Syria's hostility to the rival Ba'thist regime in Iraq resulted in periodic border closings and a shutdown in shipments of Iraqi oil through Syrian pipelines to refineries on the Mediterranean coast in the 1980s. The border was closed definitively after Syria sided with Iran during its war with Iraq and remained so in the Gulf War, as Syrian troops formed part of the coalition that drove Iraqi forces out of Kuwait. The UN sanctions on Iraq brought the two Arab neighbors closer together. The border was reopened in 1997, and the new regime removed all restrictions on travel to Iraq in 2001. In February, the two Ba'th regimes signed a free-trade agreement; and in August, Syrian prime minister Muhammed Moru made the first official visit to Baghdad of any Syrian government leader since 1979. Current trade volume between the countries is $500 million.

Assad expelled Abdullah Ocalan, leader of insurgent Kurdish forces fighting for independence from Turkey, in February 1999. The Kurdish leader had been useful to Syria as a bargaining chip with Turkey, particularly for increased allocation of water from the Euphrates River.

Syria's role as an alleged major sponsor of international terrorism has adversely affected its relations with Western countries for years. In 1986, a number of these countries broke diplomatic relations after the British discovered a Syrian-funded plot to blow up an Israeli airliner at Heathrow Airport in London, England. Syria also sent troops to support the U.S.-led Coalition in the Gulf War, despite its close economic relationship with Iraq. Following the September 11, 2001 attacks on the U.S. by al Qaeda terrorists, Syria provided intelligence information on its network. However, the Assad regime's continued support and sponsorship of anti-Israeli organizations such as Hamas and Hezbollah have tarnished its image abroad, notably in the United States. After the U.S. invasion and occupation of Iraq, Bush administration policymakers charged that its open border with that country enabled weapons and terrorists to enter and thus delay the reconstruction of Iraq. In December 2003 Congress passed the Syria Accountability Act. It bans exports of dual-use items (those which have both civilian and military applications). In March 2004 Bush ordered the imposition of economic sanctions in implementation of the act.

Syria's inclusion in the Department of State list as a state supporter of terrorism had been based on the harboring of groups engaged in violence, usually against Israel but also against Yassir Arafat's Palestinian organization. The groups included Hamas, the Popular Front for the Liberation of Palestine (PFLP), and Islamic Jihad. However, the Assad government was careful not to allow them to launch anti-Israeli operations from Syrian territory. The September 11, 2001, terrorist attacks on the United States brought a change in the equation. President Bashar denounced the terrorist attacks and criticized Osama bin Laden and his al Qaeda network for giving Islam a bad name. However, he declared that the Palestinian-Israeli conflict ultimately bore responsibility for the terrorism.

The 1993 Oslo agreements between Israel and the Palestinians and the 1994 Jordan-Israel peace treaty encouraged Assad to begin serious discussions with the Israelis for a settlement of the Golan Heights issue. Talks began with the Rabin government, but they were broken off after the election of Benjamin Netanyahu as Israel's prime minister. His defeat in the 1999 elections made possible the revival of negotiations, inasmuch as incoming prime minister Ehud Barak had stressed settlement of the Golan as part of his 15-month plan for regional peace.

Syrian and Israeli representatives met in January 2000 in the resort town of Shepherdstown, West Virginia, with then-president Bill Clinton serving as moderator. Their talks ended inconclusively, as the Syrian and Israeli positions remained far apart on such issues as the Golan Heights. However, the death of Hafez al-Assad makes the pursuit of a peace treaty with Israel less urgent for the new Syrian regime. Its priorities of necessity have concentrated on internal reform and revitalization of the economy. As one scholar observed, "[Israel] took 90 percent of Assad's energy. Israel was a legitimizing factor for the regime—Syria must always have an enemy to help create political cohesion."[7]

PROSPECTS

Hafez al-Assad died on June 10, 2000, the last of a group of autocratic rulers who had dominated the Middle East for more than a generation. His younger son, Bashar, was elected to succeed him on June 25 by the People's Assembly, confirmed by 97.5 percent of voters in a nationwide referendum. The Syrian Constitution, which precludes anyone under age 40 from serving as president, was conveniently set aside by the Ba'th Party Regional Command, since Bashar was not to attain that exalted status until 2005.

Syria's new leader was trained as an opthalmologist in Britain and had little experience in national politics before being summoned back to replace his elder brother Basil (killed in an auto accident in 1994) as the heir-apparent. His only public post was that of commander of the Republican Guard. After his election to the presidency (he was the only candidate, like his father), Bashar became head of the armed forces and of the Ba'th Regional Command.

While cynical observers joked that Syria had exchanged a dictator for an eye doctor, Bashar brought fresh air into a moribund political system and a stagnant economy. He began by enforcing the rule requiring retirement at age 60, which is mandatory for the military but had never been adhered to. As a result, many senior commanders were forced to retire. They included Hafez al-Assad's long-serving chief of staff and the head of the mukhabarat. In March 2002, the entire cabinet resigned, as the president continued to turn to new faces to strengthen political support and help liberalize the economy.

Bashar also changed the composition of the Ba'th Regional Command, bringing in younger army commanders as well as some women. In other essentially cosmetic changes, private universities were established to supplement, and revitalize, the moribund state system, and in 2002 private banks were allowed to form. By mid-2003 six such banks were in operation. According to regulations, they must be capitalized at $25.5 million with no more than 49 percent foreign ownership.

In March 2003, elections were held for a new Peoples' Assembly (Parliament). The National Progressive Front, a 7-party coalition dominated by the Ba'th, won 167 seats to 83 for independent (non-party) candidates.

A new law proposed in February 2006 would allow political parties to form outside the Ba'th. The government also announced it would restore Syrian citizenship to its Kurdish minority. And in January 2006 five prominent opposition political prisoners were released. They included Riad Seif and Mahmoud al-Homsi, former legislators who had been stripped of their parliamentary immunity due to their public demands for political reform and an end to government corruption.

These actions were in large measure Bashar's response to an opposition statement, "the Damascus Declaration" issued in October 2005. It was described as a "blueprint" for reforming the political process. Its backers included a broad range of normally fractious groups, such as Communists, Kurdish nationalists, exiles and the London-based Muslim Brotherhood. The Declaration came after the brief "Damascus Spring of 2001–2002, when "salons" were allowed to form for meetings and political discussion in members' homes. However the salons were ordered closed and human rights activists arrested in a government crackdown in 2002. As a prominent Syrian intellectual noted, "Suddenly people are not sure where the ‹red lines› on freedom of speech are any more."[8]

Continued fallout from the Hariri assassination and the resulting UN investigation into Syria's involvement has placed Bashar in an awkward position. In October 2005 the chief UN prosecutor, Detlev Mehlis, named the president's brother Maher and his brother-in-law, Asef Shawkat, who are respectively head of the presidential guard and the chief of intelligence, as prime suspects. Four Lebanese security agents formerly on the Syrian payroll were arrested, and Syria was given 6 months to surrender these officials for questioning. Failure to do so would result in punitive sanctions. As if on cue, Syrian Interior Minister Gen. Ghazi Kenaan, former head of military intelligence in Lebanon, committed suicide following publication of the investigative report.

Whether these developments would lead to regime change and more openness in Syrian political life remained an open question. In January 2006 former Vice-President Abdel-Halim Khaddam, in exile, accused Bashar of involvement in the Hariri assassination and promised to set up a rival Syrian government. He said the Bashar regime was totally discredited.

Faced with these conflicting pressures, along with domestic problems—high unemployment, lack of press and personal freedom, and the prospect that Syria's oil reserves would be exhausted in 8–10 years—it was not clear how the Syrian president would proceed. A U.S. scholar resident in Damascus noted: "He (Bashar) stands for stability. Syria is one of the safest countries in the world. Eighty percent of the people see democracy as a dangerous obstacle course, although the elites are talking democracy."[9] But the country remained unpredictable. As a journalist observed: "if we're cornered we can still destabilize the whole of the Middle East."[10]

NOTES

1. The statement is found in many chronicles of the Umayyads. See Richard Nyrop, ed., Syria, A Country Study (Washington, D.C.: American University, Foreign Area Studies, 1978), p. 13.
2. Philip Khoury, Urban Notables and Arab Nationalism: The Politics of Damascus 1860–1920 (Cambridge, England: Cambridge University Press, 1983), pp. 8–9.
3. Umar F. Abd-Allah, The Islamic Struggle in Syria (Berkeley, CA: Mizan Press, 1983), p. 39.
4. "Syrians had long seen themselves as Arabs ... who considered the Arab world as rightly a single entity." John F. Devlin, Syria: Modern State in an Ancient Land (Boulder, CO: Westview Press, 1983), p. 44.
5. He was barred from attending a cabinet meeting and then surrounded the meeting site with army units, dismissed the government, and formed his own. Ibid., p. 56.
6. Thomas L. Friedman, in From Beirut to Jerusalem, coined the phrase "Hama rules" to describe Assad's domestic political methods. "Hama rules" means no rules at all.
7. Quoted in Scott Peterson, in The Christian Science Monitor, July 12, 2000).
8. Helena Cobban, in The Christian Science Monitor, December 19, 2002.
9. Joshua Landis, in Princeton Alumni Weekly, September 14, 2005.
10. Quoted in The Economist, February 11, 2006.

Tunisia (Republic of Tunisia)

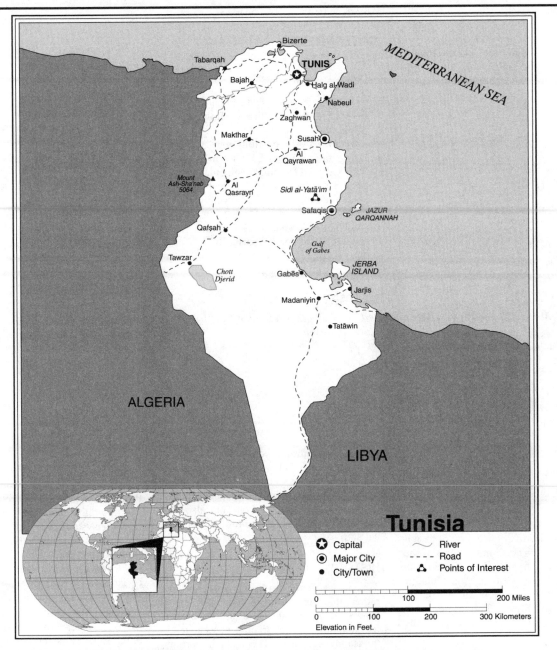

Tunisia Statistics

GEOGRAPHY

Area in Square Miles (Kilometers):
63,153 (163,610) (about the size of
Georgia)
Capital (Population): Tunis (675,000)
Environmental Concerns: hazardous-
waste disposal; water pollution; limited
fresh water resources; deforestation;
overgrazing; soil erosion; desertification
Geographical Features: mountains in
north; hot, dry central plain; semiarid
south merges into Sahara
Climate: hot, dry summers; mild, rainy
winters; desert in the south; temperate in
the north

PEOPLE

Population

Total: 10,074,951
Annual Growth Rate: 0.99%
Rural/Urban Population Ratio: 37/63
Major Languages: Arabic; French

Ethnic Makeup: 98% Arab-Berber; 1% European; 1% others

Religions: 98% Muslim; 1% Christian; less than 1% Jewish

Health

Life Expectancy at Birth: 73 years (male); 76.7 years (female)

Infant Mortality Rate: 24.77/1,000 live births

Education

Adult Literacy Rate: 74.2%

Compulsory (Ages): 6–16

COMMUNICATION

Telephones: 1,313,000 main lines

Daily Newspaper Circulation: 45 per 1,000 people

Televisions: 156 per 1,000 people

Internet Service Provider: 1 (2000)

TRANSPORTATION

Highways in Miles (Kilometers): 14,345 (23,100)

Railroads in Miles (Kilometers): 1,403 (2,260)

Usable Airfields: 32

Motor Vehicles in Use: 531,000

GOVERNMENT

Type: republic

Independence Date: March 20, 1956 (from France)

Head of State/Government: President Zine El Abidine Ben Ali; Prime Minister Mohammed Ghannouchi

Political Parties: Constitutional Democratic Rally (RCD), ruling party; others are Al-Tajdid Movement, Liberal Socialist Party (PSL), Movement of Democratic Socialists (MDS), Popular Unity Party, Unionist Democratic Union (Al-Nahda "Resistance"), Islamic fundamentalist party (currently outlawed)

Suffrage: universal at 20

MILITARY

Military Expenditures (% of GDP): 1.5%

Current Disputes: none

ECONOMY

Currency ($U.S. Equivalent): 1.24 dinars = $1

Per Capita Income/GDP: $7,100/$70.8 billion

GDP Growth Rate: 5.1%

Inflation Rate: 3%

Unemployment Rate: 13.8%

Labor Force: 6,259,000

Natural Resources: petroleum; phosphates; iron ore; lead; zinc; salt

Agriculture: olives; dates; oranges; almonds; grain; sugar beets; grapes; poultry; beef; dairy products

Industry: petroleum; mining; tourism; textiles; footwear; food; beverages

Exports: $9.9 billion (primary partners Germany, France, Italy)

Imports: $11.5 billion (primary partners France, Germany, Italy)

SUGGESTED WEB SITES

http://www.cia.gov/cia/ publications/factbook/index.html

http://www.tunisiaonline.com

Tunisia Country Report

Tunisia, the smallest of the four North African countries, is less than one tenth the size of Libya, its neighbor to the east. However, its population is nearly twice the size of Libya's.

Tunisia's long coastline has exposed it over the centuries to a succession of invaders from the sea. The southern third of the country is part of the Sahara Desert; the central third consists of high, arid plains. Only the northern region has sufficient rainfall for agriculture. This region contains Tunisia's single permanent river, the Medjerda.

DEVELOPMENT

Associate membership in the European Union has resulted in a number of advantages to Tunisia. One important one is favorable terms for its agricultural exports. Privatization of some 140 state-owned industries, a liberal investment code and tax reform have made possible a GDP growth rate averaging 4.5 to 5 percent annually.

The country is predominantly urban. There is almost no nomadic population, and there are no high mountains to provide refuge for independent mountain peoples opposed to central government. The Tunis region and the Sahel, a coastal plain important in olive production, are the most densely populated areas. Tunis, the capital, is not only the dominant city but also the hub of government, economic, and political activity.

HISTORY

Tunisia has an ancient history that is urban rather than territorial. Phoenician merchants from what is today Lebanon founded a number of trading posts several thousand years ago. The most important one was Carthage, founded in 814 B.C. It grew wealthy through trade and developed a maritime empire. Its great rival was Rome; after several wars, the Romans defeated the Carthaginians and destroyed Carthage. Later, the Romans rebuilt the city, and it became great once again as the capital of the Roman province of Africa. Rome's African province was one of the most prosperous in the empire. The wheat and other commodities shipped to Rome from North African farms were vitally needed to feed the Roman population. When the ships from Carthage were late due to storms, lost at sea, or seized by pirates, the Romans suffered hardship. Mod-

ern Tunisia has yet to reach the level of prosperity it had under Roman rule.

The collapse of the Roman Empire in the fifth century A.D. affected Roman Africa as well. Cities were abandoned; the irrigation system that had made the farms prosperous fell into ruin. A number of these Roman cities, such as Dougga, Utica, and Carthage itself, which is now a suburb of Tunis, have been preserved as historical monuments of this period.

Arab armies from the east brought Islam to North Africa in the late seventh century. After some resistance, the population accepted the new religion, and from that time on the area was ruled as the Arab-Islamic province of *Ifriqiya*. The Anglicized form of this Arabic word, "Africa," was eventually applied to the entire continent.

The Arab governors did not want to have anything to do with Carthage, since they associated it with Christian Roman rule. They built a new capital on the site of a village on the outskirts of Carthage, named Tunis. The fact that Tunis has been the capital and major city in the country for 14 centuries has contributed to the sense of unity and nationhood among most Tunisians.[1]

The original Tunisian population consisted of Berbers, a people of unknown origin. During the centuries of Islamic rule, many Arabs settled in the country. Other waves of immigration brought Muslims from Spain, Greeks, Italians, Maltese, and many other nationalities. Until recently, Tunisia also had a large community of Jews, most of whom emigrated to the State of Israel when it was founded in 1948. The blending of ethnic groups and nationalities over the years has created a relatively homogeneous and tolerant society, with few of the conflicts that marked other societies in the Islamic world.

From the late 1500s to the 1880s, Tunisia was a self-governing province of the Ottoman Empire. It was called a regency because its governors ruled as "regents" on behalf of the Ottoman sultan. Tunis was already a well-established, cosmopolitan city when it became the regency capital. Its rulers, called beys, were supported by an Ottoman garrison and a corsair fleet of fast ships that served as auxiliaries to the regular Ottoman navy. The corsairs, many of them Christian renegades, ruled the Mediterranean Sea for four centuries, raiding the coasts of nearby European countries and preying on merchant vessels, seizing cargoes and holding crews for ransom. The newly independent United States was also affected, with American merchant ships seized and cargoes taken by the corsairs. In 1799, the United States signed a treaty with the bey, agreeing to pay an annual tribute in return for his protection of American ships.

In the nineteenth century, European powers, particularly France and Britain, began to interfere directly in the Ottoman Empire and to seize some of its outlying provinces. France and Britain had a "gentleman's agreement" about Ottoman territories in Africa—the French were given a free hand in North Africa and the British in Egypt. In 1830, the French seized Algiers, capital of the Algiers Regency, and began to intervene in neighboring Tunisia in order to protect their Algerian investment.

The beys of Tunis worked very hard to forestall a French occupation. In order to do this, they had to satisfy the European powers that they were developing modern political institutions and rights for their people. Ahmad Bey (1837–1855) abolished slavery and piracy, organized a modern army (trained by French officers), and established a national system of tax collection. Muhammad al-Sadiq Bey (1859–1882) approved in 1861 the first written Constitution in the Islamic world. This Constitution had a declaration of rights and provided for a hereditary (but not an absolute) monarchy under the beys. The Constitution worked better in theory than in practice. Provincial landowners and local chiefs opposed the Constitution because it undermined their authority. The peasants, whom it supposedly was designed to protect, opposed the Constitution because it brought them heavy new taxes, collected by government troops sent from Tunis. In 1864, a popular rebellion broke out against the bey, and he was forced to suspend the Constitution.

FREEDOM

The campaign against Islamic fundamentalists have significantly reduced the civil rights normally observed in Tunisia. Foreign travel is routinely restricted, passports randomly confiscated and journalists arrested for articles that are said to "defame" the state. Tunisia was expelled from the World Press Organization in 1999 for its restrictions on press freedom.

In 1881, a French army invaded and occupied all of Tunisia, almost without firing a shot. The French said that they had intervened because the bey's government could not meet its debts to French bankers and capitalists, who had been lending money for years to keep the country afloat. There was concern also about the European population. Europeans from many countries had been pouring into Tunisia, ever since the bey had given foreigners the right to own land and set up businesses.

The bey's government continued under the French protectorate, but it was supplemented by a French administration, which held actual power. The French collected taxes, imposed French law, and developed roads, railroads, ports, hospitals, and schools. French landowners bought large areas and converted them into vineyards, olive groves, and wheat farms. For the first time in 2,000 years, Tunisia exported wheat, corn, and olive oil to the lands on the other side of the Mediterranean.

Because Tunisia was small, manageable, and primarily urban, its society, particularly in certain regions, was influenced strongly by French culture. An elite developed whose members preferred the French language to their native Arabic. They were encouraged to enroll their sons in Sadiki College, a European-type high school set up in Tunis by the French to train young Tunisians and expose them to Western subjects. After completing their studies at Sadiki, most were sent to France to complete their education in such institutions as the Sorbonne (University of Paris). The experience helped shape their political thinking, and on their return to Tunisia a number of them formed a movement for self-government that they called Destour (*Dustur* in Arabic), meaning "Constitu-

tion." The name was logical, since these young men had observed that independent countries such as France based their sovereignty on such a document. They were convinced that nationalism, "in order to be effective against the French, had to break loose from its traditional power base in the urban elite and mobilize mass support."[2] In 1934, a group of young nationalists quit the Destour and formed a new party, the Neo-Destour. The goal of the Neo-Destour Party was Tunisia's independence from France. From the beginning, its leader was Habib Bourguiba.

HABIB BOURGUIBA

Habib Ben Ali Bourguiba, born in 1903, once said he had "invented" Tunisia, not historically but in the sense of shaping its existence as a modern sovereign nation. The Neo-Destour Party, under Bourguiba's leadership, became the country's first mass political party. It drew its membership from shopkeepers, craftspeople, blue-collar workers, and peasants, along with French-educated lawyers and doctors. The party became the vanguard of the nation, mobilizing the population in a campaign of strikes, demonstrations, and violence in order to gain independence. It was a long struggle. Bourguiba spent many years in prison. But eventually the Neo-Destour tactics succeeded. On March 20, 1956, France ended its protectorate and Tunisia became an independent republic, led by Habib Bourguiba.

One of the problems facing Tunisia today is that its political organization has changed very little since independence. A Constitution was approved in 1959 that established a "presidential republic"—that is, a republic in which the elected president has great power. Bourguiba was elected president in 1957.

Bourguiba was also the head of the Neo-Destour Party, the country's only legal political party. The Constitution provided for a National Assembly, which is responsible for enacting laws. But to be elected to the Assembly, a candidate had to be a member of the Neo-Destour Party. Bourguiba's philosophy and programs for national development in his country were often called Bourguibism. It was tailored to the particular historical experience of the Tunisian people. Since ancient Carthage, Tunisian life has been characterized by the presence of a strong central government able to impose order and bring relative stability to the people. The predominance of cities and villages over nomadism reinforced this sense of order. The experience of Carthage, and even more so that of Rome, set the pattern. "The Beys continued the pattern of strong order while the French developed a strongly bourgeois, trade-oriented society, adding humanitarian and some authoritarian values contained in French political philosophy."[3] Bourguiba considered himself the tutor of the Tunisian people, guiding them toward moral, economic, and political maturity.

In 1961, Bourguiba introduced a new program for Tunisian development that he termed "Destourian Socialism." It combined Bourguibism with government planning for economic and social development. The name of the Neo-Destour Party was changed to the Destour Socialist Party (PSD) to indicate its new direction. Destourian Socialism worked for the general good, but it was not Marxist; Bourguiba stressed national unanimity rather than class struggle and opposed communism as the "ideology of a godless state." Bourguiba took the view that Destourian Socialism was directly related to Islam. He said once that the original members of the Islamic community (in Muhammad's time in Mecca) "were socialists … and worked for the common good."[4] For many years after independence, Tunisia appeared to be a model among new nations because of its stability, order, and economic progress. Particularly notable were Bourguiba's reforms in social and political life. Islamic law was replaced by a Western-style legal system, with various levels of courts. Women were encouraged to attend school and enter occupations previously closed to them, and they were given equal rights with men in matters of divorce and inheritance.

Bourguiba strongly criticized those aspects of Islam that seemed to him to be obstacles to national development. He was against women wearing the veil, polygyny, and ownership of lands by religious leaders, which kept land out of production. He even encouraged people not to fast during the holy month of Ramadan, because their hunger made them less effective in their work.

There were few challenges to Bourguiba's leadership. His method of alternately dismissing and reinstating party leaders who disagreed with him effectively maintained Destourian unity. But in later years Bourguiba's periodic health problems, the growth of Islamic fundamentalism, and the disenchantment of Tunisian youth with the single-party system raised doubts about Tunisia's future under the PSD.

The system was provided with a certain continuity by the election of Bourguiba as president-for-life in 1974, when a constitutional amendment was approved specifying that at the time of his death or in the event of his disability, the prime minister

would succeed him and hold office pending a general election. One author observed: "Nobody is big enough to replace Bourguiba. He created a national liberation movement, fashioned the country and its institutions."[5] Yet he failed to recognize or deal with changing political and social realities in his later years.

The new generation coming of age in Tunisia is deeply alienated from the old. Young Tunisians (half the population are under age 15) increasingly protest their inability to find jobs, their exclusion from the political decision-making process, the unfair distribution of wealth, and the lack of political organizations. It seems as if there are two Tunisias: the old Tunisia of genteel politicians and freedom fighters; and the new one of alienated youths, angry peasants, and frustrated intellectuals. Somehow the two have gotten out of touch with each other.

The division between these groups has been magnified by the growth of Islamic fundamentalism, which in Bourguiba's view was equated with rejection of the secular, modern Islamic society that he created. The Islamic Tendency Movement (MTI) emerged in the 1980s as the major fundamentalist group. MTI applied for recognition as a political party after Bourguiba had agreed to allow political activity outside of the Destour Party and had licensed two opposition parties. But MTI's application was rejected.

THE END OF AN ERA

In 1984, riots over an increase in the price of bread signaled a turning point for the regime. For the first time in the republic's history, an organized Islamic opposition challenged Bourguiba, on the grounds that he had deformed Islam to create a secular society. Former Bourguiba associates urged a broadening of the political process and formed political movements to challenge the Destour monopoly on power. Although they were frequently jailed and their movements proscribed or declared illegal, they continued to press for political reform.

However, Bourguiba turned a deaf ear to all proposals for political change. Having survived several heart attacks and other illnesses to regain reasonably good health, he seemed to feel that he was indestructible. His personal life underwent significant change as he became more authoritarian. He had divorced his French wife, apparently in response to criticism that a true Tunisian patriot would not have a French spouse. His second wife, Wassila, a member of the prominent Ben Ammar family, soon became the power behind the throne.

As Bourguiba's mental state deteriorated, he divorced her arbitrarily in 1986. At that time the president-for-life assumed direct control over party and government. As he did so, his actions became increasingly irrational. He would appoint a cabinet minister one day and forget the next that he had done so. Opposition became an obsession with him. The two legal opposition parties were forced out of local and national elections by arrests of leaders and a shutdown of opposition newspapers. The Tunisian Labor Confederation (UGTT) was disbanded, and the government launched a massive purge of fundamentalists.

HEALTH/WELFARE

Tunisia has overhauled its school and university curricula to emphasize respect for other monotheistic religions. They require courses on the Universal Declaration of Human Rights, democracy, and the value of the individual. The new curricula are at variance with government repressive policies, but they do stress Islamic ideals of tolerance for the school-age population.

The purge was directed by General Zine el-Abidine Ben Ali, the minister of the interior, regarded by Bourguiba as one of the few people he could trust. There were mass arrests of Islamic militants, most of them belonging to the outlawed Islamic Tendency Movement, a fundamentalist organization that Bourguiba outlawed as subversive. (It was later reorganized as a political party, Ennahda—"Renaissance"—but was again banned by Ben Ali after he became president.)

Increasingly, it seemed to responsible leaders that Bourguiba was becoming senile as well as paranoid. "The government lacks all sense of vision," said a long-time observer. "The strategy is to get through the day, to play palace parlor games." A student leader was more cynical: "There is no logic to [Bourguiba's] decisions; sometimes he does the opposite of what he did the day before."[6]

A decision that would prove crucial to the needed change in leadership was made by Bourguiba in September 1987, when he named Ben Ali as prime minister. Six weeks later, Ben Ali carried out a bloodless coup, removing the aging president under the 1974 constitutional provision that allows the prime minister to take over in the event of a president's "manifest incapacity" to govern. A council of medical doctors affirmed that this was the case. Bourguiba was placed under temporary house arrest in his Monastir villa, but he was allowed visitors and some freedom of movement within the city (after 1990).

Habib Bourguiba died in April 2001, at the age of 96. He was buried next door to this villa, in a mausoleum of white marble. The words inscribed on its door—"Liberator of women, builder of modern Tunisia"—seem an appropriate inscription for the "inventor" of his country.

NEW DIRECTIONS

President Ben Ali (elected to a full five-year term in April 1989) initiated a series of bold reforms designed to wean the country away from the one-party system. Political prisoners were released under a general amnesty. Prodded by Ben Ali, the Destour-dominated National Assembly passed laws ensuring press freedom and the right of political parties to form as long as their platforms are not based exclusively on language, race, or religion. The Assembly also abolished the constitutional provision establishing the position of president-for-life, which had been created expressly for Bourguiba. Henceforth Tunisian presidents would be limited to three consecutive terms in office.

Ben Ali also undertook the major job of restructuring and revitalizing the Destour Party. In 1988, it was renamed the Constitutional Democratic Rally (RCD). Ben Ali told delegates to the first RCD Congress that no single party could represent all Tunisians. There can be no democracy without pluralism, fair elections, and freedom of expression, he said.

Elections in 1988 underscored Tunisia's fixation on the single-party system. RCD candidates won all 141 seats in the Chamber of Deputies, taking 80 percent of the popular vote. Two new opposition parties, the Progressive Socialist Party and the Progressive Socialist Rally, participated but failed to win more than 5 percent of the popular vote, the minimum needed for representation in the Chamber. MTI candidates, although required to run as independents because of the ban on "Islamic" parties under the revised election law, dominated urban voting, taking 30 percent of the popular vote in the cities. However, the winner-take-all system of electing candidates shut them out as well.

Local and municipal elections have confirmed the RCD stranglehold on Tunisian political life; its performance was the exact opposite of that of the National Liberation Front in neighboring Algeria, where the dominant party was discredited over time and finally defeated in open national elections by a fundamentalist party. In the 1995 local and municipal council elections, RCD candidates won 4,084 out of 4,090 contested seats, with 92.5 percent of Tunisia's 1,865,401 registered voters casting their ballots.

Efforts to mobilize an effective opposition movement earlier were hampered when Ahmed Mestiri, the long-time head of the Movement of Socialist Democrats (MDS), the major legal opposition party, resigned in 1992. In the 1994 elections, the only opposition party to increase its support was the former Tunisian Communist Party, renamed the Movement for Renewal. It won four seats in the Chamber as Tunisia continued its slow progress toward multiparty democracy.

After the election the Chamber of Deputies was enlarged from the present 144 to 160 deputies. Twenty seats would be reserved for members from opposition parties. In the presidential election, Ben Ali was reelected for a third term and again in 1999 for a fourth term. In the latter election he faced modest opposition, and as a result his victory margin was a "bare" 99.44 percent.[7]

The Chamber was enlarged again in time for the 1999 elections, this time to 182 seats, to broaden representation for Tunisia's growing population. The results were somewhat different from the previous election. The RCD won 148 seats to 13 for the MDS, the largest opposition party. However, opposition parties all together increased their representation from 19 seats to 34.

THE ECONOMY

The challenge to Ben Ali lies not only in broadening political participation but also in improving the economy. After a period of impressive expansion in the 1960s and 1970s, the growth rate began dropping steadily, largely due to decreased demand and lowered prices for the country's three main exports (phosphates, petroleum, and olive oil). Tunisia is the world's fourth-ranking producer of phosphates, and its most important industries are those related to production of superphosphates and fertilizers.

Problems have dogged the phosphate industry. The quality of the rock mined is poor in comparison with that of other phosphate producers, such as Morocco. The Tunisian industry experienced hard times in the late 1980s with the drop in

global phosphate prices; a quarter of its 12,000-member workforce were laid off in 1987. However, improved production methods and higher world demand led to a 29 percent increase in exports in 1990.

Tunisia's oil reserves are estimated at 1.65 billion barrels. The main producing fields are at El Borma and offshore in the Gulf of Gabes. New offshore discoveries and a 1996 agreement with Libya for 50/50 sharing of production from the disputed Gulf of Gabes oil field have improved oil output, currently about 4.3 million barrels annually.

Tunisia became an associate member of the European Union in 1998, the first Mediterranean country to do so. The terms of the EU agreement require the country to remove trade barriers over a 10-year period. In turn, Tunisian products such as citrus and olives receive highly favorable export terms in EU countries. The EU also provides technical support and training for the government's Mise A Nouveau (Upgrading and Improvement) program intended to enhance productivity in business and industry and compete internationally.

Tunisia's political stability—albeit one gained at the expense of human rights—and its economic reforms have made it a favored country for foreign aid over the years. During the period 1970–2000 it received more World Bank loans than any other Arab or African country. Its economic reform program, featuring privatization of 140 state-owned enterprises since 1987, liberalizing of prices, reduction of tariffs and other reforms, is lauded as a model for development by international financial institutions.

The country's political stability and effective use of its limited resources for development have made it a favored country for foreign aid. Since the 1970s it has received more World Bank loans than any other Arab or African country. The funding has been equitably distributed, so that 60 percent of the population are middle class, and 80 percent own their own homes.

THE FUTURE

Tunisia's progress as an economic beacon of stability in an unstable region has been somewhat offset by a decline in its long-established status as a successful example of a secular, progressive Islamic state. Following President Ben Ali's ouster of his predecessor in 1987, he proclaimed a new era for Tunisians, based on respect for law, human rights, and democracy. Tunisia's "Islamic nature" was reaffirmed by such actions as the reopening of the venerable Zitouna University in Tunis, a center for Islamic scholarship, along with its counterpart in Kairouan. But like other Islamic countries, it has not been free from the scourge of militant Islamic fundamentalism. The fundamentalist movement Al-Nahda ("Renaissance"), which advocates a Tunisian government based on Islamic law, attacked RCD headquarters in February 1991. Many of its members, including the leader, were subsequently arrested and it was outlawed as a political party. However, the success of Osama bin Laden's al Qaeda movement is attracting young Muslims everywhere, and has emboldened those members still at large. In April 2003 a bomb attack on the ancient Jewish synagogue in Djerba, a popular tourist resort area, killed 17 people. In addition to the damage done to the important tourist industry, the attack provided further evidence of its goal of overthrowing the Ben Ali regime as "impure," one not true to Islamic law and ideals.

In subsequent years, Tunisia has become an increasingly closed society. The press is heavily censored. Bourguiba's death was not even reported in Tunisia, and the obligatory seven days of mourning were countered by instructions to banks and government offices to keep regular hours. Telephones are routinely tapped. More than 1,000 Ennahda members have been arrested and jailed without trial. In December 2000, a dozen members of another Islamic fundamentalist group were given 17-year jail sentences for forming an illegal organization; their lawyers walked out of the trial to protest the court's bias and procedural abuses.

The regime's repression of Islamic groups, even moderate nonviolent ones, has changed its former image as a tolerant, progressive Islamic country. The Tunisian League for Human Rights, oldest in the Arab world, has been shut down from time to time by the government. Until recently opposition political leaders were given no coverage in the mainstream press, and press censorship remains routine. The UN-sponsored "World Summit on the Information Society" held in Tunis in November 2005 offered opposition groups a rare opportunity to air their protests against the government's repressive policies. Leaders of the groups—the Progressive Democratic Party, Tunisian Communist Party and Human Rights League, along with the unrecognized Tunisian Journalists' Union, addressed the conference and called for freedom of the press, the right to free public association, and the release of the country's 600 political prisoners.

Following his "tainted" election victory in 1999, Ben Ali announced a new program designed to provide full employment by 2004. Called the 21–21 program, it would supplement an earlier 26–26 one that had brought the public and private sectors together to end poverty and increase home ownership, notably among the poor. In his address announcing the new program, Ben Ali stated: "Change comes from anchoring the democratic process in a steady and incremental progress aimed at avoiding setbacks or losing momentum."[8]

Some easing of repressive restrictions on political and human rights for the population was carried out in 2004–2005. Permission was granted to the International Committee of the Red Cross to inspect Tunisian prisons, and censorship of newspaper articles would no longer be automatic. Also a second, upper chamber was approved for the National Assembly. One observer noted, somewhat sardonically, that "Tunisia is the country in the region that is most able to democratize." But as the leader of the Tunisian Journalists' Union commented in launching a hunger strike by opposition leaders on the eve of the Tunis Summit, "It is the only means open to protest, because the authorities have closed all other avenues of dialogue and negotiation."[9]

NOTES

1. Harold D. Nelson, ed., *Tunisia: A Country Study* (Washington, D.C.: American University, Foreign Area Studies, 1979), p. 68.
2. *Ibid.*, p. 42.
3. *Ibid.*, p. 194. What Nelson means, in this case, by "authoritarianism" is that the French brought to Tunisia the elaborate bureaucracy of metropolitan France, with levels of administration from the center down to local towns and villages.
4. *Ibid.*, p. 196.
5. Jim Rupert, in *The Christian Science Monitor* (November 23, 1984).
6. Louise Lief, in *The Christian Science Monitor* (April 10, 1987).
7. Mamoun Fandy, in *The Christian Science Monitor* (October 25, 1999).
8. Georgie Ann Geyer, in *The Washington Post* (October 23, 1999).
9. "Hunger for Change," *The Economist*, November 12, 2005.

Timeline: PAST

264–146 B.C.
Wars between Rome and Carthage, ending in the destruction of Carthage and its rebuilding as a Roman city

800–900
The establishment of Islam in Ifriqiya, with its new capital at Tunis A.D.

1200–1400
The Hafsid dynasty develops Tunisia as a highly centralized urban state

1500–1800
Ottoman Turks establish Tunis as a corsair state to control Mediterranean sea lanes

1881–1956
French protectorate

1956
Tunisia gains independence, led by Habib Bourguiba

1974
An abortive merger with Libya

1980s
Bourguiba is removed from office in a "palace coup;" he is succeeded by Ben Ali

1990s
Tunisia's economic picture brightens; Ben Ali seeks some social modernization; women's rights are expanded

PRESENT

2000s
Human-rights abuses continue

Turkey (Republic of Turkey)

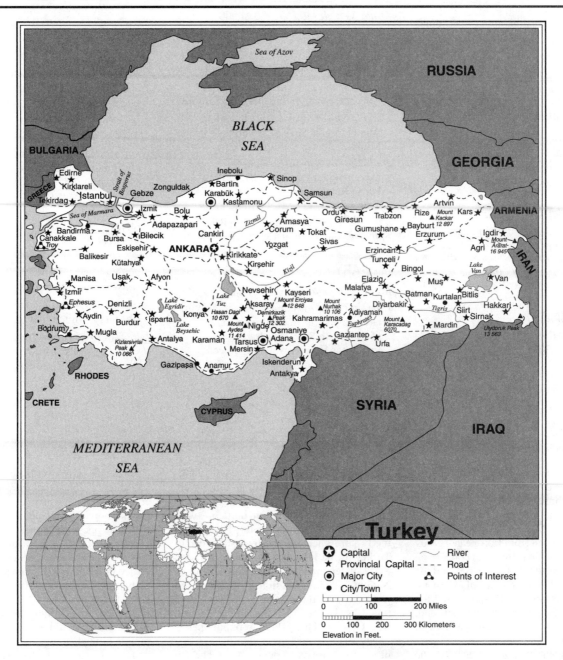

Turkey Statistics

GEOGRAPHY

Area in Square Miles (Kilometers):
301,303 (780,580) (about the size of Texas)
Capital (Population): Ankara (2,940,000)
Environmental Concerns: water and air pollution; deforestation; threat of oil spills from Bosporus ship traffic

Geographical Features: mostly mountains; a narrow coastal plain; a high central plateau (Anatolia)

Climate: temperate; hot, dry summers and mild wet winters along coasts; much drier and more extreme in temperatures in interior plateau and mountains

PEOPLE

Population

Total: 69,660,559
Annual Growth Rate: 1.09%
Rural/Urban Population Ratio: 29/71

Major Languages: Turkish; Kurdish; Arabic

Ethnic Makeup: 80% Turk; 17% Kurd; 3% others

Religions: 99% Muslim (about 79% Sunni, 20% Shia); 1% others

Health

Life Expectancy at Birth: 69 years (male); 74 years (female)

Infant Mortality Rate: 41/1,000 live births

Physicians Available (Ratio): 1/1,200 people

Education

Adult Literacy Rate: 85%

Compulsory (Ages): 6–16

COMMUNICATION

Telephones: 17,000,000 main lines

Daily Newspaper Circulation: 44 per 1,000 people

Televisions: 171 per 1,000 people

Internet Users: 22 (2000)

TRANSPORTATION

Highways in Miles (Kilometers): 237,747 (382,397)

Railroads in Miles (Kilometers): 5,336 (8,607)

Usable Airfields: 121

Motor Vehicles in Use: 4,320,000

GOVERNMENT

Type: republican parliamentary democracy

Independence Date: October 29, 1923 (successor state to the Ottoman Empire)

Head of State/Government: President Ahmet Necdet Sezer (ceremonial); Prime Minister Recep Tayyip Erdogan

Political Parties: Justice and Development Party (AKP); Majority Party; Republican Peoples' Party (CHP), True Path (DYP); Motherland Party (ANAP), principal opposition parties

Suffrage: universal at 18

MILITARY

Military Expenditures (% of GDP): 4.5%

Current Disputes: complex disputes with Greece; Cyprus question; periodic friction with Syria and Iraq over Euphrates River's water resources

ECONOMY

Currency ($U.S. Equivalent): 1,425,500 lira = $1

Per Capita Income/GDP: $7,400/$508 billion

GDP Growth Rate: 8.2%

Inflation Rate: 9.3%

Unemployment Rate: 10.8% (plus 6.1% underemployment)

Labor Force: 25,300,000, plus 1,200,000 Turks working abroad

Natural Resources: antimony; coal; chromium ore; mercury; copper; borate; sulfur; iron ore; meerschaum; arable land; hydropower

Agriculture: tobacco; cotton; grains; olives; sugar beets; pulse; citrus; livestock

Industry: textiles; food processing; automobiles; mining; steel; petroleum; construction; lumber; paper

Exports: $69.46 billion (primary partners Germany, United States, Italy)

Imports: $94.5 billion (primary partners Germany, Italy, Russia)

Turkey Country Report

Except for a small area in extreme Southeastern Europe called Thrace, the Republic of Turkey comprises the large peninsula of Asia Minor (Anatolia), which forms a land bridge between Europe and Asia. Asiatic Turkey is separated from European Turkey by the Bosporus, a narrow strait connecting the Black Sea with the Aegean Sea and the Mediterranean Sea via the Sea of Marmara. Throughout history, the Bosporus and the Dardanelles, at the Mediterranean end, have been important strategic waterways, fought over by many nations.

Except for the Syrian border, Asiatic Turkey's borders are defined by natural limits, with seas on three sides and rugged mountains on the fourth. European Turkey's frontiers with Greece and Bulgaria are artificial; they fluctuated considerably in the nineteenth and twentieth centuries before the Republic of Turkey was established.

Modern Turkey occupies a much smaller area than did its predecessor, the Ottoman Empire. The Ottoman Turks were the dominant power in the Middle East for more than five centuries. After the defeat of the empire in World War I, Turkey's new leader, Mustafa Kemal Ataturk, turned away from the imperial past, limiting the new republic to territory with a predominantly Turkish population. Since then, Turkey has not attempted to annex land beyond its natural Anatolian borders—with two exceptions. One was the Hatay, formerly a province of Syria that was ceded to Turkey by the French (who then controlled Syria under mandate from the League of Nations). The annexation was considered justified since the majority of the population was Turkish. The second exception was Cyprus. This island republic has a Greek majority in the population, but a significant minority (20 percent) are Turkish Cypriots, descended from Turkish families that settled there when Cyprus was Ottoman territory. Although it is a sovereign state, fears of violence against the Cypriot Turks led Turkish forces to occupy the northern third of the island in 1974. They have been there since then, with no agreement as yet on reunification of Cyprus. Some years ago, the Turkish government officially recognized the area under its control as the Republic of Northern Cyprus, but no other country has done so.

Asia Minor has an ancient history of settlement. Most of the peninsula is a plateau ringed by mountains. The mountains are close to the coast; over the centuries, due to volcanic action, the coastline became cracked, with deep indentations and islands just offshore. The inland plateau has an area of arid steppe with dried-up salt lakes at the center, but most of it is rolling land, well suited to agriculture. Consequently, people settled in small, self-contained villages at an early period and began to cultivate the land. Over the centuries, nomadic peoples migrated into Asia Minor, but the geographical pattern there did not encourage them to remain nomadic.

In terms of national unity, the modern Turkish state has not had the thorny problem of ethnic conflicts—with two important exceptions. One is the Armenians, an ancient Christian people who ruled over a large part of what is now eastern Turkey many centuries ago. With the outbreak of World War I the Ottoman government aligned itself with Germany against Britain, France, and its old enemy Russia. Following a declaration of war, the Czarist government invited Armenians living in Ottoman territory to revolt against the Sultan's rule. A small minority did so, and the "Young Turks," a military triumvirate that effectively governed the empire used the pretext of an "armed Armenian revolutionary uprising" to eliminate its entire Armenian population. In what is usually, and effectively, described as the first 20th century genocide, approximately 800,000 Armenians were uprooted from their towns and villages and deported to Syria and other nearby Ottoman territories. The deportations were carried out under harrowing conditions, and few survived. Those who did settled elsewhere, notably in the United States, where the Armenian community eventually reestablished itself. Nonetheless, the memories of the deportations and massacres were kept alive, passed on from generation to generation.

Since the establishment of the Turkish republic, successive governments have consistently disclaimed responsibility for the actions of its Ottoman predecessor. In the 1970s and 1980s an Armenian terrorist organization, the Secret Army for the Liberation of Armenia (ASALA), carried out a series of attacks on Turkish diplomats abroad, killing 30 of them. However, when violence proved unproductive, Armenian community leaders undertook a lobbying campaign with the U.S. government and European governments to encourage them to put pressure on Turkey for an admission of its indirect responsibility for the deportations, since they had been carried out by its predecessor. This persistent lobbying finally brought results. In March 2001, the U.S. House of Representatives approved a resolution calling on Turkey to "recognize publicly the Armenian genocide." A similar resolution was approved in April by the European Parliament, the deliberative body of the European Union.

Perhaps prodded by such actions as a resolution by the French National Assembly that the Armenian deportations constituted genocide, the Turkish government in 2005 inaugurated a first-ever national debate on its predecessor's actions toward the Armenians. The debate was given added impetus by the forthcoming talks on Turkey's possible EU membership and the need for reforms in various areas in order to meet EU membership requirements. Prime Minister Erdogan in March proposed a joint study of the issue by Turkish and Armenian historians. Although the proposal was rejected by the Armenian Republic's Foreign Ministry, a number of nongovernmental Turkish groups—academics, industrial leaders, musicians and women, among others—began meeting with their Armenian counterparts.

Many stumbling blocks remain. The publisher of the Armenian Istanbul weekly *Agos* was given a 6-year jail sentence in August for having "insulted" the Turks by urging the two peoples to stop hating each other. On its side the first Turkish scholar granted access to Armenia's national archives (because he has publicly called Ottoman policy genocidal) was arrested by Armenian authorities on charges of smuggling antique books out of the country. But as a Turkish political scientist observed, "I think Turks have come a long way even to say, We did something wrong to the Armenians."[1]

The other exception to Turkish homogeneity is the Kurds, who make up 17 percent of the population officially but may be closer to 20 percent. There are also Kurdish populations in Iraq, Syria, and Iran, but Turkey's Kurds form the largest component of this "people without a nation," one of the last ethnic peoples in the world who do not have their own indigenous government. Their clannish social structure and fierce spirit of independence have led to periodic Kurdish uprisings against the governments that rule them. In Turkey, the Ataturk regime crushed Kurdish rebellions in the 1920s, and from then on, Kurds were officially referred to as "Mountain Turks." Until the 1980s, Turkey's Kurds were considered an unimportant, albeit economically deprived, population group. Many emigrated to the cities or abroad, mainly to Germany and the Netherlands. Turkey's Kurdish population is concentrated in its southeastern region and its provincial capital, Diyarbakir. During the 15-year guerrilla war between the Turkish army and security forces and the Kurdish PKK nationalist movement, some 35,000 civilians and militants were killed and over a million villagers displaced, most fleeing to Turkish cities or such foreign countries as Germany.

DEVELOPMENT

The loans pledged by the IMF and World Bank to resolve Turkey's economic crisis require the country to maintain a 4% budget surplus, excluding interest on foreign debts. Inflation and the loss of purchasing power have made that objective almost impossible to reach. A new tracking system that requires an official personal identification number (PIN) for transactions over $3,000 or to hold a bank or stock account should reduce cheating and create a financial database. It will also provide for more equitable tax collection.

In addition to those in Cyprus, there are two other important populations of ethnic Turks outside of Turkey. They are in Bulgaria and western (Greek) Thrace. Those in Bulgaria make up about 10 percent of the population. In the 1980s, they were suppressed by the Communist Bulgarian regime as foreigners and as Muslims, although they had lived in peace with their neighbors for centuries. About one third fled to Turkey as refugees. This forced assimilation policy was reversed after the fall of the Communist regime, and most of them have now returned to their Bulgarian homes.

There are also about 120,000 ethnic Turks in Greek Thrace, left over from the forced exchange of Greek and Turkish populations in 1922. However, they have never been granted Greek citizenship and are discriminated against in various ways. In 1990, the Greek government unilaterally abrogated the 1923 Treaty of Lausanne, which, among other things, guaranteed the Turkish minority the right to choose its own religious leaders. They are mostly small farmers, forming a close-knit group mainly due to their religious separateness. Perhaps because of the republic's long-standing policy of not expanding its territory, the turks in Greek Thrace have had little contact with those living in the adjoining Turkish vilayet (province) of Edirne.

An estimated 20 percent of Turkey's population are *Alevis,* a blanket term for various Muslim communities whose Islamic rituals and beliefs differ from those of the Sunni majority. Some are Shias; others are ethnic and religious compatriots of the Alawis, who currently rule Syria, and live close by in the Hatay and other areas near the Syrian border. Other Alevis form compact communities in such small Anatolian towns as Sivas, Corum, and Kahramanras, and Istanbul has a substantial Alevi population. Alevi rituals differ from those of both Sunnis and Shias in that they incorporate music and dancing into their services. They have no religious leaders, but each Alevi community has a *dede* ("old man") who directs community affairs. One observer said of them, "They claim to live according to the inner meaning of religion (*batin*) rather than by external (*zahir*) demands ..., prayer, the fast in Ramadan, *zakat* and *hajj* are alien."[2]

A small population of Assyrian Christians, 6,000 in all, is still found in southeastern Turkey in Kurdish territory, remnants of their deportation along with the Armenians during the first World War and offshoots of the much larger Assyrian community in Iraq. In 2005, for the first time in their history under the republic, they were allowed to observe publicly their New Year celebration (Akito), presumably due to Turkey's continued efforts to polish its image and its commitment to democratic change in the run-up to its application for membership in the European Union.

In the years of the Ottoman Empire, a large Jewish population settled in Turkey's lands, invited there from Spain after their expulsion from that country by its Christian rulers. Nearly every Ottoman city had

Turkey has been populated for thousands of years by a myriad of peoples, including Hittites, Greeks, and Romans. This ancient Turkish artifact is mute evidence of one of these many bygone civilizations.

its Jewish quarter and synagogue. But with the establishment of Israel and the rise of Turkish nationalism, the great majority of Jews left Turkey. Small Jewish communities still survive in Ankara, Istanbul, and Antakya. Most of the country's synagogues have become architectural museums, beautifully designed monuments to the multiethnic, multireligious Ottoman past. Unfortunately Islamic fundamentalist violence, aimed at destabilizing secular Muslim regimes, reached Turkey in late 2003 with the suicide bombing of two Istanbul synagogues along with a branch of Britain's Barclay's Bank. The synagogues were attacked during Shabbat prayers, killing 25 worshippers and injuring several hundred more, mostly Muslims on the street outside the buildings.

HISTORY: A PARADE OF PEOPLES

The earliest political unit to develop in the peninsula was the Empire of the Hittites (1600–1200 B.C.), inventors of the two-wheeled chariot and one of the great powers of the ancient Near East. Other Asia Minor peoples made important contributions to our modern world through their discoveries and inventions. We are indebted to the Lydians for our currency system. They were great warriors but also great traders, developing a coinage based on gold and silver to simplify trade exchanges. The gold was panned from the Pactolus River near their capital of Sardis. It was then separated from other metals by melting through capellation—mixing the

particles with salt, heating in an earthenware container, and finally smelting until ready to mint. The Lydian king, Croesus (561–547 B.C.), who ruled when Lydian trade was at its peak, has become a familiar figure, "rich as Croesus" as a result of this process.[3]

Following the collapse of the Roman Empire in the fifth century A.D., Asia Minor became the largest part of the East Roman or Byzantine Empire, named for its capital, Byzantium. The city was later renamed Constantinople, in honor of the Roman emperor Constantine, after he had become Christian. For a thousand years, this empire was a center and fortress of Christianity against hostile neighbors and later against the forces of Islam.

The Ottoman Centuries[4]

Various nomadic peoples from Central Asia began migrating into Islamic lands from the ninth century onward. Among them were the ancestors of the Turks of today. They settled mostly along the borders between Christian and Islamic powers in Asia Minor and northwest Iran. Although divided into families and clans and often in conflict, the Turks had a rare sense of unity as a nation. They were also early converts to Islam. Its simple faith and requirements appealed to them more than did Christian ritual, and they readily joined in Islam's battles as *Ghazis*, "warriors for the faith." Asia Minor, having been wrested from the Greeks by the Turks, also gave the Turks a strong sense of identification with that particular place. To them it was Anadolu

(Anatolia), "land of the setting sun," a "sacred homeland" giving the Turks a strong sense of national identity and unity.

The Ottomans were one of many Turkish clans in Anatolia. They took their name from Osman, a clan leader elected because of certain qualifications considered ideal for a Ghazi chieftain—wisdom, prudence, courage, skill in battle, and justice, along with a strong belief in Islam.[5] Osman's clan members identified with their leader to such an extent that they called themselves *Osmanlis,* "sons of Osman," rather than *Turks,* a term they equated with boorish, unwashed peasants.

Although the Ottomans started out with a small territory, they were fortunate in that Osman and his successors were extremely able rulers. Osman's son, Orkhan, captured the important Greek city of Bursa, across the Sea of Marmara from Constantinople (modern-day Istanbul). It became the first Ottoman capital. Later Ottoman rulers took the title of sultan to signify their temporal authority over expanding territories. A series of capable sultans led the Ottoman armies deep into Europe. Constantinople was surrounded, and on May 29, A.D. 1453, Mehmed II, the seventh sultan, captured the great city amid portents of disaster for Christian Europe.[6]

The North African corsair city-states of Algiers, Tripoli, and Tunis, which owed nominal allegiance to the sultan but were in practice self-governing, aligned their swift fleets with his from time to time in the contest with European states for control of the

Mediterranean. On two occasions his armies besieged Vienna, and during the rule of Sultan Sulayman I, a contemporary of Queen Elizabeth I of England, the Ottoman Empire was the largest and most powerful in the world.

One reason for the success of Ottoman armies was the Janissaries, an elite corps recruited mostly from Christian villages and converted to Islam by force. Janissary units were assigned to captured cities as garrisons. Those in Constantinople enjoyed special privileges. They had their own barracks and served on campaigns as the sultan's personal guard. Invading Ottoman armies were preceded by marching bands of drummers and cymbal players, like the bagpipers who marched ahead of Scottish armies. These Janissary bands made such terrifying noises that villagers fled in terror at their arrival, while enemy forces surrendered after hearing their "fearsomely loud sounds, like an alarm."[7]

Another factor that made the Ottoman system work was the religious organization of non-Muslim minority groups as self-governing units termed *millets,* a Turkish word meaning "nations." Each millet was headed by its own religious leader, who was responsible to the sultan for the leadership and good behavior of his people. The three principal millets in Turkey were the Armenians, Greek Orthodox Christians, and Jews. Although Christians and Jews were not considered equal to freeborn Muslims, they were under the sultan's protection. Armenian, Greek, and Jewish merchants rendered valuable services to the empire due to their linguistic skills and trade experience, particularly after the wars with Europe were replaced by peaceful commerce.

The "Sick Man of Europe"

In the eighteenth and nineteenth centuries, the Ottoman Empire gradually weakened, while European Christian powers grew stronger. European countries improved their military equipment and tactics and began to defeat the Ottomans regularly. The sultans were forced to sign treaties and lost territories, causing great humiliation, since they had never treated Christian rulers as equals before. To make matters worse, the European powers helped the Greeks and other Balkan peoples to win their independence from the Ottomans.

The European powers also took advantage of the millet system to intervene directly in the Ottoman Empire's internal affairs. French troops invaded Lebanon in 1860 to restore order after civil war broke out there between the Christian and Druze communities. The European powers claimed the right to protect the Christian minorities from mistreatment by the Mus-

lim majority, saying that the sultan's troops could not provide for their safety.

One or two sultans in the nineteenth century tried to make reforms in the Ottoman system. They suppressed the Janissaries, who by then had become an unruly mob, and organized a modern army equipped with European weapons, uniforms, and advisers. Sultan Mahmud II issued an imperial decree called *Tanzimat* (literally, "reordering"). It gave equal rights under the law to all subjects, Muslims and non-Muslims alike, in matters such as taxation, education, and property ownership. Provincial governors were directed to implement its provisions. In one province, Baghdad, the governor, Miidhat Pasha, established free schools and hospitals, invited foreign missionaries to develop a Western-style curriculum for these schools, reduced taxes, and restored public security in this traditionally unruly border province.

Subsequently Midhat Pasha was appointed grand vizier (e.g., prime minister) by the new sultan, Abdul Hamid II. In 1876, prodded by Russian and British threats of a takeover, the sultan agreed to Midhat Pasha's urgings and issued a Constitution, the first such document in the empire's history. It would limit the sultan's absolute power by establishing an elected Grand National Assembly (GNA), which would represent all the classes, races, creeds, and ethnic, and linguistic groups within the empire. (The GNA survived the fall of the empire and was reborn as the Legislature of the Turkish Republic.)

However, the forces of reaction, represented by the religious leaders, the sultan's courtiers, and the sultan himself, were stronger than the forces for reform. Abdul-Hamid had no real intention of giving up the absolute powers that Ottoman sultans had always had. Thus, when the first Grand

National Assembly met in 1877 and the members ventured to criticize the sultan's ministers, he dissolved the Assembly.

The European powers became convinced that the Ottomans were incapable of reform. European rulers compared the healthy state of their economies and the growth of representative government in their countries to the grinding poverty and lack of rights for Ottoman subjects, as a healthy person looks at an ill one in a hospital bed. The European rulers referred to the sultan as the "Sick Man of Europe" and plotted his death.

But the Sick Man's death was easier to talk about than to carry out, primarily because the European rulers distrusted one another almost as much as they disliked the sultan. If one European ruler seemed to be getting too much territory, trade privileges, or control over the sultan's policies, the others would band together to block that ruler.

World War I: Exit Empire, Enter Republic

During World War I, the Ottoman Empire was allied with Germany against Britain, France, and Russia. Ottoman armies fought bravely against heavy odds but were eventually defeated. A peace treaty signed in 1920 divided up the empire into British and French protectorates, except for a small part of Anatolia that was left to the sultan. The most devastating blow of all was the occupation by the Greeks of western Anatolia, under the provisions of a secret agreement that brought Greece into the war. It seemed to the Turks that their former subjects had become their rulers.

At this point in the Turkish nation's fortunes, however, a new leader appeared. He would take it in a very different direction. This new leader, Mustafa Kemal, had risen through the ranks of the Ottoman Army to become one of its few successful commanders. He was largely responsible for the defeat of British and Australian forces at the Battle of Gallipoli, when they attempted to seize control of the strategic Dardanelles (Straits) in 1915.

Mustafa Kemal took advantage of Turkish anger over the occupation of Anatolia by foreign armies, particularly the Greeks, to launch a movement for independence. It would be a movement not only to recover the sacred Anatolian homeland but also for independence from the sultan.

The Turkish independence movement began in the interior, far from Constantinople. Mustafa Kemal and his associates chose Ankara, a village on a plateau, as their new capital. They issued a so-called National Pact stating that the "New Turkey" would be an independent republic. Its territory would be limited to areas where Turks were the majority of the population.

The nationalists resolutely turned their backs on Turkey's imperial past.

The Turkish War of Independence lasted until 1922. It was fought mainly against the Greeks. The nationalists were able to convince other occupation forces to withdraw from Anatolia by proving that they controlled the territory and represented the real interests of the Turkish people. The Greeks were defeated in a series of fierce battles; and eventually France and Britain signed a treaty recognizing Turkey as a sovereign state headed by Mustafa Kemal.

THE TURKISH REPUBLIC

The Turkish republic has passed through several stages of political development since it was founded. The first stage, dominated by Mustafa Kemal, established its basic form. "Turkey for the Turks" meant that the republic would be predominantly Turkish in population; this was accomplished by rough surgery, with the expulsion of the Armenians and most of the Greeks. Mustafa Kemal also rejected imperialism and interference in the internal affairs of other nations. He once said, "Turkey has a firm policy of ensuring [its] independence within set national boundaries."[8] Peace with Turkey's neighbors and the abandonment of imperialism enabled Mustafa Kemal to concentrate on internal changes. By design, these changes would be far-reaching, in order to break what he viewed as the dead hand of Islam on Turkish life. Turkey would become a secular democratic state on the European model. A Constitution was approved in 1924, the sultanate and the caliphate were both abolished, and the last Ottoman sultan went into exile. Religious courts were also abolished, and new European law codes were introduced to replace Islamic law. An elected Grand National Assembly was given the responsibility for legislation, with executive power held by the president of the republic.

The most striking changes were made in social life, most bearing the personal stamp of Mustafa Kemal. The traditional Turkish clothing and polygyny were outlawed. Women were encouraged to work, were allowed to vote (in 1930), and were given equal rights with men in divorce and inheritance. Turks were now required to have surnames; Mustafa Kemal took the name *Ataturk,* meaning "Father of the Turks."

Mustafa Kemal Ataturk died on November 10, 1938. His hold on his country had been so strong, his influence so pervasive, that a whole nation broke down and wept when the news came. Ataturk's mausoleum in Ankara, the Anit Kabir, is the place most frequently visited by Turkish schoolchildren. His portrait hangs in every public place, even in barber shops, and his image appears on every bank note in the new currency issued in January 2005.

Ismet Inonu, Ataturk's right-hand man, succeeded Ataturk and served as president until 1950. Ataturk had distrusted political parties; his brief experiment with a two-party system was abruptly cancelled when members of the officially sponsored "loyal opposition" criticized the Father of the Turks for his free lifestyle. The only political party he allowed was the Republican People's Party (RPP). It was not dedicated to its own survival or to repression, as are political parties in many single-party states. The RPP based its program on six principles, the most important, in terms of politics, being *devrimcilik* ("revolutionism" or "reformism"). It meant that the party was committed to work for a multiparty system and free elections. One author noted, "The Turkish single party system was never based on the doctrine of a single party. It was always embarrassed and almost ashamed of the monopoly [over power]. The Turkish single party had a bad conscience."[9]

Agitation for political reforms began during World War II. Later, when Turkey applied for admission to the United Nations, a number of National Assembly deputies pointed out that the UN Charter specified certain rights that the government was not providing. Reacting to popular demands and pressure from Turkey's allies, Inonu announced that political parties could be established. The first new party in the republic's history was the Democratic Party, organized in 1946. In 1950, the party won 408 seats in the National Assembly, to 69 for the Republican People's Party. The Democrats had campaigned vigorously in rural areas, winning massive support from farmers and peasants. Having presided over the transition from a one-party system with a bad conscience to a two-party one, President Inonu stepped down to become head of the opposition.

MILITARY INTERVENTIONS

Modern Turkey has struggled for decades to develop a workable multiparty political system. An interesting point about this struggle is that the armed forces have seized power three times, and three times they have returned the nation to civilian rule. This fact makes Turkey very different from other Middle Eastern nations, whose army leaders, once they have seized power, have been unwilling to give it up.

Ataturk deliberately kept the Turkish armed forces out of domestic politics. He believed that the military had only two responsibilities: to defend the nation in case of invasion and to serve as "the guardian of the reforming ideals of his regime."[10] Since Ataturk's death, military leaders have seized power only when they have been convinced that the civilian government had betrayed the ideals of the founder of the republic.

The first military coup took place in 1960, after a decade of rule by the Democrats. Army leaders charged them with corruption, economic mismanagement, and repression of the opposition. After a public trial, the three top civilian leaders were executed. Thus far, they have been the only high-ranking Turkish politicians to receive the death sentence. (After the 1980 coup, a number of civilian leaders were arrested, but the most serious sentence imposed was a ban on political activity for the next 10 years for certain party chiefs.)

The military leaders reinstated civilian rule in 1961. The Democratic Party was declared illegal, but other parties were allowed to compete in national elections. The new Justice Party, successor to the Democrats, won the elections but did not win a clear majority. As a result, the Turkish government could not function effectively. More and more Turks, especially university students and trade union leaders, turned to violence as they became disillusioned with the multiparty system. As the violence increased, the military again intervened, but it stopped short of taking complete control.

HEALTH/WELFARE

Improved prison conditions is an important pre-requisite for Turkey's membership in the European Union. A hunger strike by inmates protesting these conditions in 2001 led to 31 deaths. An amendment to the constitution annuals *Article 51* of the penal code, which allows reduced sentences for "honor killings," those carried out by husbands or other relatives on women who have "shamed" their families by adultery or other forms of misbehavior.

In 1980, the armed forces intervened for the third time, citing three reasons: failure of the government to deal with political violence; the economic crisis; and the revival of Islamic fundamentalism, which they viewed as a total surrender of the secular principles established by Ataturk. (The National Salvation Party openly advocated a return to Islamic law and organized huge rallies in several Turkish cities in 1979–1981.) The National Assembly was dissolved, the Constitution was suspended, and martial law was imposed throughout the country. The generals said that they would restore parliamentary rule—but not before terrorism had been eliminated.

RETURN TO CIVILIAN RULE

The military regime approved a new Constitution in 1982. It provided for a multiparty political system, although pre-1980 political parties were specifically excluded. (Several were later reinstated, notably the Republican People's Party, or RPP). Three new parties were allowed to present candidates for a new Grand National Assembly (GNA), and elections were scheduled for 1983. However, the party least favored by the generals, the Motherland Party (ANAP), ran an American-style political campaign, using the media to present its candidates to the country. It won handily. Its leader, Turgut Ozal, became the first prime minister in this phase of Turkey's long, slow progress toward effective multiparty democracy.

Ozal, an economist by profession, had served as minister of finance under the military government in 1980–1982. In that capacity, he developed a strict austerity program that stabilized the economy. But the prime ministership was another matter, especially with five generals looking over his shoulder. The Motherland Party's popularity declined somewhat in 1986–1987, a decline that owed more to a broadening of the political process than to voter disenchantment.

On September 6, 1987, the nation took a significant step forward—although some analysts viewed it as sideways—toward full restoration of the democratic process. Voters narrowly approved the restoration of political rights to about 100 politicians who had been banned from party activity for 10 years after the 1980 coup: The vote was 50.23 percent "yes" to 49.77 percent "no" in a nationwide referendum, a difference of fewer than 100,000 votes. The results surprised many observers, particularly the most prominent political exiles, former prime ministers Suleyman Demirel of the Justice Party and Bulent Ecevit, leader of the banned Republican People's Party. They had expected a heavy vote in their favor. Prime Minister Ozal's argument that the nation should not return to the "bad old days" before the 1980 coup, when a personal vendetta between these two leaders had polarized politics and paralyzed the economy and there were several dozen murders a day, had clearly carried weight with the electorate.[11]

Thus encouraged, Ozal scheduled new elections for November 1, 1987, a year ahead of schedule. But in October, the Constitutional Court ruled that a December 1986 electoral law was invalid because it had eliminated the primary system, thereby undermining the multiparty system. The elections were held on November 29 under new electoral guidelines. The Motherland Party won easily, taking 292 of 450 seats in the GNA. The Social Democratic Populist Party (SHP), a newcomer to Turkish politics, ran second, with 99 seats; while True Path (DYP), founded by Demirel to succeed the Justice Party, ran a distant third, with 59.

The Motherland Party's large parliamentary majority enabled Ozal to have himself elected president to succeed General Evren in 1989, at the end of the latter's term in office. Although the Turkish presidency is largely a ceremonial office, Ozal continued to run the nation as if it were not, with less successful results than those he had attained during his prime ministership. As a result, popular support for his party continued to erode. In the October 1991 elections for a new National Assembly, candidates of the opposition True Path Party won 180 seats to 113 for the Motherland Party, taking 27 percent of the popular vote, as compared to 24 percent for the majority party. The Social Democratic Populist Party garnered 20 percent of the vote, followed by the Islamic Welfare Party (Refah), whose growing strength was reflected in its 16 percent support from voters.

Lacking a majority in the Assembly, the DYP formed a coalition government with the SHP in November 1991. Party leader Suleyman Demirel became prime minister. The DYP-SHP coalition improved its political position in local elections early in 1992, when its candidates won a majority of urban mayorships. In July of that year, the ban on political parties existing before the 1980 military coup was lifted; most of them had been incorporated into new parties, but the Republican People's Party, founded by Ataturk, reentered the political arena. It drew a number of defections from Assembly members, due in large part to the charismatic appeal of its leader, Deniz Baykal; as a result, the coalition was left with a shaky six-vote majority in the Legislature.

President Ozal died in April 1993, abruptly ending the long political feud between him and Demirel that had weakened government effectiveness. Demirel succeeded him as president. The DYP elected Tansu Ciller, a U.S.-trained economist and university professor, as Turkey's first woman prime minister, one of two in the Muslim world (the other was Benazir Bhutto of Pakistan).

Ciller's first two years in office were marked by economic difficulties; growing tendencies toward Islamic fundamentalism, spearheaded by Refah; and intensified violence by Kurdish separatists of the Workers' Party of Kurdistan (PKK), in the southeastern region. Nevertheless, her government, a coalition of the DYP and the Republican People's Party, representing the center left and the center right, seemed to be governing effectively in at least some respects. By early 1995, the army had regained control of much of the southeast from PKK forces, and in March, agreement was reached for a customs union with the European Union. Municipal elections in June also favored the ruling coalition. It won 61.7 percent of Council seats against 17.4 percent for Refah candidates and 13.4 percent for those of ANAP.

Thus, the collapse of the coalition government in September came as a surprise to most observers. Republican People's party head Deniz Baykal had set certain terms for continuation of his party's alliance with DYP. These included repeal of a strict anti-terrorism law, which had drawn international condemnation for its lack of rights for detained dissidents; tighter controls over Islamic fundamentalists; and pay raises of 70 percent for public workers to offset inflation. When these terms were rejected, he withdrew his party from the coalition.

Elections in December 1995 brought another shock, with Refah winning 158 seats to 135 for True Path and 132 for ANAP. For the first time in modern Turkish history, an Islamic-oriented party had won more seats in the Grand National Assembly than its rivals. Refah leader Necmettin Erbakan was named Turkey's first "Islamist" prime minister, taking office in April 1996. However, his party lacked a clear majority in the Assembly. As a result, coalition government became necessary. Erbakan's cabinet included ministers from the three major parties, and Ciller became foreign minister.

The septuagenarian Erbakan initially brought a breath of fresh air into the country's stale political system. With his round face and Italian designer ties, he seemed more like a Turkish uncle than an Islamic fundamentalist. And during his year in office, his government reaffirmed traditional secularism, state socialism, and other elements of the legacy of Ataturk. The government also stressed NATO membership in its foreign policy and continued the drive for an economic and customs union with the European Community begun by its predecessors.

With Refah's victory at the polls, Turkey's military leaders believed that the party was determined to dismantle the secular state founded by Ataturk and replace it with an Islamized one. In 1997, they demanded Erbakan's resignation. Inasmuch as they have final authority over political life under the 1980 Constitution, Erbakan had no choice. After his resignation, the state prosecutor filed suit to outlaw Refah on the grounds that its programs were intended to impose Islamic law on Turkish society. The court agreed, and Erbakan was barred from politics for five years.

President Demirel then named ANAP leader Mesut Yilmaz to head a caretaker government. But he also resigned following a no-confidence resolution in the Grand National Assembly (GNA).

In the April 1999 GNA elections, however, a relatively new party, Democratic Left, surprised observers by winning a clear majority of seats. A strong pro-nationalistic party, Nationalist Action (MHP), ran second, with 18 percent of the popular vote, winning 130 seats. The Virtue Party, reformed from the ruins of Refah, finished with 102 seats and 15 percent of the popular vote.

Virtue then set out to distance itself from Islamic fundamentalism. Its members opposed the ban on wearing headscarves in university classes and government offices. Its governing board even approved the celebration of St. Valentine's Day (!) as an appropriate secular holiday.

Unfortunately, the "new image" of Virtue did not convince the country's military and civilian leaders, who are adamant in their defense of Ataturk's legacy. In July 2000, an appeals court upheld the one-year jail sentence imposed on Erbakan, and the following year the Constitutional Court, the country's highest court, banned Virtue as a political party. The action came over the objections of many political leaders, including Ecevit. Despite the ban, Virtue deputies in the GNA would be allowed to keep their seats, as independents.

ACHIEVEMENTS

The ancient Silk Road, which ran eastward from Turkey through Iran, Central Asia and Afghanistan into China, was once one of the main routes of east-west trade and versa. Today it is more likely to be carrying "black gold," oil and gas from Iran and Central Asian sources through pipelines to the West, including those that follow the road through Turkey.

In August 2001, yet another Islamic-related political party was formed, the 281st since the 1876 Constitution allowed them to form. The new party, Justice and Development (AKP), included many former Virtue leaders, including the charismatic ex-mayor of Istanbul, Recep Tayyip Erdogan. He had been banned from politics for five years in 1998 for criticizing the country's nonadherence to traditional Islam, but he was released under the 2000 amnesty law for political prisoners.

The debate between Islamists and secularists over Turkey's Islamic identity is far from

being resolved. In May 2001, the debate shifted to the presidency. Ecevit had proposed a change in the constitution to allow Suleyman Demirel, the incumbent, to run for a second term but to reduce his term to five years rather than seven. Demirel's election was seen as a sure thing, once the GNA had accepted the proposed changes. However, a majority of deputies rejected the proposal. Their candidate, Ahmet Cevdet Sezer, chief judge of the Constitutional Court, was elected to the largely ceremonial post on the third ballot by a 60 percent margin.

Sezer's election was a bitter blow to Ecevit, and when the president refused to sign a controversial measure that would cause thousands of government employees to lose their jobs if they were suspected of separatist or Islamic fundamentalist activities, there was open warfare between the two leaders. Sezer's support for the repeal of the restrictive press laws, an end to the ban on use of the Kurdish language in schools and on official documents, and civilian control over the military leadership has put him at odds with military leaders as well, although Turkey's acceptance as a member of the European Union depends on the implementation of such reforms.

One EU requirement is that of a reduction in the powers of the National Security Council. The 10-member body, composed of the president, four cabinet ministers, and the five top military commanders, sets the agenda for all important issues, even laws, before they may be debated by the GNA. The order for dismissal of government employees for their "Islamist" beliefs was originally issued as a directive to the GNA by the council. It did not go into effect because Sezer refused to sign it, not as a council member but in his capacity as president.

The November 2002 elections for the 550-member GNA resulted in another flip-flop in Turkey's seesawing political fortunes. AKP won a clear majority, 363 seats to 178 for the RPP; the remainder were spread among several minor parties. AKP leader Abdullah Gul was named prime minister of Turkey's first single-party majority government in 15 years! The GNA subsequently amended the constitution to allow Erdogan to run in a special by-election. In February, the ex-mayor took his seat in the GNA and was then named prime minister.[12]

A development that augurs well for Turkey's future as a stable nation and its future EU membership is its progress in education. In 1997 the government increased primary school attendance from 5 years to 8, and 3 out of 4 schoolchildren go on to high school. Under a World Bank grant of $250 million parents of poor families receive the equivalent of $15–$20 per month per child, the amount depending on its attendance

record. The greatest change has been in higher education. There are now 78 universities, a number of them private, and most courses are given in English. As a result more and more of the nation's future leaders are educated within the country rather than having to go abroad as was the case with previous generations.[13]

THE "KURDISH PROBLEM"

Ataturk's suppression of Kurdish political aspirations and a separate Kurdish identity within the nation effectively removed all traces of a "Kurdish problem" from national consciousness during the first decades of the republic. From the 1930s to the 1970s, the Kurdish areas were covered by a blanket of silence. Posters in Diyarbakir, the regional capital, proclaimed Ataturk's message: "Happy is he who says he is a Turk."

However, the general breakdown in law and order in Turkey in the late 1970s led to a revival of Kurdish nationalism. The Workers' Party of Kurdistan (PKK), founded as a Marxist-Leninist organization, was the first left-wing Kurdish group to advocate a separate Kurdish state. It was outlawed after the 1980 military coup; some 1,500 of its members were given jail sentences, and several leaders were executed for treason.

The PKK then went underground. In 1984, it began a campaign of guerrilla warfare. Its leader, Abdullah ("Apo") Ocalan, had won a scholarship in political science at Ankara University. While there, he became influenced by Marxist ideology and went into exile in Syria. From his Syrian base, and with Syrian support and financing, he called for a "war of national liberation" for the Kurds. Prior to the 1991 Gulf War, PKK guerrillas mounted mostly cross-border attacks into Turkey from bases in northern Lebanon, where they came under Syrian protection. But with Iraq's defeat and the establishment of an autonomous Kurdish region in northern Iraq, the PKK set up bases there to supplement their Lebanese bases.

Use of these bases posed a problem for Iraq's Kurdish leaders. On the one hand, they were committed to the cause of Kurdish sovereignty. But the cross-border raids brought Turkish retaliation, endangering their hard-won freedom from the long arm of Saddam Hussein's government. In 1992, after the raids had brought on massive Turkish counterattacks, they announced that they no longer supported the PKK and would not allow their territory to be used for its attacks on Turkish villages and police and army posts.

However, the momentum of the conflict left little maneuvering room for the groups involved. Turkey imposed martial law on its eastern provinces, and Turkish forces carried out large-scale raids on PKK bases in northern Iraq, seriously hampering PKK effectiveness. Ocalan called a unilateral cease-fire in 1993, but the Turkish government said that it would not deal with terrorists. The PKK then resumed the conflict, which by 2000 had claimed 40,000 lives, the majority of them villagers caught between security forces and the guerrillas. Some 3,000 villages had been destroyed and 2 million Kurds made refugees.

Yet despite Turkey's huge military superiority, its struggle with the PKK remained a stalemate until 1999. Syria meanwhile had expelled Ocalan after the Turks had threatened to invade its territory under the international "right to self-defense." The PKK leader went first to Italy. The Ecevit government demanded his extradition, but the Italians refused, on grounds that their Constitution forbids extradition to countries that observe the death penalty. However, by this time Ocalan had become a huge embarrassment to his hosts. He left Italy for Greece and was finally given sanctuary in the Greek Embassy in Nairobi, Kenya.

Acting on a tip from Greek intelligence, Turkish commandos abducted Ocalan from the embassy and placed him in solitary confinement on an island in the Sea of Marmara. He was then tried for treason. Although the Turkish government insisted that the trial was an internal matter, UN and European Union observers were permitted to attend it.

Testifying in his own defense, Ocalan said that he had learned his lesson. He renounced violence as a "mistaken policy" and asserted that he would work as a loyal citizen toward the goal of peace and brotherhood. "We want to give up the armed struggle and have full democracy, so the PKK can enter the political arena," he said. "I will serve the state because now I see that it is necessary."[14] The court was unconvinced, and on June 29, Ocalan was sentenced to death. The criminal appeals court, the only one in the Turkish legal system, upheld his conviction on appeal.

Ocalan's death sentence was commuted to life imprisonment in 2002, after the GNA had abolished the death penalty as a pre-requisite for EU membership. In May 2005, however, the European Court of Human Rights ruled 11–6 that his trial had not been conducted as an impartial and independent tribunal and thus violated European conventions on human rights. Since Turkey has signed this convention, defying the decision held serious implications for the country's membership application. Prime Minister Erdogan then stated that

"whether this case is opened or not, the matter is a closed one for the nation's conscience." Most Turks agreed with him.[15]

A new GNA law will limit henceforth the involvement in national politics of military generals, who have always considered themselves guardians of Ataturk's legacy to the nation. Other new laws improved prisoner rights and abolished torture, as the country continued to polish up its human rights image in order to gain admission to the European Union.

FOREIGN POLICY

Other than friction with the U.S. Congress and with France over the Armenian issue, Turkey has been consistently a Western ally in its foreign policy. Since the end of the 1991 Gulf War, it has allowed U.S. and British planes to be based at its Incirlik air base from where they protect the Kurdish and Shia populations from Saddam Hussein's regime. The agreement, code-named Operation Provide Comfort, was approved at six-month intervals by the GNA until 2003, when the overthrow of the Iraqi regime made it unnecessary.

Turkey's relations with Iraq have always been complex and have had a negative effect on the Turkish economy, with an estimated loss of $40 billion from the cutoff in Iraqi oil exports. The UN sanctions limit Turkey to 75,000 barrels per day (bpd), insufficient to meet domestic needs. A large illegal cross-border trade developed in 2000–2001, with Iraqi oil trucked to the Kurdish border and smuggled into Turkey. With the overthrow of Saddam Hussein in March 2003, the pipeline from Kirkuk to the Turkish port of Ceyhan was reopened briefly but then closed due to sabotage.

Turkey's long-established friendship with the U.S. as its major ally, maintained through two world wars and many minor conflicts, underwent a severe strain when U.S. forces invaded Iraq in 2003. During the countdown to the invasion, the GNA refused permission for American troops to enter Iraq from its territory. Despite threats that the Bush administration would reduce U.S. aid, the Turks held firm, preferring to work through the UN to force Saddam Hussein to expose and remove his presumed weapons of mass destruction. As things turned out, the aid was maintained at the same level, but Turkish troops have yet to take part in the occupation of Iraq.

Turkey's relatively independent posture in foreign policy has been marked by agreements with Iran and Israel for training of its air-force pilots. The country reached agreement with Israel in April 2001 for water deliveries, as a part of its "water for peace" program for the Middle East. Under

the terms of the agreement, Israel would receive 50 million cubic meters annually of water from the Tigris and Euphrates Rivers. Both of them rise in Turkey.

The Cyprus issue, which has divided Turkey and Greece ever since Turkish troops occupied the northern part of the island republic in 1974, moved toward a solution in 2004 as Greek and Turkish Cypriot leaders began serious negotiations under UN sponsorship. Although Turkish Cypriots approved unification in a plebiscite, the Greek Cypriot community rejected it. In February 2006 the Turkish government issued new proposals that would remove all restrictions on freedom of goods transfers, exchange of families and sharing of services in the divided island. If agreed to by the UN and the Greek community it would end the Turkish zone's economic isolation and expand Cyprus's present EU membership to a single Greek-Turkish entity.

The country has also improved its links with the newly independent Turkish-speaking nations of Central Asia. Turkey was the first country to recognize the independence of Kazakhstan and Azerbaijan. In 2000, the government signed a 15-year agreement with Azerbaijan for imports of natural gas from the Shaykh Deniz field, offshore in the Caspian Sea. Another important agreement, among Turkey, Georgia, Azerbaijan, and Kazakhstan, has initiated construction of the thousand-mile oil pipeline from Baku, Azerbaijan, to Ceyhan (Adana) on Turkey's Mediterranean coast.

THE ECONOMY

Turkey has a relatively diversified economy, with a productive agriculture and considerable mineral resources. Cotton is the major export crop, but the country is the world's largest producer of sultana raisins and hazelnuts. Other important crops are tobacco, wheat, sunflower seeds, sesame and linseed oils, and cotton-oil seeds. Opium was once an important crop, but, due to illegal exportation, poppy growing was banned by the government in 1972. The ban was lifted in 1974 after poppy farmers were unable to adapt their lands to other crops; production and sale are now government-controlled.

Mineral resources include bauxite, chromium, copper, and iron ore, and there are large deposits of lignite. Turkey is one of the world's largest producers of chromite (chromium ore). Another important mineral resource is meerschaum, used for pipes and cigarette holders. Turkey supplies 80 percent of the world market for emery, and there are rich deposits of tungsten, perlite, boron, and cinnabar, all important rare metals.

Turkey signed a customs agreement with the European Union in 1996. The agreement eliminated import quotas on Turkish textiles and slashed customs duties and excise taxes on Turkish imports of manufactured iron and steel products from the European Union.

The agreement was intended as a first step toward full membership in the EU. However, the country's poor human-rights record, its political instability, and more recently its financial crisis of 2000–2001 have delayed the process.

The "liquidity crisis" that nearly overwhelmed the Turkish economy in 2001 resulted from a combination of factors. Ironically, one of them was the economic-reform program introduced by the Ecevit government to meet EU requirements. Corruption in economic and fiscal management was another factor, while a third grew from the public dispute between Sezer and Ecevit over privatization of state-owned enterprises. The feud between the two leaders, plus the slow pace of privatization, led to a fiscal crisis in November 2000. The liquidity crisis followed, with a run on foreign-currency reserves in the Central Bank as worried Turks and foreign investors rushed to retrieve their funds. The bank lost $7.5 billion in reserves in a two-day period. The government's stopgap decision to end currency controls and allow the Turkish lira to float caused it to lose nearly 50 percent of its value. In December, 10 banks collapsed; they included Ihlas Finans, the country's largest Islamic bank. (Under Turkish banking laws, deposits in such banks, which are interest-free under Islamic prohibitions against usury, are not covered by federal deposit guarantees. Consequently, 200,000 depositors lost their life savings.)

The crisis was averted temporarily when the International Monetary Fund agreed "in principle" to provide $5 billion in emergency aid. However, the IMF's insistence on fiscal reform as a precondition brought on another crisis in March–April 2001. The government then appointed Kermal Dervis, a Turkish-born World Bank economist, as minister of the economy. He was charged with bringing economic order out of fiscal chaos. Dervis presented his reform package in April. It included a 9 percent limit on government expenditures and a hiring freeze in the bloated public sector. But despite $16 billion in loans from the IMF and the World Bank, the economy continued to slide downward, with gross domestic product down 11.8 percent for 2001.

In the last 3 years observers have described Turkey's economic growth as "stunning." Since the Dervis reform package went into effect. The GDP growth rate

Timeline: PAST

330
The founding of Constantinople as the Roman Christian capital, on the site of ancient Byzantium

1453
The capture of Constantinople by Sultan Mehmed II; the city becomes the capital of the Islamic Ottoman Empire

1683
The Ottoman Empire expands deep into Europe; the high-water mark is the siege of Vienna

1918–1920
The defeat of Ottomans and division of territories into foreign protectorates

1923
Turkey proclaims its independence

1960
The first military coup

1980s
Military coup; civilian rule later returns; the government imposes emergency rules

1990s
The Kurdish problem intensifies; Alawi and Kurdish social unrest; thousands die as earthquakes devastate Turkey

PRESENT

2000s
Serious financial crises threaten the nation
Turkey continues to try to meet requirements for EU membership

has been over 8 percent in 2004 and 2005, while inflation has dropped to the present 9.3 percent, the first in single digits since 1972. Another beneficial measure was the reduction in agricultural subsidies, from $6 billion in 2002 to $1.5 billion in 2005. Also price controls on electricity and gas were eliminated in early January 2005, while a budget surplus of 6.5 percent of GDP in 2004 has improved the country's international debt position. The one economic area which has yet to improve is privatization, mainly because of the government's failure to make state-owned businesses attractive to potential buyers.

The country does have a fairly large skilled labor force, and Turkish contractors have been able to negotiate contracts for development projects in oil-producing countries, such as Libya, with partial payment for services in oil shipments at reduced rates. The large Turkish expatriate labor force, much of it in Germany, provided an important source of revenue through worker remittances. Although some European leaders voiced nightmarish

thoughts about a vast number of unemployed Turks wandering about Europe, the combination of increased jobs at home indicated by the projected 5 percent GDP growth rate for 2006–2009 should not only enlarge the domestic labor force but also preclude wholesale immigration of Turkish workers.

NOTES

1. The New York Life Insurance Company agreed recently to pay $10 million to the heirs or relatives of Armenians killed during the deportations in claims on policies issued before 1915. The agreement resulted from a bill passed by the California state legislature that extends the statute of limitations on such claims. California has a large Armenian community, and the company chose to pay the claims rather than incur numerous lawsuits.
2. Martin van Bruinessen, "Kurds, Turks and the Alevi Revival in Turkey," *Middle East Report* (July–September 1996), p. 7.
3. John Noble Wilford, "The Secrets of Croesus' Gold," *The New York Times* (August 15, 2000) Another Asia Minor ruler, King Midas of Phrygia, was said to have the "golden touch," because everything he touched (including his daughter) turned to gold; he had angered the gods, it seemed.
4. Cf. Lord Kinross, *The Ottoman Centuries: The Rise and Fall of the Turkish Empire* (New York: William Morrow, 1977).
5. *Ibid.,* p. 25.
6. An American astronomer, Kevin Pang, advanced the proposal that the fall of the Byzantine capital was preceded by a "darkening of the skies" and other portents of doom related to the eruption of the volcano Kuwae, in the New Hebrides, in 1453. See Lynn Teo Simarski, "Constantinople's Volcanic Twilight," *Aramco World* (November/December 1996), pp. 8–13.
7. The marching bands at football games and parades in our society apparently derive from Janissary bands of drummers and cymbal players who marched ahead of invading Ottoman armies, striking terror in their enemies with their loud sounds.
8. V. A. Danilov, "Kemalism and World Peace," in A. Kazancigil and E. Ozbudun, eds., *Ataturk, Founder of a Modern State* (Hamden, CT: Archon Books, 1981), p. 110.
9. Maurice Duverger, *Political Parties* (New York: John Wiley, 1959), p. 277.
10. C. H. Dodd, *Democracy and Development in Turkey* (North Humberside, England: Eothen Press, 1979), p. 135.
11. One reason for its success at the polls was due the fact that AKP was viewed by the electorate as "clean." (AK, in Turkish, means "white" or "clean.")
12. Stephen Kinzer, in *The New York Times,* (June 1, 1999).
13. Former Prime Minister Ciller, for example, was educated in the U.S. and spoke better English than Turkish when she was elected.
14. Quoted in *The Christian Science Monitor,* date unknown.
15. Yigal Schlufer, in *The Christian Science Monitor,* May 22, 2005.

United Arab Emirates

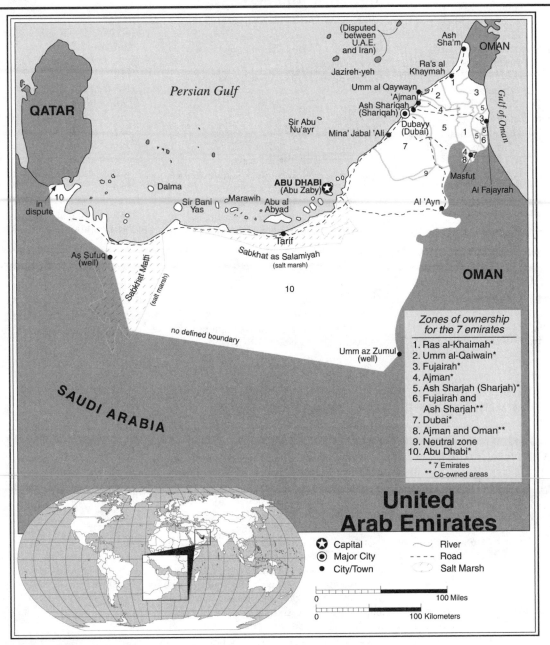

Zones of ownership for the 7 emirates

1. Ras al-Khaimah*
2. Umm al-Qaiwain*
3. Fujairah*
4. Ajman*
5. Ash Sharjah (Sharjah)*
6. Fujairah and Ash Sharjah**
7. Dubai*
8. Ajman and Oman**
9. Neutral zone
10. Abu Dhabi*

* 7 Emirates
** Co-owned areas

United Arab Emirates

★ Capital   ～ River
◉ Major City   - - - Road
● City/Town   ⬭ Salt Marsh

0   100 Miles
0   100 Kilometers

United Arab Emirates

GEOGRAPHY

Area in Square Miles (Kilometers):

31,992 (82,880) (about the size of Maine)

Capital (Population): Abu Dhabi (799,000)

Environmental Concerns: lack of natural freshwater; desertification; oil pollution of beaches and coastal waters

Geographical Features: flat, barren coastal plain merging into rolling sand dunes of vast desert; mountains in the east

Climate: hot, dry desert; cooler in the eastern mountains

PEOPLE

Population

Total: 2,563,212 (including 1,606,079 non-nationals)
Annual Growth Rate: 1.54%
Rural/Urban Population Ratio: 16/84
Major Languages: Arabic; Persian (Farsi); English; Hindi
Ethnic Makeup: 19% Emirati; 23% other Arab and Iranian; 50% South Asian; 8% East Asian and Westerner
Religions: 96% Muslim (80% Sunni, 16% Shia); 4% Hindu, Christian, and others

Health

Life Expectancy at Birth: 72.7 years (male); 77 years (female)
Infant Mortality Rate: 14.5/1,000 live births

Education

Adult Literacy Rate: 79.2%

COMMUNICATION

Telephones: 916,000 main lines
Daily Newspaper Circulation: 135 per 1,000 people
Televisions: 18 per 1,000 people

Internet Users: 1 (2000)

TRANSPORTATION

Highways in Miles (Kilometers): 3,002 (4,835)
Railroads in Miles (Kilometers): none
Usable Airfields: 35
Motor Vehicles in Use: 400,000

GOVERNMENT

Type: federation of emirates
Independence Date: December 2, 1971 (from the United Kingdom)
Head of State/Government: Shaykh Khalifa Bin Zayed al-Nuhayyan, ruler of Abu Dhabi, heads Supreme Council of rulers of the seven emirates; Prime Minister Shaykh Muhammad, ruler of Dubai
Political Parties: none
Suffrage: none

MILITARY

Military Expenditures (% of GDP): 3.1%
Current Disputes: Iranian occupation of Greater and Lesser Tunbs Islands

contested by UAE. Boundary with Oman formally demarcated in 2000

ECONOMY

Currency ($U.S. equivalent): 3.673 dirhams = $1
Per Capita Income/GDP: $25,200/$63.6 billion
GDP Growth Rate: 5.7%
Inflation Rate: 3.2%
Labor Force: 2,360,000 (74% of those in age group 15–64 are non-nationals)
Natural Resources: petroleum; natural gas
Agriculture: dates; vegetables; watermelons; poultry; dairy products; fish
Industry: petroleum; fishing; petrochemicals; construction materials; boat building; handicrafts; pearling
Exports: $69.48 billion (primary partners Japan, South Korea, Thailand)
Imports: $45.66 billion (primary partners China, India, Japan)

SUGGESTED WEBSITES

http://lcweb2.loc.gov/frd/cs/
aetoc.html

United Arab Emirates Country Report

The United Arab Emirates (U.A.E.) is a federation of seven independent states with a central governing Council located on the northeast coast of the Arabian Peninsula. The states—called emirates, from the title of their rulers—are Abu Dhabi, Ajman, Dubai, Fujairah, Ras al-Khaimah, Sharjah, and Umm al-Qaiwain. They came under British "protection" in the 1800s and were given their independence from Great Britain by treaty in 1971. At that time, they joined in the federal union. From its modest beginnings, the U.A.E. has come to play an important role in Middle East Arab affairs, because of its oil wealth.

Abu Dhabi, the largest emirate, contains 87 percent of the UAE's total land area. The federal capital is also named Abu Dhabi, but Dubai, capital of the second largest emirate, is a larger city, with a population of approximately 1 million. Dubai has the U.A.E.'s only natural harbor, which has been enlarged to accommodate supertankers, Abu Dhabi, Dubai, and Sharjah produce oil; Sharjah also has important natural-gas reserves and cement. Fujairah port is a major entrepôt for shipping. The other emirates have little in the way of resources and have yet to find oil in commercial quantities.

The early inhabitants of the area were fishermen and nomads. They were converted to Islam in the seventh century A.D., but little is known of their history before the sixteenth century. By that time, European nations, notably Portugal, had taken an active interest in trade with India and the Far East. Gradually, other European countries, particularly the Netherlands, France, and Britain, challenged Portuguese supremacy. As more and more European ships appeared in Arabian coastal waters or fought over trade, the coastal Arabs felt threatened with loss of their territory. Meanwhile, the Wahhabis, militant Islamic missionaries, spread over Arabia in the eighteenth century. Wahhabi agents incited the most powerful coastal group, the Qawasim, to interfere with European shipping. European ships were seized along with their cargoes, their crews held for ransom. To the European countries, this was piracy; to the Qawasim, however, it was defense of Islamic territory against the infidels. Ras al-Khaimah was their chief port, but soon the whole coast of the present-day U.A.E. became known as the Pirate Coast.

Piracy lasted until 1820, when the British, who now controlled India and thus dominated Eastern trade, convinced the

DEVELOPMENT

The U.A.E.'s huge oil revenues (33 percent of GDP) have enabled it overall to maintain a high per capita income and accumulate an annual trade surplus despite the uneven distribution of wealth among the emirates. With oil and gas reserves estimated to last no more than a century, Abu Dhabi and Dubai in particular have pushed non-oil sector development through such projects as Aluminium Dubai and the world's largest dry-dock. Dubai, somewhat richer and more aggressive both at home and internationally, has emphasized upscale urban development and extensive foreign investment.

principal chiefs of the coast to sign a treaty ending pirate activities. A British naval squadron was stationed in Ras al-Khaimah to enforce the treaty. In 1853, the arrangement was changed into a "Perpetual Maritime Truce." Because it specified a *truce* between the British and the chiefs, the region became known as the Trucial Coast, and the territory of each chief was termed a "trucial state." A British garrison was provided for each ruler, and a British political agent was assigned to take charge of foreign affairs. Britain paid the rulers annual subsidies; in most cases, it was all the money they could acquire. There were originally five Trucial States (also called emirates); Sharjah and Ras al-Khaimah were reorganized as separate emirates in 1966.

The arrangement between Great Britain and the Trucial States worked smoothly for more than a century, through both world wars. Then, in the 1960s, the British decided—for economic and political reasons—to give up most of their overseas colonies, including those in the Arabian Peninsula, which were technically protectorates rather than colonies. In 1968, they proposed to the Trucial Coast emirs that they join in a federation with Bahrain and Qatar, neighboring oil-producing protectorates. But Bahrain and Qatar, being larger and richer, decided to go it alone. Thus, the United Arab Emirates, when it became independent in 1971, included only six emirates. Ras al-Khaimah joined in 1972.

FREEDOM

The Supreme Council of the U.A.E. exercises overall federal authority, but rulers of the emirates have full control over their territories. In December 2005 the Supreme Council president said a democratically-elected parliament would be established "in the near future." However the combination of vast wealth and limited freedoms has generated some societal problems. The divorce rate in 2005 reached 48 percent, along with increases in alcohol usage (despite the Islamic prohibition) and a rising crime rate.

PROBLEMS OF INTEGRATION

Differences in size, wealth, resources, and population have hampered U.A.E. integration since it was formed. Another problem is poor communications. Until recently, one could travel from emirate to emirate only by boat, and telephone service was nonexistent. A combination of economic growth and technology (for example the Internet and cell phones), have produced full integration and rapid communication between the seven emirates.

Despite full integration there are certain internal disagreements among the emirates, particularly in the areas of joint economic planning and allocation of development funds for projects. One result is unnecessary duplication of ports and international airports, as each emirate wishes to have its own immediate access to the outside world. Also several borders remain undemarcated—that between Ras al-Khaimah and Umm al-Qaiwain and Sharjah's border with Oman's Musandam Peninsula.

The U.A.E. federal system is defined in the 1971 Constitution. The government consists of a Supreme Council of Rulers of the seven emirates; a Council of Ministers (cabinet), appointed by the president of the Council; and a unicameral Federal National Assembly of 40 members appointed by the ruling emirs on a proportional basis, according to size and population. A stabilizing factor in the U.A.E. system was the leadership of Shaykh Zayed of Abu Dhabi, president of the Council of Rulers from its inception until his death in 2004. The federal capital is located in Abu Dhabi, the largest and richest emirate. The ruler of Dubai, the second largest of the emirates, serves as vice-president. Other unifying features of the U.A.E. are a common commercial-law code, currency, and defense structure. The sharing of revenues by the wealthy emirates with the less prosperous ones has also helped to foster U.A.E. unity.

The 1979 Iranian Revolution, which seemed to threaten the U.A.E.'s security, accelerated the move toward centralization of authority over defense forces, abolition of borders, and merging of revenues. In 1981, the U.A.E. joined with other states in the Gulf Cooperation Council (GCC) to establish a common-defense policy toward their large and powerful neighbor. The U.A.E. also turned to the United States for help; the two countries signed a Defense Cooperation Agreement in 1994. Under the agreement, a force of several hundred U.S. military personnel is stationed in the emirates to supervise port facilities and air refueling for American planes patrolling the no-fly zone in southern Iraq.

Early in 2001, the U.A.E. joined with other GCC members in a mutual-defense pact, the first in the region. With support from the United States, the pact would increase the current rapid-deployment force from 5,000 to 22,000. Each GCC member would contribute to the force in proportion to its size and population.

The September 11, 2001, terrorist attacks on the United States intensified the importance of increased security for oil operations on the part of Abu Dhabi and Dubai, the chief oil-producing U.A.E. states. The Federal Council supported the newly formed international coalition against terrorism in a public statement, and the U.A.E. withdrew its recognition of the Taliban as the legitimate government of Afghanistan. The U.A.E. initially backed U.S. efforts in the UN to force disclosure of Iraq's alleged weapons of mass destruction. Prior to the U.S. invasion of that country, another Arab satellite all-news TV channel, Al-Arabiya, began operating in the U.A.E. to supplement Qatar's Al-Jazeera station.

The governments of the emirates themselves are best described as patriarchal. Each emir is head of his own large "family" as well as head of his emirate. The ruling emirs gained their power a long time ago from various sources—through foreign trade, pearl fishing, or ownership of lands. In recent years, they have profited from oil royalties to confirm their positions as heads of state.

HEALTH/WELFARE

The first all-female taxi service in the Gulf went into operation in the U.A.E. June 2000, as participation by women in the labor force increased. Also in 2000, U.A.E. banks set a quota system for employment of nationals to replace departed foreign workers. The quota is to be increased by 4% a year until the banks are fully staffed by U.A.E. nationals.

Disagreements within the ruling families have sometimes led to violence or "palace coups," there being no rule or law of primogeniture. The ruler of Umm al-Qaiwain came to power when his father was murdered in 1929. Shaykh Zayed deposed his brother, Shaykh Shakbut, in 1966, when the latter refused to approve a British-sponsored development plan for the protectorate. In 1987, Shaykh Abd al-Aziz, the elder brother of the ruler of Sharjah, attempted to overthrow his brother, on the grounds that economic development was being mishandled. The U.A.E. Supreme Council mediated a settlement, and Abd al-Aziz retired to Abu Dhabi. However, he continued to demand authority over Sharjah's economic policies in his capacity as minister for National Development. In 1990, the ruler dismissed him by abolishing the position.

Ajman and Umm al-Qaiwain are coastal ports with agricultural hinterlands. Ras al-Khaimah has continually disappointed oil seekers; its only natural resource is aggregate, which is used in making cement. Fujairah, although lacking in energy resources, has become a major oil-bunkering and -refining center. In 1996, the new bunkering terminal in its port went into operation. Built by the Dutch-owned Van Ommeren Tank Company, the world's largest independent operator, the facilities will eventually double the millions of tons of cargo now being handled by the port.

AN OIL-DRIVEN ECONOMY

In the past, the people of the Trucial Coast made a meager living from breeding racing camels, some farming, and pearl fishing. Pearls were the main cash crop. But twentieth-century competition from Japanese cultured pearls ruined the Arabian pearl-fishing industry.

In 1958, Shaykh Zayed, then in his teens, led a party of geologists into the remote desert near the oasis of al-Ain, following up the first oil-exploration agreement signed by Abu Dhabi with foreign oil companies. Oil exports began in 1962, and from then on the fortunes of the Gulf Arabs improved dramatically. Production was 14,200 barrels per day in 1962; by 1982, it was 1.1 million bpd, indicating how far the country's oil-driven economy had moved in just two decades. Oil reserves are approximately 98 billion barrels, while gas reserves are 205 trillion cubic feet—10 percent of global reserves. They are expected to last well into the twenty-first century at current rates of extraction.

ACHIEVEMENTS

Nearly 2 million acres of desert have been reclaimed for cultivation. The late Shaykh Zayed, who had led the effort toward achieving self-sufficiency in food through the "greening of the desert," in 1997 received the Gold Panda award from the Worldwide Fund for Nature for his services to global conservation— the first head of state to be so honored. In 2000, the U.A.E.'s first satellite went into orbit, another first for the Gulf region.

The bulk of hydrocarbon production and reserves is in Abu Dhabi. Dubai, not content with second place in U.A.E. development, launched a Strategic Development Plan in 1998, intended to increase its non-oil income to $20,000 per capita by 2010. Its government earlier had established a free-trade zone in the port of Jebel Ali. It provides 100 percent foreign ownership, full repatriation of capital and profits, and a 15-year exemption from corporate and other taxes. By 1996, more than 1,000 companies had located in the zone. In October 2000, the Dubai City Internet free-trade zone opened for business. The zone is in the process of creating a "wired economy" for the emirates, to link them with global markets and the media. Companies such as Microsoft, Compaq and IBM are helping to establish Dubai as the marketing hub of the region.[1]

The ruling family in Dubai is also quietly investing in foreign real estate, with over $1 billion invested abroad. In 2004 it purchased Madame Tussaud's Wax Works in London along with a major stake in Daimler/Chrysler, and in 2005 it acquired several large properties in New York City. However the most startling investment—at least from an American viewpoint—was acquisition by the Dubai-based Ports World of the long-established British company P & O (Peninsular and Oriental Navigation Co.). Although it was carefully vetted and the U.A.E. is a close U.S. ally in the war against terrorism, the fact that the Bush administration did not confer with Congress before the fact was mistakenly perceived as turning over security at six major U.S. ports to a foreign ARAB country. At this writing legislation has been introduced in Congress to cancel the agreement, and as one columnist noted, "this Dubai ports deal has unleashed nativist, isolationist mass hysteria, a kind of collective mania we haven't seen in decades."[3]

Although they are not blessed with the vast petroleum-based wealth of Abu Dhabi and Dubai, the other emirates do have some important economic assets. Liquefied natural gas (LNG) was discovered in Sharjah in 1992, and by 2001 its onshore Kahaif and Sajaa fields were producing 40 million cubic feet per day, sufficient to meet domestic needs. The oil refinery at Fujairah, closed from 1997 to 1999, resumed production in 2001; current output is 105,000 barrels per day of refined petroleum products. Fujairah's port was enlarged in 2000 to accommodate tankers up to 90,000 deadweight tons; it is now the largest bunkering port in the Middle East. In addition to its gas reserves, Sharjah has become the emirate of choice for small and medium-size industries; they generate 48 percent of the U.A.E.'s non-oil, industrial share of gross domestic product.

The U.A.E.'s dependence on expatriate workers, who comprise approximately 80 percent of the labor force, has been an obstacle to self-sufficiency and diversification. In October 1996, a strict new residency law governing expatriate labor was approved by the Supreme Council. The law limits both immigration numbers and length of stay; it is aimed particularly at low-level Asian workers. As a result, some 400,000 "guest workers"—approximately 15 percent of the total expatriate population—left the federation.

The U.A.E. celebrated its silver anniversary in December 1996 with a 69-ton birthday cake. An even larger one, 50 feet long by 6 feet across, was unveiled in 1998 for the country's children. Unlike other areas in the region that reverberate to the sound of bombs, in Dubai the noise comes from building construction. Like a Las Vegas writ large, but without liquor or gambling, Dubai's skyline now boasts the world's tallest skyscraper, tallest hotel, and largest theme park, Dubailand. Its facilities include a camel racetrack, golf courses, and tennis courts that regularly hold international tennis tournaments, and forty (count them) malls. Some 5.45 million tourists visited the emirate in 2004, contributing 17 percent to its economy. The legendary traveler Sinbad the Sailor would be amazed at the changes if he returned to his old town and harbor.[2]

NOTES

1. Steve Krettman, "Oil Realm Embraces a Wired Economy," *The New York Times* (June 10, 2001).
2. Ken Ringle, "The Best-kept Secret in the Middle East", *Smithsonian Magazine*, October 2003, pp. 42–46.
3. David Brooks, "Kicking Arabs in the Teeth", *The New York Times*, February 24, 2006. As other observers have pointed out, the U.A.E. was the first Middle East government to accept the U.S. Container Security Initiative to screen containers before they proceed to U.S. ports and has agreed to U.S. Energy Department requests to ban nuclear materials from shipment from its ports.

Timeline: PAST

1853, 1866
Peace treaties between Great Britain and Arab shaykhs establishing the Trucial States

1952
Establishment of the Trucial Council under British advisers, the forerunner of federation

1971
Independence

1973
The U.A.E. becomes the first Arab oil producer to ban exports to the U.S. after the Yom Kippur War

1979
Balanced federal Assembly and cabinet are established

1990s
The U.A.E. reduces its dependence on oil revenues; the free-trade zone proves a success

PRESENT

2000s
The U.A.E. joins with other GCC members in a mutual-defense pact
The U.A.E. supports the international coalition against terrorism

Yemen (Republic of Yemen)

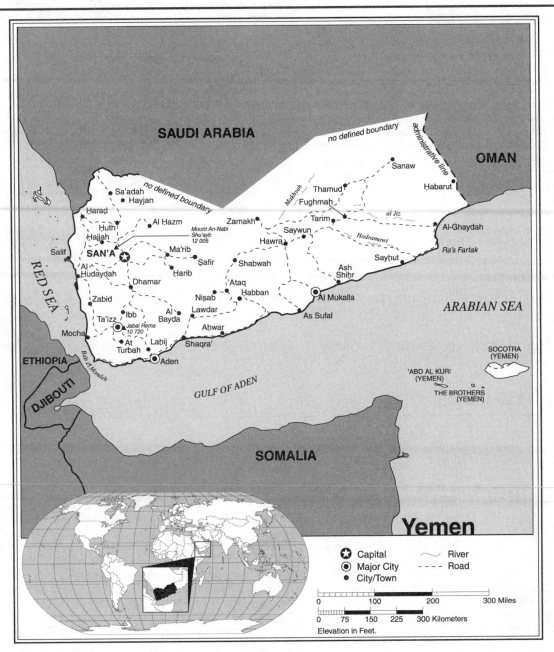

Yemen Statistics

GEOGRAPHY

Area in Square Miles (Kilometers):
203,796 (527,970) (about twice the size of Wyoming)

Capital (Population): San'a (political capital) (972,000); Aden (economic capital) (562,000)

Environmental Concerns: limited freshwater supplies; inadequate potable water; overgrazing; soil erosion; desertification

Geographical Features: a narrow coastal plain backed by hills and mountains; dissected upland desert plains in the center slope into desert

Climate: mostly desert; hot, with minimal rainfall except in mountain zones

PEOPLE

Population

Total: 20,727,063
Annual Growth Rate: 3.4%

Rural/Urban Population Ratio: 66/34
Major Language: Arabic
Ethnic Makeup: predominantly Arab;
  small Afro-Arab, South Asian, and
  European communities
Religions: nearly 100% Muslim; small
  numbers of Christians, Jews, and Hindus

Health

Life Expectancy at Birth: 59.8 years
  (male); 63.7 years (female)
Infant Mortality Rate: 61.5/1,000 live births

Education

Adult Literacy Rate: 50%
Compulsory (Ages): 6–15

COMMUNICATION

Telephones: 542,200 main lines
Televisions: 6.5 per 1,000 people
Internet Users: 1 (2000)

TRANSPORTATION

Highways in Miles (Kilometers): 37,557
  (69,263)

Railroads in Miles (Kilometers): none
Usable Airfields: 50
Motor Vehicles in Use: 510,000

GOVERNMENT

Type: republic, formed by merger of
  former Yemen Arab Republic and
  People's Democratic Republic of Yemen
Independence Date: formally united
  May 22, 1990 (date of merger)
Head of State/Government: President Ali
  Abdullah Salih (elected 1999); Prime
  Minister Abd al-Qadir Bajammal
Political Parties: 12 active; General
  Peoples' Congress serves as the majority
  party; main opposition parties are
  Islamic Reform Grouping (Islah) and
  Yemeni Socialist Party
Suffrage: universal at 18

MILITARY

Military Expenditures (% of GDP): 7.6%
Current Disputes: final boundary with
  Saudi Arabia remains to be resolved

ECONOMY

Currency ($U.S. Equivalent): 184.78 rials
  = $1

Per Capita Income/GDP: $800/$16.2 billion

GDP Growth Rate: 1.9%

Inflation Rate: 12.2%

Unemployment Rate: 30%

Natural Resources: petroleum; fish; rock
  salt; marble; small deposits of coal, gold,
  lead, nickel, and copper; fertile soil in
  west

Agriculture: grain; fruits; vegetables; qat;
  coffee; cotton; livestock; fish

Industry: petroleum; cotton textiles and
  leather goods; food processing;
  handicrafts; aluminum products; cement

Exports: $4.468 billion (primary partners
  China, Thailand, Singapore)

Imports: $3.7 billion (primary partners
  U.A.E., Saudi Arabia, China)

Yemen Statistics

The Republic of Yemen occupies the extreme southwest corner of the Arabian Peninsula. It consists of three distinct regions, which until 1990 had been separated geographically for centuries and divided politically into two states: the Yemen Arab Republic (North Yemen, or Y.A.R.) and the People's Democratic Republic of Yemen (South Yemen, or P.D.R.Y.). Until the twentieth century, the entire area was known simply as Yemen; and with the merger of the two states, it has resumed its former geographic identity. The former Y.A.R.'s territory consists of two distinct regions: a hot, humid coastal strip, the Tihama, along the Red Sea; and an interior region of mountains and high plains that shade off gradually into the bleak, waterless South Arabian Desert.

Yemeni territory also includes Socotra, a remote island 550 miles from Aden, and two other small islands, Abd al-Khuri and the Brothers, which lie off the African coast of Somalia. Socotra is the only world habitat for Dragon's Blood trees, which produce cinnabar resin, and of some 850 other plants that exist nowhere else.

The Yemeni interior is very different not only from the Tihama but also from other parts of the Arabian Peninsula. It consists of highlands and rugged mountains ranging up to 12,000 feet. At the higher elevations, the mountain ridges are separated by deep, narrow valleys, usually with swift-flowing streams at the bottom. The ample rainfall allows extensive use of terracing for agriculture. The main crops are coffee, cereals, vegetables, and qat (a shrub whose leaves are chewed as a mildly intoxicating narcotic).

This part of Yemen has been for centuries the home of warlike but sedentary peoples who have formed a stable, stratified society living in villages or small cities. These groups have been the principal support for the Shia Zaidi Imams, whose rule was the political nucleus of Yemen from the ninth century A.D. to the establishment of the republic in 1962. The Yemeni political capital, San'a, is located in these northern highlands.

The former P.D.R.Y., almost twice the size of its neighbor but less favored geographically, consists of the port and hinterland of Aden (today Yemen's economic capital); the Hadhramaut, a broad valley edged by desert and extending eastward along the Arabian Sea coast; the Perim and Kamaran Islands, at the south end of the Red Sea; and Socotra Island.

Until the recent discoveries of oil, South Yemen was believed to have no natural resources. The dominant physical feature is the Wadi Hadhramaut. It is one of the few regions of the country with enough water for irrigation. Except for Aden, the area has little rainfall; in some sections, rain may fall only once every 10 years. Less than 2 percent of the land is cultivable.

In ancient times, the whole of Yemen was known to the Greeks, Romans, and other peoples as Arabia Felix ("Happy Arabia"), a remote land that they believed to be fabulously wealthy. They knew it as the source of frankincense, myrrh, and other spices as well as other exotic products brought to Mediterranean and Middle Eastern markets from the East. In Yemen itself, several powerful kingdoms grew up from profits earned in this trade. One kingdom in particular, the Sabaeans, also had a productive agriculture based on irrigation. The water for irrigation came from the great Marib Dam, built around 500 B.C. Marib was a marvel of engineering, built across a deep river valley. The Sabaean economy supported a population estimated at 300,000 in a region that today supports only a few thousand herders.

The Sabaeans were followed by the Himyarites. Himyarite rulers were converted to Christianity by wandering monks in the second century A.D. The Himyarites had contacts with Christian Ethiopia across the Red Sea and for a time were vassals of Ethiopian kings. An Ethiopian army invaded South Arabia but was defeated by the Himyarites in A.D. 570, the "Year of the Elephant" in Arab tradition, so called because the Ethiopian invaders were mounted on elephants. (The year was also notable for the birth of Muhammad, the founder of Islam.)

DEVELOPMENT

Yemen's modest growth has been affected periodically by fluctuating world oil prices. Oil is its only significant exportable resource. The bombing of the U.S. destroyer COLE and an attack on the French tanker LIMBURG in 2002 have adversely affected commerce in Aden port, due particularly to high-risk insurance premiums. The last container line based there moved to Salalah, Oman, after the LIMBURG attack.

Sabaeans and Himyarites ruled long ago, but they are still important to Yemenis as symbols of their long and rich historical past. The Imams of Yemen, who ruled until 1962, used a red dye to sign their official documents in token of their relationship to Himyarite kings. (The word *Himyar* comes from the same root as *hamra*, "red.")

The domestication of the camel and development of an underground irrigation system of channels (*falaj*) made this civilization possible. Ships and camel caravans brought the frankincense, myrrh, and musk from Socotra and silks and spices from India and the Far East to northern cities in Egypt, Persia, and Mesopotamia. Aden was an important port for this trade, due to its natural harbor and its location at the south end of the Red Sea.

Yemenis were among the first converts to Islam. The separation of the Yemenis into mutually hostile Sunni and Shia Muslims took place relatively early in Islamic history. Those living in the Tihama, which was easily accessible to missionaries and warriors expanding the borders of the new Islamic state, became Sunnis, obedient to the caliphs (the elected "successors" of Muhammad). The Yemeni mountaineers were more difficult to reach; and when they were converted to the new religion, it was through the teachings of a follower of the Shi'at Ali, "Party of Ali," those who felt that Muhammad's son-in-law Ali and his descendants should have been chosen as the rightful leaders of the Islamic community. Yemenis in Aden and the Hadhramaut, as well as those in the Tihama, became Sunni, creating the basis for an intra-Yemeni conflict, which still exists.

Yemen was the home of a small Jewish community for at least 2,500 years, dating from the time of the Babylonian captivity. Although its members were excluded from all professions, except silversmithing, and were required to wear identifying clothing and sidecurls, the community lived side by side with Muslims without incident. After the establishment of Israel in 1948, 48,000 Jews were airlifted to the new Jewish state. Some 24,000 others left Yemen for Israel in later years. Today only 300 Jews remain in Yemen.

THE ZAIDI IMAMATE

In the late ninth century A.D., a feud among certain nominally Muslim groups in inland Yemen led to the invitation to a religious scholar living in Mecca to come and mediate in their dispute. (Use of an outside mediator was common in Arabia at that time.) This scholar brought with him a number of families of Ali's descendants who sought to escape persecution from the Sunnis. He himself was a disciple of Zaid, Ali's great-grandson. He settled the feud, and, in return for his services, he was accepted by both sides of the conflict as their religious leader, or Imam. He and his successor Imams, 111 in all, established the Zaidi Imamate in north Yemen, a theocratic state of sorts which lasted until 1962.

The first Zaidi Imam had some personal qualities that enabled him to control the unruly mountain people and bend them to his will. He was a shrewd judge of character, using his knowledge and his prestige as a member of the family of Ali to give personal favors or to give his power of *baraka* (special powers from God) to one group or withhold it from another. He had great physical strength. It was said of him that he could grind corn with his fingers and pull a camel apart barehanded. He wrote 49 books on Islamic jurisprudence and theology, some of which are still studied by modern Yemeni scholars. He was also said to bring good (or bad) fortune to a subject merely by a touch or a glance from his piercing black eyes.[1]

In a reversal of the ancient process whereby South Arabian merchants carried goods to the far-flung cities of the north, from the late 1400s on, the towns of the bleak Arabian coast attracted the interest of European seafaring powers as way stations or potential bases for control of their expanding trade with the East Indies, India, and China. Aden was a potentially important base, and expeditions by Portuguese and other Europeans tried without success to capture it at the time. In 1839, a British expedition finally succeeded. It found a town of "800 miserable souls, huddled in

huts of reed matting, surrounded by guns that would not fire," or so the American traveler Joseph Osgood described the scene.

Under British rule, Aden became an important naval base and refueling port for ships passing through the Suez Canal and down the Red Sea en route to India. For many British families bound for India, Aden was the last land, with the last friendly faces, that they would see before arriving many days later in the strange wonderland of India. The route through the Suez Canal and down the Red Sea past Aden was the lifeline of the British Empire. In order to protect Aden from possible attack by hostile peoples in the interior, the British signed a series of treaties with their chiefs, called shaykhs or sometimes sultans. These treaties laid the basis for the South Arabian Protectorates. British political agents advised the rulers on policy matters and gave them annual subsidies to keep them happy. One particular agent, Harold Ingrams, was so successful in eliminating feuds and rivalries that "Ingrams's Peace" became a symbol of the right way to deal with proud, independent local leaders.

In the 16th century A.D. the Ottoman Turks seized control of the Tihama and the port of Aden. However fierce Zaidi resistance forced them to withdraw. Subsequently an Ottoman military expedition reconquered coastal Yemen and it became an Ottoman province. However Ottoman control was tenuous at best and did not sit well with the mountain peoples. A Yemeni official told a British visitor: "We have fought the Turks, the tribes ... and we are always fighting each other. We Yemenis submit to no one permanently. We love freedom and we will fight for it."[2]

The Turkish occupation sparked a revolt. Turkish forces were unable to defeat the mountain peoples, and in 1911, they signed a treaty that recognized Imam Yahya as ruler in the highlands. In return, the Imam recognized Turkish rule in the Tihama. At the end of World War I, the Turks left Yemen for good. The British, who now controlled most of the Middle East, signed a treaty with Imam Yahya, recognizing his rule in all Yemen.

The two Yemens followed divergent paths in the twentieth century, accounting in large measure for the difficulties that they faced in incorporating into a single state. North Yemen remained largely uninvolved in the political turmoil that engulfed the Middle East after World War II. Imam Yahya ruled his feudal country as an absolute monarch with a handful of advisers, mostly tribal leaders, religious scholars, and members of his family. John Peterson notes that the Imamate "was completely dependent on the abilities of a single individual who was expected to be a competent combination of religious scholar, administrator, negotiator, and military commander."[3] Yahya was all of these, and his forceful personality and ruthless methods of dealing with potential opposition (with just a touch of magic) ensured his control over the population.

Yahya's method of government was simplicity itself. He held a daily public meeting (*jama'a*) seated under an umbrella outside his palace, receiving petitions from anyone who wished to present them and signing approval or disapproval in Himyarite red ink. He personally supervised tax collections and kept the national treasury in a box under his bed. The Imam distrusted the Ottomans, against whom he had fought for Yemeni independence, and refused to accept their coinage. He also rejected the British currency because it represented a potential foreign influence.

Yahya was determined to keep foreign influences out of Yemen and to resist change in any form. Although Yemen was poor by the industrial world's standards, it was self-sufficient, free, and fully recognized as an independent state. Yahya hoped to keep it that way. He even refused foreign aid because he felt that it would lead to foreign occupation. But he was unable to stop the clock entirely and to keep out all foreign ideas and influences.

Certain actions that seemed to be to his advantage worked against him. One was the organization of a standing army. In order to equip and train an army that would be stronger than tribal armies, Yahya had to purchase arms from abroad and to hire foreign advisers to train his troops. Promising officers were also sent for training in Egypt, and upon their return, they formed the nucleus of opposition to the Imam.

In 1948, Imam Yahya was murdered in an attempted coup. He had alienated not only army officers who resented his repressive rule but also leaders from outside the ruling family who were angered by the privileges given to the Imam's sons and relatives. But the coup was disorganized, the conspirators unsure of their goals. Crown Prince Ahmad, the Imam's eldest son and heir, was as tough and resourceful as his 80-year-old father had been.[4] He gathered support from leaders of other clans and nipped the rebellion in the bud.

Imam Ahmad (1948–1962) ruled as despotically as his father had ruled. But the walls of Yemeni isolation inevitably began to crack. Unlike Yahya, Ahmad was willing to modernize a little. Foreign experts came to design and help build the roads, factories, hospitals, and schools that the Imam felt were needed. Several hundred young Yemenis were sent abroad for study. Those who had left the country during Imam Yahya's reign returned. Many Yemenis emigrated to Aden to work for the British and formed the nucleus of a "Free Yemen" movement.

In 1955, the Imam foiled an attempted coup. Other attempts, in 1958 and 1961, were also unsuccessful. The old Imam finally died of emphysema in 1962, leaving his son, Crown Prince Muhammad al-Badr, to succeed him.

THE MARCH TO INDEPENDENCE

The British wanted to hold on to Aden as long as possible because of its naval base and refinery. It seemed to them that the best way to protect British interests was to set up a union of Aden and the South Arabian Protectorates. This was done in 1963, with independence promised for 1968. However, the British plan proved unworkable. In Aden, a strong anti-British nationalist movement developed in the trade unions among dock workers and refinery employees. This movement organized a political party, the People's Socialist Party, strongly influenced by the socialist, anti-Western, Arab nationalist programs of President Gamal Abdel Nasser in Egypt.

The party had two branches: the moderate Front for the Liberation of Occupied South Yemen (FLOSY) and the leftist Marxist National Liberation Front (NLF). About all they had in common was their opposition to the British and the South Arabian sultans, whom they called "lackeys of imperialism." FLOSY and the NLF joined forces in 1965–1967 to force the British to leave Aden. British troops were murdered; bombs damaged the refinery. By 1967, Britain had had enough. British forces were evacuated, and Britain signed a treaty granting independence to South Yemen under a coalition government made up of members of both FLOSY and the NLF.

Muhammad al-Badr held office for a week and then was overthrown by a military coup. Yemen's new military leaders formed a Revolution Command Council and announced that the Imam was dead. Henceforth, they said, Yemen would be a republic. It would give up its self-imposed isolation and would become part of the Arab world. But the Revolution proved to be more difficult to carry out than the military officers had expected. The Imam was not dead, as it turned out, but had escaped to the mountains. The mountain peoples rallied to his support, helping him to launch a counterrevolution. About 85,000 Egyptian troops arrived in Yemen to help the republican army. The coup leaders had been trained in Egypt, and the Egyptian government had not only financed the Revolution but also had encouraged it against the "reactionary" Imam.

For the next eight years, Yemen was a battleground. The Egyptians bombed villages and even used poison gas against civilians in trying to defeat the Imam's forces. But they were unable to crush the people hidden in the mountains of the interior. Saudi Arabia also backed the Imam with arms and kept the border open. The Saudi rulers did not particularly like the Imam, but he seemed preferable to an Egyptian-backed republican regime next door.

After Egypt's defeat by Israel in the 1967 Six-Day War, the Egyptian position in Yemen became untenable, and Egyptian troops were withdrawn. It appeared that the royalists would have a clear field. But they were even more disunited than the republicans. A royalist force surrounded San'a in 1968 but failed to capture the city. The Saudis then decided that the Imam had no future. They worked out a reconciliation of royalists and republicans that would reunite the country. The only restriction was that neither the Imam nor any of his relatives would be allowed to return to Yemen.

Thus, as of 1970, two "republics" had come into existence side by side. The Yemen Arab Republic was more of a tribal state than a republic in the modern political sense of the term. Prior to 1978, its first three presidents either went into exile or were murdered, victims of rivalry within the army. Colonel Ali Abdullah Saleh, a career army officer, seized power in that year and was subsequently chosen as the republic's first elected president. He was reelected in 1983 and again in 1988 for consecutive five-year terms. (With unification, he became the first head of state of all Yemen.)

Saleh provided internal stability and allowed some broadening of the political process in North Yemen. A General People's Congress (GPC) was established in

1982. A Consultative Council, elected by popular vote, was established in 1988 to provide some citizen input into legislation. Saleh displayed great skill in balancing tribal and army factions and used foreign aid to develop economic projects such as dams for irrigation to benefit highland and Tihama Yemenis alike.

SOUTH YEMEN: A MARXIST STATE

With the British departure, the South Arabian Federation collapsed. Aden and the Hadhramaut were united under Aden political leadership in 1970 as the People's Democratic Republic of Yemen. It began its existence under adverse circumstances: Britain ended its subsidy for the Aden refinery, and the withdrawal of British forces cut off the revenues generated by the military payroll.

But the main problem was political. A power struggle developed between FLOSY and the NLF. The former favored moderate policies, good relations with other Arab states, and continued ties with Britain. The NLF were leftist Marxists. By 1970, the Marxists had won. FLOSY leaders were killed or went into exile. The new government set its objectives as state ownership of lands, state management of all business and industry, a single political organization with all other political parties prohibited, and support for antigovernment revolutionary movements in other Arab states, particularly Oman and Saudi Arabia.

During its two decades of existence, the P.D.R.Y. modeled its governing structure on that of the Soviet Union, with a Presidium, a Council of Ministers, a Supreme People's Legislative Council, and provincial and district councils, in descending order of importance. In 1978, the ruling (and only legal) political party took the name Yemen Socialist Party, to emphasize its Yemeni makeup.

Although the P.D.R.Y. government's ruthless suppression of opposition enabled it to establish political stability, rivalries and vendettas among party leaders led to much instability within the ruling party. The first president, Qahtan al-Sha'bi, was overthrown by pro-Soviet radicals within the party. His successor, Salim Rubayyi Ali, was executed after he had tried and failed to oust his rivals on the party Central Committee. Abd al-Fattah Ismail, the country's third president, resigned in 1980 and went into exile due to a dispute over economic policies. Ali Nasir Muhammad, the fourth president, seemed to have consolidated power and to have won broad party support, until 1986, when he tried to purge the Central Committee of potential opponents. The peoples of the interior, who

formed Muhammad's original support base, stayed out of the fighting. After 10 days of bloody battles with heavy casualties, the president's forces were defeated. He then went into exile and was convicted of treason in absentia. He returned to Yemen in 1996 after the end of the Civil War and the reunification of the "two Yemens."

UNIFICATION

Despite their natural urge to unite in a single Yemeni nation, the two Yemens were more often at odds with each other than united in pursuing common goals. This was due in part to the age-old highland-lowland, Sunni-Shia conflict that cut across Yemeni society. But it was also due to their very different systems of government. There were border clashes in 1972, 1975, and 1978–1979, when the P.D.R.Y. was accused of plotting the overthrow of its neighbor. (A P.D.R.Y. envoy brought a bomb hidden in a suitcase to a meeting with the Y.A.R. president, and the latter was killed when the bomb exploded.)

Improved economic circumstances and internal political stability in both Yemens revived interest in unity in the 1980s, especially after oil and natural-gas discoveries in border areas promised advantages to both governments through joint exploitation. In May 1988, President Saleh and Prime Minister al-Attas of the P.D.R.Y. signed the May Unity Pact, which ended travel restrictions and set up a Supreme Yemeni Council of national leaders to prepare a constitution for the proposed unitary state.

From then on, the unity process snowballed. In 1989, the P.D.R.Y. regime freed supporters of former President Ali Nasir Muhammad. Early in 1990, the banks, postal services, ports administration, and customs of the two republics were merged, followed by the merger under joint command of their armed forces.

Formal unification took place on May 22, 1990, with approval by both governments and ratification of instruments by their legislative bodies. Ali Abdullah Saleh was unanimously chosen as the republic's first president, with a four-member Presidential Council formed to oversee the transition. A draft constitution of the new republic established a 39-member Council of Ministers headed by P.D.R.Y. prime minister al-Attas, with ministries divided equally between North and South. In a national referendum in May 1991, voters approved the new all-Yemen Constitution. (The Constitution was opposed by the newly formed Islah Party, representing the tribes and Islamic fundamentalists, on the grounds that it did not conform fully to Islamic law.)

The Constitution provides for elections to a 301-member Parliament. Elections were scheduled for November 1992 but were postponed until April 1993 after the elections committee protested that insufficient time had been allocated for voter registration, preparation of candidate lists, drawing of constituency borders, and campaigning.

The campaign itself was marred by violence, much of it directed at officials of the Yemen Socialist Party by tribal opponents of unification or others who feared that the election would result in greater influence for the more liberal, ex-Marxist Southerners in the government. In 1992, an economic crisis also hit the country; in December people took to the streets protesting price increases and a 100 percent inflation rate.

Despite the disruption, the elections were held on schedule. President Saleh's General People's Congress won 147 seats in Parliament, just shy of a majority. The elections were carried out in open democratic fashion, with women and the small Yemenite Jewish community allowed to vote, in sharp contrast to election practice in other parts of the Middle East.

A coalition government was formed in May 1993 between the General People's Congress, the Yemen Socialist Party, and Islah, which ran third in the balloting. But rivalry between the former political elites of North and South, plus differences in outlook, continued to impede progress toward full unification. Early in 1994, Ali al-Beidh, Yemen Socialist Party (YSP) leader and vice president of the ruling coalition, presented a set of 18 demands whose acceptance was a prerequisite for his return to the government. They were rejected by President Saleh, and civil war broke out in May 1994. Initially, the South Yemeni forces had the better of it, but the larger and better-equipped army of the North, moving slowly southward, surrounded Aden and captured the city after a brief siege. Vice-President al-Beidh fled into exile, effectively depriving the rebellion of its chief leader, and his Yemen Socialist Party was excluded from the governing coalition, although it was allowed to continue as a political party.

The end of the Civil War, more or less on North Yemen's terms, offered Saleh another opportunity to unify the nation. The first step would be the restoration of representative government. A 1992 law was reinstated to require political parties to have 5,000 or more members, plus offices in each governorate, in order to present candidates in the forthcoming national elections.

ACHIEVEMENTS

 Although Yemen's connections with international terrorism has caused some friction with the U.S., non-political cooperation has yielded some positive dividends. Apart from military aid, U.S. Peace Corps volunteers teach English in a number of rural villages and American specialists have been brought in to help remove the thousands of land mines left over from the civil war.

The election was held on April 27, 1993. Some 2,300 candidates vied for the 301 seats in the unicameral Yemeni national Legislature. As expected, the GPC won a large majority, 239 seats, to 62 for Islah, the main opposition party. The Yemen Socialist Party, which boycotted the elections, was shut out of legislative participation entirely.

What struck outside observers about the election was its faithful adherence to political democracy. The entire process was supervised by the Supreme Election Commission, established by law as an independent body with balanced political representation. Despite having one of the lowest literacy rates in the world, Yemenis participated with enthusiasm and in great numbers, illiterate voters being assisted by literate volunteers to mark their ballots inside the curtained polling booths. Ballots were tabulated by hand by representatives of the Supreme Election Council, prompting an American observer to ask why "they didn't use voting machines and computers. They said they would not trust such a system because it would not be transparent."[5]

A constitutional amendment approved by the electorate in 1999 allows an incumbent president to serve two consecutive terms, and President Saleh was nominated by his own party, the GPC, as well as by the opposition Islah Party. He faced opposition for the first time but won a second term with ease. A second constitutional amendment, approved in 2001, set up a 111-member Shura (Council) appointed by the president; its duties would be to advise him on basic security measures.

In September 1999, Yemen's first direct presidential election marked a milestone in the slow progress of the state—and Arab states in general—toward Western-style representative government. Prior to the election, the Constitution was amended to allow an incumbent to serve for two consecutive five-year terms. Although he was nominated by both his own party and the opposition Islah Party, President Ali Abdullah Saleh faced opposition for the first time in a presidential election. Admittedly the opposition consisted of token, unknown candidates, and Saleh won reelection with ease. However, despite its flawed nature, the election underlined not only the president's popularity among his people but also his serious commitment to representative government.

Although on the surface Yemen seems to offer fertile ground for Islamic fundamentalism due to its poverty, its high unemployment rate, and its divisions between a tribal north and a Marxist south, until recently no homegrown Islamic fundamentalist movement existed there. Following the withdrawal of Soviet troops from Afghanistan in 1989, a large number of Afghan resistance fighters (*mujahideen*) and Muslim volunteers from other countries who had gone to Afghanistan to defend Islam against atheistic Communists fled to Yemen. At the end of 1994, one mujahideen group, Aden-Abyan Islamic Jihad, carried out a number of bombings and kidnappings of foreign tourists and oil company employees. Government forces captured most of its members in 1999 and 10 were sentenced to death. They were executed in 2000.

THE ECONOMY

Discoveries of significant oil deposits in the 1980s should have augured well for Yemen's economic future. Reserves are estimated at 1 billion barrels in the Marib basin and 3.3 billion in the Shabwa field northeast of Aden, with an additional 5.5 billion in the former neutral zone shared by the two Yemens and now administered by the central government. Yemen also has large deposits of natural gas, with reserves estimated at 5.5 trillion cubic feet.

Unfortunately, the political conflicts of the 1990s had a negative effect on these rosy prospects. The Gulf War, in which Yemen supported Iraq against the UN-U.S.-Saudi coalition, caused Saudi Arabia to deport some 850,000 Yemeni workers. And the Civil War in 1994 seriously damaged the infrastructure, requiring some $200 million in repairs to schools, hospitals, roads, and power stations.

Until very recently, both Yemens were among the poorest and least-developed countries in the Middle East. This description is somewhat misleading in North Yemen's case, however, since the highland regions have traditionally supported a sizeable population, due to fertile soil, dependable and adequate rainfall, and effective use of limited arable land. South Yemen's resources were mostly unexploited during the 130 years of British rule, except for a small local fishing industry and the Aden port and refinery. During its brief period of independence, the P.D.R.Y.'s budget came mostly (70 percent) from the Soviet Union and other Communist countries. The reduction in Soviet aid from $400 million in 1988 to $50 million in 1989 was one of the economic factors that encouraged reunification from the South Yemen side.

In ancient times, Yemen, particularly in the South, had a flourishing agriculture based on monsoon rains, supplemented by a sophisticated system of small dams and canals and centered in the Wadi Hadramaut. But long neglect and two decades of disastrous Soviet-style state-farm management adversely affected agricultural production.

Since reunification, most state-owned lands in the Hadhramaut have been privatized. And oil drilling in the region has resulted in discovery of important underground water resources. As a result, and with aid from the World Bank, the region has recovered much of its former agricultural productivity. Today, date-palm groves, fields of corn, orchards, and beehives flourish.

FOREIGN RELATIONS

Prior to unification with South Yemen, North Yemen's geographical isolation and tribal social structure limited its contact with other Arab states. South Yemen's Marxist regime, in contrast, actively attempted to subvert the governments of its neighbors. Reunification has brought better and closer relations with these states. In 1995, the flags of Yemen and Oman flew side by side on their newly demarcated common border, based on a 1992 agreement to accept UN mediation.

Yemen's relations with Saudi Arabia have followed an uneven course. Yemeni workers in the kingdom were deported en masse after the Gulf War, due to Yemeni support for Iraq. The action was a severe blow to the Yemeni economy, as 20 to 25 percent of the national budget had come from expatriate-worker remittances. The relationship improved after 1995, when Syrian mediators arranged for a reconfirmation of the 1934 Treaty of Taif, which had demarcated their common border.

The resulting "Memorandum of Understanding of Taif" recognizes Saudi sovereignty over Asir, Najran, and Jizan Provinces in return for Yemeni sovereignty over the Marib oil fields.

The range of anti-U.S. terrorism reached Yemen in October 2000, when the U.S. Navy destroyer *Cole* was attacked while it was in Aden Harbor for refueling. 17 Americans were killed in the attack, which was carried out by several men in a small boat packed with explosives. A number of the attackers were later arrested but later escaped. Early in 2004 they were recaptured by security forces. One of them was Jamal Badawi, described as Yemen's most dangerous terrorist. Two turned themselves in later when offered amnesty in return for pledging to renounce violence. The remainder are still at large, including Badawi.

Following the September 11, 2001, terrorist attacks in the United States, the Yemeni government declared its willingness to join the international anti-terror coalition. It arrested 20 Yemenis suspected of having been trained in al Qaeda camps in Afghanistan before returning to their own country. To underscore Yemen's cooperation, Prime Minister Abdul-Kader Bajammal indicated that 4,000 Yemenis trained in such camps had been expelled. In January 2003 the government signed a contract with Canada's Nexen Corporation, one of the foreign firms engaged in developing its oil resources for patrol boats, surveillance equipment and anti-terrorism training for the Yemeni security services. However, the equipment arrived too late to prevent the kidnapping and murder of three American medical missionaries in Ibb province, near San'a. The murdered missionaries had founded a Baptist hospital there in the 1960s which served 40,000 poor Yemenis annually. Their attacker, after his arrest, said he had acted to "cleanse the country of infidels."

The kidnapping of an Islah Party official in January 2003, along with the missionaries' murder, marked a growing division between the people and the government, due essentially to the latter's U.S. alignment. In 2005 a previously-unknown organization called Believing Youth, formed probably in 2004, began an uprising from secret bases in the far north of the country. In addition to closing mosques and arresting militant imams the government has had some success with offers of vocational training and jobs in return for abjuring violence.

NOTES

1. Robin Bidwell, *The Two Yemens* (Boulder, CO: Westview Press, 1983), p. 10.
2. Quoted in Robert Stookey, *Yemen: The Politics of the Yemen Arab Republic* (Boulder, CO: Westview Press, 1978), p. 168.
3. John Peterson, "Nation-building and Political Development in the Two Yemens," in B. R. Pridham, ed., *Contemporary Yemen: Politics and Historical Background* (New York: St. Martin's Press, 1985), p. 86.
4. Yemenis believed that he slept with a rope around his neck to terrify visitors, that he could turn twigs into snakes, and that he once out wrestled the devil. Bidwell, *op. cit.*, p. 121.
5. William A. Rugh, "A (Successful) Test of Democracy in Yemen," *The Christian Science Monitor* (May 28, 1997), p. 19.

Lifting the Veil

Understanding the Roots of Islamic Militancy

Henry Munson

In the wake of the attacks of September 11, 2001, many intellectuals have argued that Muslim extremists like Osama bin Laden despise the United States primarily because of its foreign policy. Conversely, US President George Bush's administration and its supporters have insisted that extremists loathe the United States simply because they are religious fanatics who "hate our freedoms." These conflicting views of the roots of militant Islamic hostility toward the United States lead to very different policy prescriptions. If US policies have caused much of this hostility, it would make sense to change those policies, if possible, to dilute the rage that fuels Islamic militancy. If, on the other hand, the hostility is the result of religious fanaticism, then the use of brute force to suppress fanaticism would appear to be a sensible course of action.

Groundings for Animosity

Public opinion polls taken in the Islamic world in recent years provide considerable insight into the roots of Muslim hostility toward the United States, indicating that for the most part, this hostility has less to do with cultural or religious differences than with US policies in the Arab world. In February and March 2003, Zogby International conducted a survey on behalf of Professor Shibley Telhami of the University of Maryland involving 2,620 men and women in Egypt, Jordan, Lebanon, Morocco, and Saudi Arabia. Most of those surveyed had "unfavorable attitudes" toward the United States and said that their hostility to the United States was based primarily on US policy rather than on their values. This was true of 67 percent of the Saudis surveyed. In Egypt, however, only 46 percent said their hostility resulted from US policy, while 43 percent attributed their attitudes to their values as Arabs. This is surprising given that the prevailing religious values in Saudi Arabia are more conservative than in Egypt. Be that as it may, a plurality of people in all the countries surveyed said that their hostility toward the United States was primarily based on their opposition to US policy.

The issue that arouses the most hostility in the Middle East toward the United States is the Israeli-Palestinian conflict and what Muslims perceive as US responsibility for the suffering of the Palestinians. A similar Zogby International survey from the summer of 2001 found that more than 80 percent of the respondents in Egypt, Kuwait, Lebanon, and Saudi Arabia ranked the Palestinian issue as one of the three issues of greatest importance to them. A survey of Muslim "opinion leaders" released by the Pew Research Center for the People and the Press in December 2001 also found that the US position on the Israeli-Palestinian conflict was the main source of hostility toward the United States.

It is true that Muslim hostility toward Israel is often expressed in terms of anti-Semitic stereotypes and conspiracy theories—think, for example, of the belief widely-held in the Islamic world that Jews were responsible for the terrorists attacks of September 11, 2001. Muslim governments and educators need to further eliminate anti-Semitic bias in the Islamic world. However, it would be a serious mistake to dismiss Muslim and Arab hostility toward Israel as simply a matter of anti-Semitism. In the context of Jewish history, Israel represents liberation. In the context of Palestinian history, it represents subjugation. There will always be a gap between how the West and how the Muslim societies perceive Israel. There will also always be some Muslims (like Osama bin Laden) who will refuse to accept any solution to the Israeli-Palestinian conflict other than the destruction of the state of Israel. That said, if the United States is serious about winning the so-called "war on terror," then resolution of the Israeli-Palestinian conflict should be among its top priorities in the Middle East.

Eradicating, or at least curbing, Palestinian terrorism entails reducing the humiliation, despair, and rage that drive many Palestinians to support militant Islamic groups like Hamas and Islamic Jihad. When soldiers at an Israeli checkpoint prevented Ahmad Qurei (Abu al Ala), one of the principal negotiators of the Oslo accords and president of the Palestinian Authority's parliament, from traveling from Gaza to his home on the West Bank, he declared, "Soon, I too will join Hamas." Qurei's words reflected his outrage at the subjugation of his people and the humiliation that Palestinians experience every day at the checkpoints that surround their homes. Defeating groups like Hamas requires diluting the rage that fuels them. Relying on force alone tends to increase rather than weaken their appeal. This is demonstrated by some of the unintended consequences of the US-led invasion and occupation of Iraq in the spring of 2003.

On June 3, 2003, the Pew Research Center for the People and the Press released a report entitled *Views of a*

Changing World June 2003. This study was primarily based on a survey of nearly 16,000 people in 21 countries (including the Palestinian Authority) from April 28 to May 15, 2003, shortly after the fall of Saddam Hussein's regime. The survey results were supplemented by data from earlier polls, especially a survey of 38,000 people in 44 countries in 2002. The study found a marked increase in Muslim hostility toward the United States from 2002 to 2003. In the summer of 2002, 61 percent of Indonesians held a favorable view of the United States. By May of 2003, only 15 percent did. During the same period of time, the decline in Turkey was from 30 percent to 15 percent, and in Jordan it was from 25 percent to one percent.

Indeed, the Bush administration's war on terror has been a major reason for the increased hostility toward the United States. The Pew Center's 2003 survey found that few Muslims support this war. Only 23 percent of Indonesians did so in May of 2003, down from 31 percent in the summer of 2002. In Turkey, support dropped from 30 percent to 22 percent. In Pakistan, support dropped from 30 percent to 16 percent, and in Jordan from 13 percent to two percent. These decreases reflect overwhelming Muslim opposition to the war in Iraq, which most Muslims saw as yet another act of imperial subjugation of Muslims by the West.

The 2003 Zogby International poll found that most Arabs believe that the United States attacked Iraq to gain control of Iraqi oil and to help Israel. Over three-fourths of all those surveyed felt that oil was a major reason for the war. More than three-fourths of the Saudis and Jordanians said that helping Israel was a major reason, as did 72 percent of the Moroccans and over 50 percent of the Egyptians and Lebanese. Most Arabs clearly do not believe that the United States overthrew Saddam Hussein out of humanitarian motives. Even in Iraq itself, where there was considerable support for the war, most people attribute the war to the US desire to gain control of Iraqi oil and help Israel.

Not only has the Bush administration failed to win much Muslim support for its war on terrorism, its conduct of the war has generated a dangerous backlash. Most Muslims see the US fight against terror as a war against the Islamic world. The 2003 Pew survey found that over 70 percent of Indonesians, Pakistanis, and Turks were either somewhat or very worried about a potential US threat to their countries, as were over half of Jordanians and Kuwaitis.

This sense of a US threat is linked to the 2003 Pew report's finding of widespread support for Osama bin Laden. The survey of April and May 2003 found that over half those surveyed in Indonesia, Jordan, and the Palestinian Authority, and almost half those surveyed in Morocco and Pakistan, listed bin Laden as one of the three world figures in whom they had the most confidence "to do the right thing." For most US citizens, this admiration for the man responsible for the attacks of September 11, 2001, is incomprehensible. But no matter how outrageous this widespread belief may be, it is vitally important to understand its origins. If one does not understand why people think the way they do, one cannot induce them to think differently. Similarly, if one does not understand why people act as they do, one cannot hope to induce them to act differently.

The Appeal of Osama bin Laden

Osama bin Laden first engaged in violence because of the occupation of a Muslim country by an "infidel" superpower. He did not fight the Russians in Afghanistan because he hated their values or their freedoms, but because they had occupied a Muslim land. He participated in and supported the Afghan resistance to the Soviet occupation from 1979 to 1989, which ended with the withdrawal of the Russians. Bin Laden saw this war as legitimate resistance to foreign occupation. At the same time, he saw it as a *jihad*, or holy war, on behalf of Muslims oppressed by infidels.

When Saddam Hussein invaded Kuwait in August 1990, bin Laden offered to lead an army to defend Saudi Arabia. The Saudis rejected this offer and instead allowed the United States to establish bases in their kingdom, leading to bin Laden's active opposition to the United States. One can only speculate what bin Laden would have done for the rest of his life if the United States had not stationed hundreds of thousands of US troops in Saudi Arabia in 1990. Conceivably, bin Laden's hostility toward the United States might have remained passive and verbal instead of active and violent. All we can say with certainty is that the presence of US troops in Saudi Arabia did trigger bin Laden's holy war against the United States. It was no accident that the bombing of two US embassies in Africa on August 7, 1998, marked the eighth anniversary of the introduction of US forces into Saudi Arabia as part of Operation Desert Storm.

Part of bin Laden's opposition to the presence of US military presence in Saudi Arabia resulted from the fact that US troops were infidels on or near holy Islamic ground. Non-Muslims are not allowed to enter Mecca and Medina, the two holiest places in Islam, and they are allowed to live in Saudi Arabia only as temporary residents. Bin Laden is a reactionary Wahhabi Muslim who undoubtedly does hate all non-Muslims. But that hatred was not in itself enough to trigger his *jihad* against the United States.

Indeed, bin Laden's opposition to the presence of US troops in Saudi Arabia had a nationalistic and anti-imperialist tone. In 1996, he declared that Saudi Arabia had become an American colony. There is nothing specifically religious or fundamentalist about this assertion. In his book *Chronique d'une Guerre d'Orient*, Gilles Kepel describes a wealthy whiskey-drinking Saudi who left part of his fortune to bin Laden because he alone "was defending the honor of the country, reduced in his eyes to a simple American protectorate."

In 1996, bin Laden issued his first major manifesto, entitled a "Declaration of Jihad against the Americans Occupying the Land of the Two Holy Places." The very title focuses on the presence of US troops in Saudi Arabia, which bin Laden calls an "occupation." But this manifesto also refers to other examples of what bin Laden sees as the oppression of Muslims by infidels. "It is no secret that the people of Islam have suffered from the oppression, injustice, and aggression of the alliance of Jews and Christians and their collaborators to the point that the blood of the Muslims became the cheapest and their wealth was loot in the hands of the enemies," he writes. "Their blood was spilled in Palestine and Iraq."

Bin Laden has referred to the suffering of the Palestinians and the Iraqis (especially with respect to the deaths caused by sanctions) in all of his public statements since at least the mid-1990s. His 1996 "Declaration of Jihad" is no exception. Nonetheless, it primarily focuses on the idea that the Saudi regime has "lost all legitimacy" because it "has permitted the enemies of the Islamic community, the Crusader American forces, to occupy our land for many years." In this 1996 text, bin Laden even contends that the members of the Saudi royal family are apostates because they helped infidels fight the Muslim Iraqis in the Persian Gulf War of 1991.

A number of neo-conservatives have advocated the overthrow of the Saudi regime because of its support for terrorism. It is true that the Saudis have funded militant Islamic movements. It is also true that Saudi textbooks and teachers often encourage hatred of infidels and allow the extremist views of bin Laden to thrive. It is also probably true that members of the Saudi royal family have financially supported terrorist groups. The fact remains, however, that bin Laden and his followers in Al Qaeda have themselves repeatedly called for the overthrow of the Saudi regime, saying that it has turned Saudi Arabia into "an American colony."

If the United States were to send troops to Saudi Arabia once again, this time to overthrow the Saudi regime itself, the main beneficiaries would be bin Laden and those who think like him. On January 27, 2002, a *New York Times* article referenced a Saudi intelligence survey conducted in October 2001 that showed that 95 percent of educated Saudis between the ages of 25 and 41 supported bin Laden. If the United States were to overthrow the Saudi regime, such people would lead a guerrilla war that US forces would inevitably find themselves fighting. This war would attract recruits from all over the Islamic world outraged by the desecration of "the land of the two holy places." Given that US forces are already fighting protracted guerrilla wars in Iraq and Afghanistan, starting a third one in Saudi Arabia would not be the most effective way of eradicating terror in the Middle East.

Those who would advocate the overthrow of the Saudi regime by US troops seem to forget why bin Laden began his holy war against the United States in the first place. They also seem to forget that no one is more committed to the overthrow of the Saudi regime than bin Laden himself. Saudi Arabia is in dire need of reform, but yet another US occupation of a Muslim country is not the way to make it happen.

In December 1998, Palestinian journalist Jamal Abd al Latif Isma'il asked bin Laden, "Who is Osama bin Laden, and what does he want?" After providing a brief history of his life, bin Laden responded to the second part of the question, "We demand that our land be liberated from the enemies, that our land be liberated from the Americans. God almighty, may He be praised, gave all living beings a natural desire to reject external intruders. Take chickens, for example. If an armed soldier enters a chicken's home wanting to attack it, it fights him even though it is just a chicken." For bin Laden and millions of other Muslims, the Afghans, the Chechens, the Iraqis, the Kashmiris, and the Palestinians are all just "chickens" defending their homes against the attacks of foreign soldiers.

In his videotaped message of October 7, 2001, after the attacks of September 11, 2001, bin Laden declared, "What America is tasting now is nothing compared to what we have been tasting for decades. For over 80 years our *umma* has been tasting this humiliation and this degradation. Its sons are killed, its blood is shed, its holy places are violated, and it is ruled by other than that which God has revealed. Yet no one hears. No one responds."

Bin Laden's defiance of the United States and his criticism of Muslim governments who ignore what most Muslims see as the oppression of the Palestinians, Iraqis, Chechens, and others, have made him a hero of Muslims who do not agree with his goal of a strictly Islamic state and society. Even young Arab girls in tight jeans praise bin Laden as an anti-imperialist hero. A young Iraqi woman and her Palestinian friends told Gilles Kepel in the fall of 2001, "He stood up to defend us. He is the only one."

Looking ahead

Feelings of impotence, humiliation, and rage currently pervade the Islamic world, especially the Muslim Middle East. The invasion and occupation of Iraq has exacerbated Muslim concerns about the United States. In this context, bin Laden is seen as a heroic Osama Maccabeus descending from his mountain cave to fight the infidel oppressors to whom the worldly rulers of the Islamic world bow and scrape.

The violent actions of Osama bin Laden and those who share his views are not simply caused by "hatred of Western freedoms." They result, in part at least, from US policies that have enraged the Muslim world. Certainly, Islamic zealots like bin Laden do despise many aspects of Western culture. They do hate "infidels" in general, and Jews in particular. Muslims do need to seriously examine the existence and perpetuation of such hatred in their societies and cultures. But invading and occupying their countries simply exacerbates the sense of impotence, humiliation, and rage that induce them to support people like bin Laden. Defeating terror entails diluting the rage that fuels it.

Henry Munson is Chair of the Department of Anthropology at the University of Maine.

The Syrian Dilemma

THE RETREAT FROM LEBANON THREATENS THE VERY SURVIVAL OF THE BAATHIST REGIME.

David Hirst

Damascus

ONE ARAB NATION WITH AN ETERNAL MISSION: BAATH PARTY, SMASHER OF ARTIFICIAL FRONTIERS. Till not so long ago, this was the slogan emblazoned across a triumphal archway under which travelers passed at the Lebanese-Syrian frontier. It was a relic of that turbulent, post-independence era when revolutionary nationalist movements, bent on restoring the "Arab nation" to its former greatness, took power in various countries. No country was more central to this than Syria, the "beating heart" of Arabism, and no movement more than Syria's progeny, the Baath, or Renaissance, Party.

Since the 1950s Syria has embarked on countless, ultimately abortive unionist projects with other Arab states. None of them, oddly, involved the country that was closest to it—not at least until, in 1976, its army crossed the most "artificial" of its colonially drawn borders and, in what it portrayed as its pan-Arab duty, sought to rescue Lebanon from the civil war into which it had fallen. Thus began the overlordship, the far-reaching political, economic and institutional penetration of one Arab country by another, in which everyone—America, Israel, the Arabs—eventually acquiesced. Yet still the frontier post, odious symbol of Arab fragmentation, remained obstinately there; and eventually the decaying archway that supposedly heralded its disappearance disappeared itself, giving way to a new complex of immigration buildings and a duty-free emporium adorned not with unionist slogans but with ads for Dunkin' Donuts.

When, soon after the February assassination of former Lebanese prime minister Rafik Hariri, widely assumed to have been carried out by Syrian agents, I crossed that frontier, it had become more than just a standing reproach to Arabism; it was symbolic not merely of the Arabs' failure to unite but of the tearing asunder of the little degree to which they had. Normally teeming, it was almost deserted. Lebanese were uneasy about going to Syria. Syrians were positively fearful of going to Lebanon, where they have been insulted and assaulted, their residences attacked, a reported thirty of them murdered.

If assassinations sometimes accelerate history, Hariri's brutal, spectacular, but popularly unifying demise was surely one of them. At a stroke it unleashed, in a great and very public torrent, all the anti-Syrian sentiments that had been surreptitiously building over the years.

Ever-growing street demonstrations, unprecedented in modern Arab history, culminated in one on March 14 that drew perhaps a million people, a quarter of the population, to Beirut's Martyrs' Square. Not just the numbers were impressive; so was their composition. In this multi-confessional country, it was, if anything, a triumph over confessionalism. The people by and large stood in one trench, their Syrian-controlled rulers in another; that, not confessional antagonism, was now the fault line principally defining the course of events. True, one sect, the Shiites, was heavily underrepresented, and the Shiite resistance movement Hezbollah had earlier staged a huge—yet smaller, more regimented, essentially single-sect and tactically motivated—"pro-Syrian" rally of its own. No less true, however, the Sunni Muslim community now threw its full weight behind the hitherto mainly Christian and Druse opposition, the significance of that being that it was traditionally Sunnis, not Shiites, who chiefly stood for Lebanon's pan-Arab nationalist identity and looked to Syria to sustain them. But at bottom it was Lebanon's silent majority—of all classes, sects and stations—who had their say on March 14. And at bottom what they said was: Give us our independence, dignity and freedom back again.

For the Baathists it was surely the death throes of One Arab Nation. Here they were, its historic standard-bearers, being reviled and driven back across that "artificial" frontier from the one Arab state where they had had the means and opportunity, in their fashion, to implement it. But history may one day judge it to have marked the birth of something new. As Lebanese columnist Samir Kassir put it: "The Arab nationalist cause has shrunk into the single aim of getting rid of the regimes of terrorism and coups, and regaining the people's freedom as a prelude to the new Arab renaissance. It buries the lie that despotic systems can be the shield of nationalism. Beirut has become the 'beating heart' of a new Arab nationalism."

In its basic impulses, this was indeed a strictly Arab, and inter-Arab, affair. But where does it fit into the great debate about the degree to which America is contributing to the winds of change in the region? Certainly, at least, George W. Bush could rejoice at this timely convergence of "people power"—massive, authentic, homegrown—with his global crusade for "freedom and democracy." So could his Administration's neoconservative hawks, for whom, soulmates of Israel's Likud, the pan-Arab nationalism of the Baath is the very antithesis of Zionism and its inherent drive to keep the "Arab Nation" fragmented, weak and doomed, in the end, to make peace with Israel on Likudnik terms. The neocons have long targeted Syria as a prime candidate in their grand design for regime change throughout the region, an objective that Congress's latest "Lebanon and Syria Liberation" bill endorses in all but name. And compared with that other candidate for regime change, Iran, Syria is a temptingly "low-hanging fruit," as some in Washington put it, and probably harvestable by merely political, not military, means. No wonder Bush so smartly joined the Lebanese opposition in almost daily and peremptory demands for full and immediate withdrawal of Syrian troops and intelligence services.

This sudden and overwhelming confluence of the local and the international, the spontaneous and the long-envisaged, has shaken the Baathist regime, so much so that some in Damascus now feel that its end is only a matter of time. "Total defeat in Lebanon," said leading dissident Michel Kilo, "will mean defeat at home."

To be sure, Syria is not defeated yet. Even as its forces redeploy and some withdraw altogether, it is still sustaining Lebanon's president, Emile Lahoud, and his puppet administration without them. So far Hezbollah, now agonizingly torn between its pan-Arab, jihadist imperatives and increasingly irreconcilable Lebanese ones, remains potently, if very uncomfortably, at Syria's service. And soon after Syrian secret police departed Beirut, car bombs began to go off in Christian neighborhoods. Were these, Lebanese asked, Syria's opening shots in the manufacture of a scenario long hinted at? Namely, that if the world pushes Syria to leave Lebanon, the world will soon come begging it to return as Lebanon, sliding back into civil war, begins to look like another Iraq, another paradise for militants and terrorists of all kinds.

That remains to be seen. But even without such desperate expedients, Syria's extraordinary resolve to keep its faltering grip on Lebanon and the brutally coercive methods it has used are already evidence enough of how vitally important it deems Lebanon to be. "Along with the command economy and the apparatus of repression," said Louai Hussein, a Syrian commentator, "control of Lebanon was one of three main pillars on which [the late] President Hafez al-Assad built his power and prestige." In fact, Syria's rulers always instinctively strive for greater regional influence than the resources of Syria alone can command. They exploit their regional "cards" in a continuous quest to advance their interests—which now boil down to securing their mere survival in the new, US-dominated Middle East order. Iraq is such a card, hence the repeated recriminations over what Syria is, or perhaps isn't, doing to help the anti-American insurgency there. Palestine is another, hence persistent American charges that Syria is "unhelpful" to the peace process, or Israeli ones that Palestinian suicide bombers get their orders from Damascus.

In a long-eroding regional hand, Lebanon, and its complete and exclusive hegemony there, is Syria's only remaining trump. It is Syria's front line, its arena of proxy war, its substitute for the military confrontation with Israel that—given its vast military inferiority—it could never risk directly from its own territory. Hezbollah is the formidable instrument of this proxy war; quiescent at the moment, it is ready and waiting to offer what, in some great showdown, Iran or Syria might require of it: its jihadist zeal, its guerrilla prowess and, according to Israel, the thousands of upgraded long-range missiles it could rain down on Israeli cities.

Economically, Lebanon is Syria's milch cow, such a cornucopia of extortion, racketeering and diversion of public funds that the distribution of the spoils—authoritatively put at about a billion dollars a year—among the Baathist oligarchy is said to be a factor in the stability of the regime. Lebanon is also the place where up to a million ordinary Syrians, facing at least 20 percent (and rising) unemployment in their own country, find illicit, low-paid work, or did so until, after Hariri's murder, they started fleeing in sizable, if unknown, numbers.

Perhaps even more dangerous to Syrian Baathism than the loss of this priceless Lebanese asset would be the potential domino effect inside Syria of the Lebanese "people power" that chiefly brought it about. First there were elections in Iraq and Palestine, which, however flawed, showed Syrians the shaming fact that Arabs enjoy more electoral choice if they are occupied than if they are sovereign. Then came this huge, unscheduled outbreak of popular self-assertion in a country where an Arab sister-state, not an Israeli or American occupier, is in charge.

The Baathist order has lost all legitimacy, sunk as it is in the most cancerous corruption and abuse of law and human rights.

And could any Syrians fail to grasp that what the Lebanese were rising up against was not (despite some ugly, chauvinist side effects) Syria itself but the extension, on Lebanese soil, of what they themselves more drastically endure at home? That is to say, the oppressions of a once-revolutionary new order that has long since betrayed its three great founding principles, pan-Arab unity, freedom and socialism. Like the now-defunct Soviet-style, single-party "people's republics" on which it was largely modeled, this Baathist order has lost all true legitimacy, sunk as it is in the most cancerous corruption, minority sectarian rule, intellectual and technical backwardness, bureaucratic ossification, abuse of law and human rights, imperviousness to dissent or criticism.

Syrians also know that those who brought the Syrian presence in Lebanon to its disastrous pass are the same people—the so-called "old guard," shadowy power centers in the army and intelligence services—who have brought fear to their own lives these past forty years, as well as blocked all reform and democratization. It is no surprise that Syria's dissident intelligentsia identify entirely with Lebanon's democratic uprising; call, like it, for full Syrian withdrawal; and use the international publicity the uprising has brought to dramatize their own campaign for human rights and civil liberties, a campaign whose most visible form is small snap demonstrations, outside courts or prisons, whenever opportunity arises.

But the Syrians aren't going to rise up like the Lebanese—not yet, anyway. Long repressed, they don't have the organized opposition or the strong residue of democratic traditions that the "Syrianization" of Lebanon never snuffed out. And the barrier of fear, always much higher than in Lebanon, remains strongly in place. "Hariri's murder," said a dissident who had no doubt about its authorship, "was a savage warning to us as much as to the Lebanese." Yet weak though it may be, and still confined largely to the intelligentsia, small political groups and human rights activists, the opposition is certainly gaining ground on a regime that is in little better shape itself, rattled and insecure as it is behind the despot's characteristic facade of lofty self-confidence, loyalist street demonstrations and the portrayal of obvious reverses as great achievements in the onward people's march.

Syrians find it hard to imagine that with Lebanon and all the domestic, regional and international pressures it has unleashed, President Bashar al-Assad doesn't realize he must do something—and do it decisively—to guard his regime, or even his country, against the gathering perils. But in this aptly dubbed "dictatorship without a dictator," has he the means, or the will? Whereas his late father, Hafez, was absolute master of what he had built, Bashar often seems more like its prisoner, forever torn between two alternative courses, reform or reaction, liberalization or repression, reaching out to the people as his source of authority or falling back on his old guard. Thus, on coming to power a few years ago, he initiated the "Damascus Spring," only to rein it in when, timid though it was, he thought it was going too far.

Such alternatives now confront him more starkly than ever. He can either make a clean break with Lebanon, purge the old guard, open wide the doors to domestic reform and appease America and the world; or he can cling to Lebanon by any means, bow to the old guard, revert to full-scale repression and

defy the world. What he will probably try to do is essentially what he always has done: make no clear choice, temporize, hope that something turns up. But with his authority steadily fraying, both within his apparatus and in the country at large, how long can it be before someone, somewhere, decides it is time to rescue the regime—or overthrow it? These are the kinds of questions now being asked by Syrians, whose yearning for change is tempered only by fear of the way—liable to be tumultuous at best, civil war at worst—it might come about, and what would come after.

Not the least of the great imponderables is what America's role and objectives might be. The homily that a typical liberal, secular-modernist dissident might address to President Bush would go something like this:

> In principle we like your "freedom and democracy" and think that what you've been doing in the Middle East has, by accident or design, given a push in that direction. But the bad things your country does still so far outweigh the potentially good that the last thing reformists like us need is to be identified with you, especially if you or Israel physically attack us. For we know that whatever you do it is Israel's wishes, not ours, that concern you. And we fear that you really do take seriously your Israeli friend Natan Sharansky—the right-wing fanatic who inspires your speeches—and his preposterous theory that only when Arabs are democratic will they be ready for peace with Israel. No, we want democracy because it will serve our national interest far better than a despotic regime whose nationalism is just a cover to suppress democracy. And so long as your policies remain what they are, our national interest will be to oppose them. But in any case, if one day we do have free elections here, it won't be the likes of us who win them but—thanks largely to you—the Islamists.

David Hirst, a longtime Middle East correspondent for the Guardian, *is the author of* The Gun and the Olive Branch: The Roots of Violence in the Middle East *(Nation Books).*

A City Adorned

John Feeney

From the late ninth century well into the 15th, mosques built as prestige projects were the most spectacular buildings in Cairo. Locals, pilgrims on their way to Makkah and even Christian pilgrims were all entranced when visiting the great Cairo mosques.

Entering through immensely tall bronze doors, many inlaid with silver and gold, visitors passed into silent, imposing courtyards, then entered darkened sanctuaries lit by dozens of softly glowing glass lamps. Suspended from high ceilings on chains invisible in the darkness, the enameled and gilded lamps would have appeared to float in space, providing a soft, even light conducive to prayer, meditation and awe.

Ibn Tulun was the first ruler in the Islamic era to give back to Egypt a sense of its past importance.

Whether luxurious or humble, all of Cairo's medieval mosques served as places of meeting and of worship. But some did more. In the early Islamic city, a congregational mosque, or *masjid jami'*, endowed by its founders and well-kept by a staff of sextons, was designed to accommodate all the local inhabitants for Friday prayers. Open to all, this large building also became the locus of public education, and here all the sciences of the day were taught. The

congregational mosque also played an important social role, for it sheltered the homeless and served as a meeting place for the discussion of matters affecting the community.

In Egypt, the first congregational mosque was built about AD 641 by the conqueror 'Amr ibn al-'As in his new town of Fustat, now at the southern end of modern Cairo. The second is today the earliest mosque still standing; it was finished in 879 by Ahmad ibn Tulun for the palace-city of al-Qata'i'.

Originally from the Central Asian caravan entrepôt of Bukhara, Ahmad ibn Tulun's father rose in the service of the Abbasid caliphs of Baghdad. Ibn Tulun was educated in Tarsus, in Anatolia, and then appointed in 868 by the caliph to govern his Egyptian domains. Two years later this ambitious young man set himself up as an independent ruler intent on rivaling the regional power of his former Abbasid master.

In doing so, Ibn Tulun was the first ruler of the Islamic era to give Egypt a sense of its past importance. Just beyond the old capital of Fustat, al-Qata'i' became renowned as one of the wonders of the age. Set amid splendid gardens were palaces, barracks, a hospital, a huge *maydan* (public square)—where Ibn Tulun and his men played polo—and the mosque, of unprecedented proportions. To finance his grand project, it is said Ibn Tulun used wealth from a sudden discovery of treasure—probably a cache of Pharaonic gold.

Set on a rocky spur, the mosque was built of fired brick and modeled on the great mosques Ibn Tulun had known in the caliph's palace-city of Samarra, now in Iraq. A person coming to pray across the rock-strewn desert from nearby Fustat entered through two enormous parallel outer walls. The space between them is called the *ziada*, and was designed to keep out the heat, dust and noise of the profane exterior world. Here the teaching of theology, medicine, astrology and grammar took place. This was also where ablutions were performed before entering into the mosque proper to pray. Set into the walls were 128 finely carved, stucco-grilled windows, which filtered the sunlight through delicate tracery.

Once inside, the visitor was shaded in long aisles formed by a forest of 220 gigantic brick piers topped by pointed, slightly horseshoe-shaped arches, all decorated in the style of Samarra. (Two hundred years hence, such arches would appear for the first time in Europe's Gothic cathedrals and, later, in the Arab structures of Andalusia.)

Of all the types of monumental public architecture, the *sabil*, or fountain, was often the most modest in size.

Beyond the shaded arcades was a vast, silent, sunlit courtyard. In its center stood an elaborate fountain where visitors refreshed themselves and where, on hot summer days, they could partake of sweet lemon drinks. To ensure that plenty of fresh water was available for the mosque and all of al-Qata'i', Ibn Tulun built an aqueduct, also out of mud brick, stretching far across the desert to a spring at Basatin. The mosque was a haven for mind, body and spirit.

Beyond the courtyard, attached to the inner of the courtyard's double walls, was a minaret with a spiral staircase on the outside—again in imitation of Samarra.

Over the centuries, additions and restorations by others bore witness to the importance of Ibn Tulun's mosque. The Fatimids added a magnificent commemorative stele with boldly floral Kufic inscriptions. The Mamluks redecorated the interior of the *mihrab*, or prayer niche, and the *mimbar* (pulpit) they added is thought by many to be the most beautiful in the city. They reconstructed the minaret, and they rebuilt in stone the wooden central fountain that had been destroyed by fire at the end of Ibn Tulun's rule.

Ibn Tulun's dynasty collapsed in AD 886, and al-Qata'i' was razed. All that remained was the mosque. Over the next 1100 years, it withstood fire, flood and earthquake, and it is now surrounded by the city named by Ibn Tulun's successors *al-Qahira*, "the vanquisher"—Cairo.

The first mosque built by the Fatimids in al-Qahira was al-Azhar, "the most resplendent" or "the most blooming," founded in AD 972. It quickly took precedence over all other mosques, and for more than a thousand years, students have gathered around its columns seeking knowledge from famous scholars, making it one of the oldest universities in the world.

With their conquest of Egypt, the Fatimids also claimed the Islamic caliphate, and other large mosques were built: That of al-Hakim, outside the walls of al-Qahira, was nearly as large as Ibn Tulun's, and the stylistic resemblance can be seen. In addition to these feats of grandeur, smaller, exquisitely built prayer places were springing up everywhere, such as the

mosque of al-Aqmar, whose name means "moon-lit." It was intended as a palace chapel for the private use of the caliph and his entourage, and as a culminating point for grand Fatimid processions.

Following the fall of the Fatimid dynasty with Salah al-Din's (Saladin's) conquest of Egypt in 1171, there came with the new ruler a new religious institution, the *madrasa*, which was equivalent to a private college. This Salah al-Din introduced to reeducate the population in the four schools of Sunni jurisprudence—Shafi'i, Maliki, Hanafi and Hanbali, named after the religious leaders who founded them. Whereas the mosques were open to all, the madrasas received only limited numbers of students. Gone now were the vast enclosures built to hold all the inhabitants of the city. The layout of the madrasa was centered around a much smaller courtyard, which was often surrounded on four sides by deep alcoves (*iwans*), one for each of the schools of jurisprudence.

About this time another institution, the *khankah*, was also introduced by Salah al-Din. First founded within the city walls, khankahs—meeting places of religious brotherhoods—soon flourished in the remoteness of the surrounding deserts, where the nobility had plenty of space to build lavishly.

At first madrasas and khankahs, even when provided with minarets, were not places of public worship. But by the Mamluk era, which began in 1250, they gradually came to be fused with the congregational mosques. Within the city, large madrasa-mosque complexes were built, complete with lodgings, lecture rooms, libraries, schools for orphans and practical services such as flour mills or a cistern with a waterwheel (*sakia*) to convey running water to the surrounding district.

The largest and most luxurious of these complexes was that of Sultan Hasan.

Five hundred years passed between Ibn Tulun's construction of his mosque in the desert and the time when Sultan Hasan ibn Qala'un set out to build at the foot of Salah al-Din's 12th-century citadel. Sultan

Hasan's grandfather, Mansur Qala'un, had already set the pace for building great mosques in 1284 with his madrasa and hospital in the heart of Cairo. Sultan Hasan's father, Nasir Muhammad, who reigned for nearly 50 years, built an even vaster edifice within the walls of the citadel, capable of holding 5000 worshipers. It was during Nasir Muhammad's long reign that it became the fashion for even minor nobility to build ever more splendid mosques, palaces and public drinking fountains (*sabils*).

Though he was born into an age of architectural splendor, Sultan Hasan's years were strewn with intrigue and disaster. He only reached the throne after succeeding his seven brothers, each of whom briefly occupied it before being either murdered or deposed. First elected at the age of 11, then at 16 overthrown by one of his brothers and committed to the citadel's infamous dungeon, he languished for three years before being released and made sultan again in 1354.

Sultan Hasan's restoration coincided with a disaster: Bubonic plague, "the black death," arrived in Cairo in the autumn of 1348. Within two years, 200,000 died in Cairo alone, and large parts of the city, the historian Maqrizi tells us, were depopulated.

Sultan Hasan, however, survived, and just as Ibn Tulun, in his day, had came into sudden wealth and used it to build grandly, Sultan Hasan, too, found himself suddenly in possession of enormous wealth from the estates of plague victims. He decided to use part of the money to build another great mosque. He is said to have asked for "something impressive," as indeed he might well do if he were to stand a chance of surpassing his father's and his grandfather's colossal accomplishments.

A century of Mamluk rule culminated in the mosque-madrasa complex of Sultan Hasan.

Plans were developed for a vast mosque-madrasa complex with dormitories and accommodation for 500 teachers and students. A khankah which was to have been built in a second stage was never completed. Compared to Ibn Tulun's sprawling and essentially horizontal desert grandeur of fired brick five hundred years before, Sultan Hasan's madrasa-mosque was built out of huge blocks of stone in soaring vertical splendor.

Of unprecedented scale in the Islamic world, Sultan Hasan's mosque became the culmination of all the architectural power developed throughout a century of Mamluk rule. An impressive flight of steps leads to an entrance portal of tremendous height, its peak decorated with a *muqarnas* ceiling of stalactites. A narrow, lofty, gradually ascending passageway follows, which emerges suddenly out of darkness into the blinding sunlight of a courtyard paved in patterned marble. The Egyptian architect Hasan Fathy described the experience: "Immediately, your eyes are drawn upward into the blue sky. And as you lower your head, 'peace' with Allah's blessing of contentment descends upon you."

And as you lower your head, your gaze spreads across the patterned mosaic-paved courtyard, unprecedented in the city. Soaring up from this expanse, four enormous vaulted *iwans*, or halls, shelter the four schools of Sunni Islamic teaching. The eastern iwan is the mosque's sanctuary and the only one where the decoration begun by the Sultan

was completed, with a splendid Kufic inscription, marble paneling on the *qibla* wall (the one indicating the direction of Makkah) and a dazzling, multi-colored marble mihrab. In Sultan Hasan's day a multitude of the famous Mamluk enameled glass lamps, suspended on fine chains, glowed in the immensity of the mosque's interior space.

Of the three mosques he, his father and his grandfather gave to Cairo, Sultan Hasan's is undoubtedly the most impressive.

The mosque was meant to have four high minarets, though today there are only two: One collapsed, killing some 400 orphans; the fourth was never built. In the sultan's time, the mosque's call to prayer was announced by a chorus of 60 muezzins working in two shifts who intoned from the door of the mosque, the courtyard, inside the sanctuary, from the roof and from the high balconies of what were then the city's two tallest minarets. Unlike the bare mosque of today, in the 14th-century mosque rich carpets lay across the marble floors, while huge brass candlesticks illuminated the open pages of immense royal Qur'ans resting on carved wooden lecterns inlaid with

ivory and mother-of-pearl. Nearby were silver water bowls and censers.

Alas, in 1361, just before the completion of his mosque, Sultan Hasan lost his throne, and possibly his head, in another palace coup. No one is quite sure what happened. Some say he escaped; others say he was tortured and that he died in the citadel's dungeon. We know for certain he does not rest in the rich mausoleum he built for himself at the back of the mosque's sanctuary.

The three Qala'uns—grandfather, father and son—gave to Cairo three most impressive mosques. Of them, Sultan Hasan's is undoubtedly the most splendid. Nothing has ever surpassed it since. Six centuries after its completion, in the early 19th century, the architect Louis Sullivan was influenced by it in his design of New York's first skyscrapers. In the late 1970's, the United Nations Educational, Scientific and Cultural Organization (UNESCO) decreed Sultan Hasan's great mosque a World Heritage Site. Sultan Hasan did, at least, get his heart's desire: "Something impressive."

*Since the 1960s, **John Feeney** (NH.Phua@xtra.co.nz) has been one of* Saudi Aramco World's *most frequent contributors, and the architecture of Islamic Cairo remains one of his favorite subjects.*

Iran in Iraq's Shadow: Dealing with Tehran's Nuclear Weapons Bid

RICHARD L. RUSSELL

As the old military adage has it, no good deed ever goes unpunished. And so it would seem with American security interests in the Persian Gulf. Soon after the United States has removed a major threat to American and regional interests with the defeat of Saddam Hussein's regime, Washington has to come to terms with the looming challenge of Iran's quest for nuclear weapons. The good news is that assertive multilateral diplomacy still has some running room for negotiating a stall or derailment of Iran's nuclear weapons program. The bad news is that the prospects are dim for achieving this end without the resort to force over the coming years.

The Iraq war is the backdrop for the evolving policy debate on Iran. The Iraq situation pits competing views of American national security strategy after 11 September 2001 against one another. On one side, critics of the Iraq war are posturing that if weapons of mass destruction (WMD) failed to be a sufficient justification for waging war against Iraq, then concerns about WMD have even less merit for forcibly challenging the Iranian regime over its nuclear weapons aspirations. On the other side, the threat posed by WMD—with the associated risk that terrorists might get their hands on WMD—is emerging as a worldview to replace the grand unifying scheme of containment which governed American and Western policy during the Cold War. Those in this camp view the military campaigns in Afghanistan and Iraq as models for other policy challenges that involve WMD and potential support for terrorist groups coming from the likes of Iran and North Korea.

There are pitfalls, though, of viewing the Iran policy debate entirely through the Iraq policy prism. Just as a prism bends rays of light, Iraq and Iran, while they share many features, are distinct problems that require the modulation of policy tools. This article seeks to illuminate the commonalities and variations between past Iraq and today's Iran as well as the strengths and weaknesses of American policy options for dealing with the growing security challenge posed by Tehran's quest for nuclear weapons.

Iran's Decrepit Armed Forces and Squeezed Geopolitical Space

Iran shares with Iraq geopolitical aspirations in the Persian Gulf in which weapons of mass destruction play a critical role. Iraq's past drive for WMD was fueled by Saddam's lust for power and his will to politically and militarily dominate the Gulf. Although Iraq's behavior over the past decade captured the most international attention, Iran too has hegemonic ambitions in the Gulf. Khomeni's revolutionary goal was to remake the region in Iran's own self-image, governed by clerics and Islamic law. Iraq's 1990–91 war pushed into the far background the premier security concern of the United States and the Arab Gulf states in the 1980s—that Iran would emerge as the winner of the war with Iraq to become the dominant power capable of directly threatening Kuwait and Saudi Arabia.

Iran's geographic girth lends itself to a country with large standing armed forces, but Iran's military today is weaker than it was in the wake of the revolutionary euphoria of 1979.[1] The Iranians militarily lived off the Shah's US-provided arms and equipment to survive the Iran-Iraq War, but the war nearly exhausted their inventories and put enormous wear and tear on equipment holdings. They have managed to make due, in part, by cannibalizing American equipment to keep fewer armaments running, but these stopgap efforts are increasingly more difficult to muster to prolong the longevity of the military inventory. The Iranians also are using illicit means to bypass US restrictions on the export of military equipment to Iran.[2] Iran has been hardpressed to find direct external weapon suppliers to replace the United States. Michael Eisenstadt observes that in recent years

Russia has been Iran's main source of conventional arms, but Moscow has agreed not to conclude any new arms deals and to halt all conventional weapons transfers since September 1999.[3] The Iranians have made efforts to fill the void with indigenously produced weapons, but Tehran lacks the ability to produce high-performance conventional weapons platforms.

Tehran must have shuddered when witnessing the American military slashing through Saddam's forces in the 2003 war. Iran already had a sense of its conventional military inferiority compared to American forces. Years ago Tehran received a direct taste of that from the American re-flagging operations in the Persian Gulf during the Iran-Iraq War, when the US Navy readily destroyed much of Iran's conventional naval capabilities, leaving Iran to harass shipping with irregular hit-and-run gunboat attacks. In the spring 2003 war, American and British forces accomplished in about a month what Iranian forces had failed to do in eight years of war with Iraq between 1980 and 1988. Tehran cannot fail to appreciate that Iranian conventional forces would have little chance of resisting a US military assault.

In Iran's geopolitical landscape and strategic calculus, the United States looms large and its "demonization" remains a central feature of the cleric regime's worldview. As Anoushiravan Ehteshami observes, "Iran holds an almost paranoid and conspiratorial view of the United States' role and actions in the Middle East and sees almost every US initiative as a direct or indirect assault on Iran's regional interests."[4] Just as George Kennan in his Cold War analysis of the Soviet Union judged that the regime in Moscow needed to politically manufacture an enemy in the United States to justify its ruthless reign at home, so do the clerics in Tehran need a political opponent in the United States on which to heap the blame and deflect public attention from their own inability to deliver political freedom, basic living standards, and an adequate economic livelihood to its people. As part and parcel of its efforts to deflect domestic criticism toward outside targets, the regime portrayed numerous student demonstrations in Iran in June and July 2003—during which Tehran felt compelled to arrest about 4,000 demonstrators—as being the result of American instigation in Iranian affairs.

American military endeavors in the greater Middle East region necessitated by 9/11 have fueled Iran's insecurity and geopolitical sense of encirclement. As Ray Takeyh notes, "The paradox of the post-September 11 Middle East is that although Iran's security has improved through the removal of Saddam and the Taliban in Afghanistan, its feelings of insecurity have intensified."[5] The United States used its military presence in the Persian Gulf to support operations both in Afghanistan and Iraq, even if host-country partners were reticent about publicly discussing their support, which cut against the grain of Arab public opinion. In its campaign against al Qaeda, much to Iran's chagrin, the United States also has had hubs of military activity or transit rights in several countries in Central Asia, including Afghanistan, Pakistan, Kyrgyzstan, Uzbekistan, Kazakhstan, and Tajikistan.[6]

Glimpses of Iran's Nuclear Weapons Bid

Iran sees WMD and ballistic missiles as means to fill the void in military and deterrent capabilities. Tehran suffered under barrages of Iraqi ballistic missiles during the Iran-Iraq War and wants to have the option of using ballistic missiles that are faster and more reliable than Iran's airforce for penetrating enemy airspaces to deliver both conventional and WMD warheads. In July 2003 Iran successfully tested the Shahab-3 missile, which achieved a range of about 1,000 km. Iran is suspected of having an unspecified number of operational Shahab missiles, which are based on North Korea's No Dong-1 missile that is reportedly capable of carrying an 800 kg warhead. Iran also is working on a 2,000-km Shahab-4 based on Russian technology, as well as a 5,000-km Shahab-5 missile.[7] These missiles probably are too inaccurate to be of much military utility if armed with conventional warheads, but they would be sufficiently accurate to deliver WMD, particularly nuclear war heads.[8] According to a foreign intelligence official and a former Iranian intelligence officer, the North Koreans are working on the Shahab-4 and providing assistance on designs for a nuclear warhead.[9]

The destructive power of chemical and biological weapons pales in comparison to that of nuclear weapons, which, unfortunately, often are considered the coin of the realm for major-power status in international relations.[10] The Iranian clerics almost certainly want nuclear weapons to compensate for conventional military shortcomings to deter potential adversaries and enhance the security of their regime: "The powerful Revolutionary Guards and military strategists are convinced that only a nuclear Iran can assume its place as a major regional power and adequately deter a possible attack from the United States or Israel, said [a] policy adviser to a senior conservative cleric, who spoke on condition of anonymity."[11]

The Iranians have learned that the road to nuclear weapons is best paved with ambiguity. The Israelis, Pakistanis, Indians, and apparently the North Koreans successfully acquired nuclear weapons by cloaking their research, development, procurement, and deployment efforts with cover stories that their efforts were all geared to civilian nuclear energy programs, not to be harnessed for military applications. Tehran could not have failed to notice that once these states acquired nuclear weapons mated with aircraft and missile delivery systems, they escaped—so far, at least—military preemptive and preventive action by rival states. In marked contrast, the Iraqis suffered as the result of Israeli and American preventive military actions, in part because Baghdad was not fast enough in acquiring nuclear weapons. The Israeli strike on an Iraqi nuclear research plant in 1981 and the American wars against Iraq in 1991 and 2003 might have been deterred had Iraq managed to acquire nuclear weapons.

The Iranians therefore consistently and loudly proclaim that their pursuit of nuclear power is strictly for peaceful civilian purposes. President Muhammad Khatami, for example, said in February 2003, "I assure all peace-loving individuals in the world that Iran's efforts in the field of nuclear technology are focused on civilian application and nothing else."[12] The Iranians argue that they need electric power produced by nuclear plants to meet domestic energy needs and to free up oil for export and foreign currency. The Iranian claims have a hollow ring, however. Iran's oil industry could be modernized and made more cost-efficient and productive with the expenditure of far fewer economic resources than those needed for nuclear power, to better deliver energy to the Iranian population at lower costs while increasing production for the international market.

The Iranians are working closely with the Russians, who have an $800 million contract with the Iranians to build the 1,000-megawatt light water reactor at Bushehr.[13] Although spent nuclear fuel at Bushehr could be diverted to use in nuclear weapons, Moscow has traditionally put more weight on near-term economic interests than on longer-term strategic interests in dealing with Iran. The Russians have adapted a Keynesian approach to Iran: damn the long-run strategic threat of an Iran armed with ballistic missiles tipped with nuclear warheads hostile to Russian political interests, because in the long run we'll all be dead anyway.

The Iranians also are interested in building a heavy-water reactor, which the international community considers as more of a nuclear proliferation risk than light-water reactors such as the one at Bushehr. Tehran has announced plans to build a 40-megawatt heavy-water research reactor, and it already has a heavy-water plant at Arak that could provide heavy water to the planned research reactor. Heavy water allows a heavy-water reactor to operate with natural uranium as its fuel and to produce plutonium.[14] Spent fuel from the planned heavy-water reactor would be ideal for extracting bomb-grade plutonium. North Korea, for example, claims to have made its weapons from the plutonium-rich spent fuel of its 5-megawatt reactor.[15] Gary Milhollin, writing in a *New York Times* article, puts the planned Iranian reactor in perspective by noting that it is too small for electricity and larger than needed for research, and is the type providing fuel for nuclear weapons programs in India, Israel, and Pakistan.[16]

Iran also is developing domestic uranium production capabilities, ostensibly to fuel its "civilian-use" nuclear power plants. In February 2003, Khatami announced that Iran had begun mining uranium near Yazd.[17] The Russians, however, claim that the Bushehr contract includes "provisions for Russia to supply fresh fuel for the life of the reactor and to take spent fuel back to Russia, thus denying Iran any potential access to the plutonium contained in the spent fuel."[18] The Iranians claim that the production facility is needed for self-sufficiency to enrich uranium for nuclear power plants, but again, as with most Iranian claims regarding their ostensible "civilian" uses for nuclear power, it would be cheaper for them to purchase uranium for civilian power needs on the international market than to indigenously develop uranium production capabilities.

Perhaps most alarming are the recent international exposures of Iran's emerging uranium enrichment capabilities. The International Atomic Energy Agency (IAEA) in February 2002 discovered that Iran is building a sophisticated uranium-enrichment plant at Natanz, about 200 miles south of Tehran. The IAEA found that 160 centrifuges were installed at a pilot plant at Natanz and 5,000 more centrifuges are to be completed at a neighboring production facility by 2005. After completion of the plant, Iran will be capable of producing enough enriched uranium for several nuclear bombs per year.[19] In a June 2003 visit to Iran, moreover, the IAEA discovered traces of highly enriched, weapons-grade uranium on centrifuges at the Natanz plant and the Kalaye Electric Company, raising the international concern that Iran's centrifuges are intended to support a nuclear weapons program.[20]

Iranian uranium enrichment capabilities appear to also have benefited from Pakistani assistance. The centrifuges inspected at Natanz by IAEA officials in February 2002 were reportedly based on a Pakistani design. The now-infamous Pakistani official widely regarded as the father of Pakistan's nuclear weapons program, A. Q. Khan, reportedly traveled frequently to Tehran to share his expertise about centrifuges and nuclear weapons design. A former Iranian diplomat turned defector claims that the Iranians gave Khana villa near the Caspian Sea as a token of thanks for his support of Iran's endeavors.[21]

Some scholars and observers of Iranian politics dismiss the foregoing as evidence that Iran has embarked on a full-fledged nuclear weapons program. It is curious that they should have confidence in making such an assessment, given that the secretive regime in Tehran is not likely to publicly broadcast a decision to acquire nuclear weapons. Such a decision would be tightly held in a small circle of regime insiders. After all, many observers were surprised by the breadth, depth, and sophistication of the Iranian uranium enrichment discovered by the IAEA inspectors because the regime's decision to pursue these activities was not publicly announced. The Iranians would be foolhardy to undermine their civilian nuclear power cover story and announce their quest for nuclear weapons, only to increase their vulnerability to American and Israeli preventive military action.

Diplomatic Options for Stalling Iranian Nuclear Weapons

American diplomacy is encouraging energetic and assertive IAEA inspections of Iran under the Nuclear Non-Proliferation Treaty (NPT) regime. The specter of the US use of force against another pillar of the "axis of evil," coupled with Europe's belated doubts about the efficacy of engagement to curtail Iran's nuclear weapons program, has worked to coax Iran to accept no-notice IAEA inspections. The Europeans—the French and Germans, in particular—who had long resisted US efforts to isolate Iran and favored diplomatic and economic engagement of Tehran, were apparently taken aback by the scope of Iran's work on uranium enrichment and disregard for the NPT. The French, Germans, and British are rightly trying to exchange trade discussions for Iran's cooperation on no-notice inspections and ending its pursuit of the nuclear fuel cycle, which would give Iran the capability to pursue nuclear weapons in short order.[22] The European Union Foreign Minister declared publicly in June 2003 that if diplomatic efforts to deal with Iran's WMD should fail, coercive measures could be envisioned.[23] Obviously such bravado is in marked contrast to European opposition to the American use of force against Saddam's regime, and should push come to shove in dealing with Iran's nuclear weapons program, the Europeans may well revert to their aversion to the exercise of American military power. It is easy for the Europeans to argue theoretically that force may have to be used when that contingency appears well over the horizon, but it would be politically more unpalatable for European capitals when the concrete decision time for the resort to force beckons.

Tehran for its part probably calculates that its acceptance of the no-notice inspections will buy Iran more time to work on its clandestine nuclear weapons program by politically diffusing international support for an assertive American stance. At the same time, Tehran probably is betting that it can work on nuclear weapons undetected by IAEA inspectors. Iran has had plenty of opportunity to learn lessons on beating the IAEA inspection regime from watching Iraq and North Korea, which both cheated successfully against IAEA inspectors. Both Iraq and North Korea worked feverishly on nuclear weapons programs while officially considered "in good standing" in the eyes of the IAEA inspectors and their governing NPT. Only US intelligence was able to catch North Korea covertly working on a uranium enrichment program, which led to a chain of events that resulted in Pyongyang formally withdrawing from the NPT. The massive scope of Iraq's nuclear weapons program was revealed only after Iraq's 1991 battlefield defeat and intrusive UN weapons inspections. UN inspectors found the Iraqis to be expert in denial and deception efforts that allowed them to vigorously pursue a nuclear weapons program despite years of IAEA inspections. If IAEA inspectors were on their way to a sensitive Iranian site, Tehran's security services could manufacture all kinds of obstacles to slow the IAEA team or misdirect them, just as the Iraqis did with IAEA and UN weapons inspections.

To hedge against these potential Iranian calculations, IAEA inspectors would have to demand an unparalleled level of sustained and rapid access to Iranian facilities and personnel, with full Iranian cooperation. No-notice and intrusive IAEA inspections should be regularly and routinely mounted without international apology. IAEA inspectors should have routine, widespread, and unencumbered debriefing access to any and all Iranian scientists and technicians, who could be debriefed outside of Iran and without Iranian minders present. Such measures were only faintheartedly implemented by the United Nations under Hans Blix in the run-up to the 2003 war against Iraq.

Washington could further use international sanctions to cut Iran's trading access to the global market, particularly for oil exports, to increase pressure on Tehran to accept assertive IAEA inspections and a stoppage in Iran's nuclear fuel cycle efforts, but that course could suffer from numerous pitfalls. Sanctions would have to be sustained for a prolonged period of time before they began to hurt Iran's economy, and after that time, much like the sanctions implemented agains Saddam's regime, they would hurt the livelihood of the general populace more than regime elites. As a consequence the United States might undercut its objective of looking to the Iranian population to usher in a political change in Tehran—under the stress of such international sanctions, the population could rally around the regime rather than taking up political actions against it.

A better alternative might be for Washington to offer to sweeten the diplomatic tea with a variety of options to encourage Iran to accept an unprecedented level of intrusive IAEA inspections and to stop its nuclear fuel cycle efforts. For example, Washington could offer the resumption of diplomatic ties with Tehran severed after the 1979 revolution; the release of frozen Iranian assets in the United States; and the easing of trade sanctions that would facilitate Iranian access to the international marketplace, technology, and business, thus helping to modernize Iran's oil industry. As Takeyh observes, "The economic dimension is particularly important as, in the past decade, Tehran has grudgingly come to realize that Iran's tense relations with the United States preclude its effective integration into the global economy and access to needed technology."[24] These positive incentives, however, might still not be sufficient to reverse Iran's hostile policy toward the United States given the factions competing for power in Tehran. As Geoffrey Kemp points out, "Opponents can be counted upon to do all they can to prevent such a thing from happening, including strategic leaks designed to undermine any diplomacy in prospect."[25]

The uncertainty over the Iranian internal power structure would make it difficult for American policymakers to establish "rules of the road" in any diplomatic dialogue designed to gain a degree of confidence that Tehran could exercise responsible and stringent controls over future nuclear weapon stocks. Notwithstanding past Iranian public support for the Iranian President, the wind in Khatami's reform-minded sails is dying. And the Iranian elections in February 2004 in which conservatives barred moderates from being placed on ballots have stranded the reformers at sea without fresh water. While many in the West hope that the counterrevolutionary winds will grow stronger with public demonstrations and cast aside the conservative clerics, such a desirable course of events may await the longer run. In the short to medium terms, there are greater prospects for hard-line clerics ousting the more pragmatic clerics in the regime power struggle.

Military Options for Disrupting Iran's Nuclear Weapons Program

American diplomatic support for robust IAEA inspections is reducing widespread European and Middle Eastern criticism that the United States acts unilaterally or hegemonically in the international arena. Such criticism reached shrill heights during the lead-up to the war against Saddam's Iraq. The United States needs to work to heal those wounds to garner political support from Europe and the greater Middle East region to complement diplomacy with military force in a concerted policy to derail Iran's train ride toward nuclear weapons.

Military options could be employed to physically disrupt, delay, and destroy key components of Iran's nuclear weapons program. Such military options would be geared toward causing the Tehran regime pain and inflicting costs for Tehran's pursuit of nuclear weapons. They could be aimed at changing Tehran's strategic calculus, so that Iran views nuclear weapons not as something that enhances the security of the regime, but as a liability that increases prospects for conflict with the United States and threatens the clerical regime's hold on power.

Obviously military options would entail less risk if exercised before Iran acquires nuclear weapons. American policymakers would have to be concerned that if military options are employed after Iran acquires nuclear weapons, the Iranians could retaliate for US conventional military strikes by targeting American forces in the region with nuclear weapons or by using clandestine means to attack American civilians, perhaps via the Iranian intelligence services or collaborating transnational actors, especially Hezbollah. While such risks may not ultimately preclude the decision to use force, Iranian possession of nuclear weapons would make the decision a heavy burden.

American military superiority over Iran gives Washington a wide spectrum of military options for coercing Tehran. These options range from limited strikes against Iran's political, military, internal security, and WMD-related infrastructure. For example, the United States could target Iran's nuclear power infrastructure—to include the Bushehr nuclear power plant as well as any future nuclear power plants, heavy-water facilities, future plutonium reprocessing plants, and uranium production and enrichment plants—with cruise missiles or combat aircraft strikes. An American air campaign mounted from regional support hubs in the small Gulf Arab states could make short work of Iran's air force and air defense forces to gain air superiority for attacks against Iran's nuclear infrastructure. Such strikes could serve the practical pur-

poses of disrupting Iran's means for developing nuclear weapons as well as constituting a symbolic, political demonstration of American resolve to use whatever means are available to block Iran's nuclear weapons aspirations.

The United States would be operating with a less-than-perfect intelligence picture of Iran's nuclear weapons infrastructure, however. The Iranians cannot have escaped learning the importance of diversifying and building redundancies into their nuclear weapons program components in light of Israel's preemptive strike on Iraq's nuclear power facility. They managed to hide Iranian uranium reprocessing developments from the outside world for some time and have undoubtedly tightened security to stem further exposures of their nuclear weapons program. In the aftermath of any American air strikes against their nuclear infrastructure, Iran undoubtedly also would redouble its efforts to conceal and build redundancies into its nuclear weapons infrastructure to make follow-on American attacks more difficult.

American aircraft and cruise missiles also could target Iran's key political, security, and military infrastructures to harm the power of the regime in Tehran. Strikes could target government buildings and even the homes of clerics; facilities and compounds used by internal security and policy forces; assets of the Iranian Revolutionary Guard Corps (IRGC) and Basij forces; major army units and garrisons; and WMD delivery vehicles, such as aircraft and ballistic missiles, as well as their production facilities. Targeting internal security organs would be particularly useful because that might allow the disgruntled populace more freedom to demonstrate against the regime and substantially increase the pressure on clerics to forgo their nuclear weapons aspirations.

The threat of a US invasion of Iran should not be taken off the table, because it could be used to bolster the strength of coercive diplomacy to compel the Iranians to desist on nuclear weapons and to accept robust and intrusive international inspections to help ensure their compliance with the NPT. The most imprudent step a statesman can make is to let his adversary know what he is not prepared to do; that profoundly undermines his political leverage to achieve interests without resort to force. President Clinton made this critical mistake in the 1999 Kosovo war, in which he declared that US ground forces would not be used against Serbia.

"A nuclear-armed Tehran might fear the prospect of American and Israeli nuclear retaliation less than Western strategists would hope."

Nevertheless, the US military presence in the greater Middle East that brackets Iran would be insufficient to stage the type of massive ground campaign that would be required to occupy Iran's major cities. Iraq is a comparatively easy occupational task in comparison to Iran; it is smaller and has fewer citizens. Iraq is twice the size of Idaho and populated with about 25 million people, while Iran is nearly four times the size of Iraq with approximately 67 million people.[26] The American and British forces in neighboring Iraq are likely to be fully preoccupied with Iraq's internal security for the coming years, and without significant augmentation they would be unavailable for a cross-border invasion of Iran. US forces in Pakistan and Afghanistan are much smaller and more suited for special operations that would augment, rather than spearhead, the massive ground force campaign that would be necessitated by Iran's sheer geographic size.

American decision makers have to weigh political ends against military means as a basis for formulating strategy. The United States now has a significant portion of its total ground forces committed to Iraq and would be hard-pressed to mount a comparable or larger operation simultaneously against Iran. The United States also needs to keep its forces ready to meet contingencies elsewhere in the world, particularly in Asia where potential clashes could emerge on the Korean Peninsula or over Taiwan. The weighing of these concerns, however, would best be done in the minds of policymakers and not shared aloud in the public domain for the ears of Iran's clerics.

The domestic Iranian political fallout from American military operations could cut two ways. On one hand, US operations could undermine the regime politically as many Iranians would see them as more evidence that the nature of the regime works to prolong Iran's isolation from the world community and its economic stagnation and political retardation. On the other hand, the clerics would seize on the strikes as evidence of a hegemonic American campaign to conquer the Middle East and its oil, and use that perception as justification for repressive domestic security measures to hold onto power. In the final analysis, the United States could have to just wait and see which of these competing forces would prove to be stronger as it vigilantly monitored Iran's efforts to reconstitute its infrastructure and made follow-on strikes over a period of years to perpetually "kick the can down the road" and delay Tehran's acquisition of nuclear weapons.

As has been the case in the war against Iraq, the United States would have to ride out the international political fallout from any military actions against Iran. At first glance, Russia, China, North Korea, and Pakistan probably would politically protest "American unilateralism" out of concern over economic losses as a result of attacks on Iranian facilities that those countries are supporting. But then again, from a more cynical view, those states might work to economically exploit the situation and seek additional contracts to rebuild all that the Americans had destroyed. Military operations too would come with a tide of regional outcries against the United States. Many would accuse the United States of the all-too-familiar refrain that Washington holds a double-standard in the region by ignoring Israeli nuclear weapons while taking military actions against Muslim states such as Iraq and Iran, which were seeking to arm themselves to balance Israeli and American nuclear power. As hard as it is for American observers to appreciate, many in the region—officers, diplomats, officials, as well as the general public—harbor the view that a nuclear-armed Iran could be useful to counterbalance Israeli as well as American nuclear power.

Running Risks with Iranian Nuclear Weapons

And what if these diplomatic and military options were unsuccessful? What could Iran do with nuclear weapons? Would Iranian nuclear weapons pose a profound security challenge for the United States? Or would an Iranian nuclear weapons inventory be manageable for Washington? Could the United States accept Iranian nuclear weapons capabilities, much as Washington has accepted those possessed by Israel, Pakistan, India, and perhaps North Korea?

A grave concern is that Iran could transfer nuclear weapons to non-state actors, because for the past 20 years Tehran has consistently used non-state actors as instruments of statecraft to advance Iranian political interests and objectives. Indeed, the prospects for the transfer of nuclear weapons to non-state actors is greater in the case of Iran than it was for Saddam's regime, because Tehran has been much more active than Baghdad had been in the sponsorship of terrorist operations, particularly those orchestrated by Hezbollah, against the United States.[27] Jeffrey Goldberg reports in *The New Yorker*, "Hezbollah has an annual budget of more than a hundred million dollars, which is supplied by the Iranian government directly and by a complex system of finance cells scattered around the world."[28]

Some observers argue that the revolutionary steam has run out of Iran's regime and that Iranian sponsorship of terrorist operations against US interests has diminished. Iran's complicity and support for the 1996 bombing of Khobar Towers, the American military housing complex in Saudi Arabia, which killed 19 American servicemen, belies arguments that Iran's government has tempered its opposition to the United States, however. Former FBI Director Louis Freeh has publicly and directly linked Iran to the Khobar Towers attack: "Over the course of our investigation the evidence became clear that while the attack was staged by Saudi Hezbollah members, the entire operation was planned, funded and coordinated by Iran's security services, the IRGC and MOIS [Ministry of Intelligence and Security], acting on orders from the highest levels of the regime in Tehran."[29] More recently, Iran has shown an interest in maintaining links to al Qaeda by harboring its operatives, some of whom had fled neighboring Afghanistan and Pakistan in the midst of the October 2001 American military campaign in Afghanistan.[30]

Some observers are inclined to give the Iranian regime the benefit of the doubt regarding allegations of complicity in the Khobar Towers bombing by arguing that "rogue elements" or conservative hardliners in the regime, not President Khatami and like-minded supporters in the Ministry of Foreign Affairs and parliament, supported the operations. Conclusive evidence to bolster this argument is elusive, but even if it were found to be the case, such a fact would be of little solace to American policymakers and the public coming to terms with the potential dangers posed by Iranian possession of nuclear weapons. Policymakers would have to be concerned that hardliners in the future could control or direct transfers of nuclear weapons even if it were not the consensus policy of the regime. If an American city were to suffer from the detonation of a Hezbollah-planted Iranian nuclear weapon, it would be largely irrelevant whether or not it came about via rogue or mainstream elements of the Tehran government.

Tehran might calculate that a nuclear deterrent would give it more leeway for supporting militants in the Middle East, including Hezbollah, Islamic Jihad, and Hamas. The Iranians, even without nuclear weapons, are moving in this policy direction. As Daniel Byman observes in an article in *Foreign Affairs*, "Since the outbreak of the al Aqsa intifada in October 2000, Hezbollah has provided guerrilla training, bomb-building expertise, propaganda, and tactic altips to Hamas, Palestinian Islamic Jihad, and other anti-Israeli groups."[31]

Tehran might judge that even if its hand were revealed in supporting terrorist operations via these groups against American interests and partners among the Gulf Arab states, Iranian nuclear weapons would deter military reprisals against Iran. American and Israeli contemplation of retaliatory strikes against Iran would be substantially riskier if Iran had the means to retaliate with nuclear weapons. The Iranian clerics are not well schooled in the ins and outs of the elaborate Western strategic literature formulated during the Cold War. The clerics probably would be more influenced by their Islamic ideological world views than by a rational calculation of national interests. As George Perkovich argues, "Political leaders like Khamene'i and Rafsanjani see nuclear weapons as an almost magical source of national power and autonomy. These men are political clerics, not international strategists or technologists. They intuit that the bomb will keep all outside powers, including Israel and the US, from thinking they can dictate to Iran or invade it."[32] In short, a nuclear-armed Tehran might fear the prospect of American and Israeli nuclear retaliation less than Western strategists would hope.

The Iranians could elect to rely more heavily on integrating nuclear weapons into their war-fighting strategies. They undoubtedly have ingrained into their political and military thinking the premiset on ever again be caught in a prolonged war of attrition as was the case in the Iran-Iraq War that Tehran ultimately lost. The Iranians might come to view nuclear weapons as useful, or even essential, battlefield instruments for destroying the armed forces of an adversary, particularly those of Iraq. As Gary Sick points out, Iran's past use of unconventional hit-and-run speedboat attacks in the Persian Gulf during its war with Iran demonstrate Tehran's willingness to "use unconventional, even terrorist, methods to pursue a political and military strategy, even if that meant confronting the United States."[33] Along these lines, Tehran might be tempted to harness the threat of nuclear weapons for leverage in the political-military struggle against the United States for power and influence in the Persian Gulf.

Iranian nuclear weapons would give Tehran greater political and military prestige that could translate into leverage over the Arab Gulf states. As Kenneth Pollack warns, "Tehran appears to want nuclear weapons principally to deter an American attack. Once it gets them, however, its strategic calculus might change and it might be emboldened to pursue a more aggressive foreign policy."[34] The Arab Gulf states would be more vulnerable to Iranian political pressure to reduce security cooperation with the United States, particularly in the event of a regional contingency. Finally, an Iranian nuclear bomb also would increase the already high incentives for Arab states to procure nuclear weapons.

NOTES

The author is indebted to his Near East-South Asia Center colleagues Ray Takeyh, for his tutelage on Iranian politics, and Danielle Debroux, for her able research assistance. A thanks is also due to Henry Sokolski for comments on an earlier draft as well as for his Nonproliferation Policy Education Center's project on Iran that instigated the author's interest in the topic.

1. Tehran's regular armed forces consist of about 325,000 in the army, 18,000 in the navy, and 52,000 in the air force. It has a parallel force structure in the Revolutionary Guard Corps (IRGC) with about 125,000 soldiers, including about 100,000 ground troops, 20,000 naval, 5,000 marines, and an unknown number in an air force. Tehran also has a paramilitary force, the Popular Mobilization Army or Basij, with about 40,000 active troops. See International Institute for Strategic Studies (IISS), *The Military Balance, 2002–2003* (London: Oxford Univ. Press, 2002), p. 104.
2. In July 2003, the United States issued search warrants and grand jury subpoenas to 18 US companies in a massive raid against illegal export of American-built military components to a London front company for Iran. The front company was procuring components for the Hawk air defense system, F-14, F-4, F-5 combat aircraft, C-130 transport aircraft, and radar as well as other equipment. California police in July 2003 arrested two men trying to export military technology—including components for the F-4, F-5, and F-14 aircraft, and Hawk surface-to-air missiles—to China. These items are not in the Chinese military inventory, however, but are in the Iranian military's inventory, strongly suggesting that China is acting as a middle man for Iran's clandestine repair parts pipeline. See Christine Hanley, "Two Men Tried to Illegally Export Military Parts to China, U.S. Says," *Los Angeles Times*, 25 July 2003, p. B5.
3. Michael Eisenstadt, "Living with a Nuclear Iran?" *Survival*, 41 (Autumn 1999), 140.
4. Anoushiravan Ehteshami, "Tehran's Tocsin," in *Contemporary Nuclear Debates: Missile Defense, Arms Control,*

and Arms Races in the Twenty-First Century, ed. Alexander T. J. Lennon (Cambridge, Mass.: MIT Press, 2002), p. 152.

5. Ray Takeyh, "Iran's Nuclear Calculations," *World Policy Journal*, 20 (Summer 2003), 23.

6. See "A Survey of Central Asia: At the Crossroads," *The Economist*, 26 July 2003, p. 3.

7. Alon Ben-David, "Iran Successfully Tests Shahab 3," *Jane's Defence Weekly*, 9 July 2003, http://jdw.janes.com/.

8. See "Iran's Ballistic Missiles: Upgrades Underway," *Strategic Comments*, 9 (London: IISS, 2003).

9. Douglas Frantz, "Iran Closes in on Ability to Build a Nuclear Bomb," *Los Angeles Times*, 4 August 2003, p. A6.

10. The Iranians developed a chemical warfare program in the 1980s to match Iraq's chemical weapons capabilities demonstrated during the Iran-Iraq War and are suspected of harboring a biological warfare program. This is despite Tehran's signature on the chemical and biological weapons conventions that prohibit such programs. See Joseph Cirincione with John B. Wolfsthal and Miriam Rajkumar, *Deadly Arsenals* (Washington: The Brookings Institution, for the Carnegie Endowment for International Peace, 2002), pp. 255–56.

11. Azadeh Moaveni and Douglas Frantz, "Are Iran's Nuclear Promises Real?" *Los Angeles Times*, 21 November 2003.

12. Quoted in Nazila Fathi, "Iran Says It Has Developed Ability to Fuel Nuclear Plants But Won't Seek Weapons," *The New York Times*, 10 February 2003, p. A12.

13. David Holley, "Iran Sets Its Sights on More Reactors," *Los Angeles Times*, 3 July 2003, p. A3.

14. Douglas Frantz, "Iran Closes in on Ability to Build a Nuclear Bomb," *Los Angeles Times* 4 August 2003, p. A7. Heavy water (D_2O) is "water containing significantly more than the natural proportion … of heavy hydrogen (deuterium, D) atoms to ordinary hydrogen atoms." US Nuclear Regulatory Commission, http://www.nrc.gov/reading-rm/basic-ref/glossary/heavy-water-d2.html.

15. "Fissionable: Iran's Nuclear Program," *The Economist*, 14 June 2003, p. 24.

16. Gary Milhollin, "The Mullahs and the Bomb," *The New York Times*, 23 October 2003.

17. Fathi, p. A12.

18. Robert J. Einhorn and Gary Samore, "Ending Russian Assistance to Iran's Nuclear Bomb," *Survival*, 44 (Summer 2002), 53.

19. Joby Warrick and Glenn Kessler, "Iran's Nuclear Program Speeds Ahead," *The Washington Post*, 10 March 2003, p. A1.

20. Douglas Frantz, "Iran Discloses Nuclear Activities," *Los Angeles Times*, 24 October 2003; and Douglas Frantz, "Iran

Closes in on Ability to Build a Nuclear Bomb," *Los Angeles Times*, 4 August 2003, p. A1.

21. Frantz, "Iran Closes in on Ability to Build a Nuclear Bomb," p. A7.

22. The author is indebted to Henry Sokolski for these important points.

23. "Weapons of Mass Destruction: Europe Spies a Threat," *The Economist*, 21 June 2003, p. 27.

24. Takeyh, p. 25.

25. Geoffrey Kemp, "How to Stop the Iranian Bomb," *The National Interest*, 72 (Summer 2003), 54.

26. Central Intelligence Agency, *World Factbook*, http://www.cia.gov/cia/publications/factbook/geos/iz.html and http://www.cia.gov/cia/publications/factbook/geos/ir.html.

27. Hezbollah was responsible for the bombing of the US Marine Corps barracks in Beirut in October 1983 that killed 241 marines, the April 1983 bombing of the US Embassy in Beirut that killed 63 people, killed the CIA Beirut station chief in 1985, and killed a US Navy diver on hijacked TWA Flight 847 that landed in Beirut in 1985. For an argument against using Iraqi ties to terrorist groups as a strategic rationale for waging war against Saddam, see Richard L. Russell, "War and the Iraq Dilemma: Facing Harsh Realities," *Parameters*, 32 (Autumn 2002), 47–48.

28. Jeffrey Goldberg, "In the Party of God: Are Terrorists in Lebanon Preparing for a Larger War?" *The New Yorker*, 14 October 2002, p. 183.

29. Louis J. Freeh, "American Justice for Our Khobar Heroes," *The Wall Street Journal*, 20 May 2003, p. A18.

30. Peter Finn and Susan Schmidt, "Al Qaeda Plans a Front in Iraq," *The Washington Post*, 7 September 2003, p. A26.

31. Daniel Byman, "Should Hezbollah be Next?" *Foreign Affairs*, 82 (November/December 2003), 59.

32. George Perkovich, "Dealing with Iran's Nuclear Challenge," Carnegie Endowment for International Peace, 28 April 2003, p. 4.

33. Gary Sick, "Iran: Confronting Terrorism," *Washington Quarterly*, 26 (Autumn 2003), 87.

34. Kenneth M. Pollack, "Securing the Gulf," *Foreign Affairs*, 82 (July/August 2003), 7.

Richard L. Russell *is Professor of National Security Affairs at the National Defense University's Near East-South Asia Center for Strategic Studies. He also holds appointments at Georgetown University as Adjunct Assistant Professor in the Security Studies Program and Research Associate in the Institute for the Study of Diplomacy.*

From *Parameters*, Autumn 2004, pp. 31-45. Copyright © 2004 by Richard L. Russell. Reprinted by permission.

THE DYING OF THE DEAD SEA

THE ANCIENT SALT SEA IS THE SITE OF A LOOMING ENVIRONMENTAL CATASTROPHE

JOSHUA HAMMER

GIDON BROMBERG is nervous. On a sweltering August afternoon, the Israeli environmental activist leads me along the shore of the Dead Sea, watching every step we take. Towering sandstone mesas loom above our heads; the saline lake extends like a shimmering sheet of turquoise toward the hazy mountains of Jordan. The temperature is pushing 110 degrees, the sun beats down on my neck, and my feet crunch pieces of petrified driftwood and calcium deposits—wrinkled white sheets that bear a disturbing resemblance to human rib cages. Bromberg stops abruptly beside a gaping crater, more than 60 feet deep, and a sign that reads DANGER: OPEN PITS. "Better not walk any farther," he warns. "The ground could swallow us whole."

Up and down the Dead Sea, on the Jordanian and Israeli coasts, the shoreline is pockmarked by these sinkholes—testifying to an environmental catastrophe. The Dead Sea is shrinking, and as it recedes, the fresh water aquifers along the perimeter of the lake are receding along with it. As this fresh water diffuses into salt deposits beneath the surface of the shoreline, the water slowly dissolves the deposits until the earth above collapses without warning. More than 1,000 sinkholes have appeared in the past 15 years. In that time, sinkholes have swallowed a portion of road, date-palm fields and several buildings on the sea's northwest coast. Environmental experts believe that hotels along the shore are also in danger. "The good news is that if you get swallowed by a sinkhole, they name it after you," Bromberg deadpans.

As we trudge back along the shore, Bromberg points out the Ein Gedi Spa, built along the waterline about 20 years ago. Today the resort sits marooned on a spit of wasteland almost a mile from the water; a trolley carries guests to and from the beach along a track that must be extended every year. Driving a few miles south, past the ancient Jewish fortress of Masada, we come upon Ein Bokek, a garish strip of high-rise hotels that calls to mind Atlantic City, New Jersey. Few tourists arriving at Ein Bokek are aware of the resort's not so little secret: the shallow water in front of the hotels isn't the Dead Sea, which dried up here in the 1980s. It is a reservoir main-tained by Dead Sea Works, an Israeli company that pumps water from the northern to the southern part of the lake, where it is evaporated to extract minerals such as potash and bromide—a process hastening the sea's demise. "It's all artificial," says Bromberg. "But you're not going to hear that from the hotel management."

Bromberg was born and raised in Israel, graduated from American University's law school in Washington, D.C., and returned to Israel 17 years ago. Now he directs Friends of the Earth Middle East, the most active of several environmental groups working to galvanize concern for the dying sea. With a staff of Israelis, Palestinians and Jordanians, and with offices in Tel Aviv, Bethlehem and Amman, Friends of the Earth has become a model of regional cooperation at a time when most such ventures have all but disappeared.

During the past several years, Friends of the Earth has sponsored exhibitions of Dead Sea photographs and conducted tours for journalists and government officials. The organization has lobbied Jordan, Israel and the Palestinian Authority to nominate the Dead Sea as a United Nations World Heritage site—a designation that would mandate creation of an environmental protection plan and restrict development. The Friends have also pressured the region's governments to reform what they call "shortsighted" water policies that they say have been sucking dry the Dead Sea—and the rivers and streams that feed it—for decades.

The work has been difficult—and at times dangerous. During the most recent wave of Palestinian uprisings, beginning five years ago, simply bringing together staffers from Jordan, the Palestinian territories and Israel turned into a logistical nightmare. Islamic militants have opposed such joint ventures, accusing Arab staffers of acting as "collaborators" with Israel. Four years ago, during an intense period of Israeli-Palestinian violence, gunmen fired on Friends of the Earth's Munqeth Mehyar as he drove away from his Jordanian office in downtown Amman. (He escaped without injury.) "We thought about closing the Amman office after that incident, but the staffers here said no way," Mehyar said. "They believed that would be giving in to terror." Instead, Mehyar and his colleagues received protection from Jordanian police and intelligence. The assailants were eventually captured.

CREATED BY THE SAME shift of tectonic plates that formed the Syrian-African Rift Valley several million years ago, the Dead Sea owes its precarious state to both human and geological factors. Originally part of an ancient, much larger lake that extended to the Sea of Galilee, its outlet to the sea evaporated some 18,000 years ago, leaving a salty residue in a desert basin at the lowest point on earth—1,300 feet below sea level. Since then, this body of water, known as the Dead Sea since Greco-Roman times, has maintained its equilibrium through a fragile natural cycle: it gets fresh water from rivers and streams from the mountains that surround it and loses it by evaporation. The evaporation process, combined with its rich salt deposits, account for its extraordinary—up to 33 percent—salinity (compared with the up to 27 percent salinity of Utah's Great Salt Lake). Until the 1950s, the flow of fresh water equaled the rate of evaporation, and Dead Sea water levels held steady. Then in the 1960s, Israel built an enormous pumping station on the banks of the Sea of Galilee, diverting water from the upper Jordan, the Dead Sea's prime source, into a pipeline system that supplies water throughout the country. To make matters worse, in the 1970s Jordan and Syria began diverting the Yarmouk, the lower Jordan River's main tributary.

Since then, the Dead Sea has declined dramatically. It needs an infusion of 160 billion gallons of water annually to maintain its current size; it gets barely 10 percent of that. Some 50 miles long in 1950, the sea is about 30 miles long today. Water levels are falling at an average rate of three feet per year. According to a recent Israeli government study, the rate of evaporation will slow and the Dead Sea will reach equilibrium again in a few decades—but not before losing another third of its present volume.

Such a scenario represents an immeasurable loss. Tourists have flocked here for generations to float in the brine, soak in mineral and mud baths and take in the dramatic panorama of Israel's Judean Desert and Jordan's Moab Mountains. Sufferers from chronic skin diseases, such as psoriasis and eczema, routinely make pilgrimages, attracted by the bone-dry climate, oxygen-rich atmosphere and—some claim—the sea's miraculous healing properties.

A refuge over the millennia for messiahs, martyrs and zealots, the Dead Sea region abounds with sites sacred to Islam, Christianity and Judaism. Some Muslims believe that Moses, whom they regard as a prophet, lies buried in a hilltop mosque just off the main road from Jerusalem. Jesus Christ was said to have been baptized in the Jordan River after traveling down to the Dead Sea from Galilee. At the fortress of Masada, nearly 1,000 Israelites committed suicide en masse in A.D. 73 rather than surrender to the Romans. Fifth-century ascetics from Asia Minor re-

treated to the region's cliffside caves and built monasteries such as Mar Saba, the oldest continuously inhabited one in the world. In 1947, Bedouin shepherds, searching for a stray goat in the Judean Desert, entered a cave at Qumran near the north shore of the lake and discovered clay jars containing 2,000-year-old scripture written in Hebrew, Greek and Aramaic—the Dead Sea scrolls.

And despite its name, the Dead Sea helps support one of the world's most complex and vibrant ecosystems. Fed by fresh water springs and aquifers, a half-dozen oases along the shore harbor scores of indigenous species of plants, fish and mammals, including ibex and leopards. About 500 million birds representing at least 300 species, including storks, pelicans, lesser spotted eagles, lesser kestrels and honey buzzards, take refuge here during a biannual great migration from Africa to Europe and back again. Ein Feshka, a lush expanse of tamarisk, papyrus, oleander and pools of crystal water, was used by the late king Hussein of Jordan as a private playground in the 1950s and early '60s. But as the Dead Sea recedes, the springs that feed the oases are moving along with it; many experts believe that Ein Feshka and other oases could wither away within five years.

IN APRIL 1848, when Palestine was a desolate outpost of the Ottoman Empire, American adventurer Lt. William Francis Lynch embarked on a U.S. Navy expedition to chart the course of the Jordan River to the Dead Sea. Lynch and his party of scientists and topographers set off in three vessels from the Sea of Galilee and quickly found themselves swept up in a frothing torrent. The river was hundreds of feet wide in some places, interrupted by "frequent and most fearful rapids," Lynch wrote. "Placing our sole trust in Providence [we] plunged with headlong velocity down appalling descents." They reached the Dead Sea after seven grueling days, losing one boat, which had been battered to pieces on the rocks.

The story of the Jordan River's decline begins at the very place where Lynch launched his boats in what is no longer a roaring torrent but a pond of sluggish green water. In 1953, Israel constructed a dam, the Degania Gate, a few hundred feet south of this spot, to collect water from the Sea of Galilee for the National Water Carrier project. The dam reduced the Jordan's flow to a trickle.

ABOUT FIVE MILES south of the dam, Bromberg and I enter the Degania kibbutz, one of Israel's oldest kibbutzim, or agricultural cooperatives, founded in 1909. We bounce along a rutted dirt track through corn, tomato and avocado fields, following two giant metal pipes that siphon off some of the Jordan's water for an extensive irrigation system. Dozens of other collective farms in the area also dip into the river. After a few minutes we arrive at a small earthen dam, where the Jordan comes to a pitiful end. On one side lies a stagnant pool covered by algae. A rusted rowboat is submerged beneath the surface. On the other side of the dam, liquid gushes from two pipes and flows down the riverbed. One flow consists of raw sewage from kibbutzim in the area. The other is saline water from springs flowing into the Sea of Galilee mixed with partially treated sewage from Tiberias, captured and removed to decrease the lake's salinity. The Jordan's once annual flow of 343 billion gallons of fresh water has now been replaced by 40 billion gallons or so of mostly sewage and saline water. Irrigation "is one of the main reasons that the Dead Sea is dying," Bromberg tells me.

Another reason, according to environmentalists and various government officials, is a water policy on the part of Israel, Jordan and Syria that encourages unrestricted agricultural use. From the first years of Israel's existence as a Jewish state, for example, when collective farming transformed much of it into fertile vineyards and vegetable fields, both Labor and Likud governments have bestowed generous water subsidies on the nation's farmers. The results have been disastrous: today, agriculture accounts for just 3 percent of Israel's gross national product and uses up to half of its fresh water. Recently, Uri Sagie, chairman of Israel's national water company, told a conference of Israeli farmers that a growing and irreversible gap between production and consumption looms. "The water sources are being depleted without the deficit being restored," he warned. Jordan lavishes similar water subsidies on its farmers with similar consequences: the kingdom takes about 71 billion gallons of water a year from the Yarmouk River and channels it into the King Abdullah Canal, constructed by USAID in the 1970s to provide irrigation for the Jordan Valley; Syria takes out another 55 billion gallons. The result is near-total depletion of the lower Jordan's main source of water.

SEVERAL DAYS LATER on another outing with Bromberg, we are hiking through the Ein Gedi Nature Reserve, on a ridge 600 feet above the Dead Sea. A stream of fresh water, originating in an underground spring deep in the Judean Desert, rushes through a steep canyon dense with tamarisk, pine, birch and oleander. We ascend to the top of the canyon, where a cascade tumbles down sandstone cliffs into a cool, clear pool.

Yet not a single drop of that spring water—some 114 million gallons a year—reaches the Dead Sea. Just outside the nature reserve, the Ein Gedi kibbutz takes it, bottling some for a popular brand of mineral water and using the rest to irrigate the kibbutz grounds and botanical gardens, a sea of green amid the desert's desolation. To Bromberg and other environmentalists, kibbutz policy is rank hypocrisy. "The people of the Ein Gedi kibbutz are the first to complain about sinkholes along the shore," Bromberg says. "But they don't blame themselves for contributing to the problem."

Ein Gedi's residents deny any responsibility for the Dead Sea's plight—and lash out both at green groups such as Friends of the Earth and at the Israeli Knesset (Parliament), which recently sought to crack down on the kibbutz's water usage. "It's garbage what they're saying. If you take all water from Ein Gedi's spring, it's a small drop in the Dead Sea," Merav Ayalon, Ein Gedi's spokesperson, told me. "The problem isn't us. It's the Israeli government." Ayalon blames the Water Commission and the Agriculture Ministry for a shortsighted policy that, she says, has wrecked the local economy. "Our date palms are dying because of the sinkholes," she says. "Our farmers can't work [in some groves] because it's gotten too dangerous. People have come close to being killed. We almost had to close the kibbutz, and the government does nothing. It has no policy to save the Dead Sea."

So what is the answer? Environmental activists say that one solution is to eliminate the water subsidies altogether. "Unless water is priced at its real costs," says Ra'ed Daoud, managing director of ECO Consult, a water-use consulting firm, "there's no way you're going to reduce agriculture." But because the region's agricultural lobby is strong and the environmental movement weak, says Daoud, there has been insufficient leverage for change. Israel's water commissioner, Shimon Tal, recently spoke publicly about the need to reduce some subsidies, but he admitted that it would be a long and difficult battle. Even Prime Minister Ariel Sharon, who grows vegetables on his farm in the Negev Desert, likes the subsidies. "We desperately need to change the situation, but the agriculture lobby won't even talk about it," says Tamar Keinan, a former Israeli Water Commission official turned project manager for Friends of the Earth.

Another approach is to encourage alternate water sources. Friends of the Earth Middle East is part of a coalition of 21 environmental groups that has developed proposals to conserve household water use (about 133 billion gallons a year, as much as that used in agriculture) and to regulate the amount that can be taken out of Israel's springs. In addition, the Israeli government is promoting the building of wastewater treatment plants and desalination facilities; the first large one on the Mediterranean was completed this past August. Over the next five years, the government says, these facilities will provide as much as 106 billion gallons of fresh water annually for agricultural and domestic consumption.

Friends of the Earth is also taking its message to the farmers themselves—encouraging them to plant crops that use less water and spelling out the advantages of renewed tourism in the area. "Israeli agriculture is incredibly mismanaged," Bromberg says as we pass banana plantations along the Jordan River bank. "The farmers here could be planting olives, flowers and other crops like dates that don't require fresh water. They could be using treated sewage water and allow fresh water to flow back into the Jordan River." Friends of the Earth cites a Haifa University study that argues that current uses of the Jordan River make no sense. "The potential tourism-dollar return of a healthy river and a healthy Dead Sea outweighs the little return that agriculture offers," says Bromberg.

To see the possibilities of tourism for myself, I visit the Gesher kibbutz, which straddles the ancient trade route from the port of Akko and Jerusalem to Damascus and Baghdad. The Romans, Ottomans and British all built bridges over the Jordan at this spot; the spans remained intact until May 1948, when defending guerrillas blew them up partially to prevent 3,000 Iraqi troops from invading the newly declared state of Israel. Last year, the kibbutz put a train car on one of the bridges and restored some buildings in the area, including a 13th-century *khan*, or guesthouse, and an Ottoman-era customhouse, to lure tourists to the site.

But it remains a hard sell. The border zone, where the kibbutz is located, is one of the tensest places in the world— bristling with watchtowers, machine-gun nests and barbed wire. As we head down to the riverbank, Nirit Bagron, my tour guide from the kibbutz, halts before a military security fence covered with sensors that can detect would-be terrorist infiltrators from Jordan. Bagron, who brings tourists here by special arrangement with the Israeli Defense Forces, is quickly checked by Israeli troops and permitted to pass, as am I. As we approach the river, she points out three observation posts perched atop the rugged hills lining the Jordanian side. "They're watching us," she tells me. "We've never talked to them, but sometimes, on a very hot day, we see the Jordanian soldiers go down there to fish and even to swim."

The Jordan River, its mix of untreated sewage and saline runoff flowing below us, courses through a black basalt canyon and under the ancient Roman bridge. Bagron looks down and grimaces. "I wouldn't dive in there, not even on a hot day," she tells me. "It's very bad, bad water."

Uri Schor, the Water Commission spokesman, defends Israel's water policy and says the greens' campaign to put more water back into the Jordan River is misplaced. "These groups think about nothing but the environment, but we have other issues to consider," says Schor. In a country that has faced repeated droughts and shortages, he goes on, nature will always take a back seat to the needs of human beings. "Imagine you are stuck in a desert with one glass of water," says Schor. "You also have a potted plant. What do you do? Do you water the plant? Of course it's a bad thing to dry out the plant, but you have to live too." Schor and other Israeli officials defend the subsidies that support agriculture, arguing that farming is a vital economic and environmental resource for the country. "Agriculture made the desert bloom," says one former official in the Israeli Water Commission. "Imagine Israel without it. It would be a wasteland."

Officials at Israel's Ministry of the Environment say they doubt that the Jordan can ever be replenished. Even with new desalination and wastewater treatment plants, they say, supplies of fresh water will barely keep up with increased demand. "It would be very nice to bring the Jordan River back to where it was 100 years ago, but it's not realistic," says Miriam Haran, director general of the ministry. "We're not the only country using the river. Syria and Jordan are taking out huge amounts of water." Haran cites the report prepared for her government that says the Dead Sea can exist for another century and lose only one-third of its current volume. "We're not talking about the disappearance of the Dead Sea," she says. "The only real problem is these sinkholes, and we'll find ways to deal with them."

DURING THE PAST FOUR YEARS Tayseer al-Smadi, who was until recently director general of Jordan's Ministry of Planning and International Cooperation, was a leading advocate of the so-called Red-Dead project—a proposed 112-mile concrete conduit to carry seawater from the Red Sea to the Dead Sea. First dismissed as preposterous by international lenders and government officials alike, the notion is now taken seriously by Israel, the Palestinian Authority and Jordan. "You can always say, 'Let's go pray that the rainy season will be better next year,' but you can't rely on it," al-Smadi told me. "If we don't do something fast, the Dead Sea is going to die. [Red-Dead] is the only solution."

It's nothing if not ambitious: a six-mile-long canal would carry seawater from the Gulf of Aqaba into giant underground pipes across Jordan's desolate Wadi Araba, a desert area along the border with Israel. The water would be pumped to about 1,000 feet above sea level, then would flow downhill for about 50 miles to the Dead Sea. Some 200 billion gallons of seawater per year would be diverted to the dying sea. The same amount would be desalinated and distributed among Jordan, Israel and the Palestinian Authority. Hydroelectric stations along the route would provide 550 megawatts of power—potentially transforming the regional economy.

Red-Dead began to gather support after the historic 1994 peace agreement signed by Jordan's King Hussein and Israeli Prime Minister Yitzhak Rabin. Three years later, at the request of both countries, the World Bank conducted a preliminary study of the plan, then backed away from financing it because of the high price tag: an estimated $1 billion for the conduit, plus another $3 billion for hydroelectric power stations and desalination plants. After languishing for years, the project regained momentum at the 2002 Conference on Sustainable Development in Johannesburg, South Africa. Israel, wanting to curb attacks by Palestinian suicide bombers, had just invaded the West Bank, and as anti-Israeli protesters filled Johannesburg's streets, Jordan's minister of planning proposed the idea of a "Peace Conduit" to his Israeli counterpart. This time, the World Bank was cautiously receptive. Following discussions between then World Bank president James Wolfensohn and Jordan's King Abdullah, the bank agreed to coordinate a $15 million study once Jordan, the Palestinian Authority and Israel agreed on mutually acceptable "terms of reference." (The three did so this past May.) Bromberg says there's tremendous pressure on the World Bank to support Red-Dead because "it's the only joint project between Jordan and Israel to come out of the Palestinian intifada."

Many questions, and hurdles, remain. One of the biggest concerns is the potential effect of introducing seawater from the Gulf of Aqaba into the Dead Sea's brine. Will the sea lose its fabled buoyancy? Will dilution produce algae, which could change its color from turquoise to brown? Will mixing affect its rich mineral content? And there are thorny political issues as well. Israel, Jordan and Palestine have provisionally agreed that all three would benefit equally from the undertaking, but that remains to be seen. "This is a massive project whose impact has to be looked at very closely," says ECO Consult's Ra'ed Daoud. "We're not just building a chocolate factory here." He believes that it will be at least five years before any digging begins. Bromberg, citing cost and potential destruction to the environment, is more pessimistic. "We're very skeptical that this canal will ever be built," he says.

But what is the alternative? According to a recent study by the Technion-Israel Institute of Technology in Haifa, the Jordan could run dry within the next few years, as more dams are built on its tributaries and more raw sewage is recycled through new wastewater treatment plants in Israel, rather than dumped into the Jordan River. "We've gone from a point, five years ago, where nobody except environmentalists cared about the fate of the Dead Sea, to a point where all three governments say they value the Dead Sea and recognize that they have to rehabilitate it," Bromberg said. "But when it comes to stopping the diversion of all the water of the Jordan River, there has been zero progress."

Time is running short. Late one afternoon, Bromberg and I drive up the mountainous coast to view what could turn out to be a window into the future: an abandoned luxury hotel, built by Jordan decades ago on what used to be the Dead Sea's northern shore. After walking up a weed-choked path, we step into a marble-floored lobby, now a crumbling ruin. The wide veranda, decorated by faded murals depicting maps of the Jordan River Valley, looks out over a cracked, deserted swimming pool. "This place was once the fanciest hotel on the Dead Sea—and then it was left in the dust," Bromberg says. A mile away, across an arid moonscape, I can barely make out the glittering blue edge of the Dead Sea, receding a bit more every day.

JOSHUA HAMMER, former Newsweek *bureau chief in Jerusalem, is now the magazine's bureau chief in Cape Town, South Africa.*

THE WORLD OF HIS CHOICE

Lee Lawrence

Wilfred Thesiger often said he regretted not having lived 50 years earlier, before planes and cars shrank the globe. As it was, he came into the world in June 1910 in Addis Ababa, where his father was serving as head of the British legation in Abyssinia, today's Ethiopia. To his young eyes, it was a world of big-game hunting, horseback riding and impressive columns of Shoan warriors marching into battle. At the end of World War I, at age nine, he was sent back to Britain for 12 years of schooling, including Eton and Oxford. In those years, Ford began mass-producing automobiles, Lindbergh flew across the Atlantic, and the world of Thesiger's childhood began rapidly to disappear.

In 1930 he took a break from Oxford to return to Addis Ababa for the coronation of an old family friend as the emperor Haile Selassie. There, he snapped photographs of the ceremonies with his father's Kodak Brownie box camera. He chose not to make photographs of the air show that was one feature of the celebrations, though he mentioned it in his diary. With that choice, he gave the first hint of the directions in which he would point both his career and his camera.

Following the coronation, Thesiger undertook his first exploration up the largely uncharted Awash River, accompanied by a retinue of local guides and porters. The trip afforded him a weeklong taste of life in the wild sufficiently thrilling to prompt his return three years later, after he had completed his Oxford exams. This time, he discovered the elusive source of the Awash, high in the country inhabited by the Danakil, who violently resented the intrusions of outsiders. Thesiger apparently impressed them with sincerity and his apolitical purposes, and he passed unharmed.

From 1935 to 1939, Thesiger worked for Britain's Sudan Political Service, where he chose the most remote assignments in order to spend as much time as possible away from a desk and atop a camel. He had recently bought a Leica II, which he carried throughout his subsequent service in Sudan, Abyssinia, Syria and Palestine.

By 1945, Thesiger had read—jealously—the accounts of Bertram Thomas and Harry St. John Philby, the first westerners to cross the Rub' al-Khali, or Empty Quarter, of Saudi Arabia in 1930 and 1932, respectively. He had become fascinated with the Empty Quarter and its inhabitants, and he learned Arabic. When the Locust Research Organisation asked him to discover whether there were breeding grounds for locusts in the Arabian desert, he jumped at the offer. In 1946–1947 and again in 1948, Thesiger assembled groups of Bedouin and, taking new, more difficult routes, became the first European to cross the Empty Quarter twice.

"There was of course," he wrote in his autobiography, *The Life of My Choice*, "the lure of the unknown; there was the constant test of resolution and endurance. Yet those travels in the Empty Quarter would have been for me a pointless penance but for the comradeship of my Bedu companions." For months at a time over a five-year period, Thesiger traveled with Bedouin parties throughout the Arabian Peninsula's deserts, pushing himself to exhaustion, sipping brackish well water and risking attack by rival tribes. When he left, he knew "I should never meet their like again. I had witnessed their loyalty to each other.... I knew their pride in themselves and their tribe; their regard for the dignity of others; their hospitality when they went short to feed chance-met strangers; their generosity...their absolute honesty; their courage, patience and endurance and their thoughtfulness."

IN ABYSSINIA, THESIGER IMPRESSED THE DANAKIL PEOPLE AND WAS ABLE TO PASS THROUGH THEIR TERRITORY UNHARMED. IN THE SUDAN POLITICAL SERVICE, HE CHOSE TWO YEARS OF ASSIGNMENTS IN REMOTE DARFUR.

TO PARAPHRASE THE PHILOSOPHER VILEM FLUSSER, PHOTOGRAPHS DO NOT PRESENT THE WORLD AS IT IS; RATHER, THEY ENCHANT US INTO BELIEVING THAT THE WORLD IS A CERTAIN WAY.

In 1951, Thesiger was looking to spend a couple of weeks shooting duck, and he headed into the marshes of southern Iraq, where he was smitten with the life of the Ma'dan people. He spent the next seven winters among them. Unlike his time in Arabia, his life in the marshes had a more settled quality, and his basic knowledge of asepsis and his supply of antibiotics earned him a position in society performing circumcisions. In the summer, Thesiger left the marshes to trek in the mountains of northern Iraq, Afghanistan, Pakistan and India, and one year he joined Iran's Bakhtiari tribe for its grueling annual migration across the Zagros Mountains.

Between these ventures, Thesiger would return to England, where he visited family, and in the mid-1950's he began taking yearly summer trips around Europe, North Africa or the Middle East with his mother, Kathleen Mary, traveling by train and automobile and staying in "reasonably priced hotels," as he told biographer Michael Asher. Significantly, these trips are almost entirely absent from his photographic archive.

It was his photographs that, in the mid-1950's, prompted a literary agent to urge Thesiger to write a book, thus echoing Kathleen Mary's own persistent pleas. Finally acquiescing, Thesiger holed up in an apartment in Denmark with copies of his Royal Geographical Society reports, letters home, diaries and boxfuls of photographs. According to Alexander Maitland, now writing Thesiger's authorized biography, the diaries were "massive, but they are very workaday documents." He and others familiar with Thesiger's archives agree that the photographs were essential to his reconstructions of his Empty Quarter and later journeys. "I think the photographs certainly kept the [journey] alive for him," says Maitland. "They were the source for one of the things he did best, which was carefully honed descriptive writing." The result of Thesiger's effort was the first of 10 books Thesiger would author about his travels, the highly acclaimed *Arabian Sands*, published in 1959.

From 1968 onward, Thesiger made his base in Kenya among the Samburu and Turkana tribes. He would have liked to die there, he said, but in the mid-1990's deteriorating health forced him to return to England. A man who had sought to share the rigors of male nomad societies all his adult life, Thesiger lived out his last years in a nursing home in Surrey populated largely by women. He died on August 24, 2003.

He left a photographic legacy of thousands of prints, 75 albums and 38,000 negatives. These photos, Thesiger wrote, were his "most cherished possessions," thanks to which he could "live once more in a vanished world."

What world exactly was this? What are we, today, to make of it?

In one photograph, a man, back to the camera, looks out over rocky, rugged terrain that fills the frame of the photograph. In another shot, a Bedouin sits on his haunches, rifle planted by his side. A third image shows a chain of camels cresting a dune. In a fourth, a boy turns back toward the camera as he poles a canoe through tall reeds.

These images are the kind we have come to associate with Wilfred Thesiger, and that define the Thesiger Collection.

The great majority of photos in the archive date from his 1946 journey across the Empty Quarter of Arabia onward, for it was during that first desert crossing that he "began to consider the composition of each photograph, conscious to achieve the best result. From then on," he stated in *Visions of a Nomad*, "photography became a major interest." In his early Royal Geographical Society reports as well as in his later books, he used his photographs as simple illustrations; however, since the late 1980's, many of the photographs have themselves become the subjects of books and exhibitions. By the time he died, he was as well known in Europe for his arresting portraits of Bedouin and Marsh Arabs as he was for his writings. In both, the appeal lay at least in part in the romantic representations of times and places European audiences had never experienced.

Thesiger's albums, prints and negatives were bequeathed to the Pitt Rivers Museum in Oxford, England. Thanks to a grant from the late President Shaykh Zayed bin Al Nahyan of the United Arab Emirates, a team of three archivists has spent a year identifying, inventorying, cataloging and digitizing Thesiger's 38,000 negatives. To encourage further research, the museum is posting the catalog on its Web site along with a selection of heretofore unpublished images.

Now, for the first time, Thesiger's photographic work can be studied as a whole. Already senior curator Elizabeth Edwards has discovered a wealth of ethnographic material: series documenting the Samburus' body art in Kenya and circumcision rituals in Iraq and Kenya, as well as many shots that contain information about the way people dressed, lived and made everyday objects. "People have made esthetic judgments about his photographs," Edwards says, and now she hopes they will mine them for information, too.

More generally, the archive can reveal how Thesiger felt about the lands and peoples he photographed, and also much about the differences between viewing a single image in isolation and seeing it within the context and discernible patterns of an archive. It is important to remember that photographs are like statistics: By their very selectivity, they have a way of appearing factual even while spinning fictions. Or, to paraphrase the German–Czech–Brazilian philosopher Vilém Flusser, photographs do not present the world as it is; rather, they "enchant" us into believing that the world is a certain way. Archives, backed as they are by institutions with the power to compile, preserve and index, add a mantle of authority that cuts two ways: Sometimes it magnifies the spell of individual photographs; at other times, the succession of images in an archive helps to dispel the "enchantment" of the single photo with evidence of how it was made.

For example, the Thesiger Collection presents a chronological sequence of negatives spanning decades and continents. Since there are no significant gaps in this sequence, it is easy to assume that the archive chronicles all of Thesiger's travels. From this, it is equally easy to conjure a specific image of a man forever seeking adventure and traveling under difficult conditions in remote places. It is an image that leaves little room for the fact that Thesiger took annual trips with his mother for more than 15 years. Except for a handful of mostly architectural photographs, the Thesiger Collection hardly acknowledges that these trips took place.

Indeed, the archive presents only a world in which camels plod through sand, dhows sail into harbors, and horses clamber up mountain trails. Lone figures look out over untouched landscapes, children live in the villages of their grandfathers, and the faces of individual men and children fill frame after frame, their dark eyes often peering straight at us. On the whole, there is a static quality to the images, as few show a chance or fortuitous event unlikely ever to be repeated; when action shots occur, they portray activities we imagine people have been performing for generations: watering animals at desert wells, circumcision rituals, braiding thatch housing in the marshes, making coffee on a campfire, herding livestock across mountains or converging on a pilgrimage site.

It is a world, moreover, devoid of color. This was not always a deliberate choice on Thesiger's part, since color film was not widely available during most of his active years. Nevertheless, it is telling that, given a choice, even later on Thesiger favored black and white, tracing this to his preference for prints and drawings over paintings. "Colour," he wrote, "aims to reproduce exactly what is seen by the photographer.... With black and white, on the other hand, each subject offers its own variety of possibilities according to the use made by the photographer of light and shade."

Mary Peck, one of America's leading photographers of regional landscapes, has, like Thesiger, traveled in deserts and mountains. She helps clarify what Thesiger may have been getting at in his praise of black and white. "It is a remove from the supposed reality that color gives us," Peck explains, adding that she finds black and white film "a way to try to translate an experience" that gives her "a better chance of guiding a viewer's reaction."

This holds true in the Thesiger Collection. Were we to stand, say, in the desert, we would experience the way in which the reds, tans and oranges of dunes and the blues of the sky intensify and wane over the course of a day. Dunes would be differentiated not only by light and shadow, but also by their hues, and desert shrubs would sometimes sprout green leaves and blooms, indicating a recent rain. While color film chronicles such variations vividly, black and white film does not. The sun's journey across the sky registers only as a deepening of shadows, with the sand appearing somewhat whiter at noon and grayer at dusk. The sky remains a constant, whitish expanse occasionally interrupted by clouds. Shrubs and trees appear uniformly gray, whether they are in leaf or desiccated. The use of black and white suspends time, and it strengthens the impression of changelessness.

Moreover, just as black and white photography erases the differences in hue among the dunes, it also creates artificial similarities. The lines in some of Thesiger's mountain views in Kurdistan and Pakistan, for example, echo those in his Arabian landscapes, both favoring converging lines and balanced compositions typical of European landscape painting. In black and white, these compositional similarities become more prominent, causing the scenes to resemble one another more closely than they might in color: The various reddish shades of the dunes would set desert landscapes well apart from mountains covered in white snow. In Thesiger's images, the chromatic similarities reinforce the compositional ones to create the impression that the lands and peoples all belong to the same timeless, changeless pre-modern era—a most romantic notion.

There is yet another way in which the archive magnifies the "enchantment" of individual images. Based on a single black-and-white shot of, say, women at a well or men dressed in traditional garb, we might not assume that their world was completely untouched by modernity. But a steady succession of such scenes is more likely to enchant us into believing that the camera has recorded all there is to see. If there had been signs of modernity, we might think, they would have appeared, and thus we conclude that they were not there.

Yet this was not the case. "There were certainly cars in Salalah, where Wilfred began his journey to cross the Empty Quarter," says Asher, himself the winner of two awards for desert exploration. "Wilfred even mentions— not in his books, but in one of his reports back to the Royal Geographical Society—that they followed a motor track for some way when they got to the other side of the Empty Quarter. And you may remember also that when he and his party were arrested [during his second crossing of the Empty Quarter], they were actually taken off in a truck." It is right there, as he points out, in *Arabian Sands*.

"THOSE TRAVELS IN THE EMPTY QUARTER WOULD HAVE BEEN FOR ME A POINTLESS PENANCE BUT FOR THE COMRADESHIP OF MY BEDU COMPANIONS," THESIGER WROTE. "I KNEW THEIR PRIDE IN THEMSELVES AND THEIR TRIBE; THIER REGARD FOR THE DIGNITY OF OTHERS; THEIR HOSPITALITY WHEN THEY WENT SHORT TO FEED CHANCE-MET STRANGERS."

So are references to airplanes: In 1946, following Thesiger's first taste of the Empty Quarter, a member of his Bedouin traveling party, Musallim bin Tafl, became the first in his tribe to board a plane. A year and a half later, a favorite companion, Salim bin Ghabaisha, flew from Salalah to Hadhramawt at Thesiger's insistence in order to join the expedition in time for the second crossing. None of this could have occurred in the world the Thesiger Collection conjures up. "I would say that Wilfred was definitely setting out in his photographs to create a world that no longer existed," Asher says.

This world is one largely populated by men whose faces fill frame after frame in the archive, bearing witness to Thesiger's assertion that it was people, not places, "who offered me the most interesting subjects." In all but a handful of Thesiger's portraits, the subjects are squinting in harsh sunlight, their half-closed eyelids protecting them not only from the sun's rays but perhaps also from our gaze.

At one level, many of Thesiger's portraits follow a convention of his day whereby photographers called forth squints and their radiating wrinkles when they wanted to mark their subject as "an outdoor type" or a member of a pre-modern "primitive society." The squint also speaks of authenticity: It is proof that Thesiger was not photographing people in a studio but in their environment. At another level, the squint prevents us from looking deeply into the subjects' eyes, perhaps marking them as people whose soul a European viewer could not easily know.

At the same time, the light falls at enough of an angle to bring out the texture of skin and cloth. We see the stringy fringe of a headscarf, sand grains in matted hair, the wrinkles in newly washed cotton, the lines on individual faces. Such detail can heighten our empathy with people strong enough to withstand the harsh conditions that etch such deep wrinkles into their skin. The children portrayed may not yet bear the traces of hardship, but their juxtaposition with their elders implies that they will in time.

This is not to suggest that the portraits create a total romantic fiction. Whether in the desert, marshes or mountains, life was and often remains hard by any standard, and the faces of the men and the rugged landscapes were indeed as they appear in the photographs. Yet we know from Thesiger's and others' accounts that the Bedouin, for example, spent part of the year living in cities and with their families, as did many of the men in the Iraqi marshes. To take the archive at face value is to believe that the men portrayed lacked family lives, made their way in small bands across inhospitable lands, gathered in guest houses or stood alone contemplating trackless terrain.

Given Thesiger's admiration for the "comradeship" he found in his travels, it is perhaps surprising that images of lone men in large landscapes are such a recurring, even archetypal pattern. Such an image first appears in a 1946 shot taken in Oman; it then recurs in various settings, even in a shot of the Parthenon that is one of the rare photographs Thesiger took during his travels in Europe with his mother. Such images say more about a 19th-century convention popular with British picturesque photographers (and no less popular today) than it does about nomadic life.

In Britain such scenes were often staged using a man in city clothes who stood in the foreground to illustrate the vastness of nature and to induce viewers to appreciate and contemplate it with him. But there is a salient difference between those photographs and Thesiger's. The men in the British prototypes contemplate the landscape as outsiders in the country for a day of reflection; the men in Thesiger's photographs live in the landscape as insiders, and their contemplation can induce viewers to see it from their point of view.

In many of these photos, Thesiger showed a landscape that rose high behind the subjects. Indeed, with the exception of photographs taken in the Iraqi marshes, Thesiger's horizon line tended to remain high throughout much of the archive, so that we come away with a sense that there is no escaping the land. Whether sand, water or rock-strewn mountains, these untamed landscapes define men's lives and require that men adapt to them and not the other way around. As Edwards notes, "there is not that caressing of the landscape you see in, say, pictures of the American West. Thesiger is interested in how people survive in these landscapes."

If the archive strengthens the spell cast by individual photographs of a pre-modern world inhabited by lone men, its succession of images also serves elsewhere to diminish the power of individual images. Viewed singly, a photograph of Bin Ghabaisha might lead us to believe that Thesiger had either happened upon him or asked him to pose for a photograph. But the portrait of Bin Ghabaisha turns out to be one of five successive photographs of the young man, all taken in front of the same rock, and two more portraits follow of two other boys at the same location. Such series of portraits of various people at the same site recur frequently. One series totals 11 portraits of different Kurdish villagers, all taken at the same spot.

We thus can imagine a far different scenario: Thesiger snapping the photograph of one villager, only to find others queuing up for a turn to pose. This raises the question of who is in control of the process of representation. In the one case, the photographer (or editor, publisher or curator) selects which portraits will be published to represent a given people. Within the much broader and less selective archive, on the other hand, the individuals who stepped before the camera uninvited have as much say in who represents their group as the photographer had.

THE MEN IN THESIGER'S PHOTOGRAPHS LIVE IN THE LANDSCAPE AS INSIDERS, AND THEIR CONTEMPLATION CAN INDUCE VIEWERS TO SEE IT FROM THEIR POINT OF VIEW.

As we move chronologically through the Thesiger Collection, it becomes increasingly clear that, just as the photographer is capturing his vision of his subjects, the people portrayed are as consciously presenting themselves to his camera. Some of Thesiger's Bedouin companions relax and smile; some people he encounters in villages appear to strike poses. A Herki boy in Khazna, for example, wears the same stern expression in each of three shots Thesiger took of him, even though these were not taken in succession. Similarly, a Barzan man poses, one arm crooked, the other holding a pipe, a plume of smoke blowing sideways out of his mouth. Since this is one in a series of same-site portraits, it seems likely that the man stepped up to the camera intent on projecting a particular—presumably favorable—image of himself.

The archive also reveals another type of portrait series. For example, a photograph of Sultan, a Bedouin of the Bayt Kathir tribe, shows him from the front, looking into the camera; in the next frame, he appears in the same pose at the same spot, but this time in profile. Since Thesiger worked mostly with a standard 55mm lens, he had to photograph his subjects from about two meters' (6'6") distance to get a head-and-shoulders portrait. Looking at the archive, we imagine Thesiger, who himself stood almost two meters tall, circling his subject, leaning in periodically to take a reading with his hand-held light meter, then stepping back to take the shot while the subject collaborated by not moving, giving the frontal and profile sequences we see.

Such series are typical, particularly during Thesiger's first two years in Arabia, and even though they occur less frequently after that, the format persists through his time in Kenya in the 1980's. Those frames taken at a level or slightly downward angle recall the way colonial ethnographers documented "Orientals" in the 1800's and the way colonial police in India recorded criminals—the antecedents, in other words, of our frontal-and-profile mug shots. But it is as though Thesiger himself caught this resemblance and fought against it. In the albums, he often chose only one of the pair or, when displaying both, reversed their order so that we first see the profile followed by the frontal view.

More telling yet is that as time goes on, he varies the angle. Three-quarter views often replace frontal shots, and the angle of the camera changes from slightly downward to upward, so that he captures the subject soaring dramatically against the sky. If the closeness and the detail of the portrait invite our empathy, this dramatic upward angle creates monumentality and elicits our admiration, making us literally "look up to" these men.

As with his figures in landscapes, Thesiger's portraits are mainly rooted in colonial and picturesque precedents. Like colonial representations of non-European societies, the Thesiger Collection conjures a world suspended in time in which men live in harsh, primitive conditions. There is more than a touch of the "noble savage" in the way Thesiger eulogizes these men—and some women—who, he contended, "had no concept of any world other than their own" yet were on the verge of seeing modernity alter their way of life.

If such melancholy reeks of latent colonialism, the archive also works against such attitudes. Unlike colonial archives, the Thesiger Collection teems not with "types," but with named individuals, and when the photographer himself appears in this world, it is not as conqueror or outside observer, but as a fellow traveler, respectful of his companions and striving to be accepted as their equal. In portraits taken in Saudi Arabia, Thesiger poses the way his Bedouin subjects pose, and he dresses the way they do. The shots are taken from a similar distance, and at the same variety of angles, so were it not for his European features, there would be no distinguishing Thesiger from the Bedouin. Since Leica only introduced a self-timer in 1950, one of Thesiger's companions must have operated the camera for those shots, even if Thesiger set them up. According to a caption in *Desert, Marsh & Mountain*, in at least one case this job of photographer fell to Bin Kabina, a favorite traveling companion.

The act of handing his camera to the young man so that he, Thesiger, could place himself in Bin Kabina's world contains an obvious irony, and also poignantly highlights the yearning that infuses the Thesiger Collection. While his other portraits and landscapes work together to enchant us into believing in worlds so harsh that they sculpted men into heroes, the self-portraits assert that Thesiger not only witnessed these worlds but lived in them, and can thus claim for himself some of the heroism and admiration he bestowed on the Bedouin, the Kurds and other nomadic peoples.

At the same time, the Thesiger Collection proves by its very existence that modern technology had already encroached upon those worlds. We are thus left between the enchantment and the abolition of the enchantment, in a limbo much like the one that Thesiger himself inhabited as, try as he might to prove otherwise, his ideal pre-modern world slipped into the past.

Lee Lawrence (leeadair8@juno.com) *is a free-lance writer in Washington, D.C. who specializes in the cultures of Asia. She recently completed a master's thesis on photography in British colonial India.*

From *Saudi Aramco World*, vol. 56, no. 3, May/June 2005, pp. 24–37. Copyright © 2005 by Saudi Aramco World/PADIA. Reprinted by permission.

Freedom and Justice in the Modern Middle East

Bernard Lewis

CHANGING PERCEPTIONS

FOR MUSLIMS as for others, history is important, but they approach it with a special concern and awareness. The career of the Prophet Muhammad, the creation and expansion of the Islamic community and state, and the formulation and elaboration of the holy law of Islam are events in history, known from historical memory or record and narrated and debated by historians since early times. In the Islamic Middle East, one may still find passionate arguments, even bitter feuds, about events that occurred centuries or sometimes millennia ago—about what happened, its significance, and its current relevance. This historical awareness has acquired new dimensions in the modern period, as Muslims—particularly those in the Middle East—have suffered new experiences that have transformed their vision of themselves and the world and reshaped the language in which they discuss it.

In 1798, the French Revolution arrived in Egypt in the form of a small expeditionary force commanded by a young general called Napoleon Bonaparte. The force invaded, conquered, and ruled Egypt without difficulty for several years. General Bonaparte proudly announced that he had come "in the name of the French Republic, founded on the principles of liberty and equality." This was, of course, published in French and also in Arabic translation. Bonaparte brought his Arabic translators with him, a precaution that some later visitors to the region seem to have overlooked.

The reference to equality was no problem: Egyptians, like other Muslims, understood it very well. Equality among believers was a basic principle of Islam from its foundation in the seventh century, in marked contrast to both the caste system of India to the east and the privileged aristocracies of the Christian world to the west. Islam really did insist on equality and achieved a high measure of success in enforcing it. Obviously, the facts of life created inequalities—primarily social and economic, sometimes also ethnic and racial—but these were in defiance of Islamic principles and never reached the levels of the Western world. Three exceptions to the Islamic rule of equality were enshrined in the holy law: the inferiority of slaves, women, and unbelievers. But these exceptions were not so remarkable; for a long time in the United States, in practice if not in principle, only white male Protestants were "born free and equal." The record would seem to indicate that as late as the nineteenth or even the early twentieth century, a poor man of humble origins had a better chance of rising to the top in the Muslim Middle East than anywhere in Christendom, including post-revolutionary France and the United States.

Equality, then, was a well-understood principle, but what about the other word Bonaparte mentioned—"liberty," or freedom? This term caused some puzzlement among the Egyptians. In Arabic usage at that time and for some time after, the word "freedom"—*hurriyya*—was in no sense a political term. It was a legal term. One was free if one was not a slave. To be liberated, or freed, meant to be manumitted, and in the Islamic world, unlike in the Western world, "slavery" and "freedom" were not until recently used as metaphors for bad and good government.

The puzzlement continued until a very remarkable Egyptian scholar found the answer. Sheikh Rifa'a Rafi' al-Tahtawi was a professor at the still unmodernized al-Azhar University of the early nineteenth century. The ruler of Egypt had decided it was time to try and catch up with the West, and in 1826 he sent a first mission of 44 Egyptian students to Paris. Sheikh Tahtawi accompanied them and stayed in Paris until 1831. He was what might be called a chaplain, there to look after the students' spiritual welfare and to see that they did not go astray—no mean task in Paris at that time.

During his stay, he seems to have learned more than any of his wards, and he wrote a truly fascinating book giving his impressions of post-revolutionary France. The book was published in Cairo in Arabic in 1834 and in a Turkish translation in 1839. It remained for decades the only description of a modern European country available to the Middle Eastern Muslim reader. Sheikh Tahtawi devotes a chapter to French government, and in it he mentions how the French kept talking about freedom. He obviously at first shared the general perplexity about what the status of not being a slave had to do with politics. And then he understood and explained. When the French talk about freedom, he says, what they mean is what we Muslims call justice. And that was exactly right. Just as the French, and more generally Westerners, thought of good government and bad government as freedom and slavery, so Muslims conceived of them as justice and injustice. These contrasting perceptions help shed light on the political debate that began in the Muslim world with the 1798 French expedition and that has been going on ever since, in a remarkable variety of forms.

JUSTICE FOR ALL

AS SHEIKH TAHTAWI rightly said, the traditional Islamic ideal of good government is expressed in the term "justice." This is represented by several different words in Arabic and other Islamic languages. The most usual, *adl*, means "justice according to the law" (with "law" defined as God's law, the sharia, as revealed to the Prophet and to the Muslim community). But what is the converse of justice? What is a regime that does not meet the standards of justice? If a ruler is to qualify as just, as defined in the traditional Islamic system of rules and ideas, he must meet two requirements: he must have acquired power rightfully, and he must exercise it rightfully. In other words, he must be neither a usurper nor a tyrant. It is of course possible to be either one without the other, although the normal experience was to be both at the same time.

The Islamic notion of justice is well documented and goes back to the time of the Prophet. The life of the Prophet Muhammad, as related in his biography and reflected in revelation and tradition, falls into two main phases. In the first phase he is still living in his native town of Mecca and opposing its regime. He is preaching a new religion, a new doctrine that challenges the pagan oligarchy that rules Mecca. The verses in the Koran, and also relevant passages in the prophetic traditions and biography, dating from the Meccan period, carry a message of opposition— of rebellion, one might even say of revolution, against the existing order.

Then comes the famous migration, the *hijra* from Mecca to Medina, where Muhammad becomes a wielder, not a victim, of authority. Muhammad, during his lifetime, becomes a head of state and does what heads of state do. He promulgates and enforces laws, he raises taxes, he makes war, he makes peace; in a word, he governs. The political tradition, the political maxims, and the political guidance of this period do not focus on how to resist or oppose the government, as in the Meccan period, but on how to conduct government. So from the very beginning of Muslim scripture, jurisprudence, and political culture, there have been two distinct traditions: one, dating from the Meccan period, might be called activist; the other, dating from the Medina period, quietist.

The Koran, for example, makes it clear that there is a duty of obedience: "Obey God, obey the Prophet, obey those who hold authority over you." And this is elaborated in a number of sayings attributed to Muhammad. But there are also sayings that put strict limits on the duty of obedience. Two dicta attributed to the Prophet and universally accepted as authentic are indicative. One says, "there is no obedience in sin"; in other words, if the ruler orders something contrary to the divine law, not only is there no duty of obedience, but there is a duty of disobedience. This is more than the right of revolution that appears in Western political thought. It is a duty of revolution, or at least of disobedience and opposition to authority. The other pronouncement, "do not obey a creature against his creator," again clearly limits the authority of the ruler, whatever form of ruler that may be.

The Muslim world has long debated the merits of political obedience and rebellion.

These two traditions, the one quietist and the other activist, continue right through the recorded history of Islamic states and Islamic political thought and practice. Muslims have been interested from the very beginning in the problems of politics and government: the acquisition and exercise of power, succession, legitimacy, and—especially relevant here—the limits of authority.

All this is well recorded in a rich and varied literature on politics. There is the theological literature; the legal literature, which could be called the constitutional law of Islam; the practical literature—handbooks written by civil servants for civil servants on how to conduct the day-to-day business of government; and, of course, there is the philosophical literature, which draws heavily on the ancient Greeks, whose work was elaborated in translations and adaptations, creating distinctly Islamic versions of Plato's *Republic* and Aristotle's *Politics*.

In the course of time, the quietist, or authoritarian, trend grew stronger, and it became more difficult to maintain those limitations on the autocracy of the ruler that had been prescribed by holy scripture and holy law. And so the literature places increasing stress on the need for order. A word used very frequently in the discussions is *fitna*, an Arabic term that can be translated as "sedition," "disorder," "disturbance," and even "anarchy" in certain contexts. The point is made again and again, with obvious anguish and urgency: tyranny is better than anarchy. Some writers even go so far as to say that an hour—or even a moment—of anarchy is worse than a hundred years of tyranny. That is one point of view—but not the only one. In some times and places within the Muslim world, it has been dominant; in other times and places, it has been emphatically rejected.

THEORY VERSUS HISTORY

THE ISLAMIC TRADITION insists very strongly on two points concerning the conduct of government by the ruler. One is the need for consultation. This is explicitly recommended in the Koran. It is also mentioned very frequently in the traditions of the Prophet. The converse is despotism; in Arabic *istibdad*, "despotism" is a technical term with very negative connotations. It is regarded as something evil and sinful, and to accuse a ruler of *istibdad* is practically a call to depose him.

With whom should the ruler consult? In practice, with certain established interests in society. In the earliest times, consulting with the tribal chiefs was important, and it remains so in some places—for example, in Saudi Arabia and in parts of Iraq (but less so in urbanized countries such as Egypt or Syria). Rulers also consulted with the countryside's rural gentry, a very powerful group, and with various groups in the city: the bazaar merchants, the scribes (the nonreligious literate classes, mainly civil servants), the religious hierarchy, and the military establishment, including long-established regimental groups such as the janissaries of the Ottoman Empire. The importance of these groups was, first of all, that they did have real power. They could and sometimes did make trouble for the ruler, even deposing him. Also, the groups' leaders—tribal chiefs, country notables, religious leaders, heads of guilds, or commanders of the armed forces—were not nominated by the ruler, but came from within the groups.

Consultation is a central part of the traditional Islamic order, but it is not the only element that can check the ruler's authority. The traditional system of Islamic government is both consensual and contractual. The manuals of holy law generally assert that the new caliph—the head of the Islamic community and state—is to be "chosen." The Arabic term used is sometimes translated as "elected," but it does not connote a general or even sectional election. Rather, it refers to a small group of suitable, competent people choosing the ruler's successor. In principle, hereditary succession is rejected by the juristic tradition. Yet in practice, succession was always hereditary, except when broken by insurrection or civil war; it was—and in most places still is—common for a ruler, royal or otherwise, to designate his successor.

But the element of consent is still important. In theory, at times even in practice, the ruler's power—both gaining it and maintaining it—depends on the consent of the ruled. The basis of the ruler's authority is described in the classical texts by the Arabic word *bay'a*, a term usually translated as "homage," as in the subjects paying homage to their new ruler. But a more accurate translation of *bay'a*—which comes from a verb meaning "to buy and to sell"—would be "deal," in other words, a contract between the ruler and the ruled in which both have obligations.

Some critics may point out that regardless of theory, in reality a pattern of arbitrary, tyrannical, despotic government marks the entire Middle East and other parts of the Islamic world. Some go further, saying, "That is how Muslims are, that is how Muslims have always been, and there is nothing the West can do about it." That is a misreading of history. One has to look back a little way to see how Middle Eastern government arrived at its current state.

The change took place in two phases. Phase one began with Bonaparte's incursion and continued through the nineteenth and twentieth centuries when Middle Eastern rulers, painfully aware of the need to catch up with the modern world, tried to modernize their societies, beginning with their governments. These transformations were mostly carried out not by imperialist rulers, who tended to be cautiously conservative, but by local rulers—the sultans of Turkey, the pashas and khedives of Egypt, the shahs of Persia—with the best of intentions but with disastrous results.

Modernizing meant introducing Western systems of communication, warfare, and rule, inevitably including the tools of domination and repression. The authority of the state vastly increased with the adoption of instruments of control, surveillance, and enforcement far beyond the capabilities of earlier leaders, so that by the end of the twentieth century any tin-pot ruler of a petty state or even of a quasi state had vastly greater powers than were ever enjoyed by the mighty caliphs and sultans of the past.

But perhaps an even worse result of modernization was the abrogation of the intermediate powers in society—the landed gentry, the city merchants, the tribal chiefs, and others—which in the traditional order had effectively limited the authority of the state. These intermediate powers were gradually weakened and mostly eliminated, so that on the one hand the state was getting stronger and more pervasive, and on the other hand the limitations and controls were being whittled away.

This process is described and characterized by one of the best nineteenth-century writers on the Middle East, the British naval officer Adolphus Slade, who was attached as an adviser to the Turkish fleet and spent much of his professional life there. He vividly portrays this process of change. He discusses what he calls the old nobility, primarily the landed gentry and the city bourgeoisie, and the new nobility, those who are part of the state and derive their authority from the ruler, not from their own people. "The old nobility lived on their estates," he concludes. "The state is the estate of the new nobility." This is a profound truth and, in the light of subsequent and current developments, a remarkably prescient formulation.

Modernization in the Middle East inevitably brought Western tools of repression.

The second stage of political upheaval in the Middle East can be dated with precision. In 1940, the government of France surrendered to Nazi Germany. A new collaborationist government was formed and established in a watering place called Vichy, and General Charles de Gaulle moved to London and set up a Free French committee. The French empire was beyond the reach of the Germans at that point, and the governors of the French colonies and dependencies were free to decide: they could stay with Vichy or rally to de Gaulle. Vichy was the choice of most of them, and in particular the rulers of the French-mandated territory of Syria-Lebanon, in the heart of the Arab East. This meant that Syria-Lebanon was wide open to the Nazis, who moved in and made it the main base of their propaganda and activity in the Arab world.

It was at that time that the ideological foundations of what later became the Baath Party were laid, with the adaptation of Nazi ideas and methods to the Middle Eastern situation. The nascent party's ideology emphasized pan-Arabism, nationalism, and a form of socialism. The party was not officially founded until April 1947, but memoirs of the time and other sources show that the Nazi interlude is where it began. From Syria, the

Germans and the proto-Baathists also set up a pro-Nazi regime in Iraq, led by the famous, and notorious, Rashid Ali al-Gailani.

The Rashid Ali regime in Iraq was overthrown by the British after a brief military campaign in May–June 1941. Rashid Ali went to Berlin, where he spent the rest of the war as Hitler's guest with his friend the mufti of Jerusalem, Haj Amin al-Husseini. British and Free French forces then moved into Syria, transferring it to Gaullist control. In the years that followed the end of World War II, the British and the French departed, and after a brief interval the Soviets moved in.

The leaders of the Baath Party easily switched from the Nazi model to the communist model, needing only minor adjustments. This was a party not in the Western sense of an organization built to win elections and votes. It was a party in the Nazi and Communist sense, part of the government apparatus particularly concerned with indoctrination, surveillance, and repression. The Baath Party in Syria and the separate Baath Party in Iraq continued to function along these lines.

Since 1940 and again after the arrival of the Soviets, the Middle East has basically imported European models of rule: fascist, Nazi, and communist. But to speak of dictatorship as being the immemorial way of doing things in that part of the world is simply untrue. It shows ignorance of the Arab past, contempt for the Arab present, and unconcern for the Arab future. The type of regime that was maintained by Saddam Hussein—and that continues to be maintained by some other rulers in the Muslim world—is modern, indeed recent, and very alien to the foundations of Islamic civilization. There are older rules and traditions on which the peoples of the Middle East can build.

CHUTES AND LADDERS

THERE ARE, of course, several obvious hindrances to the development of democratic institutions in the Middle East. The first and most obvious is the pattern of autocratic and despotic rule currently embedded there. Such rule is alien, with no roots in either the classical Arab or the Islamic past, but it is by now a couple of centuries old and is well entrenched, constituting a serious obstacle.

Another, more traditional hurdle is the absence in classical Islamic political thought and practice of the notion of citizenship, in the sense of being a free and participating member of a civic entity. This notion, with roots going back to the Greek *polites*, a member of the *polis*, has been central in Western civilization from antiquity to the present day. It, and the idea of the people participating not just in the choice of a ruler but in the conduct of government, is not part of traditional Islam. In the great days of the caliphate, there were mighty, flourishing cities, but they had no formal status as such, nor anything that one might recognize as civic government. Towns consisted of agglomerations of neighborhoods, which in themselves constituted an important focus of identity and loyalty. Often, these neighborhoods were based on ethnic, tribal, religious, sectarian, or even occupational allegiances. To this day, there is no word in Arabic corresponding to "citizen." The word normally used on passports and other documents is *muwatin*, the literal meaning of which is "compatriot." With a lack of citizenship went a lack of civic representation. Al-

though different social groups did choose their own leaders during the classical period, the concept of choosing individuals to represent the citizenry in a corporate body or assembly was alien to Muslims' experience and practice.

Yet, other positive elements of Islamic history and thought could help in the development of democracy. Notably, the idea of consensual, contractual, and limited government is again becoming an issue today. The traditional rejection of despotism, of *istibdad*, has gained a new force and a new urgency: Europe may have disseminated the ideology of dictatorship, but it also spread a corresponding ideology of popular revolt against dictatorship.

The rejection of despotism, familiar in both traditional and, increasingly, modern writings, is already having a powerful impact. Muslims are again raising—and in some cases practicing—the related idea of consultation. For the pious, these developments are based on holy law and tradition, with an impressive series of precedents in the Islamic past. One sees this revival particularly in Afghanistan, whose people underwent rather less modernization and are therefore finding it easier to resurrect the better traditions of the past, notably consultation by the government with various entrenched interests and loyalty groups. This is the purpose of the Loya Jirga, the "grand council" that consists of a wide range of different groups—ethnic, tribal, religious, regional, professional, and others. There are signs of a tentative movement toward inclusiveness in the Middle East as well.

There are also other positive influences at work, sometimes in surprising forms. Perhaps the single most important development is the adoption of modern communications. The printing press and the newspaper, the telegraph, the radio, and the television have all transformed the Middle East. Initially, communications technology was an instrument of tyranny, giving the state an effective new weapon for propaganda and control.

Islam's rejection of despotism is having a profound impact on the Muslim world.

But this trend could not last indefinitely. More recently, particularly with the rise of the Internet, television satellites, and cell phones, communications technology has begun to have the opposite effect. It is becoming increasingly clear that one of the main reasons for the collapse of the Soviet Union was the information revolution. The old Soviet system depended in large measure on control of the production, distribution, and exchange of information and ideas; as modern communications developed, this became no longer possible. The information revolution posed the same dilemma for the Soviet Union as the Industrial Revolution did for the Ottoman and other Islamic empires: either accept it and cease to exist in the same manner or reject it and fall increasingly behind the rest of the world. The Soviets tried and failed to resolve this dilemma, and the Russians are still struggling with the consequences.

A parallel process is already beginning in the Islamic countries of the Middle East. Even some of the intensely and unscrupulously propagandist television programs that now infest the airwaves contribute to this process, indirectly and unintentionally, by offering a diversity of lies that arouse suspicion and questioning. Television also brings to the peoples of the Middle East a previously unknown spectacle—that of lively and vigorous public disagreement and debate. In some places, young people even watch Israeli television. In addition to seeing well-known Israeli public figures "banging the table and screaming at each other" (as one Arab viewer described it with wonderment), they sometimes see even Israeli Arabs arguing in the Knesset, denouncing Israeli ministers and policies—on Israeli television. The spectacle of a lively, vibrant, rowdy democracy at work, notably the unfamiliar sight of unconstrained, uninhibited, but orderly argument between conflicting ideas and interests, is having an impact.

Modern communications have also had another effect, in making Middle Eastern Muslims more painfully aware of how badly things have gone wrong. In the past, they were not really conscious of the differences between their world and the rest. They did not realize how far they were falling behind not only the advanced West, but also the advancing East—first Japan, then China, India, South Korea, and Southeast Asia—and practically everywhere else in terms of standard of living, achievement, and, more generally, human and cultural development. Even more painful than these differences are the disparities between groups of people in the Middle East itself.

Right now, the question of democracy is more pertinent to Iraq than perhaps to any other Middle Eastern country. In addition to the general factors, Iraq may benefit from two characteristics specific to its circumstances. One relates to infrastructure and education. Of all the countries profiting from oil revenues in the past decades, pre-Saddam Iraq probably made the best use of its revenues. Its leaders developed the country's roads, bridges, and utilities, and particularly a network of schools and universities of a higher standard than in most other places in the region. These, like everything else in Iraq, were devastated by Saddam's rule. But even in the worst of conditions, an educated middle class will somehow contrive to educate its children, and the results of this can be seen in the Iraqi people today.

The other advantage is the position of women, which is far better than in most places in the Islamic world. They do not enjoy greater rights—"rights" being a word without meaning in that context—but rather access and opportunity. Under Saddam's predecessors, women had access to education, including higher education, and therefore to careers, with few parallels in the Muslim world. In the West, women's relative freedom has been a major reason for the advance of the greater society; women would certainly be an important, indeed essential, part of a democratic future in the Middle East.

FUNDAMENTAL DANGERS

THE MAIN THREAT to the development of democracy in Iraq and ultimately in other Arab and Muslim countries lies not in any inherent social quality or characteristic, but in the very determined efforts that are being made to ensure democracy's failure. The opponents of democracy in the Muslim world come from very different sources, with sharply contrasting ideologies. An alliance of expediency exists between different groups with divergent interests.

One such group combines the two interests most immediately affected by the inroads of democracy—the tyranny of Saddam in Iraq and other endangered tyrannies in the region—and, pursuing these parallel concerns, is attempting to restore the former and preserve the latter. In this the group also enjoys some at least tacit support from outside forces—governmental, commercial, ideological, and other—in Europe, Asia, and elsewhere, with a practical or emotional interest in its success.

Most dangerous are the so-called Islamic fundamentalists, those for whom democracy is part of the greater evil emanating from the West, whether in the old-fashioned form of imperial domination or in the more modern form of cultural penetration. Satan, in the Koran, is "the insidious tempter who whispers in men's hearts." The modernizers, with their appeal to women and more generally to the young, are seen to strike at the very heart of the Islamic order—the state, the schoolroom, the market, and even the family. The fundamentalists view the Westerners and their dupes and disciples, the Westernizers, as not only impeding the predestined advance of Islam to final triumph in the world, but even endangering it in its homelands. Unlike reformers, fundamentalists perceive the problem of the Muslim world to be not insufficient modernization, but an excess of modernization—and even modernization itself. For them, democracy is an alien and infidel intrusion, part of the larger and more pernicious influence of the Great Satan and his cohorts.

Islamic fundamentalists believe democracy to be an alien and infidel intrusion.

The fundamentalist response to Western rule and still more to Western social and cultural influence has been gathering force for a long time. It has found expression in an increasingly influential literature and in a series of activist movements, the most notable of which is the Muslim Brotherhood, founded in Egypt in 1928. Political Islam first became a major international factor with the Iranian Revolution of 1979. The word "revolution" has been much misused in the Middle East and has served to designate and justify almost any violent transfer of power at the top. But what happened in Iran was a genuine revolution, a major change with a very significant ideological challenge, a shift in the basis of society that had an immense impact on the whole Islamic world, intellectually, morally, and politically. The process that began in Iran in 1979 was a revolution in the same sense as the French and the Russian revolutions were. Like its predecessors, the Iranian Revolution has gone through various stages of inner and outer conflict and change and now seems to be entering the Napoleonic or, perhaps more accurately, the Stalinist phase.

The theocratic regime in Iran swept to power on a wave of popular support nourished by resentment against the old regime, its policies, and its associations. Since then, the regime has become increasingly unpopular as the ruling mullahs have shown themselves to be just as corrupt and oppressive as the ruling cliques in other countries in the region. There are many indications in Iran of a rising tide of discontent. Some seek radical change in the form of a return to the past; others, by far the larger number, place their hopes in the coming of true democracy. The rulers of Iran are thus very apprehensive of democratic change in Iraq, the more so as a majority of Iraqis are Shiites, like the Iranians. By its mere existence, a Shiite democracy on Iran's western frontier would pose a challenge, indeed a mortal threat, to the regime of the mullahs, so they are doing what they can to prevent or deflect it.

Of far greater importance at the present are the Sunni fundamentalists. An important element in the Sunni holy war is the rise and spread—and in some areas dominance—of Wahhabism. Wahhabism is a school of Islam that arose in Nejd, in central Arabia, in the eighteenth century. It caused some trouble to the rulers of the Muslim world at the time but was eventually repressed and contained. It reappeared in the twentieth century and acquired new importance when the House of Saud, the local tribal chiefs committed to Wahhabism, conquered the holy cities of Mecca and Medina and created the Saudi monarchy. This brought together two factors of the highest importance. One, the Wahhabi Saudis now ruled the holy cities and therefore controlled the annual Muslim pilgrimage, which gave them immense prestige and influence in the Islamic world. Two, the discovery and exploitation of oil placed immense wealth at their disposal. What would otherwise have been an extremist fringe in a marginal country thus had a worldwide impact. Now the forces that were nourished, nurtured, and unleashed threaten even the House of Saud itself.

The first great triumph of the Sunni fundamentalists was the collapse of the Soviet Union, which they saw—not unreasonably—as their victory. For them the Soviet Union was defeated not in the Cold War waged by the West, but in the Islamic jihad waged by the guerrilla fighters in Afghanistan. As Osama bin Laden and his cohorts have put it, they destroyed one of the two last great infidel superpowers—the more difficult and the more dangerous of the two. Dealing with the pampered and degenerate Americans would, so they believed, be much easier. American actions and discourse have at times weakened and at times strengthened this belief.

In a genuinely free election, fundamentalists would have several substantial advantages over moderates and reformers. One is that they speak a language familiar to Muslims. Democratic parties promote an ideology and use a terminology mostly strange to the "Muslim street." The fundamentalist parties, on the other hand, employ familiar words and evoke familiar values both to criticize the existing secularist, authoritarian order and to offer an alternative. To broadcast this message, the fundamentalists utilize an enormously effective network that meets and communicates in the mosque and speaks from the pulpit. None of the secular parties has access to anything comparable. Religious revolutionaries, and even terrorists, also gain support because of their frequently genuine efforts to alleviate the suffering of the common people. This concern often stands in marked contrast with the callous and greedy unconcern of the current wielders of power and influence in the Middle East. The example of the Iranian Revolution would seem to indicate that once in power these religious militants are no better, and are sometimes even worse, than those they overthrow and replace. But until then, both the current perceptions and the future hopes of the people can work in their favor.

Finally, perhaps most important of all, democratic parties are ideologically bound to allow fundamentalists freedom of action. The fundamentalists suffer from no such disability; on the contrary, it is their mission when in power to suppress sedition and unbelief.

Despite these difficulties, there are signs of hope, notably the Iraqi general election in January. Millions of Iraqis went to polling stations, stood in line, and cast their votes, knowing that they were risking their lives at every moment of the process. It was a truly momentous achievement, and its impact can already be seen in neighboring Arab and other countries. Arab democracy has won a battle, not a war, and still faces many dangers, both from ruthless and resolute enemies and from hesitant and unreliable friends. But it was a major battle, and the Iraqi election may prove a turning point in Middle Eastern history no less important than the arrival of General Bonaparte and the French Revolution in Egypt more than two centuries ago.

FEAR ITSELF

THE CREATION of a democratic political and social order in Iraq or elsewhere in the Middle East will not be easy. But it is possible, and there are increasing signs that it has already begun. At the present time there are two fears concerning the possibility of establishing a democracy in Iraq. One is the fear that it will not work, a fear expressed by many in the United States and one that is almost a dogma in Europe; the other fear, much more urgent in ruling circles in the Middle East, is that it will work. Clearly, a genuinely free society in Iraq would constitute a mortal threat to many of the governments of the region, including both Washington's enemies and some of those seen as Washington's allies.

The end of World War II opened the way for democracy in the former Axis powers. The end of the Cold War brought a measure of freedom and a movement toward democracy in much of the former Soviet domains. With steadfastness and patience, it may now be possible at last to bring both justice and freedom to the long-tormented peoples of the Middle East.

Bernard Lewis *is Cleveland E. Dodge Professor Emeritus of Near Eastern Studies at Princeton University. This essay is adapted from a lecture given on April 29, 2004, as part of the Robert J. Pelosky, Jr., Distinguished Speaker Series at the Elliott School of International Affairs, George Washington University.*

From *Foreign Affairs*, vol. 84, no. 3, May/June 2005, pp. 36–51. Copyright © 2005 by Foreign Affairs. Reprinted by permission.

Preemption and Just War: Considering the Case of Iraq

FRANKLIN ERIC WESTER

This article demonstrates that the use of military force by the Bush Administration against the regime of Saddam Hussein does not meet the ethical criteria for "preemptive war" set forth in the classical Just War tradition. It considers ethical questions raised by the US-led attack against Iraq as part of the war against global terrorism and argues that the doctrine of preemptive war as applied in the case of Iraq fails crucial ethical tests.

Could Operation Iraqi Freedom and the global war on terrorism be as pivotal in the history of ethical decision-making as the emergence of the nation-state in the Peace of Westphalia in 1648? Do new ethics for the war on terror sever the fourth-century Augustinian roots of Just War theory and the ties to Thomas Aquinas's *Summa Theologica* 700 years later? Could the first major war of the 21st century inaugurate a revolution in ethical decision-making about warfare, justifying a new set of criteria for preemption or preventive war? Answers to these questions hinge on whether or not the doctrine of preemption matures into new ethical criteria. Such criteria would build not on foundations for constraining unavoidable human violence, but stretch toward a vision of an ideal of liberty that justifies the selective killing of some to achieve a greater good of liberty for many others. This emerging ethic installs the United States as the guardian of a universal, even transcendent, cause of freedom and the ultimate arbiter in that cause.[1]

This article applies the classic categories of Just War tradition to the doctrine of preemption as advanced by the current Administration in the justification for Operation Iraqi Freedom. It does not address the range of other explanations for and postures toward war outside the Just War tradition. Specifically, it does not develop details of three other major ways to think about war:

- Realism, the belief that war is essentially a matter of power, self-interest, and necessity, largely making moral analysis irrelevant.

- Holy War, the belief that war is an instrument of divine power and that individuals, groups, or nations apply decisions about violence to coerce or destroy those opposing divine will.

- Pacifism, the belief that all war is intrinsically evil and can never be justified.[2]

The article begins with a summary of the national security debate as expressed in the buildup to war against Iraq, including the views of policy experts and decision-makers, ethicists and academics. Second, it considers Just War ethical frameworks and definitions for two facets of warfare: justice in going to war (*jus ad bellum*) and justice in the conduct of war (*jus in bello*), focusing on the six criteria of *jus ad bellum*. In its attack on Iraq, the Bush Administration redefined criteria for preemptive and preventive war that do not satisfy the criteria established in the classical Just War tradition and may signal development of an emerging ethic.

Preemption, Prevention, and the National Security Strategy Debate

When the *National Security Strategy* was published by the Bush Administration in 2002, one of its most notable shifts specified a doctrine or principle of "preemption." Preemption—and, more notably, preventive war—exploded onto the scene of ethical debate as a major change in US security strategy. The 2002 *National Security Strategy* asserted, the "United States has long maintained the option of preemptive actions to counter a sufficient threat to our national security," and "the United States will, if necessary, act preemptively."[3]

This argument is moot if one sees the context of the war against Iraq as a continuation of the 1991 Desert Storm war, as does ethicist Thomas Nichols, Chairman of Strategy and Policy at the US Naval War College, and as the Joint Chiefs of Staff may have.[4] Professor Nichols presents an appeal for debate about how the 2003 attack against Iraq is a proper application of *jus in bello*. This argument contends that the leaders of Iraq remained at war against the coalition despite their signing the 1991 treaty at the end of Desert Storm; therefore, Operation Iraqi Freedom was not preemption but a justified use of force. The two pivotal issues in this regard are Iraq's defiance of UN resolutions and pre-war Iraqi aggression. Nichols describes the situation before the war as follows:

> In a repeating pattern, Iraq is served notice with [UN] resolutions, agrees to them, and then breaks them.... There is no longer a credible way to envision any peaceful road to Iraqi disarmament.... He [Saddam Hussein] has pledged and promised and agreed, and then reneged, so many times that only the most trusting (or cynical) diplomats would encourage him to play and win such a pointless game one more time."[5]

For Nichols, the military attack was morally right and justified based on Iraq's noncompliance with international standards. If one concedes Iraq's noncompliance, the Just War ethical question then shifts to legitimate authority, away from preemption and just cause. If Iraq was defying the previous coalition or the United Nations, then the legitimate authority for war is not a US-led "coalition of the willing," but a more clearly recognized international body—either the coalition from the 1991 war whose treaty was violated, or the United Nations, whose resolutions were ignored.

Nichols also states that Iraq did not accept the UN-imposed no-fly zones intended to prevent humanitarian abuses. "The Iraqis ... fired on coalition aircraft over 700 times since 1998 alone, in an attempt to harm those engaged in the protection of the innocent—itself an action sufficient to trigger a presumption of Just War."[6] The anti-aircraft fire of the Iraqis drew bombs and missiles from coalition aircraft, in a continuing exchange of tactical fires. As to his second argument that Operation Iraqi Freedom is not a preemptive or preventive war, Nichols summarizes, "The United States and its allies [were] already at war with the Iraqis; one cannot 'preempt' or 'preventively attack' a regime whose forces one is already attacking on a regular basis."[7]

Although Professor Nichols does not view the war against Iraq as preemptive, the Bush Administration made preemption the reason for resorting to armed conflict. The President and key presidential advisors, including the National Security Advisor and the Secretary of Defense, did not justify toppling Saddam Hussein primarily as a response to Iraq's attacks against coalition forces or the humanitarian needs of Iraqis. The principal reason for war stated by the Bush Administration to the nation and the world was the possible use of weapons of mass destruction (WMD). Disarming Iraq was the desired end, and regime change in Iraq was the only possible way to achieve that end. Preemptive military action was required, and thus justified, to prevent possible use of WMD. As expressed by Alan W. Dowd, "The Bush Doctrine's principle of preemption was tailor-made for Baathist Iraq—a country with growing ties to terror, an underground unconventional weapons program, and the means and motive to mete out revenge on the United States."[8]

President Bush addressed preemption in a major policy address at the US Military Academy on 1 June 2002. He stated, "If we wait for threats to materialize, we will have waited too long," and he declared that "our security will require all Americans ... to be ready for preemptive action when necessary to defend our liberty and defend our lives."[9] In the background of these remarks, remember that US and coalition forces were consolidating their prompt success in scattering the al Qaeda terrorists in Afghanistan and replacing the neutered Taliban regime with a new government led by Hamid Karzai. The West Point speech foreshadowed the run-up to the Iraq war of an economic, political, and by March 2003, a military coalition.[10]

Beyond the President's statements, National Security Advisor Condoleezza Rice amplified the security strategy. She used a graphic analogy to convey the increased risk of waiting and her rationale for preemptive action: "We don't want the smoking gun to become a mushroom cloud."[11]

Secretary of Defense Donald Rumsfeld, in an August 2002 interview with Fox News, argued that America could not wait for proof that Saddam had weapons of mass destruction. He compared the prelude to war against Iraq with the prelude to World War II, when the Allies appeased Hitler. The Secretary rejected alternative points of view other than war, saying, "The people who argue [against invading Iraq] have to ask themselves how they're going to feel at that point where another event occurs and it's not a conventional event, but it's an unconventional event."[12]

Secretary of State Colin Powell presented the Bush Administration's case to the United Nations Security Council for disarming Saddam Hussein's Iraq. He provided information from the core assessments made by the chief UN-appointed inspector, Dr. Hans Blix, indicating ballistic missiles were moved and hidden from inspectors and that Iraq had failed to account for biological and other weapons. Secretary Powell did not ask anything of the UN Security Council. He challenged the Security Council in a general way: "We must not shrink from the responsibilities that we set before ourselves."[13]

In contrast to the unanimous voices from within the Administration advocating war against Iraq, a former Secretary of State, one often thought of as a pragmatist and hawkish, Dr. Henry Kissinger, urged a diplomatic approach and cited a potential second-order effect of war against Iraq. He warned, "It is not in the American national interest to establish pre-emption as a universal principle available to every nation."[14] He argued not so much against the war as for the United States to use its power to shape an international response, believing that preemptive or preventive war could destroy the international order that had prevailed since Westphalia.

> *"Preemptive strikes and preemptive war have a recognized historic and narrowly defined place in the Just War tradition."*

Interestingly, the 2002 *National Security Strategy* indirectly acknowledges the Just War ethic. Logic in the document relies on the special case of preemption based on "imminent threat,"recognizing that Just War tradition makes room for arresting or resisting "imminent threat" as an extension of legitimate self-defense. However, the *National Security Strategy* goes on to assert, "We must adapt [that is, change] the concept of imminent threat to the capabilities and objectives of today's adversaries."[15] How to change a concept like "imminent threat" or the moral reasoning associated with the Just War ethic is not specified. The *National Security Strategy* assesses that the United States faces a new threat from the convergence of rogue states, failed states, and terrorists operating with potential access to weapons of mass destruction. This combination makes measuring the imminent nature of threats so difficult that both preemptive war and preventive war are justified. In addition, the *National Security Strategy* concludes that against such adversaries with such weapons, deterrence is no longer possible.

With these introductory aspects of the ethical issues as background, we now turn to some practical problems in the case of Iraq. The first relates to the compressed time in moving toward and implementing the current military doctrine, Rapid Decisive Operations. Military actions connect concepts and ideas; military decision-making connects ethical decision-making and outcomes. The question may then be asked, is preemption just another way to take "rapid" to the next level? A second practical problem is the risk of a wrong decision. Given the potential threats cited in the 2002 *National Security Strategy* and elsewhere, the stakes for either action or inaction are very high. Just as the doctrine of Mutually Assured Destruction kept the world in a precarious balance, a doctrine of preemption elevates the risks of premature action or useless inaction and increases the danger of mistakes. A nation preempting another nation or group may win a battle against a specific threat, but lose the war of acting rightly. Right actions include obtaining victory while addressing the moral duty to prevent destruction of vital resources (e.g., oil) and Western ways of life (universal human values, inalienable rights), as well as preventing unnecessary casualties among noncombatants.

A third practical problem is that of having inaccurate or incomplete information which then becomes "actionable intelligence." Information overload and faulty patterns of selecting information can create stepping stones to incorrect decision-making. In the case of the war against Iraq, significant complications with human intelligence emerged—relying on people with much to gain from regime change in Iraq, a shortage of human intelligence sources, and poor translation of human reports. All conspired to weaken a critical link in building a case for preemptive or preventive war.

How much accuracy in intelligence is needed? In the case of the US use of an inaccurate white paper from Great Britain reporting Niger's providing yellow-cake uranium as a component for weapons of mass destruction, more information or a better analysis was clearly needed. And in the post hoc analysis, the findings by Dr. David Kay, chief UN inspector, confirmed prewar reports presented by Hans Blix to the UN:

- Iraq's nuclear weapons program was dormant.

- No evidence suggested that Iraq possessed chemical or biological weapons.

- But Iraq was attempting to develop missile capability exceeding the UN-mandated limit of 93 miles.[16]

Views of Ethicists

Noting these practical problems, were turn to some of the ethical difficulties. With preemption included in the *National Security Strategy*, the reaction of ethicists and academics in social science, law, religion, and philosophy was prompt, if not widely noticed. Part of the debate occurred in the public square of newspaper opinion pages and magazines. Even more discussion percolated among academics, and in the *Chronicle of Higher Education* on 23 September 2002, 100 scholars made a one-sentence declaration: "As Christian ethicists, we share a common moral presumption against a pre-emptive war on Iraq by the United States."[17] Signatories ranged from Duke University pacifist professor Dr. Stanley Hauerwas, to Dr. Shaun Casey, a just-war ethicist from Wesley College in Washington, D.C.

Some ethicists and religious leaders endorsed military action. Some saw it as a continuation of the 1991 conflict, while others saw the action in 2003 as moral. For example, one theologian, the Reverend Richard Land, president of the ethics and religious-liberty commission of the Southern Baptist Convention, endorsed military action on grounds of self-defense: "I believe we are defending ourselves against several acts of war by a man who did not keep treaties and who has already used weapons of mass destruction."[18]

In the press, futurist and military commentator Ralph Peters applauded the war and advanced the position that Operation Iraqi Freedom was changing the criteria for imminent threat described in the *National Security Strategy*. As previously noted, the strategy document explained we must adapt the concept of imminent threat to the capabilities and objectives of today's adversaries. Peters said, "We have cast off old, failed rules of warfare" for a new paradigm "that makes previous models of warfare obsolete."[19]

Responding to Peters, Dr. John Brinsfield (Colonel, USA Ret.), a former professor at the Army War College and retired Army chaplain, challenged this view:

> Adopting pre-emptive strikes (followed by bombing more massive than anything since World War II) should never be a *normative* part of our ethical thinking about war. To embrace pre-emptive strikes as normal policy rather than a very narrowly defined exception to the rules of civilized warfare is not to advance to a position of "waging just wars humanely" (quoting Peters) but rather to retreat to barbarism, waging war whenever we think "might makes right."[20]

Another ethicist, Paul Schroeder, summarizing the Administration's justification for preemptive war, noted the arguments went largely unexamined and concluded the rationale will quickly prove unacceptable:

> The Bush Administration's case for preemptive war asserts: the dangers and costs of inaction far outweigh those of acting now. Saddam Hussein, an evil despot,

a serial aggressor, an implacable enemy of the United States, and a direct menace to his neighbors must be deposed before he acquires weapons of mass destruction that he might use or let others use against Americans or its allies and friends. A few thousand Americans died in the last terrorist attack; many millions could die in the next one. Time is against us; once Hussein acquires such weapons, he cannot be overthrown without enormous losses and dangers. Persuasion, negotiation, and conciliation are worse than useless with him. Sanctions and coercive diplomacy have failed. Conventional deterrence is equally unreliable. Preemptive action to remove him from power is the only effective remedy and will promote durable peace in the region.

Do we have a right to wage preemptive war against Iraq to overthrow its regime? Would this be a necessary and just war? What long-range effects would it have on the international system? On these questions the Administration won by default. The assumption that a war to overthrow Hussein would be a just war and one that, if it succeeded without excessive negative side effects, would serve everyone's interests went unchallenged in the mainstream. The Administration's claim of a right to overthrow regimes it considers hostile is extraordinary—and one the world will soon find intolerable.[21]

Preemptive strikes and preemptive war have a recognized historic and narrowly defined place in the Just War tradition. In the section that follows, the four criteria for anticipatory self-defense will be considered, noting the specific cases of the *Caroline* in 1837, precipitating events for World War I, and Operation Iraqi Freedom. Preventive war has no legal or ethical sanction, because the threat is neither clear nor present.

Definitions: Preemptive Strike, Preemptive War, and Preventive War

At this point, let us define and differentiate the terms preemptive strike, preemptive war, and preventive war. A preemptive strike is a tactical activity, intended to have a strategic effect. Preemptive strikes may be actions in war or discrete acts that one nation takes against another apart from war. Ethically, a preemptive strike in war is evaluated in the category of *jus in bello* and is a way to seize the initiative. A preemptive strike may be preceded by warnings and is not necessarily a "sneak attack." The Confederate attack against Fort Sumter illustrates a preemptive strike. In a more recent example, perhaps the most well-known modern preemptive strike not associated with war is the long-distance attack by Israel in 1984 against Iraq. Israel successfully destroyed a nuclear power plant in Iraq based on the suspicion that eventually Saddam Hussein would have the means for a missile attack with a nuclear weapon against Israel.

In comparison, a preemptive war is associated with one aspect of the just cause standard of going to war (*jus ad bellum*).

If attack is imminent, with a clear and present danger, a nation is right to defend itself. With Egypt's tanks on Israel's border in the Sinai as a clear and present danger, Israel launched the 1967 war. Also, the act of proceeding to war before actual attack is moral when the threat is real and so near at hand that launching war could be considered self-defense. A nation or nations also may rightly intercede to prevent humanitarian abuses, even inside the boundaries of another sovereign nation.

"*Preventive war has no legal or ethical sanction, because the threat is neither clear nor present.*"

In contrast, a preventive war is started well before the imminent threat or humanitarian crisis, when the balance of forces is the primary consideration. As noted above, a preemptive war is launched at a time close to a documented or presumed threat, when the forces initiating war retain tactical, operational, or strategic advantage. Preventive war, on the other hand, is built on a sheer calculation of advantage—nation X can gain an advantage by acting now to attack nation Y, regardless of the threat. By launching a war now, a later conflict—more costly in human life, national resources, or even lost victory—is avoided. The justification for such a war must withstand the critique of a just intent standard.[22]

Six Criteria for Jus ad Bellum

The next two sections summarize the criteria of Just War ethics and apply these six criteria to the case of Iraq. By definition, ethical discourse about preventive or preemptive war fits in the category of *jus ad bellum* (justice in going to war).[23] The six criteria are as follows.

- *Legitimate authority.* Different countries assign different legitimate authorities for declaring war. In the United States, though the Constitution specifies Congress as the agent to declare war, the unresolved tension between the President wielding the War Powers Act and the control of appropriations by the Congress has functioned sufficiently to legitimize war by US forces. In cases of international forces, recognized organizations and institutions have formal procedures for legitimizing military power.
- *Public declaration.* National leaders or leaders of international organizations or institutions are called on to announce intentions to pursue war and to provide the conditions for avoiding or ending conflict.[24]
- *Just intent.* A general rule for just intent, or just cause in going to war, is to restore the status quo ante bellum, a return to international relations when war was not pursued. Other facets of just intent are to protect the innocent, recover something wrongly taken, punish evil, or defend against wrongful attack.[25]

• *Proportionality*. This criterion focuses on restraint and precision in the use of force. Warfare presents notorious difficulty in predicting its costs—both human and economic—yet the application of military force is legitimate only to the degree it takes account of such effects and outcomes.

• *Last resort*. This criterion presents a logical conundrum. In theory, something else can always be done. The point of this specification is to clarify that force is justified only as a sad necessity after other good faith ways to avoid or resolve conflict have failed.

• *Reasonable hope of success*. Leaders make a morally grave decision to commit the lives of their military forces, and those of innocent civilians, to death for the hope of reversing the cause of going to war. Only conflict with some expectation of restoration to an acceptable status quo is usually ethical. Revenge and "suicide stands" are not moral choices in cases where there is no hope whatever of success.

Asserting the Bush Administration Did Not Meet the Just War Criteria

• *Legitimate Authority*. The first test of legitimate authority includes determining the legitimacy of the person as a rightful office-holder. In the case of President Bush, there is certainty that he is the legitimate national leader acting properly in his role in the United States. Gerald Bradley, law professor at Notre Dame, asserts that the morally upright leadership of the President of the United States in concert with credible advisors is recognized,[26] and that assertion is uncontested here.

A second test of legitimate authority focuses on whether the office-holder is in a position to properly authorize war. In the case of the war against Iraq, in which President Bush led a "coalition of the willing," this aspect is in doubt. Consider the proper authority criterion along a continuum of legitimacy. From the most to the least credible authority for taking military action against a nation-state or national leader, the continuum would be: (1) unanimous international commitment, (2) a United Nations decision, (3) a UN Security Council decision, (4) other regional or international alliance, (5) an ad hoc coalition, (6) unilateral action. Viewing these frameworks as being from most to least legitimate and compelling, the US-led coalition is in the range of 5 to 6. Even the strongest advocates for military action against Iraq express the legitimate authority for the war as being based on a loose coalition of nations. Critics describe the military action as being a case of a limited number of nations trailing along behind the United States out of obligation, or, more bluntly, describe the war as unilateral action by the United States.

If the reason for going to war was based on violations of UN sanctions by Iraq, then the United Nations would be the legitimating authority. If the war was based on international humanitarian concern for the victims of the Hussein regime in Iraq, then some international body, such as the North Atlantic Treaty Organization or the UN, would be the legitimating authority. To act on the authority of a "coalition of the willing" relies on vague ethical criteria. US leaders indicated they possessed persuasive information that an attack against the United States or US interests using weapons of mass destruction was possible, and that Iraq was advancing terrorism. Both the assertion of possible attack with WMD or conventional means and the involvement of Iraq with terrorism (specifically al Qaeda) have since come under considerable dispute, to say the least.

• *Public Declaration*. The second criterion, a public declaration of an ultimatum, was met by the President on 19 March 2003 when he demanded that Saddam Hussein and his sons disarm and step down from power in Iraq within 48 hours. Public communication of intentions for the use of military force by the US-led coalition is clearly recognized.

• *Just Intent*. The war against Iraq presents problems, however, regarding the just intent of the US-led coalition. Conventional Just War logic uses high standards to grant legitimacy for going to war. National self defense and overcoming a grave evil, such as the suffering of innocents, are typical.

Just intent in Operation Iraqi Freedom focused on the preemptive use of force to disarm Saddam Hussein and replace his regime, based on defending against an imminent threat. Dr. Michael Walzer, Professor of Social Science at the Institute of Advanced Study, in Princeton, N.J., argues that the intent to disarm Iraq was just, based on the failed inspections linked to monitoring compliance with weapons restrictions. The intended end—a disarmed Iraq—met with international support. However, many nations preferred the ways of sanctions and inspections, not war and regime change. To Paul J. Griffiths, Professor of Catholic Studies at the University of Illinois, the definition of imminent has not changed: "It means the gun is at your head." And in the case of Iraq, "We just don't have that." He states that redefining imminent offers "well-intentioned support for US foreign policy, but it's not defensible in terms of traditional Just War theory."[27] Professor Griffiths proposes two standards for a just, preemptive attack in the framework of Just War:

- Knowledge that the threat is in place; that nuclear, biological, or chemical weapons are armed and ready to be used.
- Knowledge that the weapons are aimed at the nation proposing preemption in self-defense.

The intended end of a disarmed Iraq was founded on the conviction that the regime of Saddam Hussein had both motive and means to launch biological and chemical attacks and would very soon have a capability for nuclear attack or the threat of nuclear attack. Bush Administration proponents for preemption believed both standards proposed by Professor Griffiths were met. Two practical questions intrude on this ethical discussion: How does one determine the threat from a regime continuously demonstrating no willingness to comply with treaties, international sanctions, or standards of truth-telling, and what culpability rests on the preempting nation if and when the rationale for preemption is proven false?

Another facet of the just intent for war with Iraq was the idea of protecting innocent people from humanitarian abuses. International law has developed rapidly along these lines in recent interventions in Africa and even more explicitly in the former Yugoslavia. In 1999, when "ethnic cleansing" in Kosovo exceeded an unspecified threshold for public and international tolerance, the United States led NATO in an intervention, employing air attacks and the evacuation of noncombatants, to bomb the Serbian forces into stalemate and halt the slaughter of Albanian Muslims. This was one of the most vivid examples of an international intervention within the boundaries of a sovereign state to protect innocents from humanitarian abuses. In Saddam's Iraq, there were graphic reports and references to widespread humanitarian abuses, including the use of chemical weapons against separatist, civilian populations in the Kurdish areas of northern Iraq and the religiously distinct minority in southern Iraq. But humanitarian intervention was never seriously advanced by either the Clinton or Bush administrations, until it was included as part of the justification for ousting the regime of Saddam Hussein. In this instance, protecting victims from humanitarian abuse is probably most accurately described as an *ex post facto* reason for war.

In the case of the war on Iraq, regime change was a way, not an end, and the end of a disarmed Iraq was determined by the Bush Administration to be achievable only by regime change. Regime change as a "morally desirable side-effect"[28] of disarming an aggressor is consistent with the Just War ethic. Regime change as the end or intent falls outside the recognized standards of Just War logic. Regime change was incorporated explicitly in the justification for Operation Iraqi Freedom. Regime change is not a status quo ante bellum and commits the nation far beyond military application to postwar responsibility, building a new, politically functioning nation. This is a commitment shouldered determinedly, so far, by the United States, while most other nations and the UN still search for a morally and politically acceptable role. Regarding the ethic of regime change, in his letter to President Bush on 13 September 2002, Bishop Wilton Gregory, president of the US Catholic Conference of Bishops, asked, "Should not a distinction be made between efforts to change unacceptable behavior of a government and efforts to end that government's existence?"[29] This echoes the debate about the appropriate ends and leads to questions of appropriate ways—destroying Iraq, containing Iraq, isolating Iraq.

The vortex of the debate about just cause for this war swirls here: Was disarming Iraq morally required to prevent a destruction or endangerment of the core human value of freedom? Was disarming Iraq possible only through military victory over Iraq and replacing the government of Saddam Hussein? In the case of Operation Iraqi Freedom, the US-led coalition of nations acted to destroy the military and political forces of Iraq, and sought to replace the regime to protect all freedom-loving people, extend freedom to Iraq, and increase stability in the region.

"The Bush doctrine contends that preemption is right, just, and different from aggression."

The argument for this just cause, or just intent, points toward the central dimension of an emerging paradigm shift in the ethics of war. Preemptive war to prevent a potential threat

through regime change using military force exemplifies the change proposed in the 2002 *National Security Strategy*. Disarming and restructuring a nation using preemptive or preventive war is driven by an ideal future vision, not defense of or return to the status quo. This model or framework for action employs military force to improve the lot of citizens in a foreign land while eliminating a real or potential threat to the territory of the United States, allies, and US political or other interests. This ethic asserts an idealist, universal, God-given liberty as the bedrock for decision-making. This freedom is to be advanced by the United States with or without coalition partners, not as a model nation or political persuader.

• *Proportionality.* Proportionality aims to limit the cost and damage of war. In the discussion of Iraq, proportionality relates to the validity of the threats. A strong aspect of the Bush Administration's case for war is the worst-case risks associated with weapons of mass destruction. If tens of thousands—or millions—of people were killed by attacks with WMD, or should the entire Western way of life and the values of liberty and self-determination be obliterated, obviously the charge of negligence would rest against those who failed to act.

In the case of protecting innocents, proportionality becomes more tangible. Proportionality is intended to engage both restraint in the use of deadly force and precision in employing such force. In both regards, the coalition invasion of Iraq is viewed as acceptable in the damage inflicted to disarm Saddam.

Critics of the case for war argue that Operation Iraqi Freedom violated the Just War ethic because Iraq had not attacked nor threatened to attack the United States or other nations, so no proportional response was indicated. If the threat was genuinely the possible use of weapons of mass destruction, then disarming Iraq's regime seems proportional. Regime change, however, exceeds that measure.

• *Last Resort.* As noted above, the standard of using war as a last resort is logically difficult. Something other than going to war can always be done. However, the prudential test is whether or not all reasonable options have been exhausted prior to launching military action. In the case of Iraq, was military action the sad but necessary next step? Some of the harshest criticisms of the Bush Administration have focused on this criterion. Opponents of military action argue that the UN's inspection regime should have been given more time. But when does more time become too much time?

The Bush Administration's emphasis on the threat of WMD was an attempt to move the threshold of last resort forward and to discount the applicability of deterrence. In the book *Structure for Scientific Revolution*[30] a paradigm shift is said to occur when old scientific data can no longer explain new facts. The 2002 *National Security Strategy* argues that we face a threat that cannot be managed within the old paradigm, so a new paradigm is created. Preemption based on partial but sufficient evidence that we face a clear but not necessarily present danger is this new paradigm. In this view, the danger is so unpredictable and volatile that we must act immediately rather than waiting to act only as a last resort.

In the case of Iraq, supporters of military action sooner rather than later point to the habitual deceit and evasion of Saddam

Hussein—signing a 1991 treaty he consistently violated, perpetually shooting at coalition aircraft in the UN "no fly zone," and agreeing to inspections of weapon sites only to create obstacles and diversions with inspection teams. Should we have sought more inspections? Continued sanctions? Such steps are always alternatives, but the pragmatics of choosing when to consider war as a last resort are based on the prudential political and diplomatic perceptions of restraint, not on a theoretical last act.

- *Reasonable hope for success.* The reasonable hope for success is a criterion intended to prevent the pointless use of military forces that have no chance at victory. In effect, this protects military personnel and nations from authorities who recklessly steer a nation into armed conflict. In the case of the US-led coalition against Iraq, there is little doubt of the military superiority of the allied forces. Questions about success became more complex in two areas. First, would the application of military force provoke attacks with weapons of mass destruction that otherwise would have been unlikely?[31] Second, reasonable hope for military success does not necessarily lead to reasonable hope for success in other domains of Iraq's future. Is there reasonable hope for success in building a secure and free Iraq? Will military success against Iraq provoke other regional conflict? And following a successful military campaign to change the regime, does the US-led coalition have there sources and endurance to effect the changes necessary in creating a stable government?

Focus on Imminent Threat as a Just Cause

Anticipatory self-defense is recognized in international law and Just War tradition as a just cause. The first test for this is whether one's action is self-defense or aggression. Professor William Galston teaches at the University of Maryland and is Director of the Institute for Philosophy and Public Policy. He summarized much legal and philosophical argument by specifying four criteria of preemption as self-defense. These criteria are not dichotomous, either/or categories, but each presents a continuum of possibilities for consideration: the severity of the threat, the degree of probability of the threat, the imminence of the threat, and the cost of delay.[32]

Historically, these four specifications emerged from the British military attack against an American ship, the *Caroline*, on 29 December 1837, when British forces crossed the Niagara River to the US side to intercept assistance reportedly being provided to Canadian revolutionaries by US nationals. Daniel Webster subsequently argued in a letter to Lord Ashburton that a British case for self-defense was untenable, because the risks to Britain did not meet the four criteria.

Self-defense has been used sparingly in justifying preemption in the West over many centuries. Even in the post hoc analysis of World War II, preemption of Nazism by Allied power has not been advanced.[33] In an earlier case, historian David Fromkin analyzes the causes of World War I: "The archduke's murder was the excuse Austria-Hungary wanted to declare war on Serbia. Things might have ended there. But then Germany, using Austria's war as its excuse, declared war on Russia on the

possibility that Russia might interfere in the Serbian conflict."[34] This was aggression, not anticipatory self-defense.

Applying the four standards of preemption, one can reconsider "imminent threat" and interpret the proposal in the *National Security Strategy* for "adapting the concept of imminent threat to the capabilities and objectives of today's adversaries."[35] Professor Galston applied these four criteria to the case for war against Iraq as follows.

- *Threat.* In the analysis of the US national leadership, the threat was high. In the worst case, Iraq had nuclear weapons (or would have them within a near-term) that could be used against US citizens, allies, or friends, or transferred to terrorists.
- *Probability.* The probability of the threat was certain, based on analysis by US national leadership. This issue of probability is the decisive point of contention in international discussions after the fact. As certain as the Bush Administration was about the probability of Iraq's motives and capability to use weapons of mass destruction, France, Germany, Russia, and the majority of the rest of the world were not convinced (or at least would not act on the evidence presented).
- *Imminence.* No persuasive case was argued that the threat was imminent, at least in any conventional definition of imminent.
- *The cost of delay.* This appeared to be low; that is, delay was rejected by the United States and other willing coalition partners based primarily on the analysis of the past in transigence of Saddam, not the risk of delay. In fact, most other nations argued in favor of delay, to allow inspections and sanctions to work.

Anticipatory self-defense, though technically justified, is an unusual case in military ethics and is not supported in this instance. Consequently we are left with the revolutionary idea of redefining imminent threat and just cause according to the Bush doctrine. The Bush doctrine contends that preemption is right, just, and different from aggression, transcends imperialism, and is based on a vision of the future achievable through preemptive or preventive war. The Bush doctrine builds on a vision of extending liberty and an open-market economy, and authorizes invading a sovereign nation to topple a regime through preemptive war against an enemy, using criteria such as these:

- The enemy is controlled by a despotic ruler with a record of aggression.
- The enemy threatens, directly or through proxies, neighboring people or nations.
- The enemy seeks weapons of mass destruction to use against or threaten others.
- The enemy is stoppable now, and delaying attack only increases the cost of action later.
- The enemy is unresponsive to persuasion, negotiation, deterrence, or conciliation.

Conclusions

In the case of achieving the end of a disarmed Iraq, the ethically preferred way would have been to use coercive inspections based on a significantly larger coalition, preferably an established alliance of nations. Even stronger legitimacy would have been based on a majority, if not a consensus, of member states of the United Nations to implement coercive inspections.[36]

As stated in the introduction, a new paradigm for right thinking and right acting in war may be emerging. Preempting a nuclear, chemical, or biological attack (or any attack) against innocent people is a morally desirable action. In the case of weapons of mass destruction, the questions remain, "When is imminent, and when is too late?" But these questions can be framed in Just War language. The Bush doctrine of preemption builds on a conviction that using armed force is just when based on partial but sufficient evidence of a clear, unpredictable, and volatile danger that threatens the security of US citizens, liberty for people in other nations, and the rational structure of nation-states. Redefining imminent, as called for in the *National Security Strategy*, may mean a clear danger is not necessarily a present danger.

The 2002 *National Security Strategy*, then, is used to justify selective killing of some to achieve a greater good of liberty for many others, driven by an idealistic approach for universal human freedom. This is a strategic move oriented toward future vision and away from the recent, realist historical politics of a balance of power or a balance of terror. In the ethical framework of increased good for the most people and balancing ethical ends and means, the Bush doctrine advances democracy at the tip of a spear. This application of military force presents a moral dilemma.

The case of Operation Iraqi Freedom is a catalyst for further thinking about this new ethic regarding the use of force, particularly the leaner, rapid, decisive, lethal new force of the US military. The pivotal issue about retaining "imminent threat" but redefining its criteria for preemptive self-defense is driven by the threat of terrorists with the means and motives to employ weapons of mass destruction. Until further moral thinking provides a more substantial ethical framework for decision-making, however, it seems proper to withdraw references to preemption as a doctrinal element of the *National Security Strategy*. Further, even should nations proceed toward a new ethic for launching war preemptively or preventively, consideration of second-order effects is important. Even accepting the benefit of preventing Iraq from obtaining nuclear weapons, will other nations pursuing nuclear weapons move more quickly and covertly, believing the United States would attack during their pursuit of WMD (such as Iraq) but not nations possessing them (such as North Korea)?

Recommended Areas for Further Study

The application of preventive war breaks new ground in the ethics of going to war. In the age of a so-called "clash of cultures," how can nations inheriting the Just War tradition find ethical frameworks which intersect with, if they don't correlate with and complement, the ethical perspectives of nations without this Just War legacy?

In the case of Operation Iraqi Freedom, the United States and other coalition nations were involved with military forces enforcing a no-fly zone. Diplomatic efforts to monitor, if not diminish, Iraq's weapons-related activities meant warfare overarched Iraq's interaction with the rest of the world. Lacking a clear and certain view of Iraq's capabilities, especially for weapons of mass destruction, the issue of imminent threat was blurred. This blurring applied to the practical matter at hand, with questions about what weapons Saddam Hussein might have and when and where he might use them. On a theoretical level, the case of Iraq's possible possession of WMD raised the question to be further explored regarding an imminent threat: How does imminence apply in cases where time and space before attack are not clearly discernible? In other words, when is it timely and when is it too late to act?

Another recommended area for further investigation involves distinguishing between preemption against non-state or "stateless" terrorists and their organizations[37] and sovereign states. Distinguishing among nations which harbor or financially support terrorists gives a focal point for conventional military, diplomatic, and economic planning. An additional area for investigation is the assertion that non-state actors and some states (specifically Iraq, though this will certainly apply in other future cases) are beyond deterrence. For those doubtful about deterrence, this method was linked and limited by the Cold War standoff between the United States and the former Soviet Union. Deterrence works in a framework of rational thought and a balance of power. However, even in hypothetical cases of holy war, which was not precisely a factor between the adversaries in Operation Iraqi Freedom, are deterrence and containment possible when dealing with so-called "rogue" states? How, besides deterrence, can one explain Iraq's not using its weapons of mass destruction against its avowed enemy who was within range of its delivery systems—Israel?

Finally, regime change was based on analysis that the only way to disarm Saddam Hussein and release the grip of his tyranny was by going to war. This is a significant departure from Just War standards where the status quo ante bellum is normative. This may, once again, indicate the importance of developing a new ethic when dealing in cases of conflict across the cultural divides separating nations inheriting and relying on a Just War tradition and those nations where the status quo represented a source of the conflict in the first place.

Given the decade of broken promises and provocative pinpricks by Iraq toward the United Nations and the coalition of the early 1990s, perhaps Operation Iraqi Freedom is not a "pure" case of preemptive or preventive war. But if military forces are to be used to extend liberty, democracy, and free enterprise based on a universal principle of liberty, and if these forces will be called on by national leaders to wage war preemptively, or preventively, then it is imperative to probe for ethical models on which military personnel and legitimate political leaders can launch such wars.

NOTES

1. James W. Skillen, "Iraq, Terrorism, and the New American Security Strategy," *Public Justice Report*, First Quarter 2003, http://www.cpjustice.org/stories/storyReader$933.

2. David Blankenhorn, moderator, "Iraq and Just War: A Symposium," Washington, D.C., Elihu Root Room, The Carnegie Endowment for International Peace, 30 September 2002, http://pewforum.org/events/index.php?EventID=36.

3. George W. Bush, *The National Security Strategy of the United States of America* (Washington: The White House, 2002), p. 15, http://www.whitehouse.gov/nsc/nss.html. Hereinafter, *National Security Strategy*.

4. "Military Chiefs Stand by War," Associated Press, 11 February 2004, http://www.bvvinc.org/WebPages/Miltary.com%20images/standbywar.htm. The article notes that in joint testimony to the Senate Armed Services Committee on 10 February 2004, Admiral Vernon Clark, Chief of Naval Operations, read a declaration on the morality of Operation Iraqi Freedom in his statement: "It was my belief that this cause was just. That was my position then [during the war] and that's what I believe today." In reviewing part of a letter he wrote to Secretary of Defense Donald Rumsfeld, Admiral Clark stated: "For some this is about WMD. For others, this is about al Qaeda. For us, it's about all these and more. Iraq has been shooting at our aircraft for over five years."

5. Thomas M. Nichols, "Just War, Not Prevention," Carnegie Council on Ethics and International Affairs 2004, http://www.carnegiecouncil.org/viewMedia.php/prmTemplateID/8/prmID/867.

6. Ibid.

7. Ibid.

8. Alan W. Dowd, "Thirteen Years: The Causes and Consequences of the War in Iraq," *Parameters,* 33 (Autumn 2003), 48.

9. George W. Bush, "Graduation Speech at West Point," 1 June 2002, http://www.nti.org/e_research/official_docs/pres/bush_wp_prestrike.pdf.

10. Richard Cheney, "The Vice President Delivers Remarks to the National Association of Home Builders," Washington, D.C., 6 June 2002, http://www.whitehouse.gov/vicepresident/news-speeches/speeches/vp20020606.html. Vice President Cheney spoke bluntly about preempting adversaries: "Wars are not won on the defensive. We must take the battle to the enemy—and, where necessary, preempt grave threats to our country before they materialize."

11. Condoleezza Rice, "Search for the Smoking Gun," CNN television interview by Wolf Blitzer, 8 September 2002.

12. Cited by Greta Knutzen, "Between Iraq and A Hard Place," From The Wilderness Publications, 27 August 2002, http://www.fromthewilderness.com/free/ww3/082702_iraq.html. Also, DOD News Transcript, 19 August 2002, http://www.dod.gov/transcripts/2002/t09122002_t0819foxanew.html.

13. Colin Powell, Address to the U.N. Security Council, 5 February 2003, www.whitehouse.gov/news/releases/2003/02/20030205-1.html.

14. Henry Kissinger, "Consult and Control," *The Washington Post*, 12 August 2002, sec. B, p. 1. His call to action and view of second-order effects were: "America's special responsibility, as the most powerful nation in the world, is to work toward an international system that rests on more than military power—indeed, that strives to translate power into cooperation. Any other attitude will gradually isolate and exhaust America."

15. Bush, *National Security Strategy*, p. 19.

16. Bill Nichols, "U.N.: Iraq Had No WMD after 1994," *USA Today*, 3 March 2004, p. A1.

17. Scott McLemee, "100 Christian Ethicists Challenge Claim that Pre-emptive War on Iraq Would Be Morally Justified," *Chronicle of Higher Education*, web daily, 23 September 2002, http://maxspeak.org/gm/archives/chron.htm.

18. "Support for Bush in 'Just War,'" The Ethics & Religious Liberty Commission of The Southern Baptist Convention, 4 October 2002, http://www.christianity.com/partner/Article_Display_Page/0,,PTID314166|CHID597896|CIID1554306,00.html.

19. Ralph Peters, "Revolutionizing Warfare on the Fly," *Atlanta Journal-Constitution*, 23 March 2003, p. D6.

20. John Brinsfield, "Going Against Rules No Reason to Rejoice," *Atlanta Journal-Constitution*, 30 March 2003, p. C2.

21. Paul W. Schroeder, "Iraq: The Case Against Preemptive War," *American Conservative*, 21 October 2002, http://www.amconmag.com/10_21/iraq.html. In trying to put preemption and preventive war in context, Schroeder goes on to say: A more dangerous, illegitimate norm and example can hardly be imagined. As could easily be shown by history, it completely subverts previous standards for judging the legitimacy of resorts to war, justifying any number of wars hitherto considered unjust and aggressive. It would, for example, justify not only the Austro-German decision for preventive war on Serbia in 1914, condemned by most historians, but also a German attack on Russia and/or France as urged by some German generals on numerous occasions between 1888 and 1914. It would in fact justify almost any attack by any state on any other for almost any reason. This is not a theoretical or academic point. The American example and standard for preemptive war, if carried out, would invite imitation and emulation, and get it. One can easily imagine plausible scenarios in which India could justly attack Pakistan or vice versa, or Israel any one of its neighbors, or China Taiwan, or South Korea North Korea, under this rule that suspicion of what a hostile regime might do justifies launching preventive wars to overthrow it.

22. Jeffrey Record, "The Bush Doctrine and War With Iraq," *Parameters,* 33 (Spring 2003), p. 19. In his conclusion, Record recounts an illuminating historical case of proposed preventive war by a US military leader when the Cold War was heating up: In the earliest years of the Cold War, before the Soviet Union exploded the first atomic bomb, there were calls in the United States for preventive war against another evil dictator. The call continued even after the Soviets detonated their first bomb in 1949. Indeed, in the following year,

the Commandant of the Air Force's new Air War College publicly asked to be given the order to conduct a nuclear strike against fledgling Soviet atomic capabilities. "And when I went to Christ," said the Commandant, "I think I could explain to Him why I wanted to do it now before it's too late. I think I could explain to Him that I had saved civilization. With it [the A-bomb] used in time, we can immobilize a foe [and] reduce his crime before it happened." President Truman fired the Commandant, preferring instead a long, hard, and, in the end, stunningly successful policy of containment and deterrence.

23. US Army, "Apply Just War Tradition to Your Service as a Leader and the Profession of Arms," Training Support Package (TSP) 158-C-1131, Center for Army Leadership, US Army Command and General Staff College, Fort Leavenworth, Kans., http://155.217.58.58/cgi-bin/atdl.dll/cctsp/158-c-1131/158-c-1131.htm. This TSP presents a thorough introduction to the topic and the role of officers as leaders and professionals in thinking about foundations for and key references to the Just War tradition.

24. *The Law of Land Warfare*, requires a "previous and explicit warning" as a declaration of war or an ultimatum with a conditional declaration of war, based on the 1907 Hague III, Article 1. US Department of the Army, *The Law of Land Warfare*, Field Manual 27-10 (Washington: GPO, 18 July 1956, with Change 1, 15 July 1976), ch. 2, sec. 1, para. 20.

25. James Turner Johnson, "Threats, Values, and Defense: Does Defense of Values by Force Remain a Moral Possibility?" *Parameters*, 15 (Spring 1985), pp. 14–15.

26. Gerald Bradley, "Iraq and Just War: A Symposium," The Carnegie Endowment for International Peace, 30 September 2002.

27. Paul J. Griffiths, quoted in Joe Feuerherd, "Preemption, Aggression, and Catholic Teaching: Iraq War Highlights Problems of Seeking Justice Through Force," *National Catholic Reporter*, 25 October 2002, p. 25, http://www.findarticles.com/p/articles/mi_m1141/is_1_39/ai_94079348.

28. Bradley, "Iraq and Just War: A Symposium."

29. Joe Feuerherd, "Preemption, Aggression, and Catholic Teaching."

30. Thomas Kuhn, *The Structure of Scientific Revolutions* (Chicago: Univ. of Chicago Press, 1962).

31. Central Intelligence Agency, *Iraq's Weapons of Mass Destruction Programs* (Washington: GPO, October 2002).

32. William Galston, "Iraq and Just War: A Symposium," The Carnegie Endowment for International Peace, 30 September 2002. Also see Schroeder, p. 7.

33. Ibid. Schroeder considers, "the superficially plausible idea that a preventive war launched against Hitler's Germany in 1936 at the time of Germany's reoccupation of the Rhineland or in 1938 at the annexation of Austria would have prevented all the horrors of World War II and the Holocaust."

34. Cited by Malcolm Jones in his review of David Fromkin, *Europe's Last Summer* (New York: Knopf Publishing Group, 2004) in *Newsweek*, 5 April 2004, p. 60.

35. Bush, *National Security Strategy*, p. 19.

36. Michael Walzer, "Iraq and Just War: A Symposium," The Carnegie Endowment for International Peace, 30 September 2002. Walzer's proposal is an excellent approach. In order to conduct coercive inspections, both a strong, international case would be needed (and provide legitimate authority) and military force would be moved into position to express the will of the international community, protect the inspectors, and expand the diplomatic strength and military capability for future action, should coercive inspections be resisted.

37. Martin Cook, "Ethical and Legal Dimensions of the Bush 'Preemptive' Strategy," paper presented at the Marshall Center International Law Conference, Garmish-Partenkirchen, Germany, 16 September 2003 (and forthcoming in the *Harvard Journal of Law and Public Policy*), proposes that the contemporary setting is more akin to a "pre-Westphalian" world than a world of sovereign states: "In that respect, at least if one measures the contemporary situation by al Qaeda's aspirations, the contemporary Global War against Terrorism bears much more striking resemblance to Augustine's circumstance than it does to Westphalian states responding to aggression committed against them by other Westphalian states."

Chaplain (Colonel) **Franklin Eric Wester**, *US Army Reserve, is assigned as the Command Chaplain for the US Army Reserve Command, Fort McPherson, Georgia. As a Lutheran minister, he has served as an Army Chaplain since 1982 in the active component, the Army National Guard, and the US Army Reserve. He was the Joint Task Force Chaplain for Operation Provide Refuge, an interagency mission aiding Kosovar refugees. He holds master's degrees from Trinity Lutheran Seminary, New Brunswick Theological Seminary, and the US Army War College.*

Does Israel Belong in the EU and NATO?

RONALD D. ASMUS AND BRUCE P. JACKSON

*O*VER THE COURSE of the past year, a debate has started over whether Israel should rethink its relationship with the core institutions of the Euro-Atlantic community, namely NATO and the EU, and if so, how. The impetus for this rethinking has originated both in Israel and on both sides of the Atlantic. At first blush, an outside observer might ask: Why are we having this debate—and why now? The answer to that question has several parts.

First, the Euro-Atlantic community itself has undergone a profound process of transformation since the end of the Cold War, shifting its strategic focus east and south toward the wider Middle East. That shift started with the epochal fall of the Berlin Wall on November 9, 1989, the ensuing collapse of communism in Central and Eastern Europe, and the eventual unraveling of the former Soviet Union. The delayed aftershocks of that geopolitical earthquake are still being felt today, as we can see in the dramatic events unfolding in Ukraine.

This revolutionary set of events has led to a dramatic strategic response by the West. Since the early 1990s, NATO and the EU community have expanded across the eastern half of the continent, nearly doubling their size and membership to help consolidate democracy and security across the new Europe. They have intervened beyond their borders to stop ethnic wars in the Balkans and have developed into pan-European institutions stretching from the Baltic Sea in the north to the Black Sea and Turkey in the south. Today they have stretched their borders to the northern edge of the wider Middle East and are assuming new responsibilities across this wider security space.

The other seminal event reshaping the Euro-Atlantic community was September 11, 2001 and Al Qaeda's terrorist attacks against the United States. Those attacks accelerated the strategic shift of the West away from an insular focus on threats in Europe to those emanating from beyond it. They drove home the fact that the greatest threats to Euro-Atlantic security may well originate from regions such as the wider Middle East. To be sure, there is still unfinished business in

Europe and Eurasia. A dictator remains in power in Belarus, Russia is moving in an anti-democratic direction, Ukraine's democratic future must still be consolidated, and a final settlement remains outstanding in the Balkans. Integrating Turkey into the EU and developing a strategy for the wider Black Sea region remain major challenges.

The strategic contours of a Euro-Atlantic geopolitical system are beginning to settle.

While complex questions of policy still confront Washington and Brussels, the strategic contours of a new Euro-Atlantic geopolitical system are beginning to settle. That system has now anchored Central and Eastern Europe and the Baltics; it is working to consolidate peace and security in the Balkans and is starting to reach out across the Black Sea region. It now stretches across the European continent—possibly to include a democratizing Ukraine now seeking to turn west. Despite the many painful debates that lie in front of it, the EU has decided to embrace the full integration of Turkey, which in turn will consolidate the Euro-Atlantic community's border on the northern edge of the Middle East. Finally, the rift across the Atlantic and within Europe created by the Iraq war is gradually being overcome and the strategic unity of the West laboriously reestablished.

One strategic question remaining from the twentieth century is the relationship of Israel to a Euro-Atlantic community that is coming closer and closer to its borders. Closely related to the process of strategic redefinition of the Euro-Atlantic community is what we would term the perishability of revolutionary time. For the past decade there has been a plastic or malleable quality to the process of reshaping the Euro-Atlantic community. This window would seem to offer the United States, Europe, and Israel an unprecedented opportunity to reshape their own relations in ways inconceivable in previous periods. It is difficult to see how these quite extraordinary circumstances will persist indefinitely. Therefore, a compelling reason to address this question now is that we may not have the window of opportunity to address it again in the foreseeable future.

As important as the residual challenges of securing peace in Europe are, the deadly threat to Western societies posed by the nexus of new anti-Western fundamentalist ideologies, terrorism, and the possible use of weapons of mass destruction in the wider Middle East is pulling the Euro-Atlantic community into this region. That is why NATO has embraced its first modest missions in Afghanistan and, to a lesser degree, in Iraq. Under American prodding, the West is debating whether and how to pursue a long-term strategy aimed at the transformation and democratization of the region as a whole, and the U.S.-European agenda is increasingly dominated by how to cooperate on questions ranging from Iran to Middle East peace. The old compartmentalization between a European and Middle Eastern security space is crumbling, and in this context, the question of whether and how Israel relates to and is included in broader Western strategy has inevitably arisen.

NATO *has embraced its first modest missions in Iraq and Afghanistan.*

The second part of the answer to why the issue of rethinking and upgrading Israel's relations with NATO and the EU is now being raised has to do with events in Israel and the region. With the collapse of the Oslo peace process and the second intifada, the vision of Israel successfully integrating itself into a new and transforming Middle East was dealt a severe setback. The vision of closer integration between Israelis and Palestinians has been supplanted by a desire on both sides for separation. The prospect of Iran obtaining nuclear weapons—possibly encouraging other countries like Egypt to consider the same—would pose a very real threat to Israeli security. While Israel could be a great beneficiary of a Western strategy aimed at transforming and democratizing the region, should such a strategy backfire or fail, Israel would be one of the first countries to feel the consequences.

We want to be clear on one point. Much of the recent discussion in the West about Israel and NATO has focused on a possible peacekeeping or monitoring role for Alliance forces in connection with a possible Israeli-Palestinian peace agreement. But what some Israeli strategic thinkers are starting to discuss—and what we are addressing here—is something different, namely an upgraded strategic relationship between Israel and Euro-Atlantic institutions like NATO and the EU that would lead to increasingly closer ties and could include eventual membership. Such an upgraded relationship could become a crucial part of an overall package aimed at securing a peace settlement as well as

a part of an overall reassessment of NATO and EU ties in the region. It would not exclude NATO and/or the EU from assuming some role in a future peace settlement. But the strategic purpose would be different, namely to bring Israel closer to and anchor it in the Euro-Atlantic community.

These are the issues and questions that some far-sighted Israelis have also started to pose. What seems remarkable to us, however, is not that they are now being posed, but rather that Israelis have not been more curious and assertive in exploring such opportunities for enhancing Israel's security and long-term viability. At a minimum, both Israel and the West will need to review what kind of relationship does or does not make sense as the European and Middle Eastern security spaces increasingly overlap. Should Israel seek closer relations with NATO and the EU? Or should Israel remain outside? We believe Israel may have a unique window in which it can seek to realign itself vis-à-vis Euro-Atlantic institutions, and we turn now to a sketch of the reasons why such a realignment is in the interests of Israel, the United States, and Europe.

What's in it for Israel?

*T*HE PROPER PLACE to start such an analysis is Israel. After all, if Israelis are not interested in seeking an upgraded[1] relationship with the Euro-Atlantic community, then there is little point in this exercise. Why might Israel be interested in such a step? It is, of course, up to the Israelis themselves to determine their national interest. Yet an outsider might offer the following thoughts for consideration.

First, at a minimum, Israel should want to have closer ties with NATO and the EU simply because they are actors who are coming closer to Israel geographically and who are developing strategies to shape the Middle Eastern neighborhood in which Israel lives. Israel should aspire to have the closest possible relations with the actors and institutions setting those policies.

Second, a new and upgraded relationship between Israel and the Euro-Atlantic community could become a critical element in helping to provide the security Israel will need to take steps to make peace with a Palestinian state in the Middle East. Anchoring Israel more closely with NATO and the EU can reduce the sense of isolation that Israel feels. In a post-Oslo political environment, such a step could be especially important in convincing a skeptical Israeli public to support such a settlement.

Third, an upgrading of Israel's relations with the institutions of the Euro-Atlantic community could play an important role in ending Israel's political and diplomatic isolation and strengthen Israel's position vis-à-vis other parts of the world, including its adversaries in the Middle East.

Last but perhaps most important, the American connection is a necessary but not necessarily a sufficient condition for Israel's long-term survival and viability. It is and will remain *the* key Western anchor for Israel, but it is also clear that the country would benefit from a second European or Euro-Atlantic anchor as well. This is especially true if one views Israel's needs in a broader strategic sense extending beyond military security and including economic markets, access to technology, currency stability, etc. Developing closer relations with the Euro-Atlantic community can also serve as an insurance policy in case Israel is ever faced with a rapidly deteriorating security situation in the region. In such a scenario, Israel might feel the need to seek closer strategic relations with the West. It would make sense to lay the foundation for such an option in advance and before such a crisis.

Anchoring Israel more closely with NATO *and the* EU *can reduce the sense of isolation.*

We would be the last people to question the importance of Israel's American connection. Like most Americans, we are proud of our country's track record of supporting Israel. Yet it is not fanciful to raise the question of what might happen in, for example, 20 years' time if the U.S. were embroiled in a conflict in the Pacific and then also faced with a Middle East conflict, one in which protecting Israel could expose us to terrorist attacks in the American homeland. Even if we agree that Americans see themselves as defenders of Israel (for reasons of history, faith, and cultural values), it is hard to see why Israel should rely exclusively on America's assurance forever. Few states in history have relied upon a single alliance and an informal one at that. Most have sought to construct a web of interlocking relationships as a strategic insurance policy. It seems only prudent for Israel to seek a multilateral complement to a strong bilateral relationship.

This list of potential benefits should be matched by what some Israelis could view as the possible downsides or "costs" of such a move. One set of concerns centers on Israel's deeply rooted belief in the need for political and strategic self-reliance and its reluctance to rely on allies. Related to this is Israel's own negative history with and distrust of multilateral institutions, especially the United Nations. Israel will think hard about whether closer relations with the EU and NATO could constrain its freedom of maneuver on core issues central to its security.

A second set of concerns has to do with Israel's own identity and its relationship with Europe. The question of national identity is a vast subject. Suffice it to mention several key issues: Do Israelis today see themselves as a democratic Jewish state whose values are fundamentally the same as those of the Euro-Atlantic community? Or do they view themselves as a people essentially betrayed by Europe? If the answer is the former, then there is no reason why Israel should not seek a close relationship with and

perhaps even inclusion in those institutions created to defend and sustain those values. If the answer is the latter, however, it is hard to see why Israelis would see a strategy of returning Israel to European institutions as desirable.

As Americans, we sympathize with feelings of Israeli exceptionalism. Yet, as Euro-Atlanticists, it strikes us as a bit odd to argue that in terms of value Israel is so distinct that it cannot fit in the broader Euro-Atlantic community, while Erdogan's Turkey can and does. It also strikes us as curious that Israel sees itself as a close American ally yet at times is nervous about developing close relationships with other close American allies. The reason is clear: Israel's political relations with the United States are excellent, but with Europe they are troubled. Many Israelis today doubt Europe's commitment to Israel and are concerned about growing anti-Semitic currents on the continent. Israelis fear that closer ties with Europe will generate greater pressure for a peace settlement on unfavorable terms. These issues and fears in Israel need to be faced and resolved, something that will happen only in a real dialogue with Europe.

A third and final set of doubts has to do with the viability and cohesion of the Euro-Atlantic community itself. After all, why should Israel make a major move to get close to the Euro-Atlantic community if that community itself is in danger of falling apart? Are Americans and Europeans capable of overcoming the divisions of recent years and will they undertake this kind of strategy? Even if the Euro-Atlantic community regains its footing and comes back together again, many Israelis would ask whether such an upgrade in Israel's relations with the West is really on offer.

Many of these concerns are real and need to be discussed and addressed at length. Even this brief survey suggests that there is a compelling case for Israelis to explore the option of such an upgrade and its potential benefits and downsides. But such a cost-benefit analysis also requires us to look at American and European interests and views.

What about the U.S.?

*a*AS THE MAIN supporter of Israel, the United States shares many of the interests and benefits listed above. This is a case where the interests of both sides potentially dovetail, as Washington clearly would benefit from a strategy that would make Israel more secure and that would enhance its long-term viability as a country and nation. In addition, the following considerations should be taken into account:

First, the U.S. would acquire partners and assistance in sharing the burden of helping to secure Israel and anchor her to the West. To be sure, Europe, by assuming more burden and responsibility, would also gain more potential influence. But it is unlikely to displace the United States as the senior partner and friend of Israel in any meaningful way barring some radical crisis in U.S.-Israeli relations.

Americans can afford to be relaxed. There is no danger of American influence with Israel being marginalized.

Second, the transatlantic rift over how to deal with Israel would presumably be narrowed significantly if not overcome, thus eliminating one of the current sources of tension in U.S.-European relations. One way to help narrow the gap between the United States and Europe is to force both sides to work together in developing a more common approach. It is noteworthy how deep differences often suddenly narrow when one has to share responsibility and contemplate joint action.

Third, a common Euro-Atlantic policy toward Israel would also mean that the Arab world would be less able to play on differences between the United States and Europe. Over time, this could increase the U.S. negotiating leverage and position in the Arab world.

The opening to Israel would help resolve the moral and strategic contradictions that chafe within U.S. policy.

To be sure, there will be Euro-skeptical voices in the United States who will question such an approach. They will argue that ensuring Israel's security through a bilateral relationship with the United States is easier, more flexible, and perhaps even advantageous. They would claim that the United States would be making a mistake by "allowing" Europeans to acquire a more important voice and greater influence in Israel and in the Middle East. Yet how can we assert that Israel is part of the "West" but also insist that developing Israel's ties with the core institutions of the West is somehow too hard or complicated? At the end of the day, if Israel makes it clear that it desires a closer relationship with Europe, then such voices are likely to be muted and limited in their impact.

As Americans, we discuss U.S. policy toward Israel frequently with our friends and colleagues. In our view, it is clear that the opening of Euro-Atlantic institutions to Israel would help resolve the moral and strategic contradictions that chafe within U.S. policy. For example, the U.S. proposes to launch the greatest democracy program for the wider Middle East ever conceived but cannot define the role of a democratic Israel in that program. NATO has upgraded a "26 plus 1" relationship with Russia because Moscow can assist the West on terror and proliferation, but not with Israel in spite of its obvious potential contribution in these areas. Americans are overwhelmingly convinced that Turkey is an integral member of the Euro-Atlantic community but unsure or vague about whether Israel is or should be. We believe that U.S. policymakers should welcome a closer Israeli relationship with key Euro-Atlantic institutions and that such a step would help resolve these contradictions.

A good deal of political legwork would undoubtedly be required on the American side as well to make this official U.S. policy. Yet, arguably, the United States would have the fewest problems adopting such a strategy. It will not be the obstacle if Israel wants to move forward.

And what about Europe?

*T*HE REAL QUESTION lies in Europe and in European attitudes. In many ways, this is the key issue, since Europeans not only run the EU and have a decisive voice in NATO as well, but also have a more troubled relationship with Israel. Yet here, too, there are arguably several ways in which Europe could benefit from such an upgrade:

First, if such an upgrade was part and parcel of a move toward peace in the Middle East, the EU would move from the sidelines to center stage in the peace process and Middle Eastern politics more generally. If Europeans truly believe that achieving Middle East peace is critical, this is one way in which they can contribute to this goal. It could acquire the kind of major role many European leaders have long aspired to have—and give an enormous boost to European diplomatic credibility and standing in the region and beyond.

Second, Europe's own strained relationship with Israel could be mended. The current situation, in which the EU has extremely close economic and other ties with Israel but almost no meaningful political or strategic dialogue, could be overcome. A Europe that is more engaged on the ground is also likely to be a more responsible one, including in Israeli eyes.

Third, obviously some in Europe may fear that such a move would mean abandoning Europe's policy of being "even-handed" in the Israeli-Palestinian conflict and undercut Europe's standing in the Arab world. One should question whether that needs to be the case. If handled correctly, such a step might actually lead pro-Western moderate Arab states to seek their own closer ties with the Euro-Atlantic community as well, something we should welcome.

There are three big questions about the feasibility of Europeans making such a leap of strategic imagination to embrace such a bold strategy. The first is whether European leaders have the vision and courage to take such a step and whether it is domestically sustainable given the kind of critical sentiments toward Israel one finds today in many parts of Europe. As we have debated this issue over the last year, the initial response of many European colleagues has been that Israel and Palestine must first make peace, and then and only then should we discuss bringing Israel closer to and perhaps into our Euro-Atlantic institutions. We would suggest Europeans need to move beyond this static and reactive approach, flip or reverse this logic, and think in terms of what they and we can do and offer in advance of or parallel to moves toward peace in order to reinforce that process. Indeed, it is only by thinking in these more dynamic terms that Europe can acquire the sort of role and influence it wants.

The second and perhaps equally important question for many Europeans will be whether Europe can find a way to upgrade its relations with Israel yet sustain what it views as its special commitment to the Palestinians as well as to key Arab states. As mentioned earlier, it would be simplistic to assume that such a move would automatically lead to deterioration in Europe's relations with the Arab world. Indeed, if handled properly, one could argue that such a move would enhance Europe's prestige and influence in the Arab world. But this underscores that European countries will be more comfortable in upgrading Israel's relations with Europe if that step can be embedded in a broader regional approach that also contains opportunities to step up outreach to key Arab states. This is a question of packaging.

Finally, there is the question of whether the EU will be willing to assume the kind of added responsibility such a strategic shift would entail—and whether it would be willing to do so in partnership with the United States. Many Europeans could be concerned that they are being drawn into potential conflicts and assuming new risk in the region. Yet at the end of the day, it may be far easier for Europe to mend its relations with Israel in a trans-Atlantic framework. Many Europeans are also aware that the problematic relationship between Israel and Europe also creates a long-term strain on U.S.-European relations, which manifests itself in doubt about the reliability of the U.S.-European partnership in the Middle East. Establishing a better Israeli-European relationship would not only serve to enhance Israeli's security, but also mitigate those doubts.

Where to start

𝓕OR THE REASONS laid out above, we believe there is a compelling strategic argument why Israel should explore the option of building closer ties to the Euro-Atlantic community. As noted, we are living in a moment of strategic fluidity—both across the Atlantic and in the Middle East. The future contours of the Euro-Atlantic community are likely to settle in the years ahead. The question is whether they will come to an end on the northern edge of the wider Middle East and stop with Turkey and the Black Sea region—or whether they will reach down to embrace a democratic country like Israel as well. In the Middle East itself, we may be entering a new phase of strategic fluidity as well—in connection with Prime Minister Sharon's disengagement plan for Gaza, the election of a new Palestinian leadership, and in the region more broadly.

For all these reasons, this is the right moment for Israel to decide whether it wants to pursue a Euro-Atlantic upgrade. Both Europe and Israel need to participate equally with the United States in such a rethinking. Movement will be required on both the European and Israeli sides to make progress. As a first step, it is nevertheless Israel that needs to decide that it wants to seek a new and expanded relationship with the Euro-Atlantic community.

The instruments or tools to do so already exist. The recent Istanbul NATO summit has for the first time opened the door to creating a separate bilateral Israeli-NATO relationship outside of and in addition to the Mediterranean Dialogue. Israel has a friend in the current secretary general of NATO, Jaap de Hoop Scheffer. And the NATO format would seem more manageable in political terms since the U.S. is also involved and it has a greater focus on security issues that are of immediate concern.

Israel today actually has a much closer relationship with the EU than it has with NATO—arguably closer than any other non-EU member—but that relationship is non-strategic, politically stunted, and very much limited to trade, technology, and science. Yet here, too, the offer from the Essen summit of building a special relationship between the EU and Israel is still on the table. Moreover, the EU's European Neighborhood Policy (ENP) offers a broader framework within which the EU can deepen ties with both Israel and other countries in the region. The history of the past decade in terms of Euro-Atlantic outreach has shown that it is possible for the receiving country to fill initiatives like the ENP with more substance than its drafters may originally have intended. From the bottom up, both the EU and NATO would have to start to build a political and strategic relationship that could grow over time as well.

All long journeys start with small steps, and a strategic reorientation of the kind discussed is no different. It would require a top-down and bottom-up component. At the top there are a number of political issues—largely but not exclusively between Israel and Europe—that would have to be resolved and would undoubtedly take time. Progress toward a peace settlement with the Palestinians and clarity on Israel's final borders undoubtedly are at the top of that list.

The scope of what is imaginable or possible is wide. It will depend upon the interest of the NATO nations as well as Israel. Israel can start by turning to those NATO nations that it considers to be friends and that are likely to be most interested in developing this relationship. They in turn can take the lead in creating opportunities for Israel to deepen its relationship through the plethora of existing partnership mechanisms—or by working with Israel in a subgroup of NATO allies. Over time, Israel might aspire to develop the kind of close partnership relationship that countries like Sweden or Finland have developed over the past decade and enjoy today—a very close political relationship, close military interoperability, and the de facto yet unspoken option to join if the strategic environment ever makes such a move necessary.

Note

1. A point on language: In this essay we use words like "upgrading" Israel's relations with the Euro-Atlantic community or "anchoring" Israel to the West. These words include a spectrum of relationships ranging from closer ties up to and including possible membership. We reserve judgment at this stage on the exact form such an upgraded relationship would or should take. What we are talking about is the creation of a new and much closer relationship in which Israel both sees itself as part of the West and aspires to have the closest possible relationship with the Euro-Atlantic community—and one in which the United States and Europe think of and include Israel as a close partner and what might be termed a member of the Euro-Atlantic community's extended family.

Ronald D. Asmus *is executive director of the German Marshall Fund of the United States' Transatlantic Center in Brussels.* ***Bruce P. Jackson*** *is president of the Project on Transitional Democracies. They have been the Democratic and Republican heads of the U.S. Committee on NATO. This essay grew out of a U.S.-European-Israeli dialogue supported by the German Marshall Fund of the United States.*

IRAQ'S RESILIENT MINORITY

Shaped by persecution, tribal strife and an unforgiving landscape, Iraq's Kurds have put their dream of independence on hold—for now

ANDREW COCKBURN

In THE SAVAGE HEAT of summer on the Mesopotamian plain, where the temperature regularly tops 110 degrees, Baghdadis crave the cool mountains and valleys of Kurdish Iraq, where the wild landscape climbs up to the rugged borders of Iran and Turkey. Even amid this dramatic scenery, the rocky gorge of Gali Ali Beg stands out as a spectacular natural wonder, and it was there one day last August that I encountered Hamid, an engineer from Baghdad, happily snapping photographs of his family against the backdrop of a thundering waterfall.

Hamid had just arrived with his wife, sister, brother-in-law and four children. By his account, the dangerous nine-hour drive from Baghdad—much of the ongoing Iraq War is fought on the highways—had been well worth it. Excitedly, he reeled off a long list of Kurdish beauty spots he planned to visit before heading home.

Given that Kurds have vivid memories of genocidal onslaughts by Saddam Hussein and his Baath Party henchmen, and are currently wary of attacks by Arab Sunni insurgents, I was surprised to see Hamid here. Was he nervous? Were the Kurdish people friendly? The 30-year-old Hamid, who earns a prosperous wage working for a major American corporation in Baghdad, looked puzzled. "Why not?" he replied, "it's all the same country. It's all Iraq."

"They still don't get it," hissed a Kurdish friend as we walked past a line of cars with Baghdad plates in a parking lot. "They still think they own us."

Kurds like to tell people that they are the largest nation in the world without a state of their own. There are roughly 25 million of them, predominantly non-Arab Muslims practicing a traditionally tolerant variant of Islam. Most live in the region where Iraq, Turkey and Iran meet. They claim to be an ancient people, resident in the area for thousands of years, an assertion not necessarily accepted by all scholars. Until the 20th century, they were largely left to themselves by their Persian and Ottoman rulers.

As nationalism spread across the Middle East, however, Kurds, too, began to proclaim a common bond as a nation, even though they remained riven by tribal feuds and divisions. The British, after defeating the Ottomans in World War I, briefly considered the creation of an independent Kurdish state. Instead, in 1921, Great Britain opted to lump what was called southern Kurdistan into the newly minted Iraqi state, ruled by Arabs in Baghdad. Successive Iraqi governments broke agreements to respect the Kurds' separate identity, discouraging, for example, the teaching of Kurdish in schools. The Kurds protested and periodically rebelled, but always went down to defeat. In the 1980s, Saddam Hussein sought to solve the Kurdish problem by eliminating them in vast numbers; as many as 200,000 died on his orders, often in chemical weapons attacks. Thousands of villages were destroyed. Survivors who had lived by farming were herded into cities where they subsisted on government handouts.

Briefing

"We live in a tough neighborhood," says a war-weary official in the Kurdish Regional Government of Iraq. Scattered across the Middle East, some 25 million Kurds live as minorities in Turkey, Iraq, Iran, Syria and neighboring countries (shaded pink). Most are non-Arab Muslims who speak a distinctive Indo-European language with Persian roots. Although separated by national boundaries and internal bickering, few Kurds can forget the cycles of rebellion and defeat that have defined their long history. Arabs first conquered Kurdish lands in the seventh century. Since then the Kurdish story has been written in blood and disappointment, with control of their homelands changing hands no less than seven times. Thousands of Kurds died when they rebelled at Turkish and Iranian rule in the 1920s and 30s. As many as 200,000 Iraqi Kurds died in the late 1980s when Saddam Hussein cracked down on them. They achieved a respite in the early 1990s, when a U.S.-backed "no-fly" zone kept Saddam from renewing his assaults. Kurds were happy to help topple their old nemesis and his Baathist regime as the United States and allies invaded Iraq in 2003.

Today, however, Iraqi Kurdistan appears in shining contrast to the lethal anarchy of occupied Iraq. Kurds provide their own security and, with some bloody exceptions, have deflected the strife raging around them. The economy is comparatively prosperous. Exiles who escaped to the West are returning to invest and make a living, as are Christian Iraqis now fleeing the embattled cities to the south. The electricity works most of the time (still a distant dream in Baghdad). Iraqi Kurds can now celebrate the outward symbols of independent statehood, from flags to national anthems. The agreement they have negotiated with the groups that dominate the rest of the country allows them to run their own affairs in return for remaining part of a federated Iraq. As the slogan of Kurdistan Airlines proclaims: "Finally a dream comes true." Yet despite these hopeful signs, Kurds are still at the mercy of unfriendly neighbors who will not even let the tiny Kurdish airline service land in their countries. And the past rivalries that so plagued Kurdistan have not gone away. Despite outward appearances, the Kurds remain very much divided.

But at least Saddam has gone. "My age is 65 years, and in my life I have witnessed this village destroyed and burned four times," a Kurdish farmer named Haji Wagid announced to me outside his very modest stone house, in the village of Halawa, tucked away in a mountain valley at the southern end of the Zagros range. "The first time was in 1963, the last time was in 1986." As his wife sorted sunflower seeds in the shade of a mulberry tree, he explained how after the last onslaught, the whole area had been declared a closed military zone. "Four people were

taken away, and to this day we do not know what happened to them," said a neighbor who had sauntered over from his house to invite me for tea and watermelon, "and they killed so many livestock." The villagers were herded off to the city of Irbil, a few hours away on the dusty plain, where it would be easier for authorities to keep an eye on them.

Most of the outside world learned of the Kurdish predicament only in March 1991. Following Saddam's defeat in the Gulf War, the Kurds launched a revolt throughout Kurdistan, briefly securing most of the territory, only to flee in terror when the Iraqi army counterattacked. Suddenly, more than a million men, women and children poured across the Turkish and Iranian frontiers and onto the world's TV screens. The United States, backed by the United Nations and pressured by public opinion, forced Saddam to withdraw from much of Kurdistan. Refugees returned to live more or less independently under the protection of allied fighter jets, which patrolled a newly established "no-fly" zone over Kurdistan. When U.S. ground forces invaded Iraq in 2003, the Kurds were eager to assist in the destruction of their nemesis, contributing troops and providing territory as a staging ground for the assault. The United States has hardly been consistent in its dealings with Kurds, however. Having cheered resistance to Saddam, the United States now discourages all manifestations of Kurdish independence—to preserve Iraqi unity and to avoid offending America's allies in Turkey. Kurds complain that the United States takes them for granted.

I visited Kurdistan for the first time shortly after the Iraqi withdrawal of 1991, driving across the bridge over the Habur River that marks the major crossing at the Turkish border. The former Iraqi immigration and customs post was deserted, and the ubiquitous official portraits of Saddam had in every case been destroyed or defaced. Blackened swaths marked where entire villages had been wiped off the face of the earth. There was no electricity, hardly any traffic and precious little food, but the atmosphere was one of amazed and euphoric relief. Everywhere there were cheerful *peshmerga*, Kurdish fighters with AK-47 rifles and their distinctive baggy pants and turbans. Sometimes whole groups burst into song as they marched through the devastated countryside.

Fourteen years later, the Kurdish end of the Habur Bridge has sprouted a crowded passport control office, complete with flag, a "Welcome to Kurdistan" sign and a bureaucracy demanding proof of Iraqi accident insurance coverage. The guards have abandoned their dashing traditional garb in favor of drab camouflage fatigues. Almost everyone carries a cellphone, and the smooth highway, framed by rich wheat fields on either side, runs thick with traffic.

Approaching Hawler, to use the Kurdish name for Irbil, capital of the Kurdish region, the traffic grew heavier, and eventually halted in an impenetrable jam. In the gathering dusk, firelight flickered all along the mountainside, for it was Friday night and the city folk had streamed out of town for family barbecues.

At the time, Kurdish politicians in Baghdad were negotiating the new Iraqi constitution, one that they hope will guarantee them control of Kurdish affairs. Most important, the Kurdish leaders want most of the revenues from any new oil fields struck in their territory, calculating that if they have an independent income, they will truly be free. Until then, they must rely on money from Baghdad to run the Kurdish Regional Government, which is supposed to get about $4 billion a year, 17 percent of Iraq's national revenues. But Kurdish officials grumble that Baghdad always shortchanges them, passing along a fraction of the amount due. "It's not a favor they're doing us by sending money," a minister complained to me. "We have the right. They should be grateful that we are staying in Iraq."

Meanwhile, because most of Iraqi Kurdistan has been effectively autonomous since 1991, young people cannot remember ever living under anything but Kurdish authority. To them, the horrors of the past are the stuff of legend.

"What happened to your families when the Baathists were here?" I asked a classroom of teenagers in Sulaimaniyah, Kurdistan's second-largest city. A few hands rose. "My father was a nationalist, and he was put in prison," said a boy named Darya. Two students had visited Kirkuk while it was still controlled by the Baathists and had been harassed and kicked by police. Silwan, sitting at the next desk, has a friend whose family was showered with chemical weapons by the Iraqi air force. "His brothers and sisters died." Berava, three rows back, had had a brother imprisoned.

"How many of you think Kurdistan should be an independent country?" I asked.

All of the 13 young people raised their hands.

Only three of them know any Arabic, once a required subject in school. Since 1991 a generation of students has graduated speaking only Kurdish. "That is why," one Kurd remarked to me, "there is no going back."

Each member of the class had paid $52 for an introductory course in English, as offered in the brightly painted premises of the Power Institute for English Language. The school itself, founded in July 2005 by Raggaz, a young Kurd who had grown up in the London suburb of Ealing, is something of an advertisement for the new Kurdistan. Following the 2003 war, Raggaz returned to Sulaimaniyah, the hometown he barely remembered, and saw that Kurdish youths were eager to learn English. He borrowed $12,500 from an uncle, set up the new school and was turning a profit after just three months.

Despite the billions pledged for the reconstruction of Baghdad, all of the cranes visible on that city's skyline are rusting memorials of Saddam's time. The major cities of Kurdistan, by contrast, feature forests of cranes towering over construction sites. Part of this prosperity can be accounted for by money from Baghdad—even the central government's parsimonious contribution helps some. In addition, Kurdistan's comparative peace has attracted investors from abroad and from Arab Iraq. Driving out of Sulaimaniyah early one morning, I passed a long line of laborers toiling at road repairs in 100-degree heat. "Arabs, bused in from Mosul," explained a businessman. "There's 100 percent employment in Sulaimaniyah. You have to wait ages for a Kurdish worker, and Arabs are 40 percent cheaper anyway."

But they're not welcome everywhere. "We don't employ any Arabs, as a security measure," said another returned exile, named Hunar. A year after arriving home from Sweden, he is security director for 77G, the most successful manufacturer in Kurdistan. Tucked away on the outskirts of Irbil, the company claims to make every one of the huge free-standing concrete slabs designed to deflect the blast from the heaviest suicide car bomb or rocket. The company's structures, rising up to 12 feet, have become the symbol of the new Iraq, where any building of consequence is encircled by 77G's long gray walls—including the American Embassy in Baghdad, according to the company. The bunker monopoly is very profitable. Desperate customers have paid as much as $700 per 12-foot-long section—producing roughly 30 percent profit for an enterprise operated by Kurds.

"When Arabs apply to work here, we can't do a detailed background check, so we don't employ them," Hunar explained offhandedly. "It's not discrimination; it's just that we don't trust them. Why? We have to fight our way through to make deliveries in Baghdad—we are always under attack. Arabs have killed six of our guys—but we killed *more*!"

Recounting a typically Kurdish life story of upheaval, persecution and exile, Hunar insisted that the Kurds have no future as part of the Iraqi nation. Semi-seriously, he posited the notion of fencing all of Kurdistan with 77G products: "We could do it. We could seal off all our borders."

Such overconfidence may be dangerous, says David McDowall, a scholar of Kurdish history. "The Kurds should remember that Washington may come and go, but Baghdad is there forever. One day Baghdad will be strong again, and that could lead to a day of reckoning."

Pending that, the Kurds face persistent problems on their borders. "It's hard for our people to understand the difficulties we face," says Falah Mustafa Bakir, minister of state in the Kurdish Regional Government. "None of our neighbors are happy with a strong Kurdistan. When the foreign ministers of Turkey, Iran and Syria, who in reality hate each other, get together, at least they can agree about the 'problem' of Kurdistan. For the Turks, the Kurdistan at the other end of the Habur Bridge does not exist, even though they are looking at it. That's why it's impossible for Kurdistan Airways to get permission to fly to Istanbul."

Turkish attitudes toward Kurdistan are molded by perennial distrust of its own 14 million Kurds, who constitute 20 percent of the population. Irked by discrimination, the Turkish Kurds fought a brutal guerrilla war against Turkey in the 1980s and '90s. Fighting flared up again this year.

A proudly independent Kurdistan just across their border is anathema to the Turks, an attitude most bluntly expressed in the line of fuel tankers stretching back as far as 20 miles into Turkey from the Habur River crossing. They are carrying the gasoline much needed in Kurdistan, which is rich in oil but short on refining capacity. But the Turks feel little inclination to speed the flow. Kurds must wait for their fuel while hapless drivers sleep in their trucks for days or even weeks. "Every now and then the price of gas soars here, because the Turks feel like tightening the screws a little bit by slowing border traffic further," one businessman told me. "Then you see people lining up for 24 hours to get gas, sleeping in their cars."

There is little prospect that Kurdish identity will be subsumed by allegiance to any other nation. "There is more of Kurdistan in Iran," asserted Moussa, whom I encountered in Tawela, a remote mountain village near the Iranian border. About the same number of Kurds—five million—live in Iraq and Iran each. Moussa's sentiment was firmly endorsed by the crowd gathered in the cobbled street.

"Should all Kurds be together as one country?" I asked. "Yes," came the thunderous reply from the group gathered around me. "It *has* to be."

Meantime, the villagers get by as they always have, farming, smuggling and taking jobs with the police.

Kurds, scattered across international borders, have traditionally been well positioned for smuggling. In northeastern Iraq, where the landscape is dominated by soaring mountainsides dotted with the black tents of nomadic shepherds, I encountered an unattended horse trotting along with a bulging pack strapped to its back. This was one of the *aeistri zirag*, or "clever horses," trained to travel alone across the frontier with loads of contraband, such as alcohol, into Iran.

From 1991 to 2003, when Iraqi Kurdistan offered a way around the U.N. trade embargo, a good smuggler-horse was worth as much as a car. At that time, the roads leading to Habur were slick with oil leaking from the tanks on thousands of trucks smuggling crude to Turkey. Kurds at the Habur River checkpoint levied millions of dollars in fees each month. Happy to see the Kurds support themselves, Western powers winked at this flagrant sanction-busting.

In addition, anyone with good connections to powerful Kurds and the ruling elite in Baghdad made huge amounts of money smuggling such basic commodities as cigarettes from Turkey shipped across Kurdish territory to Baghdad. These fortunes may account for much of the frenetic construction activity around Kurdish cities.

Tribal alliances still bring money and power to their adherents. The Barzani clan, headed by Massoud Barzani, dominates the Kurdistan Democratic Party, or KDP. The Patriotic Union of Kurdistan, or PUK, is led by an energetic intellectual named Jalal Talabani. The two groups fought side by side in the 1991 uprising that followed Saddam's defeat in the Gulf War. Then both Kurdish factions came home to rule under the shelter of American air power in the respective areas they controlled, Barzani in the northwestern corner of Iraqi Kurdistan, Talabani to the east.

Rivalry turned to civil war in 1994, over land disputes and, some say, spoils from oil smuggling. The fighting raged on and off through the summer of 1996, when Talabani enlisted military support from Iran and soon had Barzani on the ropes. Desperate, Barzani made a deal with the devil himself—Saddam Hussein—who sent Talabani's forces reeling.

In 1998, the U.S. government persuaded the two parties to sign a peace agreement. They cooperated—with each other and with the United States—through the 2003 war and the negotiations on the Iraqi constitution. Barzani agreed that Talabani could become president of Iraq. Meanwhile, Barzani was given authority as president of the Kurdish Regional Government.

The two sides no longer shoot it out, though there have been scattered and unpublicized armed clashes as recently as this past February. But divisions remain deep and persistent. The city of Irbil is festooned exclusively with portraits of the Barzani family, while portraits of Talabani watch over the streets of Sulaimaniyah, the PUK capital. Barzani's Irbil is somewhat dour, with the few women visible on the streets almost invariably clad in enveloping black *abayas*. Talabani's Sulaimaniyah appears more vibrant, with a lively literary and musical scene and some of its women in Western fashions.

"Sulaimaniyah is the cultural heart of Kurdistan," said Asos Hardi, the crusading editor of *Hawlati*, a weekly newspaper based in the city. "It's relatively new, founded only 200 years ago. Irbil is 9,000 years old, and very traditional. No one has ever seen Barzani's wife. Talabani's wife is very active and visible, the daughter of a famous poet."

Like many Kurds, Hardi, known to his youthful staff as "the old man," despite being only 42, shares the common distrust of the Arab Iraqis who ruled here for so long. "If we can live in this country with proper rights, why not?" he said. "But who can guarantee our future?"

Founded in 2000, Hardi's muckraking journal, whose name means citizen, enjoys the largest circulation of any Kurdish paper. It is clearly doing its job; each of Kurdistan's major political parties has, from time to time, boycotted the paper, each party charging that it is financed by the other's secret police. Hardi conceded that there have never been any physical threats against him or his staff. Nevertheless, he is critical of Kurdistan's current rulers.

"Since 2003 they've been forced to show unity vis-à-vis Baghdad," he remarked, "but there's no real practicable agreement. Although they all talk about democracy, no party accepts being number two for a while."

To maintain an uneasy peace, the two parties have carved up their territory. So Kurdistan has two prime ministers, two ministers of finance, interior, justice, agriculture and so on down the line. They have two chiefs of peshmerga, two secret police forces—even two cellphone companies. Travelers passing from the land of the KDP to the land of the PUK mark their passage by tugging out their cellphones and changing the memory cards, an irritating but revealing fact of life in the new Kurdistan. Asia Cell, which covers PUK territory, was licensed in 2003 by authorities in Baghdad to serve northern Iraq. This arrangement cut little ice in Irbil, where local officials refused to switch from Korek Telecom, a monopoly that existed before the fall of Saddam.

The dominant Barzani family has blessed other entrepreneurs in its part of Iraq, such as the fast-expanding Ster Group. Motorists entering Iraq at the Habur River crossing are required to buy an accident policy from Ster's insurance subsidiary—the fee ranges from $5 to $80, depending on who is collecting the money or talking about the practice. Most travelers who make it to Irbil stay in a shiny high-rise hotel owned principally by the Ster Group. Salah Awla, Ster's fast-talking general manager, gave me a summary of the group's impressive penetration of local business, starting with the new hotel where we were chatting. "We own 60 percent," he said, going on to describe his company's interest in oil wells, shopping centers, gas stations, bottling plants and tourist sites. There seemed no part of the economy immune from Ster's influence—including the lucrative realm of government contracts. "We lend more than $10 million to each ministry," Awla explained cheerfully, "for 'goodwill.' In this way the minister has to give us projects." But he left little doubt about a bright economic future for Kurdistan, especially for those with the right contacts.

Meanwhile, in a fold in the mountains, the village of Halawa, destroyed four times since 1963, has been once more rebuilt. It probably does not look that different now, apart from the smart little mosque financed by a Saudi charity and a school built by UNICEF. The Kurdish administration, said locals, had not offered any help, but even so, one villager mused: "It would be better if Kurdistan were independent. Then everything will be under our control." On the long drive back to Turkey, I had to make wide detours to avoid cities like Mosul where the Iraq War laps at Kurdish borders. And at the Turkish border, the line of immobile trucks and tankers was as long as ever.

ANDREW COCKBURN regularly covers the Middle East. He is co-author of Out of the Ashes, *a biography of Saddam Hussein.*

Originally appeared in *Smithsonian*, vol. 36, no. 9, December 2005, pp. 42–58. Copyright © 2005 by Andrew Cockburn. Reprinted by permission.

The Seas of Sindbad

Paul Lunde

I went down to Basra with a group of merchants and companions, and we set sail in a ship upon the sea, and at first I was seasick because of the waves and the motion of the vessel, but soon I came to myself and we went about among the islands, buying and selling."

—*The Tale of Sindbad the Sailor, from* The 1001 Nights

Besides sailing across the Indian Ocean, there was another sea route from Arabia to India, the oldest of them all. It was not dependent on the monsoon and could be sailed without knowledge of the stars. The Arabian Gulf was the natural corridor between Mesopotamia and India, and the voyage could be made in small boats simply by hugging the coast, always keeping land in sight. Maritime contacts between Mesopotamia and India through Gulf waters go back to the very beginnings of urban civilization in the third millennium BC, when Sumer on the Tigris and Euphrates Rivers was in touch with Harappa on the Indus.

Unlike the Red Sea, whose reef-filled waters and complex wind regime required skilled pilotage, the Arabian Gulf was relatively easy to navigate. While the shores of the Red Sea were sparsely inhabited and almost waterless, the headwaters and eastern shore of the Gulf were home to ancient civilizations. Along its coasts have been found the scattered evidence of some five millennia of trade: fragments of pre-Sumerian al-'Ubaid pottery from the third millennium BC, Chinese celadon and early Islamic glazed jars, Indian bangles, Gujarati carnelian beads, 19th-century coffee cups, Roman coins and the occasional Chinese cash.

On the western shore, in the second century BC, at around the time that Hippalos discovered the secret of the monsoon, the caravan city of Gerrha flourished in what is now the Eastern Province of Saudi Arabia. Strabo, quoting a source named Artemidorus, says: "From their trafficking both the Sabaeans and the Gerrhaeans have become richest of all; and they have a vast equipment of both gold and silver articles, such as couches and tripods and bowls, together with drinking vessels and very costly houses; for doors and walls and ceilings are variegated with ivory and gold and silver set with precious stones."

Although no certain traces of Gerrha have yet been found, it has been identified tentatively with the site known as Thaj, while its name may be preserved in that of the little port of al-'Uqair not far away.

At the head of the Gulf was the port the Greeks called Apologos and the Arabs called al-Ubulla; it was conquered in the year 636 by 'Utba ibn Ghazwan, who, in a letter to the Caliph 'Umar informing him of his victory, called it "the port of Bahrain, Oman and China." We see that in the five hundred or so years since the composition of the *Periplus*, a significant change had taken place: The sea route to China, a country known to the classical world only by hearsay, had been discovered.

Basra, a military camp near al-Ubulla in 638, quickly grew into the largest and most ebullient city in the Islamic world, its population exploding from zero to 200,000 in three decades. The streets were thronged with Arabs, East Africans, Persians, Indians and Malay-speakers from "Zabaj" (Indonesia). A clearinghouse for information, it was here that practical knowledge of Southeast Asia and China began to reach the Arabic-speaking peoples and gradually find its way into both sober works of geography and entertaining wonder books. Basra's most famous writer, the ninth-century polymath al-Jahiz, reflected the excitement of the city's intellectual world: Our sea is worth all the others put together," he wrote, "for there is no other into which God has poured so many blessings. It flows into the Indian Ocean, which extends for an unknown distance."

During the half millennium between the *Periplus* and the founding of Basra, most of the references to trade with the Far East emanate from Nestorian Christian sources. Persecuted in Byzantine lands for their unorthodox beliefs, the Nestorians took refuge in Sasanian Persia, where the Sasanians, inveterate enemies of the Byzantines, welcomed them. The Nestorians were active prose-lytizers, and, like other marginalized minorities in largely agrarian Asian states, they turned to trade. They followed the overland Silk Roads from Persia into Central Asia and ultimately to China. Nestorian communities were established on the west coast of India—where they were encountered by the Portuguese a millennium later—as well as in Sri Lanka, Socotra and Yemen.

From Basra, traders could sail to the Malabar coast of India without ever losing sight of land.

To the southwest of the Arabian Peninsula, across the Red Sea, lay the Ethiopian Christian kingdom of Aksum, with its port of Adulis. Aksum exported gems, spices—including cassia—incense and gold to Byzantium, India, Sri Lanka and Persia. The gold came from the African interior and the incense was grown in Aksumite territory, but the cassia and other spices almost certainly came from India and what is now the Indonesian archipelago.

The Ethiopic liturgical language, Ge'ez, was derived from one of the non-Arabic Semitic languages of South Arabia. Its alphabet is the only Semitic alphabet to indicate vowels, and the system by which it does this is elsewhere found only in the various writing systems of India—further evidence of the extent to which ideas as well as spices were trafficked across the Indian Ocean.

Between 540 and 570, two events occurred that had profound effects on Arabian society. The great dam of Ma'rib in Yemen, which had supported the South Arabian agrarian kingdoms of Saba and Himyar, burst, either through neglect or because of some unrecorded natural disaster. The huge area it had irrigated was laid waste. The fall of these South Arabian kingdoms was to Arab tradition what the fall of Rome was to the early Christians: the end of an old era and the beginning of a new one.

The other important event was the Sasanian Persian invasion of Yemen, which was at that time ruled by Ethiopia. At the request of Sayf ibn Dhi Yazan, scion of an old Yemeni aristocratic house, the Sasanian fleet landed at Aden and the Ethiopians were exterminated. The fabulous palace of Ghumdan at San'a, with its ceilings of translucent alabaster so thin that the shapes of flying birds could be discerned through it, was destroyed. The historic connection between the African and Arabian shores of the Red Sea was severed, and the wealth of Aksum was no more. The port of Adulis vanished. The Christians of Ethiopia withdrew to the highlands, where they preserved their distinctive culture in isolation, inadvertently giving rise to the legend of the Kingdom of Prester John.

Shortly after the death of the Prophet in 632, the Islamic conquests put an end to the Sasanian dynasty. In the year 711, Muslim armies invaded both Spain in the far West and the province of Sind, roughly corresponding to modern Pakistan, in the East. In 751 the Muslims defeated a Chinese army at Talas, north of the Oxus River, in what is now Kyrgyzstan. The Byzantine Empire, too, suffered: It lost its provinces in Syria, Egypt and North Africa to the Muslims.

Byzantium, India and China were suddenly neighbors of a new world empire, with a new religion—Islam—and a unifying language—Arabic. Persia, the historic enemy of Greece and Rome, was reduced to a province of an Arab empire.

In pre-Islamic times, Sasanian Persia had blocked direct contact between China and the Byzantine empires. Chinese silks and ceramics had reached the West overland almost exclusively through Persia. Exactly when maritime trade with China began is unknown. A Chinese work called the *Chhien Han Shu* mentions trade with the "South Seas" in the first century BC and speaks of voyages lasting a year. Whether these voyages terminated in India or crossed the Indian Ocean to the Arabian Gulf or Red Sea is unknown, but there is little doubt that Chinese ships at this time reached Malaya, the traditional halfway house between what was in pre-Islamic times the Chinese-dominated Pacific and the Indian-dominated Indian Ocean.

That there were Arab and Persian merchants domiciled in Chinese ports in the immediately pre-Islamic period is indicated by the words used to refer to them in Chinese annals. Persians were called *Po-ssu*, Arabs *Ta-shih*. Po-ssu is obviously an attempt to render the word *Pars*, which gave rise to the Greek *Perses*, the Latin *Parthia* and the Arabic *Fars*. Ta-shih is derived from the Aramaic name of the Arab tribe of Tayy, which must have reached China via Aramaic-speaking Nestorian merchants from the region of al-Hira, where this tribe was dominant.

The traditional date for the introduction of Nestorian Christianity to China is 636, the date given on the beautiful black stone stele erected in 781 at the Tang capital of Xi'an in central China. The text consists of 1900 Chinese characters that give an account of the event, while a short inscription in Syriac, the language used by the Nestorians, gives a list of 70 Nestorian missionaries. Only a few years after the stele was raised, the Tang annals speak of 2000 Arab and Persian merchants trading in Canton: Clearly, the sea route between the Middle East and China was wide open.

The discovery of the sea route between the Arabian Gulf and China was an event equal in importance to the discovery by the Portuguese of the sea route to India. It was one thing to cross the Indian Ocean with the monsoon to Gujarat or the Malabar coast, or even to sail south of Sri Lanka and turn north to the Bay of Bengal or east to Malaya—but it was quite another to make the far longer voyage to Canton through a lesser-known sea with its own pattern of winds, to say nothing of the perpetual danger of piracy and the typhoons of the South China Sea. Yet in early Islamic times, direct sailing to Canton via the Gulf seems to have been common practice.

Looking for Waqwaq

In a passage about the sea route to China in his *Kitab al-Masalik wa 'l-Mamalik (Book of Roads and Kingdoms)*, Ibn Khurradadhbih gives an estimate of the size of the Indian Ocean: "The length of this sea, from Qulzum [at the head of the Red Sea] to Waqwaq, is 4500 *farsakhs*." He also states that the distance from Qulzum to the Mediterranean port of Farama is 25 farsakhs. The latter distance, he writes, corresponds to the length of one degree on the meridian; thus, the 4500-farsakh distance to Waqwaq corresponds to 180 degrees. Therefore Waqwaq lies exactly halfway around the world from Qulzum. With its outlandish name and incredible distance eastward, Waqwaq seems to belong to legend rather than commercial geography.

Yet Ibn Khurradadhbih clearly thought Waqwaq was a real place. He mentions it twice more: "East of China are the lands of Waqwaq, which are so rich in gold that the inhabitants make the chains for their dogs and the collars for their monkeys of this metal. They manufacture tunics woven with gold. Excellent ebony wood is found there." And again: "Gold and ebony are exported from Waqwaq."

So where *was* Waqwaq?

Ibn Khurradadhbih's first European editor, the Dutch scholar Michael Jan de Goeje, noted that one of the Chinese names for Japan was *wo-kuo*, "the country of Wo." In the Cantonese dialect, which Arab merchants would have heard, this is pronounced *wo-kwok*. The mystery was solved: "Waqwaq" was an almost perfect rendering of a Chinese name for Japan.

This solution was doubly satisfying, for it solved the mystery of Waqwaq and proved, for the first time, that the Arabs knew of Japan. The trouble with de Goeje's identification, however, is that nothing Ibn Khurradadhbih says about Waqwaq seems to have anything to do with Japan. Although "tunics woven with gold" are barely possible, it is difficult to imagine monkeys and dogs with golden collars in the sophisticated, austere society of ninth-century Japan. Nor does Japan export ebony.

Stories of a land where people grew on trees like fruit endured until the late 17th century.

"Waqwaq" was also the name of an unusual tree. The earliest reference to it (though without the name) occurs in a Chinese source, the *T'ung-tien* of Ta Huan, written before 801. Ta Huan was told the story by his father, who had lived in Baghdad for 11 years as a prisoner of war after the Battle of Talas. He claimed to have heard the following story from Arab sailors:

The king of the Arabs had dispatched men who boarded a ship, taking with them their clothes and food, and went to sea. They sailed for eight years without coming to the far shore of the ocean. In the middle of the sea, they saw a square rock; on this rock was a tree with red branches and green leaves. On the tree had grown a number of little children; they were six or seven thumbs in length. When they saw the men, they did not speak, but they could all laugh and move. Their hands, feet and heads were fixed to the brancehes of the trees.

The same story occurs repeatedly in Arabic sources, where the tree is identified as "the waqwaq tree," and is later embellished by turning the little children into beautiful young women, suspended from the branches by their hair. The classic account, written in 12th-century al-Andalus, says the women "are more beautiful than words can describe, but are without life or soul.... This is a wonder of the land of China. The island is at the end of the inhabited world...."

Two accounts, however, do not fit with the others. One describes a fairly advanced culture: "I have been told by some people that they met a man who had traveled to Waqwaq and traded there. He described the large size of their towns and their islands. I do not mean by this their area, but the size of their population. They look like Turks. They are very industrious in their arts and everywhere in their country they try to improve their ability."

The other is much more intriguing:

In the year 945 the people of Waqwaq sailed with 1000 ships to attack Qanbaluh [on the coast of East Africa, opposite Zanzibar].... When asked why they attacked them, rather than some other city, they answered that it had things needed in their own country and also in China, such as ivory, tortoiseshell, panther skins and ambergris; besides, they wanted to capture men of Zanj, who are strong and able to stand hard labor. They said their voyage lasted a year.... If these men were telling the truth when they said they had sailed for a year, then Ibn Lakis was right when he says the islands of Waqwaq are opposite China.

De Goeje, who knew this text and was still convinced that Waqwaq was Japan, tried without success to find historical evidence of a Japanese naval assault on East Africa in 945. The French scholar Gabriel Ferrand, who first identified Waqwaq with Madagascar, then with Sumatra, wondered with more reason if this were not an account of an Indonesian attack on Madagascar and the East African coast, or even a memory of the aggressive migration of speakers of Austronesian languages from the Indonesian archipelago to Madagascar.

Al-Biruni, who wrote his wonderful book *Kitab al-Hind (The Book of India)* in AD 1000 based largely on Sanskrit sources, mentions a country where people are born from trees and hang suspended from the branches by their navels. Perhaps the waqwaq tree too goes back to a Sanskrit source, and the Arab tales of Waqwaq are themselves a faint memory of a time when the Indonesian archipelago was in the cultural orbit of Hindu–Buddhist culture.

The story of the waqwaq tree traveled westward, like many other oriental stories, appearing in at least one of the surviving manuscripts of the 14th-century traveler Friar Odoric and in one of the many medieval French romances of Alexander the Great. Its final appearance dates from 1685, when all the mysteries of the Indian Ocean had long faded in the light of pragmatic European accounts. It occurs in the *Safinat Sulayman (The Ship of Solomon)*, an account of a Persian embassy to Siam (now Thailand) written by a scribe who accompanied the mission. He says he heard it from a Dutch captain:

Once on our way to China we dropped anchor in the bay of an island to avoid a heavy storm. There was a strange collection of people inhabiting the island who only barely resembled human beings. Their feet were three cubits long and just as wide and they were completely nude and had very long hair. At night they all climbed to the top of their own trees in the jungle, even the women, who bore their children with them under their arms. Once up in the tree they would tie their hair to a branch and hang there all night resting.

Nothing shows the medley of cultures of the Indian Ocean so well as the story of the waqwaq tree: It probably originated in a Sanskrit Hindu text, was told in the eighth century to a Chinese envoy by an Arab sailor, was brought to Europe by a Franciscan friar and was retold by a Dutch sea captain to a Persian envoy to the king of Siam.

The *bahr al-hind*, the "Sea of India," or the *bahr al-sin*, the "Sea of China," as the Indian Ocean was often called, was not a single entity. Those who sailed it said it was made up of seven different seas, each with its own characteristics—the traditional division from which we derive our expression for far-ranging travel, "to sail the seven seas."

Here is how al-Ya'qubi, who died in 897, describes it, obviously following oral tradition:

Whoever wants to go to China must cross seven seas, each one with its own color and wind and fish and breeze, completely unlike the sea that lies beside it. The first of them is the Sea of Fars, which men sail setting out from Siraf. It ends at Ra's al-Jumha; it is a strait where pearls are fished. The second sea begins at Ra's al-Jumha and is called Larwi. It is a big sea, and in it is the island of Waqwaq and others that belong to the Zanj. These islands have kings. One can only sail this sea by the stars. It contains huge fish, and in it are many wonders and things that pass description. The third sea is called Harkand, and in it lies the island of Sarandib, in which are precious stones and rubies. Here are islands with kings, but there is one king over them. In the islands of this sea grow bamboo and rattan. The fourth sea is called Kalah-bar and is shallow and filled with huge serpents. Sometimes they ride the wind and smash ships. Here are islands where the camphor tree grows. The fifth sea is called Salahit and is very large and filled with wonders. The sixth sea is called Kardanj; it is very rainy. The seventh sea is called the sea of Sanji also known as Kanjli. It is the sea of China; one is driven by the south wind until one reaches a freshwater bay, along which are fortified places and cities, until one reaches Khanfu [Canton].

The names of these seven seas derive from the languages of the peoples who lived on their shores. The names preserve, as if in amber, the varied linguistic and cultural backgrounds of the seafarers who first explored these waters, and they sounded just as exotic to Arabic-speakers in the ninth and 10th centuries. These are the seas sailed by Sindbad in *The 1001 Nights*, for the stories of his adventures almost certainly date from the ninth or 10th century; they give a vivid picture of the dangerous and mysterious world of these early mariners. The little book by Buzurgh ibn Shahriyar, *'Aja 'ib al-Hind (The Wonders of India)*, written in the 10th century, is a compendium of seafaring tales remarkably similar in tone to the adventures of Sindbad.

The most important surviving document on international trade in the ninth century is a brief account by Ibn Khurradadhbih, a director of the government postal service in Baghdad, called *Kitab al-Masalik wa 'l-Mamalik (Book of Roads and Kingdoms)*. It describes the overland and maritime routes that linked the Abbasid Empire to the world, including a description of the sea routes to India, Malaya, Indonesia and China. Most interestingly, Ibn Khurradadhbih's account describes regular, organized long-distance trade between western Europe and China long before the days of Marco Polo.

One of his most interesting passages describes a group of international traders called the Radhaniyya, who were Jewish merchants from the "land of the Franks," that is, the Carolingian Empire. Their name may be derived from the Latin name of the river Rhône, and their most probable home port is Venice:

These merchants speak Arabic, Persian, Greek, Latin, Frankish, Spanish and Slavic, [Ibn Khurradadhbih wrote.] They travel from West to East and from East to West, sometimes by land, sometimes by sea. From the West they bring eunuchs, female slaves, young boys, brocades, beaver, marten and other furs and swords. They set sail from the land of the Franks, on the Western Sea [the Mediterranean], and make for al-Farama [on the Isthmus of Suez]. There they transfer their merchandise to camels and go overland to [the Red Sea port of] Qulzum, a distance of 25 farsakhs. From there they set sail on the [Red Sea] and make for al-Jar and Jiddah. Then they sail to Sind, India and China. On their return from China, they bring musk, aloeswood, camphor, cinnamon and other products of the East. They return to Qulzum, then back to al-Farama, where they take ship once again on the [Mediterranean] Sea. Some sail to Constantinople, to sell their merchandise to the Greeks; others go to the capital of the king of the Franks to sell their goods. Occasionally these Jewish merchants sail from the land of the Franks to Antioch. From there they go to al-Jabiyah, three days overland. There they embark on the Euphrates, making for Baghdad. Then they go down the Tigris to al-Ubulla. From al-Ubulla they set sail for Oman, Sind, India and China.

Of the four routes of the Radhaniyya merchants he mentions, two are overland and two are maritime. They coincide with trade routes described in other Arabic sources.

But some scholars have been skeptical about the existence of a "Radhaniyya network." Other sources, most notably the correspondence of Jewish merchants from Egypt preserved in the Cairo Genizah, a fundamental source for our knowledge of medieval commerce, show that the great majority of merchants confined their activities to their own regions, seldom making long journeys themselves. Long-distance trade consisted of a chain of interregional trips rather than the long and inevitably perilous journeys described by Ibn Khurradadhbih. This is certainly true for the period after about 1000, when the trade networks were denser and more organized; however, we have already seen that earlier maritime trade with China was indeed direct. The *Wonders of India* tells of a ship's captain from Siraf who made seven voyages to Canton and back, all before the middle of the 10th century.

The 10th-century Baghdadi historian al-Mas'udi, in his *Muruj al-Dhahab (Meadows of Gold)*, explains how direct voyages to China were replaced with a system in which merchants stopped at a halfway house in either Sri Lanka or Malaya, where they purchased goods brought there from China:

China was prosperous because of the justice with which the country was administered, as it always had been under its former kings, until the year 878. From that year until our own days [943], events have occurred which upset order and overturned

Of Cowry Shells and Coir

Beyond the mouth of the Gulf, past Ra's al-Hadd, the Sea of Larwi begins. Beware, the way is barred by sea monsters upon whose backs grass and seashells grow. Mariners have taken them for islands and come to grief. They blow water into the air, high as a minaret, and when the sea is calm, sweep whole schools of unwary fish into their gaping mouths with their tails. Sailors beat wooden clappers at night to keep them at bay.

These monsters are not the same as the wal, which is 20 cubits long. When caught and slit open, a smaller version of itself is found inside, and this in turn contains another. Despite its size, the wal is preyed upon by a fish called laskh, no more that a cubit long. When the wal misbehaves, maltreating smaller fish, the laskh penetrates its ear and kills it. The laskh attaches itself to the hulls of ships; such ships are safe, for no wal dares approach. Then there are the flying fish with human faces, called mij. When they fall back into the sea, they are devoured by the fish called 'anqatus. For all fish eat each other.

The Sea of Larwi is separated from the Sea of Harqand by an archipelago of 1900 islands. They are closely spaced, the distance between islands being two, three or four farsakhs. They are small, but inhabited and ruled by a queen. The coconut palm thrives; pieces of ambergris the size of a house are thrown up on shore from the bottom of the sea. Wealth is measured in cowry shells, which are stored up in the royal treasury. The local name for them is kastaj. The people of these islands are expert weavers. They make long chemises with sleeves, side panels and pockets, all woven in a single piece. Their houses, boats and other things are made with equal skill.

This is a paraphrase from the oldest firsthand Arabic account of India, island Southeast Asia and China, *Akhbar al-Sin wa'l-Hind (Notes on China and India)*, which dates from 851.

The ninth-century reader would have been struck no less than we are by the number of outlandish words in these three short paragraphs, and the use of all these exotic words is the author's literary device to prefigure the unfamiliarity of the region he was describing.

The voyage described in the *Notes* begins in the mysteriously named Sea of Larwi and immediately enters a stranger and more threatening environment than the familiar waters of the Arabian Gulf. The first landfall is an archipelago—its name is later given as Dibajat—that is ruled by a woman, where wealth is counted not in gold or silver but in cowry shells, a gift of the sea. Just as Dibajat is outside the patriarchal system of Islamic lands, it is outside the bimetallic monetary system. Yet the inhabitants are far from primitive; they are skilled craftsmen and weavers, though they flourish far from cities. A civilization without cities is a contradiction in terms, and the author is warning the reader that his mental categories must be adjusted.

Even the naïve anecdotes about sea monsters make a point about appearance and reality, warning us not to accept a foreign environment at face value: What appears to be an island may turn out to be the back of a sea monster. The story of the wal, that when opened contains smaller replicas of itself, illustrates a different point of the author's: We are continually faced with reflections of our own beliefs. The vulnerability of the wal to the little lashk shows the fragility of power—something to be borne in mind when considering the vulnerability of the small trading emporia perched along the coasts of the great land-based, agricultural empires of India and China.

With the force and simplicity of a proverb, a predator-prey model of marine life is presented: "For all fish eat each other."

The reader would have been expected to draw the parallel with human society.

The writer's "archipelago of 1900 islands" is in fact the Laccadives and Maldives, 10 days' sail west of Calicut; their actual 2000-plus coral atolls stretch 1500 kilometers (900 mi) north to south, like a net deployed to catch ships. Five hundred years after the author of the *Notes* wrote down his metaphorical and fantastic tale, the Maldives and their waters were well integrated into the system of Indian Ocean trade. Ibn Battuta called them "one of the wonders of the world," and remarked on their annular form and proximity to each other: "A hundred or so are arranged in a circle like a ring, with an opening at one point to form a passage; ships may reach the islands only through this passage.... They are so close together that when leaving one, the tops of the palm trees on the next are visible."

In historical times, the Maldives have successively been Hindu, Buddhist and Muslim. Unusually, we have a precise date for the introduction of Islam there—July 1153—confirmed by an inscription in the Friday Mosque on Male, the principal island, and by an 18th-century Arabic history of the Maldives, the *Ta'rikh Islam Diba Mahal* by the jurist Hasan Taj al-Din. The latter states that the Buddhist ruler, Siri Bavanaditta Maha Radun, embraced Islam as the result of a miracle performed by a pious Muslim who was visiting the islands.

"A strange thing about these islands," wrote Ibn Battuta, "is that their ruler is a woman." So by a curious chance, a queen held power there again, 500 years after the *Akhbar al-Sin wa 'l-Hind* was written. Ibn Battuta calls her Sultana Khadija, but her Maldivian name was Rehendi Kambadikilage; she was 19th in the line of the ruling Theemuge dynasty. She ruled from 1347 to 1362, and her grandfather was Sultan of Bengal, the huge province in northeast India that stretched from the foothills of the Himalayas to the Ganges delta. This connection was of great importance: Bengal, until well into the 19th century, used cowry shells for its currency, and the leading exporter of cowries in the Indian Ocean was the Maldives. In exchange for their cowries, the Maldives imported their staple food, Bengali rice.

The Maldivians farmed their cowries by floating branches of coconut palms in the sea, to which the shells attached themselves. Ibn Battuta described the next step: "They gather this animal in the sea and then put them in holes in the ground until the flesh rots, leaving the white shell.... They exchange [the shells] for rice with the people of Bengal, who also use them as currency. They also sell them to the people of Yemen, who ballast their ships with them instead of with sand. These cowries are also used in the lands of the blacks. I saw them being sold in Mali and Gawgaw at a rate of 1150 per dinar."

In the Maldives, the exchange rate at that time was 400,000 cowries to the dinar, or more when the cowry was weak. This was 1/350 of the Malian rate, a proportion that gives an idea of the profits possible in the cowry trade if the shells could be transported far enough from their place of origin. And they were transported great distances: After Yemeni ships, Portuguese, Dutch and English ships also carried them as ballast, and huge quantities were auctioned to slavers in Amsterdam and London in the 18th century.

continued

continued

The other essential product of the Maldives was coir, the fiber of the dried coconut husk. Cured in pits, beaten, spun and then twisted into cordage and ropes, coir's salient quality is that it is resistant to saltwater. It stitched and rigged the ships that plied the Indian Ocean. Maldivian coir was exported to India, China, Yemen and the Gulf.

"It is stronger than hemp," wrote Ibn Battuta, "and is used to sew together the planks of Indian and Yemeni boats, for this sea abounds in reefs, and if the planks were fastened with iron nails, they would break into pieces when the vessel hit a rock. The coir gives the boat greater elasticity, so that it doesn't break up."

The Maldives were the first landfall for traders sailing to India from the Arabian Gulf, South Arabia or the Red Sea. In the Maldives, ships could take on fresh water, fruit and the delicious, basket-smoked red flesh of the black bonito, a delicacy exported to India and China and Yemen. The people of the archipelago were gentle, civilized and hospitable. They produced brass utensils as well as fine cotton textiles, exported in the form of sarongs and turban lengths. These local industries must have depended on imported raw materials, although it is possible cotton was grown on some of the islands.

Although the archipelago produced not a single spice or exotic wood, nor any of the precious things commonly associated with the eastern trade, the inhabitants generated considerable wealth for themselves by developing their own resources.

the authority of the law. A rebel named Huang Ch'ao...ravaged the cultivated lands of the kingdom and set up his camp before Canton an important city situated on a river larger and more important than the Tigris. This river flows into the Sea of China, six or seven days' journey from Canton. Ships from Basra, Siraf and Oman and other kingdoms sail up it with their merchandise and cargoes. Huang Ch'ao besieged Canton and chopped down the mulberry groves which surrounded the city and had fed the silkworms that produced the silk that was exported to Islamic countries. Silk production and export came to a halt. He took Canton and slaughtered 200,000 Muslims, Christians, Jews and Mazdaeans. This estimate could not have been made except for the custom of Chinese kings of keeping census lists of their subjects and neighboring nations tributary to them. They have officials charged with making the census, for they wish to have an up-to-date idea of the number of their subjects.

Everything in al-Mas'udi's account is confirmed by Chinese annals—though the number of casualties among the merchants is clearly much exaggerated—and it serves to give an idea of the importance of Canton in international trade at the time. Its note of Jews among the merchants killed gives credence to Ibn Khurradadhbih's account of the Radhaniyya.

As a result of Huang Ch'ao's sack of Canton and the political uncertainty in southern China in the wake of his rebellion, the pattern of maritime trade changed. Al-Mas'udi's informant on Chinese affairs, the Sirafi merchant Abu Zayd, told him, "Today the city of Kalah [in Malaya] is the terminus for Muslim vessels from Siraf and Oman. Here they meet the ships from China. But this was not so in the past. Formerly, ships from China sailed directly to Oman, Siraf, the coast of Persia and Bahrain, al-Ubulla and Basra, and ships from these places sailed directly to China. It is only since people could no longer trust in the justice of governments and in their good intentions and since the state of China has become what we have described that merchants meet at this intermediate point."

These political events in China reverberated in the Arabian Gulf. The sack of Canton occurred in the very years that all of lower Iraq was in the throes of the Zanj Rebellion, the 20-year insurgency of slaves from East Africa who worked in the nitrate beds in the marshes of Lower Iraq. During this period, Basra, al-Ubulla and Abadan were dangerous places. Just as traders had moved away from Canton, so they moved south down the Gulf out of harm's way. They settled in Siraf, a small town on the Persian side controlled by men from Oman. Siraf became the main port for the eastern trade, and Basra never reclaimed its former position.

The classical pattern of Asian maritime trade was now set. First Siraf, then a series of other Gulf ports on both the Persian and Arabian sides, ending with Hormuz, became major emporia. The Malay-speaking peoples of Malaya and the Indonesian archipelago became the intermediaries between Islam and China, as Kalah and other ports became the nuclei of local dynasties that in time became Muslim. East of the Strait of Malacca, the seas were dominated by Chinese shipping. Indian shipping continued to play a most important role, and ships from Gujarat, Malabar and Bengal plied the waters of the Indian Ocean. But the days of direct sailing between China and the Arabian Gulf were over.

A ninth-century rebellion in China changed the pattern of maritime trade, making such ports as Malacca important intermediaries between China and the Islamic world.

The new system lasted for nearly five centuries, until the coming of the Portuguese. Although it began as a reaction to unstable political situations, other factors were at work as well. One was speed: Moving ports ever closer to the margins of the monsoon winds reduced sailing time, diminished the ever-present danger of shipwreck and allowed more goods to flow. By establishing emporia halfway to China, in ports like Kalah and later Malacca, the year-long voyage to China could be halved. Reducing the turnaround time made it possible to meet increasing demand, as markets expanded in both the Muslim world and in Europe. These emporia often enjoyed semi-independent or even fully independent status, also a major attraction to merchants.

The key figure in the system remained the small trader who, like Sindbad, risked his life and capital to set sail upon the Indian Ocean, to "go about among the islands, buying and selling."

From *Saudi Aramco World,* col. 56, no. 4, July/August 2005, pp. 20–37. Copyright © 2005 by Saudi Aramco World/PADIA. Reprinted by permission.

[LETTER FROM IRAN]

A NEW DAY IN IRAN?

The regime may inflame Washington, but young Iranians say they admire, of all places, America

AFSHIN MOLAVI

The police officer stepped into the traffic, blocking our car. Tapping the hood twice, he waved us to the side of the road. My driver, Amir, who had been grinning broadly to the Persian pop his new speaker system thumped out, turned grim. "I don't have a downtown permit," he said, referring to the official sticker allowing cars in central Tehran at rush hour. "It could be a heavy fine."

We stepped out of the car and approached the officer. He was young, not more than 25, with a peach fuzz mustache.

"I'm a journalist from America," I said in Persian. "Please write the ticket in my name. It's my fault."

"You have come from America?" the officer asked. "Do you know Car … uh … Carson City?"

Carson City? In Nevada?

He crinkled his eyebrows. The word "Nevada" seemed unfamiliar to him. "Near Los Angeles," he said.

It's a common reference point. The city hosts the largest Iranian diaspora in the world, and homes across Iran tune in to Persian-language broadcasts from "Tehrangeles" despite regular government efforts to jam the satellite signals. The policeman said his cousin lives in Carson City. Then, after inspecting my press pass, he handed it back to me and ripped up the traffic ticket. "Welcome to Iran," he beamed. "We *love* America."

Back in the car, Amir popped in a new tape, by the American rapper Eminem, and we continued on our way to the former U.S. Embassy. It was there, of course, 25 years ago last November, that radical Iranian students took 52 Americans hostage for 444 days, sparking one of the gravest diplomatic crises in U.S. history. The former embassy compound—now a "university" for Iran's most elite military unit, the Revolutionary Guards—was an important stop on my itinerary. I'd gone to Iran to peel back some of the layers of its shifting, sometimes contradictory relations with the United States. America has played an outsized role in Iran over the past century, and is locking horns with Tehran once again over the country's nuclear program.

Perhaps the most striking thing about anti-Americanism in Iran today is how little of it actually exists. After the September 11 attacks, a large, spontaneous candlelight vigil took place in Tehran, where the thousands gathered shouted "Down with terrorists." Nearly three-fourths of the Iranians polled in a 2002 survey said they would like their government to restore dialogue with the United States. (The pollsters—one a 1970s firebrand and participant in the hostage-taking who now advocates reform—were arrested and convicted in January 2003 of "making propaganda against the Islamic regime," and they remain imprisoned.) Though hard-line officials urge "Death to America" during Friday prayers, most Iranians seem to ignore the propaganda. "The paradox of Iran is that it just might be the most pro-American—or, perhaps, least anti-American—populace in the Muslim world," says Karim Sadjadpour, an analyst in Tehran for the International Crisis Group, an advocacy organization for conflict resolution based in Brussels.

He is hardly alone. Traveling across Iran over the past five years, I've met many Iranians who said they welcomed the ouster of the American-backed Shah 26 years ago but who were now frustrated by the revolutionary regime's failure to make good on promised political freedoms and economic prosperity. More recently, I've seen Iranians who supported a newer reform movement grow disillusioned after its defeat by hard-liners. Government mismanagement, chronic inflation and unemployment have also contributed to mistrust of the regime and, with it, its anti-Americanism. "I struggle to make a living," a Tehran engineer told me. "The government stifles us, and they want us to believe it is America's fault. I'm not a fool."

Amir, who is 30, feels the same way. "In my school, the teachers gathered us in the playground and told us to chant 'Death to America.' It was a chore. Naturally, it became boring. Our government has failed to deliver what we want: a normal life, with good jobs and basic freedoms. So I stopped listening to them. America is not the problem. *They* are." It's increasingly apparent that Iran's young are tuning out a preachy government for an alternative world of personal Web logs (Persian is the third most commonly used language on the Internet, after English and Chinese), private parties, movies, study, and dreams of emigrating to the West. These disenchanted "children of the revolution" make up the bulk of Iran's population, 70 percent of which is under 30. Too young to remember the anti-American sentiment of the '70s, they share little of their parents' ideology. While young Iranians of an earlier generation once revered Che Guevara and romanticized guerrilla movements, students on today's college campuses tend to shun politics and embrace practical goals such as getting a job or admission into a foreign graduate school. Some 150,000 Iranian professionals leave the country each year—one of the highest rates of brain drain in the Middle East. Meanwhile, Iranian intellectuals are quietly rediscovering American authors and embracing values familiar to any American civics student—separation of church and state, an independent judiciary and a strong presidency.

But intellectuals are not running the show, and the government continues to clash with the United States. In a January interview, Vice President Dick Cheney said Iran was "right at the top of the list" of potential trouble spots. The most recent crisis is Iran's alleged nuclear weapons program. At issue is whether Iran has the right to enrich uranium—important for a civilian nuclear energy program, but also crucial to creating an atomic bomb.

Recent news reports suggest the Bush administration has not ruled out military action, including an airstrike on the nuclear facility by Israeli or American forces. It wouldn't be the first in the region—in 1981, Israeli jets bombed a nuclear reactor at Osirak in Iraq, prompting condemnation from the U.N. and the United States. Iranian president Mohammad Khatami described the idea of an American strike in Iran as "madness," noting that Iran had "plans" to defend itself. A strike would likely provoke Iran's government to retaliate, possibly against Americans in nearby Iraq or Afghanistan, setting off a cycle of violence with uncertain consequences. One thing's for sure: Iran's government would use an attack as an excuse to crack down once again, perhaps even declaring martial law.

AFTER A FEW DAYS IN TEHRAN, I headed for Tabriz, known for its cool mountain air, succulent stews and reformist politics. It was a homecoming for me: I was born in Tabriz in 1970, when thousands of American businessmen, teachers, Peace Corps volunteers and military contractors called Iran home. I left with my parents for the United States when I was almost 2 years old. It wasn't until the late 1990s that I got to know the place again—first while reporting for Reuters and the *Washington Post*, then while researching a book on contemporary Iran. I was the only "American" that many Iranians had ever met. "Why do the Americans hate us?" they often asked me. After my book was published in 2002, I received dozens of letters from Americans who'd worked in Iran before the 1979 revolution and who remembered the country and its people with deep fondness. Clearly, there remained a lot of goodwill as well as misunderstanding between Iranians and Americans.

Situated on the northern route from Tehran to Europe, Tabriz has long been an incubator for new ideas. In the late 19th century, intellectuals, merchants and reformist clergy in both Tehran and Tabriz had begun openly criticizing Iran's corrupt Qajar monarchs, who mismanaged the state's resources and gave large concessions to foreign powers. Iran was a vital piece in the geopolitical struggle between Russia and Britain to gain influence in Asia, and the two powers carved the country into spheres of influence in a 1907 agreement. At the time, Iranian reformers, frustrated by royal privilege and foreign interference, advocated a written constitution and a representative Parliament, and they sparked Iran's Constitutional Revolution of 1906–11.

The affection that many liberal Iranians have for America has roots in Tabriz, where a Nebraskan missionary named Howard Baskerville was martyred. Baskerville was a teacher in the American School, one of many such institutions created by the American missionaries who'd worked in the city since the mid-19th century. He arrived in 1908, fresh out of Princeton and, swept up in the revolutionary mood, fought a royalist blockade that was starving the city. On April 19, 1909, he led a contingent of 150 nationalist fighters into battle against the royalist forces. A single bullet tore through his heart, killing him instantly nine days after his 24th birthday.

Many Iranian nationalists still revere Baskerville as an exemplar of an America that they saw as a welcome ally and a useful "third force" that might break the power of London and Moscow in Tehran. Yet I found few signs of America's historic presence in Tabriz. One day, I tried to pay a visit to Baskerville's tomb, which is at a local church. Blocking my way was a beefy woman with blue eyes and a red head scarf. She told me I needed a permit. Why? "Don't ask me, ask the government," she said, and closed the door.

I WENT TO AHMAD ABAD, a farming town 60 miles west of Tehran, to meet the grandson of Mohammad Mossadegh, whose legacy still towers over U.S.-Iran relations nearly 40 years after his death.

PERSIAN POINTS IN TIME 1900–2005

1906 Constitutional Revolution establishes Iran's first Parliament.

1921 Military commander Reza Khan topples Ahmad Shah of the Qajar dynasty that has ruled for almost 150 years.

1925 Reza Khan names himself Shah of Iran, first of the Pahlavi dynasty.

1935 Country's name officially changes from Persia to Iran.

1941–1945 Allies deem Reza Khan sympathetic to Nazis and force him to abdicate to his son Mohammad Reza. Allies use Iran as a supply line to Russia.

1951 Prime Minister Mohammad Mossadegh nationalizes oil industry. Britain retaliates with a trade embargo, leading to economic collapse.

1953 The Mossadegh government is overthrown in a CIA-British-backed coup; Mohammad Reza Shah, who had fled, returns to power.

1963 The Shah begins the "White Revolution," an ambitious—and controversial—modernization program. Opponents are repressed.

1979 The Shah is toppled in an Islamic revolution, establishing the Islamic Republic of Iran as a theocracy under the rule of Ayatollah Ruhollah Khomeini. Fifty-two Americans are taken hostage at the U.S. Embassy.

1980 Attempted rescue of hostages fails; eight U.S. servicemen die.

1980 Iraq invades Iran, sparking an eight-year war that kills an estimated 500,000. America backs Iraq with military intelligence but supplies weapons to both sides.

1981 American hostages are released after 444 days in captivity.

1985 Secret arms deals conducted between the U.S. and Iran later becomes known as the Iran-Contra affair.

1988 Iran and Iraq sign a cease-fire agreement.

1995 Accusing Iran of sponsoring terrorism, the United States bans all trade with the nation.

1997 Reformist Mohammad Khatami is elected president by a landslide.

2000 Reformists win 70 percent of Parliament seats. Hard-line judiciary begins crackdown on newspaper editors and dissidents, leading to newspaper closures and arrests.

2001 President Khatami is reelected in a landslide victory.

2002 President George W. Bush lists Iran, together with North Korea and Iraq, as part of the "Axis of Evil."

2004 Conservatives regain control of Parliament after thousands of reformist condidates are disqualified. The International Atomic Energy Agency (IAEA) rebukes Iran for failing to cooperate with nuclear inspections.

2005 Iran allows IAEA inspections in January.

Mossadegh, a Swiss-educated descendant of the Qajar dynasty, was elected prime minister in 1951 on a nationalist platform, and he soon became a hero for defying the British, whose influence in Iran had aroused resentment and anger for more than half a century. The Anglo-Iranian Oil Company, which monopolized Iran's oil production, treated Iranians with imperial disdain, regularly paying more in taxes to the British government than they did in royalties to Iran. Mossadegh, after fruitless attempts to renegotiate the terms of the oil concession, stood up in Parliament in 1951 and declared that he was nationalizing Iran's oil industry. Overnight he emerged as a paragon of resistance to imperialism. *Time* magazine celebrated him as 1951's "Man of the Year," describing him as a "strange old wizard" who "gabbled a defiant challenge that sprang out of a hatred and envy almost incomprehensible to the west."

Mossadegh's move so frightened the United States and Britain that Kermit Roosevelt, grandson of President Theodore Roosevelt and FDR's distant cousin, turned up in Tehran in 1953 on a secret CIA mission to overthrow the Mossadegh government. Together with royalist generals, Iranian merchants on London's payroll and mobs for hire, Roosevelt organized a coup that managed to overwhelm Mossadegh's supporters in the army and among the people in a street battle that ebbed and flowed for several days. Mohammad Reza Shah, only the second shah in the Pahlavi dynasty, had fled to Rome when the fighting began. When it stopped, he returned to Tehran and reclaimed his power from Parliament. The coup, which Iranians later learned had been engineered by the United States, turned many Iranians against America. It was no longer viewed as a bulwark against British and Russian encroachment but the newest foreign meddler. Mossadegh was tried for treason in a military court, and in 1953 was sentenced to three years in jail. He remained under house arrest in Ahmad Abad, quietly tending his garden, until his death in 1967.

In the 1960s, the Shah began an aggressive, U.S.-backed modernization effort, from antimalaria programs to creating the SAVAK, the country's feared internal security service. As Britain pulled out of the region in the 1960s, Iran became the guardian of the Persian Gulf. Iran-U.S. relations were never better. Yet while Iran's economy boomed, democracy withered. The Shah stifled all political opposition, dismissing or repressing opponents as enemies of the state. The 1979 revolution, led by religious fundamentalists, took him by surprise. Today, Iranians look back on the Shah's era with a mingling of nostalgia, regret and anger. "He certainly ran the economy better than these mullahs," one Tehran resident told me. "But he was too arrogant and too unwilling to share political power."

Mossadegh, in contrast, was more of a democrat at heart. Even though his reforms were modest, he is respected today for his nationalism and tough stance against foreign interlopers. Today, his admirers regularly make the trek (some call it a pilgrimage) to his tomb. I went there early one Friday morning with Ali Mossadegh, the prime minister's great-grandson. As we toured the worn, creaking house, I asked Ali, who is in his late 20s, what he considered his great-grandfather's legacy. "He showed Iranians that they, too, deserve independence and democracy and prosperity," he said. He then led me to an adjoining annex where Mossadegh's tombstone rests amid a mound of Persian carpets. The walls were covered with photographs of the prime minister: making fiery speeches in Parliament; defending himself in a military court after the coup; gardening in Ahmad Abad. Ali pointed to an inscription taken from one of Mossadegh's speeches: "If, in our home, we will not have freedom and foreigners will dominate us, then down with this existence."

THE HIGH WALL surrounding the former U.S. Embassy, which occupies two Tehran blocks, bears numerous slogans. "On that day when the U.S. of A will praise us, we should mourn." "Down with USA." The seizing of the hostages here in 1979 was only the beginning of a crisis that shook American politics to its core.

After a six-month standoff, President Jimmy Carter authorized a rescue mission that ended disastrously after a helicopter collided with a transport plane in the Dasht-e-Kavir desert in north-central Iran, killing eight Americans. Secretary of State Cyrus Vance, who had opposed the operation, resigned. Carter, shaken by the failure, was defeated in the 1980 election by Ronald Reagan. The hostages were freed on the day of Reagan's inauguration. Still, Iran was regarded by the United States and others as an outlaw state.

Adjacent to the compound, a bookstore sells religious literature, anti-American screeds and bound copies of American diplomatic files painstakingly rebuilt from shredded documents. The place is usually empty of customers. When I bought a series of books entitled *Documents from the U.S. Espionage Den*, the chador-clad woman behind the desk looked surprised. The books were covered with a thin film of dust, which she wiped away with a wet napkin.

Mohsen Mirdamadi, who was a student in Tehran in the 1970s, was one of the hostage-takers. "When I entered university in 1973, there was a lot of political tension," he told me. "Most students, like me, were anti-Shah and, as a result, we were anti-American, because the U.S. was supporting the Shah's dictatorship." I asked him if he regretted his actions. "Clearly, our actions might have hurt us economically because it led to a disruption of relations, but I don't regret it," he said. "I think it was necessary for that time. After all, America had overthrown one Iranian government. Why wouldn't they try again?"

Bruce Laingen, who was the chargé d'affaires at the U.S. Embassy when he was taken as a hostage, said he had no orders to work to destabilize the new government, contrary to what the revolutionaries alleged. "Quite the contrary," the now-retired diplomat told me. "My mandate was to make clear that we had accepted the revolution and were ready to move on." One hostage-taker, he remembers, told him angrily: "You complain about being a hostage, but your government took an entire country hostage in 1953."

The passage of time has cooled Mirdamadi's zeal, and today he is an informal adviser to Iranian president Mohammad Khatami, who inspired Iranians in 1997 with his calls for greater openness. Elected by landslides in both 1997 and 2001 despite clerics' efforts to influence the outcome, Khatami has lost much of his popularity as religious conservatives have blocked his reforms. In any event, Khatami's power is limited. Real authority is wielded by a group of six clerics and six Islamic jurists called the Guardian Council, which oversaw the selection of Ayatollah Ali Khamenei as the country's supreme spiritual leader in 1989. The council has the power to block the passage of laws as well as prevent candidates from running for the presidency or the Parliament. Mirdamadi, like Khatami, says Iran deserves a government that combines democratic and Islamic principles. "We need real democracy," he told me, "not authoritarian dictates from above." He advocates the resumption of dialogue with the United States, though specifics are unclear. His reformist views won him a parliamentary seat five years ago, but in the 2004 elections he was among the 2,500 candidates the Guardian Council barred.

A PRESIDENTIAL ELECTION is scheduled for June, and social critics in Iran as well as international analysts say a free and fair contest is unlikely. With many Iranians expected to stay away from the polls in protest, a conservative victory is almost guaranteed. But what flavor of conservative? A religious hard-liner close to current supreme leader Khamenei? Or someone advocating a "China-style" approach, with limited cultural, social and economic liberalization coupled with continued political repression? No matter what, neither is likely to share power with secular democrats or even Islamist reformers like Mirdamadi. And the clerics' grasp on power is firm: Reporters Without Borders, Human Rights Watch, Amnesty International and the U.S. State Department have all sharply criticized Iranian officials for their use of torture and arbitrary imprisonment.

There's ample evidence that many ordinary Iranians are fed up with the involvement of Muslim clerics in government. "During the Constitutional Revolution, we talked about the separation of religion and state, without really knowing what that means," historian Kaveh Bayat told me in his book-filled Tehran study. "Our understanding today is much deeper. Now we know that it is neither in our interests nor the clergy's interest to rule the state." Or, as a physician in Tehran put it to me: "The mullahs, by failing, did what Ataturk could not even do in Turkey: secularize the populace thoroughly. Nobody wants to experiment with religion and politics anymore."

Ramin Jahanbegloo, one of Iran's leading secular intellectuals, agrees. "I am constantly being invited by university students to speak at their events," he told me over mounds of saffron-flecked rice and turmeric-soaked chicken at a Tehran cafeteria. "Just a few years ago they invited predominantly religious reformers. Now, they want secular democrats."

In Qom, Iran's holy city and home of the largest collection of religious seminaries in Iran, I spoke with a shopkeeper who sold religious trinkets and prayer stones just outside the stunning blue-tiled mosque of Hazrat-e-Masoumeh. He was a religious man, he said, and that's precisely why he felt religion should stay out of politics. "Politics is dirty," he said. "It only corrupts people."

I browsed several seminary bookstores in Qom, where I spotted titles ranging from Islamic jurisprudence to Khomeini's legacy. A bookstore owner told me that the ideas of reformist clergy are much more popular than the pronouncements of conservative mullahs. And translated American self-help books by the likes of motivational guru Anthony Robbins outsell political tracts. But the owner keeps the hottest commodities discreetly in a back corner. There I saw technical texts on sex and female anatomy. He just smiled sheepishly and shrugged his shoulders.

IRAN TODAY is at a turning point. Either the Islamic revolution must mellow and embrace political change, or face a reckoning down the road when hard-line clerics come into conflict with the secular, democratic ideals of the younger generation. But though the influence of religion in politics is under assault in Iran, national pride remains a potent force. In a recent poll of dozens of countries published in *Foreign Policy* magazine, 92 percent of Iranians claimed to be "very proud" of their nationality (compared with 72 percent of Americans).

To get a glimpse of raw Iranian patriotism, a good place to go is a soccer stadium. Back in Tehran, I went to a Germany-Iran exhibition game at the Azadi stadium with my friend Hossein, a veteran of Iran's brutal 1980–88 war with Iraq, and his sons and brother. The atmosphere gave me a new appreciation for Iran's reality: a fierce tension between a populace ready for change and a regime so shackled by ideological zeal and anti-American sentiment it can't compromise.

Hossein, like many Iranians who served in the war, resents America for supporting Iraq in the conflict: Washington provided Saddam Hussein's regime with satellite images of Iranian troop movements and cities, looked the other way as Iraq used chemical weapons on Iranian sol-diers and, in 1983, sent then businessman Donald Rumsfeld as a presidential envoy to Iraq, where he greeted Saddam Hussein with a handshake. But Hossein, who served as a frontline soldier, said he's willing to forgive and forget "as long as America does not attack Iran."

In the traffic jam leading to the stadium, young men leaned out of car windows and chanted "Iran! Iran! Iran!" Once inside, several doors to the arena were blocked. Crowds grew antsy, and a few hurled insults at police patrols. When a group of bearded young men—members of the Basij volunteer militia, linked to conservative religious figures—sauntered to the front of the line and passed through the gate, the crowd roared its disapproval. (I saw this frustration again later, when a parking attendant outside the stadium demanded a fee. "You are killing us with your fees!" Hossein's brother shouted at the man. "Don't the mullahs have enough money?")

Finally, the gates flew open and we stampeded into the stadium, clutching Hossein's young sons by the hands. At halftime, the chairman of the German football federation presented a check to the mayor of Bam, a city in southeastern Iran devastated by an earthquake that killed 30,000 people in 2003. "That will help the mayor pay for his new Benz," one man near me joked.

Throughout the game, which Germany won, 2-0, large loudspeakers blasted government-approved techno music. The mostly young men filling the 100,000 seats swayed to the beat. A small group near us banged on drums. The music stopped, and an announcer recited from the Koran, but most people continued chatting with one another, appearing to ignore the verses. When the music came back on, the crowd cheered.

AFSHIN MOLAVI *is a fellow at the New America Foundation, a nonpartisan public policy think tank, and the author of* Persian Pilgrimages: Journeys Across Iran. *He lives in Washington, D.C.*

Glossary of Terms and Abbreviations

A'ayan In Arabic, collective name for noble families in Palestine during the British Mandate.

Abd In Arabic, "slave" or "servant" (of God) commonly used in personal names (Gamal Abd al-Nasir).

Abu In Arabic, "father of," commonly used in names.

Alawi (Nusayri), a minority Muslim community in Syria, currently in power under the Assad family. They are nominally Shia but have separate liturgy and secret rites, with some non-Muslim festivals. The Alevi in Turkey are unrelated but follow some of the same rituals.

Aliyah In Hebrew, "ascent" or "rising up," term used to describe the return of Jews to Palestine from the Diaspora.

Allah God, in Islam.

Al-Qaeda In Arabic, "the Base," name of the Islamic terrorist organization founded by Usama bin Ladin.

Ashkenazi In Hebrew/German, name for Jews who emigrated from northern/eastern Europe to Palestine/Israel.

'Ashura In Arabic, the 10th day of the Islamic month of Muharram, marking for Shia Muslims the death and martydom of Hussain, Muhammad's grandson.

Ayatollah "Sign of God," the title of highest rank among the Shia religious leaders in Iran and elsewhere.

Baraka ("blessing"), given by God or a holy person.

Ba'th In Arabic, full name; (in English), Arab Socialist Resurrection Party-the dominant and ruling party in Syria and formerly in Iraq. In the latter country it was the only legal party from 1968 to 2003.

Bey In Turkish, a commander in the Ottoman army; also used for the heads of the Ottoman Regency of Algiers and Tunisia prior to the French conquest.

Bilad al-Makhzan In Arabic, in the Maghrib (North Africa) land under the control of a central authority (e.g. the sultans of Morocco).

Bilad al-Siba´ In Arabic land of dissidence, not under central control.

Caliph In Arabic, *khalifa;* agent, representative, or deputy; in Sunni Islam, the line of successors to Muhammad.

chador In Arabic, a full-length body covering for women, usually in black, which completely conceals the female form. In the Arab Gulf states it is usually called an abaya.

Colon Settler, colonist (French), a term used for the French population in North Africa during the colonial period (1830–1962).

Dar al-Islam "House of Islam," territory ruled under Islam. Conversely, *Dar al-Harb,* "House of War," denotes territory not under Islamic rule.

Diaspora In Hebrew, the dispersal of Jews outside Palestine on a more or less permanent basis, beginning with the destruction of Jerusalem by the Roman emperor Titus in 70 C.E.

Druze (or Druse) An offshoot of Islam that has developed its own rituals and practices and a close-knit community structure; Druze populations are found today in Lebanon, Jordan, Syria, and Israel.

Emir (or Amir) A title of rank, denoting either a patriarchal ruler, provincial governor, or military commander. Today it is used exclusively for rulers of certain Arabian Peninsula states.

Faqih In Arabic, jurist specializing in Islamic law.

Fatwa A legal opinion or interpretation delivered by a Muslim religious scholar-jurist; a religious edict.

Fida'i (plural Fida'iyun, also Fedayeen, cf. Mujahideen) Literally, "fighter for the faith"; a warrior who fights for the faith against the enemies of Islam.

fiqh In Arabic, Islamic law or jurisprudence.

GCC (Gulf Cooperation Council) Established in 1981 as a mutual-defense organization by the Arab Gulf states. Membership: Bahrain, Kuwait, Oman, Qatar, Saudi Arabia, and United Arab Emirates. Headquarters: Riyadh.

Gush Emunim In Hebrew, "Bloc of the Faithful," a fundamentalist Zionist movement.

Hadith "Traditions" of the Prophet Muhammad, the compilation of sayings and decisions attributed to him that serve as a model and guide to conduct for Muslims.

Hajj Pilgrimage to Mecca, one of the Five Pillars of Islam.

Halakhah In Hebrew, the legal system in Judaism, based on the Talmud and Torah.

Halal In Arabic, any action approved by fiqh (q.v.)

Haluzin "Pioneers" in Hebrew, early Zionist settlers there.

Hamas "zeal" In Arabic, acronym for Movement of the IslamicResistance, a radical organization founded 1988, which currentlyholds the majority in the Palestinian legislature elected 2006

Hanbali, Hanafi In Arabic, two of the four schools of legal interpretation in Islam.

Haram al-Sharif In Arabic, collective name for the Dome of the Rock and al-Aqsa Mosque in Jerusalem

Haredim In Hebrew, "Tremblers," Ultra-Orthodox Jews. The term is derived from a verse in Isaiah.

Hijab In Arabic, head scarf usually with a veil, worn by women and covering the entire hair. Oobligatory in such Islamic countries as Iran.

Hijrah (Hegira) The Prophet Muhammad's emigration from Mecca to Medina in A.D. 622 to escape persecution; the start of the Islamic calendar.

hizb In Arabic, party.

Ibadi A militant early Islamic group that split with the majority (Sunni) over the question of the succession to Muhammad. Their descendants form majorities of the populations in Oman and Yemen.

Ihram In Arabic, the seamless white robe worn by all Muslims making the hajj.

Ijma' In Arabic, consensus (of the community), used in decision-making when neither Qur'an nor hadith seem to apply.

Imam In Arabic, religious leader, prayer-leader of a congregation. When capitalized it refers to the descendants of Ali who are regarded by Shia Muslims as the rightful successors to Muhammad.

Intifada In Arabic, "resurgence," referring to the uprising of Palestinians against Israeli occupation of the West Bank and Gaza Strip.

Islam In Arabic, "submission," i.e. to the Will of God as revealed to Muhammad in the Koran.

Jahiliyya The "time of ignorance" of the Arabs before Islam. Sometimes used by Islamic fundamentalists today to describe secular Muslim societies, which they regard as sinful.

Jama'a The Friday communal prayer, held in a mosque (*jami'*). By extension, the public assembly held by Muslim rulers for their subjects in traditional Islamic states such as Saudi Arabia.

Jamahiriyya Popular democracy (as in Libya).

Jihad In Arabic, "struggle"; the term basically refers to efforts by individuals to reform bad habits or shortcomings in the practice of the faith, either within themselves or the larger Islamic community. In the broader sense, war waged in the defense of Islam.

Kabbala In Hebrew, traditions of mysticism in Judaism.

Khan A title of rank in eastern Islam (Turkey, Iran, etc.) for military or clan leaders.

Khedive Viceroy, the title of rulers of Egypt in the nineteenth and twentieth centuries who ruled as regents of the Ottoman sultan.

Kibbutz A collective settlement in Israel.

Koran (Arabic Qur'an) "recitation," the book compiled from God's revelations to Muhammad via the Angel Gabriel that form the basis for Islam.

League of Arab States (Arab League) Established in 1945 as a regional organization for newly independent Arab countries.

Maghrib "West," the hour of the sunset prayer; in Arabic, a geographical term for North Africa.

Mahdi "The Awaited One"; the Messiah, who will appear on earth to reunite the divided Islamic community and announce the Day of Judgment. In Shia Islam he is the Twelfth and Last Imam (al-Mahdi al-Muntazir) who disappeared 12 centuries ago but is believed to be in a state of occultation (suspended between heaven and earth.)

Majlis (Meclis, in Turkish) literally, "assembly," used traditionally for a ruler's weekly public meetings with subjects to hear complaints. When capitalized, refers to a national legislature, such as Turkey's Buyuk Millet Meclisi or Iran's Majlis.

Maliki The third school of Islamic legal interpretation.

Mandates An arrangement set up under the League of Nations after World War I for German colonies and territories of the Ottoman Empire inhabited by non-Turkish populations. The purpose was to train these populations for eventual self-government under a temporary occupation by a foreign power, which was either Britain or France.

Millet "Nation," a non-Muslim population group in the Ottoman Empire recognized as a legitimate religious community and allowed self-government in internal affairs under its own religious leaders, who were responsible to the sultan for the group's behavior.

Muezzin A prayer-caller, the person who announces the five daily obligatory prayers from the minaret of a mosque.

Mufti A legal scholar empowered to issue fatwas. Usually one mufti is designated as the Grand Mufti of a particular Islamic state or territory.

Mujahideen (*see* Fida'i)—A common term for resistance fighters in Afghanistan and opposition militants in Iran.

mullah In Arabic, Farsi, a religious cleric, especially in Iran.

Muslim (*see* Islam)—One who submits (to the will of God).

OAPEC (Organization of Arab Petroleum Exporting Countries) Established in 1968 to coordinate oil policies—but not to set prices—and to develop oil-related inter-Arab projects, such as an Arab tanker fleet and dry-dock facilities. Membership: all Arab oil-producing states. Headquarters: Kuwait.

OIC (Organization of the Islamic Conference) Established in 1971 to promote solidarity among Islamic countries, provide humanitarian aid to Muslim communities throughout the world, and provide funds for Islamic education through construction of mosques, theological institutions of Islamic learning, etc. Membership: all states with an Islamic majority or significant minority. Headquarters: Jiddah.

OPEC (Organization of Petroleum Exporting Countries) Established in 1960 to set prices and coordinate global oil policies of members. A majority of its 13 member states are in the Middle East. Headquarters: Vienna.

Polisario A national resistance movement in the Western Sahara that opposes annexation by Morocco and is fighting to establish an independent Saharan Arab state, the Sahrawi Arab Democratic Republic (SADR).

PSD (Parti Socialiste Destourien) The dominant political party in Tunisia since independence and until recently the only legal party.

Qadi In Arabic, also kadi. An Islamic judge, one who administers the Islamic law (shari'a)

Qanat In Arabic, An underground tunnel used for irrigation.

qiyas In Arabic, reasoning by analogy, one of the four sources of decision-making by the Islamic community, after the Koran, hadith and ijma'.

Quraysh The group of clans who made up Muhammad's community in Mecca.

Shafi'l In Arabic, the fourth school of Islamic legal interpretion.

Shari'a "The Way," the corpus of the sacred laws of Islam as revealed to Muhammad in the Koran. The sacred law is derived from three sources, the Koran, Sunna (q.v.), and hadith (q.v.).

Sharif "Holy," a term applied to members of Muhammad's immediate family and descendants through his daughter Fatima and son-in-law Ali.

Shaykh In Arabic, also sheikh, sheik; title of a tribal leader, religious group leader or prominent Muslim personality, especially in the Gulf states.

Shia Commonly, but incorrectly, *Shiite*. Originally meant "Party," i.e., of Ali, those Muslims who supported him as Muhammad's rightful and designated successor. Today, broadly, a member of the principal Islamic minority.

Glossary of Terms and Abbreviations

Shura an advisory council appointed by a ruler to advise on national issues. Commonly used in the Arab Gulf states.

Sunnah In Arabic, also sunna; "custom," the habits and practices of the Prophet Muhammad as recorded by his companions and family. Collectively they constitute the ideal behavior enjoined on Muslims.

Sunnis The majority in the Islamic community.

Suq (Souk) In Arabic, a public weekly market in Islamic rural areas, always held in the same village on the same day of the week, so that the village may have the word incorporated into its name. Also refers to a section of an Islamic city devoted to the wares and work of potters, cloth merchants, wood workers, spice sellers, etc.

sura In Arabic, a chapter in the Koran.

Taliban "Students"; originally used for students in Islamic madrasas (schools), the term acquired political significance with the rise to power in Afghanistan of this extreme Islamist movement.

Talmud In Hebrew, "study," Jewish scripture containing the options and statements of the rabbis of Palestine and Babylonia in the 1st–5th centuries C.E. and their interpretations.

Taqiyya Dissimulation, concealment of one's religious identity or beliefs (as by Shia under Sunni control) in the face of overwhelming power or repression.

tawhid In Arabic, belief in the absolute oneness of God; the central tenet in Wahhabism (q.v.)

Torah In Hebrew lit., "teaching", the Pentateuch, the first 5 books of Jewish scripture (Old Testament) plus the law of Moses.

U.A.R. (United Arab Republic) The name given to the abortive union of Egypt and Syria (1958–1961).

Ulema The corporate body of Islamic religious leaders, scholars, and jurists.

Umma The worldwide community of Muslims.

UNHCR (United Nations High Commission for Refugees) Established in 1951 to provide international protection and material assistance to refugees worldwide. UNHCR has several refugee projects in the Middle East.

UNIFIL (United Nations Interim Force in Lebanon) Formed in 1978 to ensure Israeli withdrawal from southern Lebanon. After the 1982 Israeli invasion, UNIFIL was given the added responsibility for protection and humanitarian aid to the people of the area. Headquarters: Naqoura.

United Nations Peacekeeping Forces Various military observer missions formed to supervise disengagement or truce agreements between the Arab states and Israel. They include UNDOF (United Nations Disengagement Observer Force). Formed in 1974 as a result of the October 1973 Arab-Israeli War and continued by successive resolutions. Headquarters: Damascus.

UNRWA (United Nations Relief and Works Agency for Palestine Refugees) Established in 1950 to provide food, housing, and health and education services for Palestinian refugees who fled their homes after the establishment of the State of Israel in Palestine. Headquarters: Vienna. UNRWA maintains refugee camps in Lebanon, Syria, Jordan, the occupied West Bank, and the Gaza Strip. UNRWA has also assumed responsibility for emergency relief for refugees in Lebanon displaced by the Israeli invasion and by the Lebanese Civil War.

Bibliography

CRADLE OF ISLAM

Karen Armstrong, *The Battle for God* (New York: Knopf, 2000). Chapters 2–4 devoted to Islam.

Rexa Aslan, *No God But God: The Origins, Evolution, and Future of Islam* (New York: Random House, 2005).

James Bill & John A. Williams, *Roman Catholics and Shia Prayer, Passion and Politics* (Chapel Hill: University of North Carolina Press, 2002).

Jonathan Bloom & Sheila Blair, *Islam: A Thousand Years Of Faith and Power* (New Haven: Yale University Press, 2002). Textual companion to the PBS Documentary *Islam: Empire of Faith.*

Richard Bulliett, *The Case for Islamic-Christian Civilization* (New York: Columbia Univ. Press, 2004).

Edmund Burke III & David Yaghoubian, *Struggle and Survival in the Modern Middle East* (Berkeley: University of California Press, 2006, 2nd ed.).

Patricia Crone, *God's Rule: Government and Islam, Six Centuries of Medieval Islamic Political Thought* (New York: Columbia Univ. Press, 2004).

Farhad Daftary, *The Assassins: Legends, Myths of the Ismailis* (London: I.B. Tauris, 1997).

Lawrence Davidson, *Islamic Fundamentalism: An Introduction* (Westport CT: Greenwood Press, Rev. ed, 2003).

Harm De Blij, *Why Geography Matters: Three Challenges Facing America* (New York: Oxford, 2005).

Faisal Devji, *Landscapes of the Jihad: Militancy, Morality, Modernity* (Ithaca: Cornell Univ. Press, 2005).

John Esposito, *Unholy War: Terror in the Name of Islam* (New York: Oxford University Press, 2002).

M.A.S. Abdel Haleem, *The Qur'an: A New Translation* (Oxford, UK: Oxford Univ. Press, 2004).

Robert Irvin, *The Alhambra* (Cambridge: Harvard University Press, 2004).

Hugh Kennedy, *When Baghdad Ruled the Muslim World* (Cambridge: Perseus Books Group, 2004).

Gilles Kepel, *Jihad: The Trail Of Political Islam* (Cambridge: Harvard University Press, 2002).

Ibn Khaldun, *The Muqaddimah: An Introduction to History*, translated by Franz Rosenthal, abridged and edited by N.J. Dawood with Introduction by Bruce Lawrence (Princeton: Princeton Univ. Press, 2005).

Paul Kunitzsch, *Stars and Numbers: Mathematics in the Medieval Arab and Western Worlds* (London: Ashgate Variorum, Collected Studies Series, n.d.) Articles by a German historian of science.

Ira M. Lapidus, *A History of Islamic Societies* (Cambridge: Harvard Univ. Press, 2004).

Jacob Lassner, *The Middle East Remembered: Forged Identities, Competing Narratives* (Ann Arbor: University of Michigan Press, 2002).

Bernard Lewis, *The Crisis of Islam: Holy War and Unholy Terror* (New York: Modern Library, 2003).

____, *What Went Wrong? Western Impact and Middle Eastern Response* (New York: Oxford University Press, 2002).

Makers of the Muslim World, A monograph series: Maribel Fierro, *Abd al-Rahman III, The First Cordoban Caliph*; Chase Robinson, *'Abd al-Malik*; Shahzad Bashir, *Fazlallah Astarabadi and the Hurufis*; Usha Sanyal, *Ahmad Riza Khan Barelwi: In the Path Of the Prophet*; Sunil Sharma, *Amir Khusraw: The Poet of Sultans And Sufis* (New York: One World/Oxford Univ. Press, 2005).

Mansoor Moaddel, *Islamic Modernism, Nationalism and Fundamentalism* (Chicago: Univ. of Chicago Press, 2005).

Maria Rosa Menocal, *The Ornament of the World* (Boston: Little, Brown, 2002) the story of Islamic Spain and its capital, the "ornament," Cordoba.

Seyyed Hossein Nasr, *The Heart of Islam: Enduring Values for Humanity* (New York: HarperCollins, 2002).

____, *Islam: Religion, History and Civilization* (New York: HarperCollilns, 2003).

F. E. Peters, *Muhammad and the Origins of Islam* (Albany, NY: SUNY Press, 1994).

Ahmad Rashid, *Taliban: Militant Islam* (New York: New York University Press, 2001).

Lawrence Rosen, *The Culture of Islam* (Chicago: University of Chicago Press, 2002).

Oliver Roy, *Globalized Islam* (London: Hurst, 2004).

Michael Sells, *Approaching the Qur'an: The Early Revelations* (Ashland OR: White Cloud Press, 1999).

Tamara Sonn. *A Brief History of Islam* (London: Blackwell, 2004).

Richard L. Smith, *Ahmad al-Mansur, Islamic Visionary* (New York: Pearson/Longman, 2006).

THEATER OF CONFLICT

Israaeli-Palestinian Conflict

Roy R. Anderson, Robert F. Seibert & John G. Wagner, *Politics and Change in the Middle East...* (Pearson/Prentice-Hall, 2004, 7th edition.

Rick Atkinson, *An Army at Dawn: The War in North Africa, 1942–1943* (New York: Henry Holt, 2002).

Warren Bass, *Support Any Friend: Kennedy's Middle East and the Making of the U.S.–Israel Alliance* (New York: Oxford University Press, 2003).

Yossi Beilin, *The Path to Geneva: The Quest for a Permanent Agreement, 1996–2004* (New York: RDV Books/Akaskic Books, 2004).

Ian Bicherton & Carla L. Klausner, *A Concise History of The Arab-Israeli Conflict* (Englewood Cliffs, NJ: Prentice-Hall, 2002, 4th edition).

Robert Bowker, *Palestinian Refugees: Mythology, Identity and the Search for Peace* (Boulder CO: Lynne Rienner, 2003).

Noam Chomsky, *Fateful Triangle: The U.S., Israel and the Palestinians* (Cambridge: South End Press, 1999, 2nd edition).

Joyce Davis, *Martyrs: Innocence, Vengeance and Despair In the Middle East* (New York: Palgrave MacMillan, 2003).

Charles Enderlin, *Shattered Dreams: The Failure of the Peace Process in the Middle East, 1995–2002* (New York: Other Press, 2003).

Avner Falk, *Fratricide in the Holy Land...* (Madison: University of Wisconsin Press, 2003).

Fawaz Gerges, *The Far Enemy: Why Jihad Went Global* (Cambridge UK: Cambridge University Press, 2005).

Ahmad Nizar Hamzeh, *In the Path of Hizbullah* (Syracuse: Syracuse University Press, 2004).

Mehran Kamrava, *Democracy in the Balance: Culture and Society in the Middle East* (New York: Chatham House, 1998).

Ilan Pappe, *The Modern Middle East* (London: Routledge, 2005), an introductory textbook.

Christopher Reuter, *My Life as a Weapon: A Modern History of Suicide Bombing* (Princeton: Princeton University Press, 2004).

Dennis Ross, *The Missing Peace...* (New York: Farrar, Straus & Giroux, 2004).

Gilad Sher, *Just Beyond Reach: Israeli-Palestinian Negotiations 1999–2001* (Tel Aviv: Frank Cass, 2005).

A.J. Sherman, *Mandate Days: British Lives in Palestine, 1918–1948* (Baltimore: Johns Hopkins University Press, 1997).

Clayton Swisher, *The Truth About Camp David...* (New York: Nation Books, 2004).

Other Conflicts

Rick Atkinson, *An Army at Dawn: The War in North Africa, 1942–1943* (New York: Henry Holt, 2002).

Thomas Barnett, *The Pentagon's New Map: War and Peace In the 21st Century* (New York: Putnam, 2004).

Jason Burke, *Al-Qaeda* (London: I.B. Tauris, 2003).

Richard Clarke, *Against All Enemies: Inside America's War On Terror* (New York: Free Press, 2004).

Fred Halliday, *The Middle East in International Relations: Power, Politics and Ideology* (Cambridge, UK: Cambridge Univ. Press, (2005).

Rashid Khalidi, *Resurrecting Empire: Western Footprints and America's Perilous Path in the Middle East* (Boston: Beacon Press, 2004).

Walter Laqueur, *A History of Terrorism* (New Brunswick, NJ: Transaction Books, 2002).

Matthew Levitt, *Targeting Terror: U.S. Policy Toward Middle East State Sponsors and Terrorist Organizations* (Washington DC: Washington Institute for Near East Policy, 2002).

Hugh Miles, *Al-Jazeera, The Inside Story* (New York: Grove Press, 2005).

Michael O'Hanlon, *Defense Strategy for the post-Saddam Era* (Washington DC: Brookings Inst. Press, 2005).

Michael Oren, *Six Days of War: June 1967 and the Making of the Modern Middle East* (New York: Oxford University Press, 2002).

Lawrence G. Potter & Gary Sick, eds., *Iran, Iraq and the Legacies of War* (New York: Palgrave MacMillan, 2004).

Michael Scheuer, *Imperial Hubris...* (Washington DC: Brassey's, 2004).

Shelby Telhami, *The Stakes: America and the Middle East* (Boulder CO: Westview Press, 2002).

Steven Yetiv, *Explaining Foreign Policy: U.S. Decision-making and the Persian Gulf War* (Baltimore: Johns Hopkins Univ. Press, 2004).

ALGERIA

Kay Adamson, *Political and Economic Thought and Practice in 19th Century France and the Colonization of Algeria* (New York: Edwin Mellen Press, 2002).

Ali Aissaoui, Algeria: *The Political Economy of Oil and Gas* (New York: Oxford University Press, 2001).

Matthew Connelly, *A Diplomatic Revolution: Algeria's Fight for Independence and the Making of the Modern Middle East* (New York: Oxford University Press, 2002).

Ted Morgan, *My Battle for Algiers* (Washington D.C.: Smithsonian Books, 2005).

Irwin Wall, *France, The U.S. and the Algerian War* (Berkeley: University of California Press, 2001).

BAHRAIN

Fred H. Lawson, *Bahrain: The Modernization of Autocracy* (Boulder, CO: Westview Press, 1989).

Mahdi A. al-Tajir, *Bahrain, 1920–1945: Britain, The Shaykh and the Administration* (London: Croom Helm, 1987).

EGYPT

Genevieve Abdo, *No God But God: Egypt and the Triumph Of Islam* (New York: Oxford U.P., 2002).

Leila Ahmed, *A Border Passage* (New York: Farrar, Strauss & Giroux, 1999).

John B. Alterman, *Egypt and American Assistance 1932–1958* (New York: Palgrave, 2002).

Beth Baron, *Egypt as a Woman: Nationalism, Gender and Politics* (Berkeley: Univ. of California Press, 2005).

Ninette Fahmy, *The Politics of Egypt: State-Society Relationship* (London: Routledge Curzon, 2002).

Anthony Gorman, *Historians, State & Politics in 20th Century Egypt* (New York: Routledge, 2002).

Zachary Karabell, *Parting the Desert: The Creation of the Suez Canal* (New York: Knopf, 2003).

Timothy Mitchell, *Rule of Experts: Egypt-Technopolitics And Modernity* (Berkeley: Univ. of California Press, 2002).

Max Rodenbeck, *Cairo: The City Victorious* (New York: Vintage Departures, 2000).

Carrie R. Wickham, *Mobilizing Islam: Religion, Activism and Political Change in Egypt* (New York: Knopg, 2003).

IRAN

Genevieve Abdo & Jonathan Lyons, *Answering Only to God: Faith and Freedom in Twentieth-Century Iran* (New York: Henry Holt, 2003).

Ali Ansari, *Iran, Islam and Democracy* (London: Royal Institute of International Affairs, 2000).

Daniel Brumberg, *The Struggle for Reform in Iran* (Chicago: Univ. of Chicago Press, 2001).

Mark J. Gasiorowski & Malcolm Byrne, eds., *Mohammed Mosaddeq and the 1953 Coup in Iran* (Syracuse: Syracuse Univ. Press, 2004).

David Harris, *The Crisis—The President, The Prophet, and The Shah...* (Boston: Little, Brown, 2004).

Stephen Kinzer, *All the Shah's Men* (New York: John Wiley, 2003).

Christine Marschall, *Iran's Persian Gulf Policy: From Khomeini to Khatami* (London: Routledge Curzon, 2003).

Abbas Milani, *The Persian Sphinx: Amir Abbas Hoveyda and the Riddle of the Iranian Revolution* (Washington, D.C.: Mage Publications, 2000).

____, *Lost Wisdom: Rethinking Modernity in Iran* (Washington DC: Mage Publishers, 2004).

Negan Nabavi, *Intellectuals and the State in Iran* (Gainesville: University Press of Florida, 2003).

Azar Nafisi, *Reading Lolita in Tehran: A Memoir in Books* (New York: Random House, 2003).

Kenneth Pollack, *The Persian Puzzle Palace: The Conflict Between Iran and America* (New York: Random House, 2004).

Elaine Sciolino, *Persian Mirrors: The Elusive Face of Iran* (New York: Free Press, 2000).

Brenda Shaffer, *Borders and Brethren: Iran and the Challenge of Azerbaijani Identity* (Boston: MIT Press, 2003).

Robin Wright, *The Last Great Revolution: Turmoil and Transformation in Iran* (New York: Knopf, 2000).

IRAQ

Said K. Aburish, *Saddam Hussain: The Politics of Revenge* (North Pomfret, VT: Bloomsbury Press, 2001).

Ofra Bengio, *Saddam's Word: Political Discourse in Iraq* (New York: Oxford University Press 2000).

Anthony Cordesman, *Iraqi Military Capabilities in 2002: A Dynamic Net Assessment* (Washington DC: Center for Strategic and International Studies, 2002).

Sandra Mackey, *The Reckoning: Iraq and the Legacy of Saddam Hussein* (New York: Norton, 2002).

Phoebe Marr, *The Modern History of Iraq* (Boulder CO: Westview Press, 2004, 2nd ed.).

Kenneth Pollack, *The Threatening Storm: The Case for Invading Iraq* (New York: Random House, 2002).

Anthony Shadid, *Night Draws Near: Iraq's People in the Shadow of America's War* (New York: Henry Holt, 2005).

Charles Tripp, *A History of Iraq* (Cambridge UK: Cambridge University Press, 2002).

ISRAEL

Asher Arian, *Politics in Israel: The Second Republic* (Washington DC: CQ Press, 2005, 2nd ed.).

Yossi Beilin, *The Path to Geneva: The Quest for a Permanent Agreement, 1996–2004* (New York: RDV Books/Akashic Books, 2004).

Aaron Bornstein, *Crossing the Green Line: Between Palestine and Israel* (Philadelphia, PA: University of Pennsylvania Press, 2001).

Robert Bowker, *Palestinian Refugees: Mythology, Identity and the Search for Peace* (Boulder CO: Lynne Rienner, 2003).

Francis A. Boyle, *Palestine, Palestinians and International Law* (Atlanta, GA: Clarity Press, 2004).

Avraham Brichta, *Political Reform in Israel* (Portland, OR: Sussex Academic Press, 2001).

Ronet Chacham, *Breaking Ranks: Refusing to Serve in the West Bank and Gaza Strip* (New York: Other Press, 2003).

Yoel Cohen, *The Whistle-blower of Dimona: Israel, Vanunu and the Bomb* (New York: Other Press, 2003).

Asad Ghanem, *The Palestine–Arab Minority in Israel 1948–2000: A Political Study* (Albany, NY: State University of New York Press, 2001).

Martin Gilbert, *Israel: A History* (New York: Morrow, 1998).

Daphne Golan-Agnon, *Next Year in Jerusalem: Everyday Life in a Divided Land* (New York: New Press, 2005).

Baruch Kimmerling and Joel S. Migdal, *The Palestinian People: A History* (Cambridge: Harvard University Press, 2003). Updated and revised version of an earlier book, *Palestinians: The Making of a People* (Harvard, 1994).

David Kretzmer, *The Occupation of Justice: The Supreme Court of Israel and the Occupied Territories* (Albany, NY: State University of New York Press, 2001).

Joel Migdal, *Through the Lens of Israel: Explorations in State and Society* (Albany, NY: State University of New York Press, 2001).

Ilan Pappe, *A History of Modern Palestine: One Land, Two Peoples* (Cambridge UK: Cambridge University Press, 2004).

Peter Rodgers, *Herzl's Nightmare: One Land, Two Peoples* (New York: Nation Books, 2005).

Avi Shalim, *The Iron Wall: Israel and the Arab World* (Scranton PA: W.W. Norton, 2001).

Virginia Tilley, *The One-State Solution: A Breakthrough Plan for Peace in the Israeli-Palestinian Deadlock* (Ann Arbor: Univ. of Michigan Press, 2004).

Barbara Victor, *Army of Roses: Inside the World of Palestinian Women Suicide Bombers* (New York: Rodale Press/t. Martin's, 2003).

Bernard Wasserstein, *Divided Jerusalem: Struggle for the Holy City* (New Haven, CT: Yale University Press, 2001).

JORDAN

George Joffe, ed., *Jordan in Transition, 1990–2000* (London: Hurst, 2002).

Beverly Milton-Edwards & Peter Hinchcliffe, *Jordan—A Hashemite Legacy* (London: Routledge, 2001).

Benny Morris, *The Road to Jerusalem: Glubb Pasha, Palestine and the Jews* (London: I.B. Tauris, 2002).

Timothy J. Paris, *Britain, the Hashimites and Arab Rule 1920–1925: The Sherifian Solution* (London: Frank Cass, 2002).

Philip Robins, *A History of Jordan* (Cambridge: Cambridge University Press, 2004).

Curtis Ryan, *Jordan in Transition: From Hussein to Abdullah* (Boulder CO: Lynne Rienner, 2002).

KUWAIT

Anthony Cordesman, *Kuwait: Recovery and Security After the Gulf War* (Boulder, CO: Westview Press, 1997).

Robert Jarman, *Sabah al-Salim Al-Sabah, Ruler of Kuwait 1965–1977: A Political Biography* (London: Centre of Arab Studies, 2002).

Miriam Joyce, *Kuwait 1945–1996: An Anglo-American Perspective* (Boston: Woburn Press, 2000).

Ben J. Slot et al, *Kuwait: The Growth of a Historic Identity* (London: Arabian Publishing Company, 2003).

Mary Ann Tetrault, *Stories of Democracy* (New York: Columbia University Press, 2000).

LEBANON

Carol Dagher, *Bring Down the Walls: Lebanon's Postwar Challenge* (New York: St. Martin's Press, 2000).

Kail C. Ellis, ed., *Lebanon's Second Republic: Prospects for The 21st Century* (Gainesville: University Press of Florida, 2002).

David Grafton, *The Christians of Lebanon: Political Rights In Islamic Law* (London: I.B.Tauris, 2003).

Simon Haddad, *The Palestinian Impasse in Lebanon: The Politics of Refugee Integration* (Brighton, UK: Sussex Academic Press, 2003).

Ahmad Nizar Hamzeh, *In The Path of Hizballah* (Syracuse: Syracuse University Press, 2004).

Bayan N. Al-Hout, *Sabra and Shatila, September 1982* (London: Pluto Press, 2004).

Samir Khalaf, *Civil and Uncivil Violence in Lebanon: A History of the Internationalization of Communal Conflict* (New York: Columbia University Press, 2002).

Samir Makdisi & Richard Sadaka, *The Lebanese Civil War, 1975-1990* (Beirut: A.U.B. Institute of Financial Economics, 2003).

Elizabeth Picard, *Lebanon: A Shattered Country* (New York: Holmes & Meier, 2001).

Naim Qassem, *Hizbullah: The Story from Within,* translated from the Arabic by Dalia Khalil. (London: Saqi Books, 2005).

Amal Saad-Ghorayeb, *Hizbullah: Politics and Religion* (London: Pluto Press, 2002).

Nawaf Salem, ed., *Options for Lebanon* (London: I.B. Tauris, 2004).

Raghid El-Solh, *Lebanon and Arabism: National Identity and State Formation* (London: I.B. Tauris, 2004).

LIBYA

Guy Arnold, *The Maverick State: Libya and the New World Order* (London: Cassell Academic Press, 1997).

Judith Gurney, *Libya: The Political Economy of Oil* (New York: Oxford University Press, 1996).

Amar Obeidi, *Political Culture in Libya* (Leonia, NJ: Curzon Press, 2000).

Dirk Vandewalle, *Libya Since Independence* (Ithaca, NY: Cornell University Press, 1998).

MOROCCO

Rahma Bourqia and Susan Miller, eds., *In the Shadow of the Sultan: Culture, Power and Politics in Morocco* (Cambridge, MA: Harvard University Press, 1999). No. 31, Harvard Middle Eastern Monographs.

Vivian Mann, ed., *Morocco: Jews and Art in a Muslim Land* (New York: Merrell Publishers, 2000).

David A. McMurray, *In and Out of Morocco: Smuggling and Migration in a Frontier Boomtown* (Minneapolis, MN: University of Minnesota Press, 2001).

James Miller and Jerome Bookin-Weiner, *Morocco: The Arab West* (Boulder, CO: Westview Press, 1998).

Kitty Morse, *The Scent of Orange Blossoms: Sephardic Cuisine From Morocco* (Santa Barbara, CA: Ten Speed Press, 2001).

C. R. Pennell, *Morocco Since 1830: A History* (New York: New York University Press, 2001).

Gregory White, *Comparative Political Economy of Tunisia and Morocco* (Albany, NY: State University of New York Press, 2001).

OMAN

Nicholas Clapp, *The Road to Ubar* (Boston, MA: Houghton Mifflin, 2000).

Isam al-Rawas, *Early Islamic Oman: A Political History* (Chicago: Garnet/Ithaca, 1998).

Carol A. Riphenberg, *Oman: Political Development in a Changing World* (Westport, CT: Greenwood Press, 1998).

Raghid al-Solh, *The Sultanate of Oman, 1914–1918* (Chicago: Garnet/Ithaca, 1999).

QATAR

Jill Crystal, *Oil and Politics in the Gulf: Rulers and Merchants in Kuwait and Qatar, 2nd ed.* (Cambridge, England: Cambridge University Press, 1995).

Steven Dorr and Bernard Reich, *Qatar,* 2nd ed. (Boulder, CO: Westview Press, 2000).

SAUDI ARABIA

Hamid Algar, *Wahhabism: A Critical Essay* (Oneonta NY: Islamic Publications International, 2002).

Anthony Cave Brown, *Oil, God and Gold: The Story of Aramco and the Saudi Kings* (Boston, MA: Houghton Mifflin, 1999).

Anthony Cordesman, *Saudi Arabia Enters the Twenty-first Century* (New York: Praeger, 2003, 2 vols).

Dore Gold, *Hatred's Kingdom: How Saudi Arabia Supports the New Global Terrorism* (Washington DC: Regnery, 2003).

J.E. Peterson, *Saudi Arabia and the Illusion of Security* (Oxford UK: Oxford Univ. Press, 2002).

Madawi al-Rasheed, *A History of Saudi Arabia* (Cambridge: Cambridge University Press, 2002).

Stephen Schwartz, *The Two Faces of Islam: The House of Sa'ud from Tradition to Terror* (New York: Doubleday/Random House, 2002).

Alexei Vasiliev, et al., *The History of Saudi Arabia* (New York: New York University Press, 2000).

SUDAN

Francis Deng & Mohammed Khalil, *Sudan's Civil War—The Peace Process Before and After Machakos* (Lansing: Michigan State Univ. Press, 2004).

Amir Idris, *Sudan's Civil War: Slavery, Race and Formational Identities* (Lewiston, NY: Edwin Mellen Press, 2001).

Douglas Johnson, *The Root Causes of Sudan's Civil War* (Bloomington: Indiana University Press, 2002).

Mansour Khalid, *War and Peace in Sudan: A Tale of Two Countries* (New York: Kegan Paul International, 2003.) The author is a former Sudanese Foreign minister.

A. Rolandson, *Guerrilla Government: Political Changes in The Southern Sudan during the 1990s* (Nordic Africa Institute, 2005).

Abdel Salam Sidahmed & Alsis Sidahmed, *Sudan* (New York: Routledge, 2004).

SYRIA

Raymond Hinnebusch, *Syria: Revolution From Above* (San Diego, DA: Gordon and Breach, 2000).

Flynt Leverett, *Inheriting Syria: Bashar's Trial By Fire* (Washington, DC: Brookings Institution Press, 2005).

Norman Lewis, *Nomads and Settlers in Syria and Jordan, 1800–1980* (New York: Cambridge Univ. Press, 1987).

Volker Perthes, *Syria under Bashar al-Assad: Modernization and the Limits of Change* (New York: Oxford Univ. Press, 2004).

Patrick Seale, *Asad of Syria: The Struggle for the Middle East* (Berkeley, CA: University of California Press, 1989).

Ghada Hashem Talhami, *Syria and the Palestinians: The Clash of Nationalisms* (Gainesville, FL: University Presses of Florida, 2001).

Lisa Wedeen, *Ambiguities of Domination: Politics, Rhetoric and Symbols in Contemporary Syria* (Chicago: University of Chicago Press, 1999).

Eyal Ziser and Itamar Rabinovitch, *Asad's Legacy, Syria in Transition* (New York: New York University Press, 2000).

TUNISIA

Eva Bellini, *Stalled Democracy: Capital, Labor and the Paradox of State-Sponsored Development* (Ithaca: Cornell University Press, 2002).

Mohamed Elhachmi, *The Politicization of Islam: A Case Study of Tunisia* (Boulder CO: Westview Press, 1998).

Derek Hopwood, *Habib Bourguiba of Tunisia: The Tragedy of Longevity* (New York: St. Martin's Press, 1992).

Kenneth J. Perkins, *The Historical Dictionary of Tunisia*, 2nd ed. (Metuchen, NJ: Scarecrow Press, 1997).

Azzam Tamimi, *Rachid Ghannouchi: A Democrat Within Islamism* (New York: Oxford University Press, 2001).

Gregory White, *Comparative Political Economy of Tunisia and Morocco* (Albany, NY: State University of New York Press, 2001).

TURKEY

Taner Akcan, *From Empire to Republic: Turkish Nationalism and the Armenian Genocide* (London: Zed Books, 2004).

Roger Crowley, *1453: The Holy War for Constantinople...* (New York: Hyperion, 2005. Published in Britain by Faber & Faber as *Constantinople: The Last Great Siege, 1453*).

Selim Deringil, *The Well-Protected Domain: Ideology and the Legitimation of Power in the Ottoman Empire, 1876–1909* (London: I. B. Tauris, 2000).

Jason Goodwin, *Lords of the Horizon: A History of the Ottoman Empire* (New York: Henry Holt, 1998).

M. Sukru Hanilogu, *Preparation for a Revolution: The Young Turks, 1902–1908* (New York: Oxford, 2002).

Christopher Houston, *Islam, the Kurds and the Turkish Nation-State* (New York: Oxford, 2001).

Stephen Larrabee, *Turkish Foreign Policy in an Age of Uncertainty*, With Ian O. Lesser. (Santa Monica CA: Rand, 2002).

Andrew Mango, *Ataturk: The Biography of the Founder of Modern Turkey* (New York: Overlook Publishing, 2000).

Bruce Masters, *Christians and Jews in the Ottoman Arab World: The Roots of Sectarianism* (New York: Cambridge University Press, 2001).

UNITED ARAB EMIRATES

Frank A. Clements, *The United Arab Emirates,* rev. ed. (Santa Barbara, CA: ABC-Clio, 1998).

Joseph Kechichian, ed., *A Century in Thirty Years: The United Arab Emirates* (Washington, D.C.: Middle East Policy Council, 2000).

Peter Lienhardt and Ahmed Al-Shahi, eds., *Shaikhdoms of Eastern Arabia* (New York: St. Martin's Press, 2001).

YEMEN

Robin Bidwell, *The Two Yemens* (Boulder, CO: Westview Press, 1983).

Paul Dresch, *A History of Modern Yemen* (Cambridge: Cambridge Univ. Press, 2000).

Caesar Farah, *The Sultan's Yemen: Nineteenth Century Challenges to Ottoman Rule* (New York: I.B. Tauris, 2002).

Ulrike Freitag and William Clarence-Smith, *Hadhrami Traders, Scholars and Statesmen in the Indian Ocean, 1750s to 1960s* (Leiden, the Netherlands: E. J. Brill, 1997).

E.G.H. Joffe et al., *Yemen Today: Crisis and Solution* (London: Caravel, 1997).

Leila Ingrams, *Yemen Engraved: Foreign Travellers to the Yemen, 1496–1890* (London: Kegan Paul International, 2000).

REGIONAL STUDIES

Ibrahim M. Abu-Rabi', *Contemporary Arab Thought: Studies in Post-1967 Arab Intellectual History* (London: Pluto Press, 2003).

Edmund Burke III & David Yaghoubian, eds., *Struggle and Survival in the Modern Middle East* (Berkeley: Univ. of California Press, 2005, 2nd ed.).

Anthony Cordesman, *Bahrain, Oman, Qatar & the UAE: Challenges of Security* (Boulder, CO: Westview Press, 1997).

Adeed Dawisha, *Arab Nationalism in the Twentieth Century* (Princeton: Princeton University Press, 2003).

Raymond Hinnebusch and Anoushirvan Enteshami, eds., *The Foreign Policies of Middle Eastern States* (Boulder, CO: Lynne Rienner, 2002).

Maya Jasanoff, *Edge of Empire: Lives, Culture and Conquest In the East, 1750–1850* (New York: Knopf, 2005).

Mehran Kamrava, *Democracy in the Balance: Culture and Society in the Middle East* (New York: Chatham House, 1998).

Robert W. Olson, *Turkey's Relations With Iran, Syria, Israel and Russia 1991–2000: The Kurdish and Islamist Questions* (Costa Mesa, CA: Mazda Publishers, 2001).

Robert Rabil, *Embattled Neighbors: Syria, Israel and Lebanon* (Boulder, CO: Lynne Rienner, 2003).

Barry Rubin, *The Tragedy of the Middle East* (Cambridge, UK: Cambridge University Press, 2003).

John Waterbury, *The Nile Basin National Determinants and Collective Action* (Princeton: Princeton University Press, 2002).

Richard Zachs, *The Pirate Coast* (New York: Hyperion, 2005) Deals with the U.S. in North Africa.

Rosemarie Said Zahlan, *Making of the Modern Gulf States* (Chicago, IL: Garnet/Ithaca, 1999).

WOMEN'S STUDIES

Mounira M. Charrad, *States and Women's Rights: The Making of Post-Colonial Tunisia, Algeria and Morocco* (Berkeley: University of California Press, 2003).

Andra Dworkin, *Scapegoat: The Jews, Israel and Women's Liberation* (Glencoe, NY: Free Press, 2000).

Jennifer Heath, *The Scimitar and the Veil: Extraordinary Women of Islam* (Hidden Spring Press, 2004).

Norma Khouri, *Honor Lost: Love and Death in Present-Day Jordan* (New York: Atria Press, 2003).

Rene Melammed, *Heretics or Daughters of Israel? The Crypto-Jewish Women of Castile* (New York: Oxford University Press, 2002).

Azadeh Moaveni, *Lipstick Jihad: A Memoir of Growing Up Iranian in America and American in Iran* (New York: Public Affairs Press/ Perseus Books, 2005).

Haya al-Mughni, *Women in Kuwait: The Politics of Gender* (London: Al-Saqi, 2001).

Judith E. Tucker, *In the House of the Law: Gender and Islamic Law in Ottoman Syria and Palestine* (Berkeley, CA: University of California Press, 1998).

LITERATURE IN TRANSLATION

Suad Amiry, *Sharon and My Mother-in-Law: Ramallah Diaries* (New York: Pantheon Books, 2004).

Ahmad Faqi, ed., *Libyan Stories* (London: Kegan Paul, 2001).

Elias Khoury, *Gate of the Sun,* translated by Humphrey Davies (New York: Archipelago Books, 2005) (stories told by Palestinians in refugee camps).

Yeshayahu Koren, *Funeral at Noon* (South Royalton, VT: Steerforth Press, 1996). English version of *Levayah ba-tsohorayin.*

Bernard Lewis, *Music of a Distant Drum* (Princeton: Princeton University Press, 2002).

Naguib Mahfouz, *The Dreams,* translated from the Arabic by Raymond Stack (Cairo: A.U.C. Press., 2004).

Amos Oz, *A Tale of Love and Darkness,* translated from the Hebrew by Nicholas de Lange (New York: Harcourt Brace, 2004).

Holly Payne, *The Weaver's Knot* (New York: Penguin Books, 2002). A novel about a famous femake weaver in Lycia, Turkey.

Elise Salem, *Constructing Lebanon: A Century of Literary Narratives* (Gainesville: University Press of Florida).

CURRENT EVENTS

To keep up to date on rapidly changing events in the contemporary Middle East and North Africa, the following materials are especially useful:

Africa Report Bimonthly, with an "African Update" chronology for all regions.

Africa Research Bulletin (Exeter, England) Monthly summaries of political, economic, and social developments in all of Africa, with coverage of North–Northeast Africa.

The Christian Science Monitor 1 Norway St., Boston, MA; weekly Mon-Fri, extensive regional coverage with periodic in-depth articles and reports.

Current History, A World Affairs Journal At least one issue per year is usually devoted to the Middle Eastern region.

The Economist U.S. edition, 1111 West 57th St., New York, NY; weekly except Sunday.

Middle East Economic Digest (London, England) Weekly summary of economic and some political developments in the Middle East–North African region generally and in individual countries. Provides special issues from time to time.

The Nation 33 Irving St., New York, NY; weekly news magazine, liberal-leftist in approach.

SaudiAramco World Published in Houston, TX by Saudi American Oil Company. Bimonthly, with extensive coverage of social-cultural life in the Islamic lands of the Middle East and elsewhere.

Tikkun 22 Shattuck Lane, Berkeley, CA, bimonthly magazine of the Jewish left, edited by Michael Lerner, distinguished scholar-rabbi.

World Press Review Shrub Oak, NY; monthly magazine with articles from the world press, translated into English.

PERIODICALS

The Economist 25 St. James's Street London, England.

The Middle East and North Africa Europa Publications 18 Bedford Square London, England A reference work, published annually and updated, with country surveys, regional articles, and documents.

The Middle East Journal 1761 N Street, NW Washington, D.C. 20036 This quarterly periodical, established in 1947, is the oldest one specializing in Middle East affairs, with authoritative articles, book reviews, documents, and chronology.

New Outlook 9 Gordon Street Tel Aviv, Israel A bimonthly news magazine, with articles, chronology and documents. Reflects generally Israeli leftist peace-with-the-Arabs views of the movement Peace Now with which it is affiliated.

Index

Index